P9-DBL-973

MEDICAL INFORMATICS

Knowledge Management and Data Mining in Biomedicine

INTEGRATED SERIES IN INFORMATION SYSTEMS

Series Editors

Professor Ramesh Sharda
Oklahoma State University

Prof. Dr. Stefan Voß
Universität Hamburg

Other published titles in the series:

E-BUSINESS MANAGEMENT: *Integration of Web Technologies with Business Models*/ Michael J. Shaw

VIRTUAL CORPORATE UNIVERSITIES: *A Matrix of Knowledge and Learning for the New Digital Dawn*/ Walter R.J. Baets & Gert Van der Linden

SCALABLE ENTERPRISE SYSTEMS: *An Introduction to Recent Advances*/ edited by Vittal Prabhu, Soundar Kumara, Manjunath Kamath

LEGAL PROGRAMMING: *Legal Compliance for RFID and Software Agent Ecosystems in Retail Processes and Beyond*/ Brian Subirana and Malcolm Bain

LOGICAL DATA MODELING: *What It Is and How To Do It*/ Alan Chmura and J. Mark Heumann

DESIGNING AND EVALUATING E-MANAGEMENT DECISION TOOLS: *The Integration of Decision and Negotiation Models into Internet-Multimedia Technologies*/ Giampiero E.G. Beroggi

INFORMATION AND MANAGEMENT SYSTEMS FOR PRODUCT CUSTOMIZATION/ Blecker, Friedrich, Kaluza, Abdelkafi & Kreutler

MEDICAL INFORMATICS

Knowledge Management and Data Mining in Biomedicine

edited by
Hsinchun Chen
Sherrilynne S. Fuller
Carol Friedman
William Hersh

Hsinchun Chen
The University of Arizona, USA

Sherrilynne S. Fuller
University of Washington, USA

Carol Friedman
Columbia University, USA

William Hersh
Oregon Health & Science Univ., USA

Library of Congress Control Number: 2005044125

ISBN-10: 0-387-24381-X (HB) ISBN-10: 0-387-25739-X (e-book)
ISBN-13: 978-0387-24381-8 (HB) ISBN-13: 978-0387-25739-6 (e-book)

Printed in the United States of America.

9 8 7 6 5 4 3 2

springer.com

TABLE OF CONTENTS

UNIT II: Information and Knowledge Management

UNIT III: Text Mining and Data Mining

EDITORS' BIOGRAPHIES

Hsinchun Chen is the McClelland Professor of Management Information Systems (MIS) at the Eller College of the University of Arizona. He received his Ph.D. degree in Information Systems from New York University. He is the author of more than nine books and 200 articles covering medical informatics, knowledge management, homeland security, semantic retrieval, and Web computing in leading information technology publications. He serves on the editorial boards of *Journal of the American Society for Information Science and Technology, ACM Transactions on Information Systems, IEEE Transactions on Systems, Man, and Cybernetics, IEEE Transactions on Intelligent Transportation Systems,* and *Decision Support Systems.* He is a scientific counselor/advisor of the Lister Hill Center of the National Library of Medicine (NLM/USA) and the National Library of China. Dr. Chen is the director of the University of Arizona's Artificial Intelligence Lab (40+ researchers). Since 1990, Dr. Chen has received more than $17M in research funding from various government agencies and major corporations. He has been a PI of the NSF Digital Library Initiative Program and the NIH/NLM's Biomedical Informatics Program. His group has developed advanced medical digital library, data mining, and text mining techniques for gene pathway and disease informatics analysis and visualization. Dr. Chen's work also has been recognized by major US corporations and been awarded numerous industry awards including: AT&T Foundation Award in Science and Engineering, SAP Award in Research/Applications, and Andersen Consulting Professor of the Year Award. Dr. Chen has been heavily involved in fostering digital library, medical informatics, knowledge management, and intelligence informatics research and education in the US and internationally. Dr. Chen was conference co-chair of ACM/IEEE Joint Conference on Digital Libraries (JCDL) 2004 and has served as the conference/program co-chair for the past seven International Conferences of Asian Digital Libraries (ICADL). Dr. Chen is also conference co-chair of the IEEE International Conference on Intelligence and Security Informatics (ISI) 2003, 2004, and 2005. He has been a frequent advisor for major US and international research programs. (Email: hchen@eller.arizona.edu; URL: http://ai.bpa.arizona.edu/hchen/)

Sherrilynne Fuller currently serves as Director, Health Sciences Libraries and Information Center, University of Washington. Her other responsibilities at the University of Washington include: Director, National Network of Libraries of Medicine, Pacific Northwest Region, and Assistant Director of Libraries. She is Professor, Division of Biomedical and Health Informatics, Department of Medical Education and Biomedical Informatics, School of Medicine; Professor, Information School and Adjunct Professor, Department of Health Services, School of Public Health and Community Medicine. Dr. Fuller has a BA degree in Biology, a Master's in Library Science from Indiana University, and a Ph.D. in Library and Information Science from the University of Southern California. Dr. Fuller's areas of research include: developing new approaches to represent and map the results of scientific research; design and evaluation of information systems to support decision making at the place and time of need; and integrated health sciences information systems design.

Dr. Fuller serves as Principal Investigator of the Health Sciences Libraries and Information Center contract from the NLM to serve as the Regional Medical Library for the Pacific Northwest (Alaska, Idaho, Montana, Washington and Oregon); Principal Investigator, Telemakus: Mining and Mapping Research Findings to Promote Knowledge Discovery in Aging funded by the Ellison Medical Foundation; Co-Investigator of Biomedical Applications of the Next Generation Internet (NGI): Patient-centric Tools for Regional Collaborative Cancer Care Using the NGI funded by the National Library of Medicine; Co-Investigator of an International Health and Biomedical Research and Training grant from the Fogarty International Center; and advisor to a Health Services Research Administration (HRSA) grant to explore models of Faculty Leadership in Interprofessional Education to Promote Patient Safety.

Dr. Fuller has served as a member of the President's (White House) Information Technology Advisory Committee and the Board of Regents of the National Library of Medicine and on the Boards of the American Medical Informatics Association and the Medical Library Association. She is an elected fellow of the American College of Medical Informatics. (Email: sfuller@u.washington.edu ; URL: http://faculty.washington.edu/sfuller/)

Dr. Carol Friedman is a Professor of Biomedical Informatics at Columbia University. Dr. Friedman has a B.S. degree in mathematics from the City University of New York, an M.A. degree and a Ph.D. in Computer Science from New York University. She has been involved in NLP research for several decades, starting with the pioneering Linguistic String Project. Dr. Friedman's other areas of research include knowledge representation, database design, object-oriented design, and information visualization. Initially her research focused on the clinical domain, and use of NLP for clinical applications. In the last few years she has been involved in research in the biological domain as well. She is known for the development of the MedLEE NLP system, which extracts and encodes information occurring in clinical reports. It is being used operationally at Columbia University Medical Center, where it has been shown to improve patient care. She is involved in development of two other NLP systems based on adaptations of MedLEE, GENIES and BioMedLEE, which process scientific text in the biological domain. Dr. Friedman has received more than $10M in research funding from various corporations and government agencies including the National Library of Medicine, The National Science Foundation, the New York State Office of Science and Technology, and the Research Foundation of the City University of New York. Dr. Friedman is involved in advancing biomedical informatics, text mining, and knowledge management research and education. Dr. Friedman was a conference co-chair for the Natural Language Track of the 2002 Pacific Symposium in Bioinformatics, the Workshop in Biomedicine in the 2002 and 2003 Association for Computational Linguistics Conferences, and the 2004 BioLink Workshop in the Human Language Techology Conference of the North American Chapter of the Association for Computational Linguistics. Dr. Friedman is a member of the Board of Scientific Counselors of the National Library of Medicine, is a member-at-large of the Executive Board of the American College of Medical Informatics, is on the Editorial Boards of the Journal of Biomedical Informatics and the Journal of the Association of Medical Informatics, and is a reviewer for numerous journals associated with bioinformatics. Dr. Friedman has been a guest editor of special issues of the Journal of Biomedical Informatics, has published over 100 articles on NLP, has co-authored a book on natural language processing (NLP), and is the author of various chapters on NLP.

(Email: friedma@dbmi.columbia.edu; URL:
http://www.dbmi.columbia.edu/~friedma/)

William Hersh, M.D. William Hersh, M.D. is Professor and Chair of the Department of Medical Informatics & Clinical Epidemiology in the School of Medicine at Oregon Health & Science University (OHSU) in Portland, Oregon. He also has academic appointments in the Division of General Internal Medicine of the Department of Medicine and in the Department of Public Health and Preventive Medicine. Dr. Hersh obtained his B.S. in Biology from the University of Illinois at Champaign-Urbana in 1980 and his M.D. from the University of Illinois at Chicago in 1984. After finishing his residency in Internal Medicine at University of Illinois Hospital in Chicago in 1987, he completed a Fellowship in Medical Informatics at Harvard University in 1990. Dr. Hersh has been at OHSU since 1990, where he has developed research and educational programs in medical informatics. He is internationally recognized for his contributions to the field. He is a Fellow of the American College of Medical Informatics and of the American College of Physicians. Dr. Hersh also recently served as Secretary of the American Medical Informatics Association. He is currently co-chair of the Working Group on Education of the International Medical Informatics Association. Dr. Hersh's research focuses on the development and evaluation of information retrieval systems for biomedical practitioners and researchers. The majority of his research funding comes from the National Library of Medicine, the Agency for Healthcare Research and Quality, and the National Science Foundation. He has published over 100 scientific papers and is author of the book, *Information Retrieval: A Health and Biomedical Perspective* (Second Edition, Springer-Verlag, 2003), which has an associated Web site, www.irbook.info. Dr. Hersh has also served on the Editorial Board of five scientific journals. He is also a member of the program committee of the Text Retrieval Conference (TREC) and currently chairs TREC's Genomics Track. Dr. Hersh also serves as Associate Director of the OHSU Evidence-Based Practice Center funded by the Agency for Healthcare Research and Quality. Dr. Hersh's work in medical informatics education is equally well-known. He serves as Director of OHSU's educational programs in biomedical informatics. He also teaches medical informatics to medical students, nursing students, and internal medicine residents. (Email: hersh@ohsu.edu; URL: http://medir.ohsu.edu/~hersh/)

AUTHORS' BIOGRAPHIES

Daniel Berleant, PhD, received the B.S. degree in 1982. After practicing in the software engineering field, he received the MS (1990) and PhD (1991) degrees from the University of Texas at Austin. He then developed a research program in text mining and interaction and in inference under severe uncertainty. In 1999 he accepted a position at Iowa State University where he continues to pursue research on text mining and text interaction, as well as uncertainty quantification and software engineering. He has advised or co-advised 21 master's theses and six PhD students who have either graduated or are in progress. He has authored over 50 refereed papers and book chapters. (Email: berleant@iastate.edu ; URL: http://class.ee.iastate.edu/berleant/home/)

Dr. Olivier Bodenreider, MD, PhD, is a Staff Scientist in the Cognitive Science Branch of the Lister Hill National Center for Biomedical Communications at the National Library of Medicine. He obtained the MD degree from the University of Strasbourg, France in 1990 and a PhD in Medical Informatics from the University of Nancy, France in 1993. Before joining NLM, he was an assistant professor for Biostatistics and Medical Informatics at the University of Nancy, France, Medical School. His research interests include terminology, knowledge representation, and ontology in the biomedical domain, both from a theoretical perspective and in their application to natural language understanding, reasoning, information visualization, and interoperability. (Email: olivier@nlm.nih.gov)

Dr. Anita Burgun, MD, PhD, is an associate professor at the University of Rennes I, School of Medicine (France), where she conducts research on knowledge representation and ontology in the biomedical domain. She is involved in several projects addressing semantic heterogeneity issues for information integration.
(Email: anita.burgun-parenthoine@univ-rennes1.fr)

Michael Chau, PhD, is currently Research Assistant Professor in the School of Business at the University of Hong Kong. He received his PhD degree in Management Information Systems from the University of Arizona and a Bachelor degree in Computer Science (Information Systems) from the University of Hong Kong. He was an active researcher in the Artificial Intelligence Lab at the University of Arizona, where he participated in several research projects funded by NSF, NIH, NIJ, and DARPA. (Email: mchau@business.hku.hk ; URL: http://www.business.hku.hk/~mchau/)

Christopher G. Chute, MD, DrPH, received his undergraduate and medical training at Brown University, internal medicine residency at Dartmouth, and doctoral training in Epidemiology at Harvard. He is Board Certified in Internal Medicine, and a Fellow of the American College of Physicians, the American College of Epidemiology, and the American College of Medical Informatics. He became Head of the Section of Medical Information Resources at Mayo Foundation in 1988 and is now Professor and Chair of Biomedical Informatics. As a career scientist at Mayo, Dr. Chute's NIH and AHCPR/AHRQ funded research in medical concept representation, clinical information retrieval, and patient data repositories have been widely published. He is Vice-chair of the ANSI Health Information Standards Board, Convener of Healthcare Concept Representation WG3 within the ISO Health Informatics Technical Committee, chair-elect of the US delegation to ISO TC215 for Health Informatics, co-chair of the HL7 Terminology Committee and a past member of the NIH Medical Informatics Study Section. He has chaired International Medical Informatics Association WG6 on Medical Concept Representation since 1994. (Email: chute@mayo.edu ; URL: http://mayoresearch.mayo.edu/mayo/research/staff/chute_cg.cfm)

Jeff Collmann, PhD, Associate Professor, Department of Radiology, Georgetown University, obtained his PhD in Social Anthropology from the University of Adelaide, Adelaide, South Australia. He completed a Postdoctoral Fellowship in Clinical Medical Ethics, Department of Philosophy, University of Tennessee and worked as a health care administrator. He joined the Department of Radiology, Georgetown University in January 1992 where he now conducts research, writes, consults and lectures widely on organizational dimensions of health information assurance. He serves as a medical ethicist for the Telemedicine and Advanced Technology Research Center, US Army Medical Research and Materiel Command, Ft. Detrick, Maryland. He functions as an advisor to the HIPAA compliance effort of the Department of Defense and the US Air Force Surgeon General. He also teaches courses at Georgetown University in the anthropology of medicine, science and technology and Australian culture. (Email: collmanj@georgetown.edu ; URL: http://www.isis.georgetown.edu/)

Ted Cooper, MD, Clinical Associate Professor, Department of Ophthalmology, Stanford University, received his MD and completed a residency in ophthalmology at the George Washington University. He is a fellow of the American College of Medical Informatics and the American Academy of Ophthalmology. As National Director for Confidentiality and Security he helped guide Kaiser Permanente's response to HIPAA. He has lectured widely on information assurance. He has participated in a number of health informatics activities including director and chairperson of the Computer-based Patient Record Institute. He is currently the chairperson of the Health Information and Systems Society Privacy and Security Steering Committee, a member of the Health Information and Systems Society Electronic Health Record Steering Committee, and chairperson of *The CPRI Toolkit: Managing Information Security in Health Care* Work Group. (Email: tcooper@stanford.edu ; URL: http://med.stanford.edu/school/eye/)

Julie Dickerson, PhD, received her B.S. degree from the University of California, San Diego and her MS and PhD degrees from the University of Southern California. She is currently an Associate Professor of Electrical and Computer Engineering at Iowa State University. Dr. Dickerson designed radar systems for Hughes Aircraft Company and Martin Marrietta while getting her PhD Her current research activities are intelligent systems, bioinformatics, pattern recognition, and data visualization. She is a Carver Fellow in the Virtual Reality Applications Center and a member of the Baker Center for Bioinformatics in the Plant Sciences Institute at Iowa State University. (Email: julied@iastate.edu ; URL: http://clue.eng.iastate.edu/~julied/)

Jing Ding, is currently a PhD candidate in the department of Electrical and Computer Engeering, and the interdepartmental program of Bioinformatics and Computational Biology at Iowa State University. He received a MS (Computer Engineering) in 2003, and a MS (Toxicology) in 2000, both from Iowa State University. His research interests are in text-mining and knowledge representation. (Email: dingjing@iastate.edu)

Pan Du received the BS and MS degrees in Electrical Engineering from National University of Defense Technology, Changsha, China, in 1995 and 1998, respectively. He is currently a co-major PhD student in Electrical Engineering major and Bioinformatics and Computational Biology major at Department of Electrical and Computer Engineering, Iowa State University. His research interests include systems biology, genetic network modeling and inference, microarray data analysis, signal processing and pattern recognition. (Email: dupan@iastate.edu)

Shauna Eggers is a computer programmer at the University of Arizona Artificial Intelligence Lab. She earned a B.S. in Computer Science and a B.A. in Linguistics and German Studies from the University of Arizona in May 2004. Her research interests include natural language processing for biomedical applications and knowledge visualization.
(Email: seggers@email.arizona.edu ; URL: http://ai.eller.arizona.edu/people/shauna/index.htm

Millicent Eidson, MA, DVM, DACVPM (Epidemiology) is State Public Health Veterinarian and Director of the Zoonoses Program, New York State Department of Health. She is also an Associate Professor in the Department of Epidemiology, University at Albany School of Public Health. Dr. Eidson previously served as an Epidemic Intelligence Service (EIS) Officer with the Centers for Disease Control and Prevention based at the National Cancer

Institute, and as the Environmental Epidemiologist and State Public Health Veterinarian with the New Mexico Department of Health. Dr. Eidson currently is the President of the Epidemiology Specialty, American College of Veterinary Preventive Medicine, and President-Elect of the National Association of State Public Health Veterinarians. She has been inducted as a member into the Delta Omega honorary public health society and as a Honorary Diplomate into the American Veterinary Epidemiology Society. Her research interests focus on surveillance and control of zoonotic diseases. (Email: mxe04@health.state.ny.us)

Marcelo Fiszman, MD, PhD, is currently a postdoctoral fellow at the National Library of Medicine. He received a MD from the State University of Rio de Janeiro and a PhD in medical informatics from the University of Utah. His research interests are in developing, applying, and evaluating natural language processing techniques in the biomedical domain. At the University of Utah, he concentrated on processing clinical text, emphasizing information extraction from chest x-ray to support the automation of pneumonia guidelines and from computerized tomography reports for quality improvement applications. At the National Library of Medicine, his interests focus on semantic analysis in the biomedical research literature. He recently developed a module to interpret hypernymic propositions and integrated it into a general semantic processor. He is also involved in a project that uses semantic interpretation for automatic abstraction summarization. (Email: fiszman@nlm.nih.gov)

Carol Foster, PhD, received a B.S. in Biology and a Secondary Teaching Credential from the University of Iowa. After working as a high school biology teacher, Dr. Foster earned an M.A. in Biology from California State University-Fresno and her PhD in Plant Physiology from Iowa State University. As a research associate at Iowa State University, her interests include plant growth and development, plant-microbe interactions, and effects of photoperiod and abiotic stresses on acetyl-CoA utilization and starch metabolism in Arabidopsis. (Email: cmfoster@iastate.edu)

Ivan Gotham holds a PhD in Biology from the State University at Albany and subsequently served as Research Fellow at the Wadsworth Public Health Laboratory at the New York Data Department of Health. He is currently Director of the Bureau of HEALTHCOM network systems Management at the NYS Department of Health and responsible for development, management and operation of the Agency's network, computer architecture and integarated health information sytems. He is a co-principal investigator on the NYSDOH/CDC NEDSS Charter Site grant and lead for Focus Area E Health informatics in the State's Bioterrorism Grant. He is also an Associate Professor at the School of Public Health at the State University of Albany, NY and a faculty member of the computer science department at Siena College, Loudonville, NY. (Email: ijg01@health.state.ny.us)

Zan Huang is a PhD candidate in Management Information Systems at the University of Arizona and is a research associate in the Artificial Intelligence Lab. He earned his B.Eng. in Management Information Systems from Tsinghua University. His research interests include data mining in biomedical and business applications, recommender systems, and mapping knowledge domains.
(Email: zhuang@eller.arizona.edu ; URL: http://eller.arizona.edu/~zhuang/Zan/ and http://ai.eller.arizona.edu/people/zan/

Catherine A. Larson is associate director, Artificial Intelligence Lab, at the University of Arizona. She received the B.A. from the University of Illinois at Urbana-Champaign in 1980, with majors in Spanish and Anthropology, and a minor in Portuguese. She received the MS in Library and Information Science from UIUC in 1986. She has been with the Lab since 2003. Previous positions include team leader of the fine arts and humanities team at the University of Arizona Library, and head of the preservation department at the University of Iowa Libraries. (Email: cal@eller.arizona.edu)

Gondy Leroy, PhD, is an assistant professor in the School of Information Science at Claremont Graduate University. She is director of the Intelligent Systems Lab. Her research seeks to evaluate information systems that are user-friendly, dynamic, and learn from and for the user. (Email: gondy.leroy@cgu.edu ; URL: http://beta.cgu.edu/faculty/leroyg/)

Ling Li, received her Bachelor of Science and Master of Science in Peking University in P.R.China in 1997 and 2000 respectively. She is currently a PhD candidate in the Wurtele lab in the Genetics, Development, Cell and Biology department at Iowa State University. (Email: liling@iastate.edu)

Bisharah Libbus has a PhD in genetics from the University of Missouri. After postdoctoral work at Johns Hopkins University in reproductive biology (supporting meiotic differentiation of early spermatocytes and early mouse embryogenesis in culture), he served as chair of the science department at Haigazian College and was on the faculty at the American University Medical Center in Beirut, Lebanon, and director of the cytogenetics laboratory at the National Unit of Human Genetics. At the University of Vermont, he investigated the process of fiber tumorigenesis and cancer cytogenetics, which remained the focus of his research for a number of years. More recently his interests have shifted to bioinformatics, in particular, searching for non-coding RNA genes. He is currently a visiting scientist at the National Library of Medicine investigating the application of natural language processing techniques to harvesting information about genes and disease from the research literature. (Email: libbus@nlm.nih.gov)

Chienting Lin, PhD, is assistant professor, Information Systems Department at Pace University. He has a PhD in Management Information Systems (MIS) with a minor in Computer Engineering from the University of Arizona. His research interests center on technology acceptance issues in Digital Government and E-Commerce Applications, Information Assurance and Network Security, Knowledge Management Systems, and implementation of Enterprise Systems. His research work has appeared in the *Journal of Management Information Systems* (JMIS), *Decision Support Systems* (DSS), *Social Science Computer Review* (SSCORE), *Journal of the American Society for Information Science* (JASIS), and *IEEE Transactions on Pattern Analysis and Machine Intelligence* (PAMI). He is a member of ACM, AIS, DSI, ISSA, and IEEE Computer Society. (Email: clin@pace.edu)

Cecil Lynch, MD, MS,is an Assistant Professor and Chair of the Graduate Group in Medical Informatics at the University of California, Davis. He received his MD from UCLA and an MS in Medical Informatics from the University of California, Davis and now restricts his professional activities to Informatics. His current projects include the NCI Cancer Bio-Informatics Grid project (caBIG) where he serves in the Vocabulary and Common Data Elements workspace and as a liaison to the Architecture and Clinical Trials workspaces. In addition to the NCI activities, he serves as a Consultant to the California Department of Health Services for the Public Health Informatics system architecture design, works on Public Health surveillance activities with the Centers for Disease Control and Department of Homeland Security, and actively participates on the HL7 Vocabulary Technical Committee. He is actively involved in the BioPortal infectious disease surveillance system design, concentrating on an International Foot and Mouth Disease registry. (Email: colynch@ucdavis.edu)

Daniel McDonald is a PhD candidate in the Department of Management Information Systems at the University of Arizona. He is a research associate in the Artificial Intelligence Lab. His research interests include natural language processing and web computing. (Email: dmm@eller.arizona.edu; URL: http://eller.arizona.edu/~dmm and http://ai.eller.arizona.edu/people/dan/)

Mark Minie holds a PhD in Immunology from the University of California, Berkeley, was a Senior Staff Fellow in the National Institutes of Health (NIH) Laboratory for Molecular Biology, and has worked in both the biotechnology and information technology sectors, most recently at Corbis Corporation before joining the University of Washington Health Sciences Library as their Basic Biosciences Liaison in 2002. He also consults for the National Center for Biotechnology Information (NCBI) and trains library based information specialists in bioinformatics for them nationwide. Additionally he does research on the use of telepresence technologies for distance learning as part of project LARIAT. He is an active member of the Seattle biomedical research community, and consults professionally for research scientists, entrepreneurs, and educators in that region. His research interests are in the regulation of eukaryotic gene expression, in silico biology, and DNA-based nanotechnology. Dr. Minie also works with several prominent science fiction writers, most recently with Greg Bear on "Darwin's Radio" and "Darwin's Children," and was an invited speaker at the Science Fiction Writers of America 2004 Nebula Awards Meeting. (Email: meminie@u.washington.edu ; URL: http://healthlinks.washington.edu/hsl/liaisons/minie)

Snehasis Mukhopadhyay, PhD, is an Associate Professor in the Department of Computer and Information Science and Associate Director (Bioinformatics) in the School of Informatics at Indiana University Purdue University Indianapolis. Dr. Mukhopadhyay is a holder of degrees from Jadavpur University, India and the Indian Institute of Science, India as well as a Master of Science and Doctorate in Electrical Engineering from Yale University. His research interests include Intelligent Systems, Neural Networks, Multi-Agent Systems, Intelligent Information Filtering, and Bioinformatics. He is a National Science Foundation CAREER Award recipient in 1996. (Email: smukhopa@iupui.edu)

Mathew Palakal, PhD, is a Professor and Chair of the Department of Computer and Information Science and Director of Informatics Research Institute at Indiana University Purdue University Indianapolis. He received his Ph. D. in computer science from Concordia University, Montreal in 1987. His primary research interests are in pattern analysis and machine intelligence. He is working on problems related to information management using information filtering and text mining approaches and structural health monitoring and smart diagnostics based on intelligent computational methods. One of his current primary areas of interest is in intelligent systems applied to biomedical literature mining. (Email: mpalakal@cs.iupui.edu; URL: http://www.cs.iupui.edu/~mpalakal/)

Asraa Rashid is a masters student in the Department of Management Information Systems at the University of Arizona and is a research assistant in the Artificial Intelligence Lab. She received her bachelors degree in Computer Science (2002) from the University of Arizona. Her research interests include information extraction and content analysis. (Email: asraa@eller.arizona.edu)

Debra Revere, MLIS, MA is Research Coordinator of the Telemakus Project in the Department of Medical Education and Biomedical Informatics and holds a Clinical Faculty position in the School of Public Health and Community Medicine at the University of Washington in Seattle. Her varied informatics background includes work designing survey and observation tools to assess clinicians' information needs, qualitative data analysis, and assessment of computer-generated outpatient behavioral interventions. Her research interests include examining the utility, application, and potential of spatial-semantic navigation tools for information seeking and retrieval within unstructured document collections. (Email: drevere@u.washington.edu)

Thomas Rindflesch has a PhD in linguistics from the University of Minnesota and is currently principal investigator for the Semantic Knowledge Representation project at the National Library of Medicine. His research interests concentrate on linguistic algorithms for natural language processing, and he is developing general methods that exploit symbolic, rule-based techniques to extract usable semantic information from biomedical text. The methods being pursued emphasize the interaction of domain knowledge (such as the Unified Medical Language System) and English syntactic structure. The goal of this research is to use semantic interpretation as the basis for building innovative biomedical information management applications. (Email: tcr@nlm.nih.gov)

Larry Smith, PhD, worked as systems analyst/developer in the pharmaceutical and agrichemicals sector for 10 years before earning a PhD in mathematics from Indiana University. In 1998 he joined a research group headed by Dr. John Weinstein at the National Cancer Institute where he developed statistical and graphical software used in the analysis of the gene expression study of the NCI60 cell lines. His recent research, as a consultant to the National Center for Biotechnology Institute under Dr. John Wilbur, involves natural language phenomena underlying search and summarization of scientific literature. (Email: smith@ncbi.nlm.nih.gov)

Matthew Stephens is currently a research associate at the Center for Medical Genomics at Indiana University School of Medicine. He received his MS in Computer Science from Purdue University at Indianapolis in 2002 and a B.S. in chemistry from Indiana University at Bloomington in 1999. He currently works on problems related to microarray data analysis, presentation, and integration with other informatic resources. At present, he is pursuing a Medical Degree at Indiana University Medical School. (Email: mastephe@iupui.edu)

Hua Su, PhD, is a postdoctoral researcher in the Artificial Intelligence Lab. She earned her PhD in Plant Sciences and MS in Management Information Systems from the University of Arizona. Her research interests lie in biomedical data mining, knowledge integration, and their applications in genomics. (Email: hsu@eller.arizona.edu; URL: http://ai.eller.arizona.edu/people/hsu/)

Lorraine Tanabe, PhD, holds a B.S. in Molecular Biology from San Jose State University, and a PhD in Computational Sciences and Informatics from George Mason University. She received a Cancer Research Training Award Fellowship at the National Cancer Institute, where she was the lead developer of MedMiner, a biomedical text mining application, and co-developer of EDGAR, a system for extracting genes, cell lines, drug names, and their interactions from the biomedical literature. She is currently a research scientist at the National Center for Biotechnology Information. Her interests include semantic tagging, information extraction and text data mining. (Email: tanabe@ncbi.nlm.nih.gov)

Peter Tarczy-Hornoch, MD, is an Associate Professor in the Department of Pediatrics and in the Department of Medical Education and Biomedical Informatics and an Adjunct Associate Professor in the Department of Computer Science and Engineering at the University of Washington. Within the Department of Medical Education and Biomedical Informatics, he is the Head of the Division of Biomedical and Health Informatics. (Email: pth@u.washington.edu; URL: http://faculty.washington.edu/pth)

George R. Thoma, PhD, is Chief of the Communications Engineering Branch of the Lister Hill National Center for Biomedical Communications, a research and development division of the U.S. National Library of Medicine. In this capacity, he directs R&D programs in document image analysis and understanding, biomedical image processing, image compression, automated document image delivery, digital x-ray archiving, animated virtual books, and high speed image transmission. He earned a B.S. from Swarthmore College, and the MS and PhD from the University of Pennsylvania, all in electrical engineering. He is the recipient of the NIH Merit Award, NLM Regents Award, and Federal Computer Week's *Federal 100 Award* among many others. He has served on the Maryland Governor's Task Force on High Speed Networks and similar panels, and currently serves on the Internet2 Applications Strategy Council. Dr. Thoma is a Fellow of the SPIE, the International Society for Optical Engineering. (Email: thoma@nlm.nih.gov ; URL: archive.nlm.nih.gov)

Chun-Ju Tseng is a software engineer for the Artificial Intelligence Lab at the University of Arizona. Chun-Ju's work has included search engine construction, visualization interfaces, and architecting software for reuse. His research interests include data visualization and and language processing. (Email: chunju@email.arizona.edu; URL: http://ai.eller.arizona.edu/people/lu/index.htm)

W. John Wilbur, MD, PhD, is a Senior Scientist in the Computational Biology Branch of the National Center for Biotechnology Information. He is a principal investigator leading a research group in the study and development of statistical text processing algorithms. He obtained a PhD in pure mathematics from the University of California at Davis and an MD from Loma Linda University. While at NCBI he has developed the algorithm that produces PubMed related documents and the algorithm that in PubMed allows fuzzy phrase matching. More recently his group has authored algorithms for phrase identification in natural language text that are used in NCBI's electronic textbook project and allow for easy reference from MEDLINE documents to related textbook material. (Email: wilbur@ncbi.nlm.nih.gov)

Peter Winkelstein, MD, is a general pediatrician with the Department of Pediatrics at the University at Buffalo. His experience in informatics predates his medical training and includes several years of professional computer programming. In addition, he holds a Master of Science in Astronomy. For the Department of pediatrics, he has been Medical Director of two primary care outpatient health centers and Director of Medical Informatics. He is currently chair of the American Medical Informatics Association's (AMIA) Ethics Committee, past chair of the AMIA Ethical, Legal and Social Issues Working Group and past chair of the Children's Hospital of Buffalo Ethics Committee.
(Email: Peter.Winkelstein@Eclipsys.com)

Eve Syrkin Wurtele, PhD, received the B.S. degree in Biology from U.C. Santa Cruz, CA, USA, 1971, and the PhD degree in Biology from U.C. Los Angeles, CA, USA, 1980. She was a Postdoctoral Fellow at Department of Biochemistry, U.C. Davis from 1980 to 1983, Senior Research Scientist at Cell Biology Division, NPI, Inc. from 1983 to 1988. In 1988, she joined the department of Botany and Food Technology as an Affiliate Assistant Professor, Iowa State University. She became an Assistant Professor and Associate Professor at Department of Botany, Iowa State University in 1990 and 1995, respectively. Since 1998, she has been the Professor at Department of Development & Cell Biology, Iowa State University. Dr. Wurtele organized the International Symposium of Metabolic Networking in Plants, April, 1999. She received the Herman Frasch Foundation Award, American Chemical Society in 1997. She is also the Co-Organizer of the Third International Congress on Plant Metabolomics, Iowa State University, June, 2004. (Email: mash@iastate.edu ; URL: http://www.public.iastate.edu/~mash/MetNet/homepage.html)

Jennifer J. Xu is a doctoral candidate in Management Information Systems (MIS) at the University of Arizona, where she is a member of the Artificial Intelligence Lab. Her research interests include social network analysis, computer mediated communication, and information visualization.
(Email: jxu@eller.arizona.edu; URL: http://eller.arizona.edu/~jxu; http://ai.arizona.edu/people/jxu/

Lijun Yan is currently a Masters student in the Computer Science Department at the University of Arizona. She received her Bachelors degree in Computer Science from East China University of Science and Technology, Shanghai, China. Her research interests include Systems and Application programming, information retrieval in knowledge domains. (Email: lijunyan@email.arizona.edu)

Terry S. Yoo, PhD, is a Computer Scientist in the Office of High Performance Computing and Communications, National Library of Medicine, NIH, where he heads the Program for 3D Informatics. His research explores the processing and visualizing of 3D medical data, interactive 3D graphics, and computational geometry. He is also the project officer who conceived and managed the development of ITK, the Insight Toolkit, under the Visible Human Project. Previously, as a professor of Radiology, he managed a research program in Interventional MRI with the University of Mississippi. Terry holds an A.B. in Biology from Harvard, and an MS and PhD in Computer Science from UNC Chapel Hill. (Email: yoo@nlm.nih.gov ; URL: http://visual.nlm.nih.gov)

Daniel Zeng, PhD, received the MS and PhD degrees in industrial administration from Carnegie Mellon University, Pittsburgh, PA, and the B.S. degree in economics and operations research from the University of Science and Technology of China, Hefei, China. Currently, he is an Assistant Professor and Honeywell Fellow in the Department of Management Information Systems at the University of Arizona. He is also currently directing four National Science Foundation (NSF)-funded research projects as PI or co-PI. His research interests include security informatics, infectious disease informatics, spatio-temporal data analysis, software agents and their applications, computational support for auctions and negotiations, and recommender systems. He has co-edited two books and published about 60 peer-reviewed articles in Management Information Systems and Computer Science journals, edited books, and conference proceedings. He also serves on the editorial boards of two Information Technology-related journals. (Email: zeng@eller.arizona.edu ; URL: http://eller.arizona.edu/~zeng/)

PREFACE

The field of medical informatics has grown rapidly over the past decade due to the advances in biomedical computing, the abundance of biomedical and genomic data, the ubiquity of the Internet, and the general acceptance of computing in various aspects of medical, biological, and health care research and practice. This book aims to be complementary to several other popular introductory medical informatics textbooks. The focus of this book is on the new concepts, technologies, and practices of biomedical knowledge management, data mining, and text mining that are beginning to bring useful "knowledge" to biomedical professionals and researchers. The book will serve as a textbook or reference book for medical informatics, computer science, information systems, information and library science, and biomedical, nursing, and pharmaceutical researchers and students. Biomedical professionals and consultants in the health care industry will also find the book a good reference for understanding advanced and emerging biomedical knowledge management, data mining, and text mining concepts and practices.

Readers of this book will learn the new concepts, technologies, and practices developed in biomedical informatics through the comprehensive review and detailed case studies presented in each chapter. Students and researchers will broaden their knowledge in these new research topics. Practitioners will be able to better evaluate new biomedical technologies in their practices.

SCOPE AND ORGANIZATION

The book is grouped by three major topic units. Unit I focuses on the critical foundational topics of relevance to information and knowledge management including: bioinformatics challenges and standards, security and privacy, ethical and social issues, and biomedical knowledge mapping. Unit II presents research topics of relevance to information and knowledge management including: representations of biomedical concepts and relationships, creating and maintaining biomedical ontologies, genomic information retrieval, public access to anatomic images, 3D medical informatics, and infectious disease informatics. Unit III presents emerging biomedical text mining and data mining research including: semantic parsing and analysis for patient records, biological relationships, gene pathways and metabolic networks, exploratory genomic data analysis, and joint learning using data and text mining.

We have compiled a list of interesting and exciting chapters from major researchers, research groups, and centers in medical informatics, focusing on emerging biomedical knowledge management, data mining, and text mining research. In particular, the three topic units consist of the following chapters, organized in a logical sequence:

Unit I: Foundational Topics in Medical Informatics

- Knowledge Management, Data Mining, and Text Mining in Medical Informatics: The chapter provides a literature review of various knowledge management, data mining, and text mining techniques and their applications in biomedicine.
- Mapping Medical Informatics Research: The chapter presents an overview of key medical informatics researchers and research topics by applying knowledge mapping techniques to medical informatics literature and author citation data between 1994 and 2003.
- Bioinformatics Challenges and Opportunities: The chapter presents a number of exciting biomedical challenges and opportunities for biologists, computer scientists, information scientists, and bioinformaticists.
- Managing Information Security and Privacy in Health Care Data Mining: The chapter explores issues in managing privacy and security of health care information used to mine data by reviewing their fundamentals, components, and principles, as well as relevant laws and regulations.
- Ethical and Social Challenges of Electronic Health Information: The chapter explores ethical and social challenges of health care information including implications from biomedical data mining.

Unit II: Information and Knowledge Management

- Medical Concept Representation: The chapter presents an overview of biomedical concept characteristics and collections.
- Characterizing Biomedical Concept Relationships: The chapter examines innovative approaches utilizing biomedical concept identification and relationships for improved information retrieval and analysis.
- Biomedical Ontologies: The chapter discusses challenges in creating and aligning biomedical ontologies and examines compatibility issues among several major biomedical ontologies.
- Information Retrieval and Digital Libraries: The chapter presents information retrieval and digital library techniques of relevance to

biomedical research.

- Modeling Text Retrieval in Biomedicine: The chapter presents current challenges and example document retrieval systems that help improve biomedical information access.
- Public Access to Anatomic Images: The chapter presents an overview and case study of several systems that provide Internet access to high resolution Visual Human Images and other associated anatomic documents and knowledge.
- 3D Medical Informatics: The chapter describes the emerging discipline of 3D medical informatics and suggests some of the future research challenges.
- Infectious Disease Informatics and Outbreak Detection: The chapter provides an overview of the emerging infectious disease informatics field and describes relevant system design and components for information sharing and outbreak detection.

Unit III: Text Mining and Data Mining

- Semantic Interpretation for the Biomedical Research Literature: The chapter discusses several semantic interpretation systems being developed in biomedicine and presents two applications that exploit semantic information in MEDLINE citations.
- Semantic Text Parsing for Patient Records: The chapter focuses on semantic methods that map narrative patient information to a structured coded form.
- Identification of Biological Relationships from Text Documents: The chapter describes computational problems and their solutions in automated extraction of biomedical relationships from text documents.
- Creating, Modeling, and Visualizing Metabolic Networks: The chapter presents the FCModeler and PathBinder systems for metabolic network modeling, creation, and visualization.
- Gene Pathway Text Mining and Visualization: The chapter describes techniques that automatically extract gene pathway relationships from biomedical text and presents two case studies.
- The Genomic Data Mine: The chapter focuses on the genomic data mine consisting of text data, map data, sequence data, and expression data, and concludes with a case study.
- Exploratory Genomic Data Analysis: The chapter describes approaches to exploratory genomic data analysis, stressing cluster analysis.
- Joint Learning Using Multiple Types of Data and Knowledge: The chapter discusses joint learning research in biomedical domains and

presents two representative case studies in protein function classification and regulatory network learning.

CHAPTER STRUCTURE

The book aims to present its chapters in a manner understandable and useful to general IT and biomedical students and professionals. Each chapter begins with an overview of the field to allow readers to get a quick grasp of the research landscape. Selected case studies are then provided to allow readers to get a closer look at the implementation challenges and opportunities.

Each chapter follows a consistent structure to ensure uniformity:

- Title
- Authors and Affiliations
- Abstract and Key Words
- Introduction: Introduces the importance and significance of the topic.
- Literature Review/Overview of the Field: A coherent and systematic review of related works in the topic area suitable for non-experts.
- Case Studies/Examples: One or two detailed case studies or examples of selected techniques, systems, implementations, and evaluations.
- Conclusions and Discussion
- Acknowledgement and References
- Suggested Readings: A list of essential readings (books or articles) for readers who wish to gain more in-depth knowledge in this topic area.
- Online Resources: A list of online resources that are relevant to the topic, e.g., web sites, open source software, datasets, testbeds, demos, ontologies, benchmark results, (evaluation) golden standards, etc.
- Questions for Discussion: A list of questions that are important to the topic and that would be suitable for classroom discussions or future research.

AUDIENCE

Most medical informatics departments in the United States and international universities will be able to use this book as a senior-level or graduate-level textbook. Selected medical, nursing, and pharmaceutical schools in the United States and internationally will be able to use our book in related health computing courses. Selected computer science and information systems departments could use this book in biomedical computing or data mining courses. Information and library science departments can also use the book in graduate-level digital library, information retrieval, or knowledge management courses.

The book could serve as a textbook or reference book for medical informatics, computer science, information and library science, and information systems students; medical, nursing, and pharmaceutical researchers; and bioinformatics/biomedical practitioners in the health care industries. Biomedical professionals and consultants in the health care industry including biotech companies will find the book a good reference for understanding advanced and emerging biomedical knowledge management, data mining, and text mining concepts and practices. Most of the medical libraries and/or science/engineering libraries in the United States and other countries will find our book a must-have for their patrons.

UNIT I

Foundational Topics in Medical Informatics

Chapter 1

KNOWLEDGE MANAGEMENT, DATA MINING, AND TEXT MINING IN MEDICAL INFORMATICS

Hsinchun Chen[1], Sherrilynne S. Fuller[2], Carol Friedman[3], and William Hersh[4]

[1]Management Information Systems Department, Eller College of Management, University of Arizona, Tucson, Arizona 85721; [2]University of Washington, Biomedical and Health Informatics, Seattle, Washington 98195-7155; [3]Columbia University, Department of Biomedical Informatics New York, New York 10032; [4]Oregon Health and Science University, Medical Informatics and Clinical Epidemiology, Portland, Oregon 97239-3098

Chapter Overview

In this chapter we provide a broad overview of selected knowledge management, data mining, and text mining techniques and their use in various emerging biomedical applications. It aims to set the context for subsequent chapters. We first introduce five major paradigms for machine learning and data analysis including: probabilistic and statistical models, symbolic learning and rule induction, neural networks, evolution-based algorithms, and analytic learning and fuzzy logic. We also discuss their relevance and potential for biomedical research. Example applications of relevant knowledge management, data mining, and text mining research are then reviewed in order including: ontologies; knowledge management for health care, biomedical literature, heterogeneous databases, information visualization, and multimedia databases; and data and text mining for health care, literature, and biological data. We conclude the paper with discussions of privacy and confidentiality issues of relevance to biomedical data mining.

Keywords

knowledge management; data mining; text mining

1. INTRODUCTION

The field of biomedical informatics has drawn increasing popularity and attention, and has been growing rapidly over the past two decades. Due to the advances in new molecular, genomic, and biomedical techniques and applications such as genome sequencing, protein identification, medical imaging, and patient medical records, tremendous amounts of biomedical research data are generated every day. Originating from individual research efforts and clinical practices, these biomedical data are available in hundreds of public and private databases, which have been made possible by new database technologies and the Internet. The digitization of critical medical information such as lab reports, patient records, research papers, and anatomic images has also resulted in large amounts of patient care data. Biomedical researchers and practitioners are now facing the "info-glut" problem. Currently, the rate of data accumulation is much faster than the rate of data interpretation. These data need to be effectively organized and analyzed in order to be useful.

New computational techniques and information technologies are needed to manage these large repositories of biomedical data and to discover useful patterns and knowledge from them. In particular, knowledge management, data mining, and text mining techniques have been adopted in various successful biomedical applications in recent years. *Knowledge management* techniques and methodologies have been used to support the storing, retrieving, sharing, and management of multimedia and mission-critical tacit and explicit biomedical knowledge. *Data mining* techniques have been used to discover various biological, drug discovery, and patient care knowledge and patterns using selected statistical analyses, machine learning, and neural networks methods. *Text mining* techniques have been used to analyze research publications as well as electronic patient records. Biomedical entities such as drug names, proteins, genes, and diseases can be automatically extracted from published documents and used to construct gene pathways or to provide mapping into existing medical ontologies.

In the following sections, we first survey the background of knowledge management, data mining, and text mining research. We then discuss the use of these techniques in emerging biomedical applications.

2. KNOWLEDGE MANAGEMENT, DATA MINING, AND TEXT MINING: AN OVERVIEW

Knowledge management, data mining, and text mining techniques have been widely used in many important applications in both scientific and business domains in recent years.

Knowledge management is the system and managerial approach to the gathering, management, use, analysis, sharing, and discovery of knowledge in an organization or a community in order to maximize performance (Chen, 2001). Although there is no universal definition of what constitutes knowledge, it is generally agreed there is a continuum of data, information, and knowledge. Data are mostly structured, factual, and oftentimes numeric, and reside in database management systems. Information is factual, but unstructured, and in many cases textual. Knowledge is inferential, abstract, and is needed to support decision making or hypothesis generation. The concept of knowledge has become prevalent in many disciplines and business practices. For example, information scientists consider taxonomies, subject headings, and classification schemes as representations of knowledge. Consulting firms also have been actively promoting practices and methodologies to capture corporate knowledge assets and organizational memory. In the biomedical context, knowledge management practices often need to leverage existing clinical decision support, information retrieval, and digital library techniques to capture and deliver tacit and explicit biomedical knowledge.

Data mining is often used during the knowledge discovery process and is one of the most important subfields in knowledge management. *Data mining aims to analyze a set of given data or information in order to identify novel and potentially useful patterns* (Fayyad et al., 1996). These techniques, such as Bayesian models, decision trees, artificial neural networks, associate rule mining, and genetic algorithms, are often used to discover patterns or knowledge that are previously unknown to the system and the users (Dunham, 2002; Chen and Chau, 2004). Data mining has been used in many applications such as marketing, customer relationship management, engineering, medicine, crime analysis, expert prediction, Web mining, and mobile computing, among others.

Text mining aims to extract useful knowledge from textual data or documents (Hearst, 1999; Chen, 2001). Although text mining is often considered a subfield of data mining, some text mining techniques have originated from other disciplines, such as information retrieval, information visualization, computational linguistics, and information science. Examples of text mining applications include document classification, document clustering, entity extraction, information extraction, and summarization.

Most knowledge management, data mining, and text mining techniques involve learning patterns from existing data or information, and are therefore built upon the foundation of *machine learning* and *artificial intelligence*. In the following, we review several major paradigms in machine learning, important evaluation methodologies, and their applicability in biomedicine.

2.1 Machine Learning and Data Analysis Paradigms

Since the invention of the first computer in the 1940's, researchers have been attempting to create knowledgeable, learnable, and intelligent computers. Many knowledge-based systems have been built for various applications such as medical diagnosis, engineering troubleshooting, and business decision-making (Hayes-Roth and Jacobstein, 1994). However, most of these systems have been designed to acquire knowledge manually from human experts, which can be a very time-consuming and labor-intensive process. To address this problem, machine learning algorithms have been developed to acquire knowledge automatically from examples or source data. Simon (1983) defined machine learning as "*any process by which a system improves its performance.*" Mitchell (1997) gives a similar definition, which considers machine learning to be "*the study of computer algorithms that improve automatically through experience.*" Although the "machine learning" term has been widely adopted in the computer science community, in the context of medical informatics, "data analysis" is more commonly used to represent "*the study of computer algorithms that improve automatically through the analysis of data.*" Statistical data analysis has long been adopted in biomedical research.

In general, machine learning algorithms can be classified as supervised learning or unsupervised learning. In supervised learning, training examples consist of input/output pair patterns. Learning algorithms aim to predict output values of new examples based on their input values. In unsupervised learning, training examples contain only the input patterns and no explicit target output is associated with each input. The unsupervised learning algorithms need to use the input values to discover meaningful associations or patterns.

Many successful machine learning systems have been developed over the past three decades in the computer science and statistics communities. Chen and Chau (2004) categorized five major paradigms of machine learning research, namely probabilistic and statistical models, symbolic learning and rule induction, neural networks, evolution-based models, and analytic learning and fuzzy logic. We will briefly review research in each of these areas and discuss their applicability in biomedicine.

2.1.1 Probabilistic and Statistical Models

Probabilistic and statistical analysis techniques and models have the longest history and strongest theoretical foundation for data analysis. Although it is not rooted in artificial intelligence research, statistical analysis achieves data analysis and knowledge discovery objectives similar to machine learning. Popular statistical techniques, such as regression analysis, discriminant analysis, time series analysis, principal component analysis, and multi-dimensional scaling, are widely used in biomedical data analysis and are often considered benchmarks for comparison with other newer machine learning techniques.

One of the more advanced and popular probabilistic models in biomedicine is the *Bayesian model.* Originating in pattern recognition research (Duda and Hart, 1973), this method was often used to classify different objects into predefined classes based on a set of features. A Bayesian model stores the probability of each class, the probability of each feature, and the probability of each feature given each class, based on the training data. When a new instance is encountered, it can be classified according to these probabilities (Langley et al., 1992). A variation of the Bayesian model, called the *Naïve Bayesian model,* assumes that all features are mutually independent within each class. Because of its simplicity, the Naïve Bayesian model has been adopted in different domains (Fisher, 1987; Kononenko, 1993). Due to its mathematical rigor and modeling elegance, Bayesian learning has been widely used in biomedical data mining research, in particular, genomic and microarray analysis.

A machine learning technique gaining increasing recognition and popularity in recent years is the *support vector machines* (SVMs). SVM is based on statistical learning theory that tries to find a hyperplane to best separate two or multiple classes (Vapnik, 1998). This statistical learning model has been applied in different applications and the results have been encouraging. For example, it has been shown that SVM achieved the best performance among several learning methods in document classification (Joachims, 1998; Yang and Liu, 1999). SVM is also suitable for various biomedical classification problems, such as disease state classification based on genetic variables or medical diagnosis based on patient indicators.

2.1.2 Symbolic Learning and Rule Induction

Symbolic learning can be classified according to its underlying learning strategy such as rote learning, learning by being told, learning by analogy, learning from examples, and learning from discovery (Cohen and Feigenbaum, 1982; Carbonell et al., 1983). Among these, *learning from examples* appears to be the most promising symbolic learning approach for knowledge discovery and data mining. It is implemented by applying an algorithm that attempts to induce a general concept description that best describes the different classes of the training examples. Numerous algorithms have been developed, each using one or more different techniques to identify patterns that are useful in generating a concept description. Quinlan's ID3 decision-tree building algorithm (Quinlan, 1983) and its variations such as C4.5 (Quinlan, 1993) have become one of the most widely used symbolic learning techniques. Given a set of objects, ID3 produces a decision tree that attempts to classify all the given objects correctly. At each step, the algorithm finds the attribute that best divides the objects into the different classes by minimizing entropy (information uncertainty). After all objects have been classified or all attributes have been used, the results can be represented by a decision tree or a set of production rules.

Although not as powerful as SVM or neural networks (in terms of classification accuracy), symbolic learning techniques are computationally efficient and their results are easy to interpret. For many biomedical applications, the ability to interpret the data mining results in a way understandable to patients, physicians, and biologists is invaluable. Powerful machine learning techniques such as SVM and neural networks often suffer because they are treated as a "black-box."

2.1.3 Neural Networks

Artificial neural networks attempt to achieve human-like performance by modeling the human nervous system. A neural network is a graph of many active nodes (neurons) that are connected with each other by weighted links (synapses). While knowledge is represented by symbolic descriptions such as decision trees and production rules in symbolic learning, knowledge is learned and remembered by a network of interconnected neurons, weighted synapses, and threshold logic units (Rumelhart et al., 1986a; Lippmann, 1987). Based on training examples, learning algorithms can be used to adjust the connection weights in the network such that it can predict or classify unknown examples correctly. Activation algorithms over the nodes can then be used to retrieve concepts and knowledge from the network (Belew, 1989; Kwok, 1989; Chen and Ng, 1995).

Many different types of neural networks have been developed, among which the *feedforward/backpropagation model* is the most widely used. Backpropagation networks are fully connected, layered, feed-forward networks in which activations flow from the input layer through the hidden layer and then to the output layer (Rumelhart et al., 1986b). The network usually starts with a set of random weights and adjusts its weights according to each learning example. Each learning example is passed through the network to activate the nodes. The network's actual output is then compared with the target output and the error estimates are then propagated back to the hidden and input layers. The network updates its weights incrementally according to these error estimates until the network stabilizes. Other popular neural network models include Kohonen's *self-organizing map* and the *Hopfield network*. Self-organizing maps have been widely used in unsupervised learning, clustering, and pattern recognition (Kohonen, 1995); Hopfield networks have been used mostly in search and optimization applications (Hopfield, 1982). Due to their performances (in terms of predictive power and classification accuracy), neural networks have been widely used in experiments and adopted for critical biomedical classification and clustering problems.

2.1.4 Evolution-based Algorithms

Evolution-based algorithms rely on analogies to natural processes and Darwinian *survival of the fittest*. Fogel (1994) identifies three categories of evolution-based algorithms: *genetic algorithms*, *evolution strategies*, and *evolutionary programming*. Among these, genetic algorithms are the most popular and have been successfully applied to various optimization problems. Genetic algorithms were developed based on the principle of genetics (Holland, 1975; Goldberg, 1989; Michalewicz, 1992). A population of individuals in which each individual represents a potential solution is first initiated. This population undergoes a set of genetic operations known as *crossover* and *mutation*. Crossover is a high-level process that aims at exploitation while mutation is a unary process that aims at exploration. Individuals strive for survival based on a selection scheme that is biased toward selecting fitter individuals (individuals that represent better solutions). The selected individuals form the next generation and the process continues. After some number of generations the program converges and the optimum solution is represented by the best individual. In medical informatics research, genetic algorithms are among the most robust techniques for feature selection problems (e.g., identifying a subset of genes that are most relevant to a disease state) due to their stochastic, global-search capability.

2.1.5 Analytic Learning and Fuzzy Logic

Analytic learning represents knowledge as logical rules and performs reasoning on such rules to search for proofs. Proofs can be compiled into more complex rules to solve similar problems with a smaller number of searches required. For example, Samuelson and Rayner (1991) used analytic learning to represent grammatical rules that improve the speed of a parsing system.

While traditional analytic learning systems depend on hard computing rules, there is usually no clear distinction between values and classes in the real world. To address this problem, *fuzzy systems* and *fuzzy logic* have been proposed. Fuzzy systems allow the values of False or True to operate over the range of real numbers from 0 to 1 (Zedah, 1965). Fuzziness has been applied to allow for imprecision and approximate reasoning. In general, we see little adoptation of such approaches in biomedicine.

2.1.6 Hybrid Approach

As Langley and Simon (1995) pointed out, the reasons for differentiating the paradigms are "more historical than scientific." The boundaries between the different paradigms are usually unclear and many systems have been built to combine different approaches. For example, fuzzy logic has been applied to rule induction and genetic algorithms (e.g., Mendes et al., 2001), genetic algorithms have been combined with neural network (e.g., Maniezzo, 1994; Chen and Kim, 1994), and because neural network has a close resemblance to probabilistic model and fuzzy logic they can be easily mixed (e.g., Paass, 1990). It is not surprising to find that many practical biomedical knowledge management, data mining, and text mining systems adopt such a hybrid approach.

2.2 Evaluation Methodologies

The accuracy of a learning system needs to be evaluated before it can become useful. Limited availability of data often makes estimating accuracy a difficult task (Kohavi, 1995). Choosing a good evaluation methodology is very important for machine learning systems development.

There are several popular methods used for such evaluation, including *holdout sampling*, *cross validation*, *leave-one-out*, and *bootstrap sampling* (Stone, 1974; Efron and Tibshirani, 1993). In the holdout method, data are divided into a training set and a testing set. Usually 2/3 of the data are assigned to the training set and 1/3 to the testing set. After the system is trained by the training set data, the system predicts the output value of each instance in the testing set. These values are then compared with the real output values to determine accuracy.

In cross-validation, a data set is randomly divided into a number of subsets of roughly equal size. Ten-fold cross validation, in which the data set is divided into 10 subsets, is most commonly used. The system is trained and tested for 10 iterations. In each iteration, 9 subsets of data are used as training data and the remaining set is used as testing data. In rotation, each subset of data serves as the testing set in exactly one iteration. The accuracy of the system is the average accuracy over the 10 iterations. Leave-one-out is the extreme case of cross-validation, where the original data are split into n subsets, where n is the size of the original data. The system is trained and tested for n iterations, in each of which $n-1$ instances are used for training and the remaining instance is used for testing.

In the bootstrap method, n independent random samples are taken from the original data set of size n. Because the samples are taken with replacement, the number of unique instances will be less than n. These samples are then used as the training set for the learning system, and the remaining data that have not been sampled are used to test the system (Efron and Tibshirani, 1993).

Each of these methods has its strengths and weaknesses. Several studies have compared them in terms of their accuracies. Hold-out sampling is the easiest to implement, but a major problem is that the training set and the testing set are not independent. This method also does not make efficient use of data since as much as 1/3 of the data are not used to train the system (Kohavi, 1995). Leave-one-out provides the most unbiased estimate, but it is computationally expensive and its estimations have very high variances, especially for small data sets (Efron, 1983; Jain et al., 1987). Breiman and Spector (1992) and Kohavi (1995) conducted independent experiments to compare the performance of several different methods, and the results of both experiments showed ten-fold cross validation to be the best method for model selection.

In light of the significant medical and patient consequences associated with many biomedical data mining applications, it is critical that a systematic validation method be adopted. In addition, a detailed, qualitative validation of the data mining or text mining results needs to be conducted with the help of domain experts (e.g., physicians and biologists), and therefore this is generally a time-consuming and costly process.

3. KNOWLEDGE MANAGEMENT, DATA MINING, AND TEXT MINING APPLICATIONS IN BIOMEDICINE

Knowledge management, data mining, and text mining techniques have been applied to different areas of biomedicine, ranging from patient record management to clinical diagnosis, from hypothesis generation to gene clustering, and from spike signal detection to protein structure prediction. In this section, we briefly survey some of the relevant research in the field, covering the applications of learning techniques in knowledge management, and data mining and text mining in biomedicine. More exhaustive and detailed reviews and discussions of selected knowledge management, data mining, and text mining techniques and applications in biomedicine can be found in the subsequent chapters in this book.

3.1 Ontologies

Before we examine different biomedical applications, it is important to understand the role of ontologies in knowledge management and knowledge discovery, especially for text mining applications. *An ontology is a specification of conceptualization. It describes the concepts and relationships that can exist and formalizes the terminology in a domain* (Gruninger and Lee, 2002). Ontologies are often used to facilitate knowledge sharing between people, information processing, data mining, communication between software agents, or other knowledge processing applications.

Many ontologies have been developed in the biomedical field. The *Unified Medical Language System* (UMLS), supported by the National Library of Medicine (NLM), is a major resource for facilitating computer programs to process and manage biomedical documents (McCray et al., 1993; Humphreys et al., 1993; Campbell et al., 1998; Humphreys et al., 1998). The UMLS offers three knowledge sources: the Metathesaurus, the Semantic Network, and the Specialist Lexicon. The Metathesaurus is a large multilingual controlled vocabulary database for biomedicine that allows users to map biomedical names and textual terms to concepts (i.e., controlled vocabulary terms), or to identify a set of different terms that are associated with a single concept. The Metathesaurus is formed by integrating about 100 different controlled vocabularies including the Medical Subject Headings (MeSH), a controlled vocabulary, and SNOMED-CT, a controlled clinical vocabulary established by the College of American Pathologists. The Semantic Network provides the categorization of the concepts in the Metathesaurus and also the relationships among the concepts. The Specialist Lexicon, designed to facilitate natural language processing for biomedical text, is a lexicon containing syntactic definitions for both biomedical terms and general English terms. These resources provide a framework and ontology for knowledge representation in biomedicine. UMLS resources have been widely used in biomedical language processing (Baclawski et al., 2000; Bodenreider and McCray, 2003; Perl and Geller, 2003; Rosse and Mejino, 2003; Zhang et al., 2003; Caviedes and Cimino, 2004). Several studies have investigated the mapping of concepts from the Metathesaurus to the Semantic Network (Cimino et al., 2003; Rindflesch and Fiszman, 2003).

Besides biomedical documents, it is also important for researchers and computers to understand the different terminologies for genes and proteins. The *Gene Ontology* (GO) project is an effort to address the need for consistent descriptions of gene products in different databases (The Gene Ontology Consortium, 2000). Aiming to produce a dynamic, controlled vocabulary of genes that can be applied to all eukaryotes, the project includes many databases, including FlyBase (Drosophila), the Saccharomyces Genome Database (SGD), the Mouse Genome Database (MGD), and several other major genome databases. GO consists of three structured ontologies that describe genes and gene products. GO terms are also cross-referenced with indexes from other databases. Similarly, the *Human Genome Nomenclature* (HUGO) specifies the standard, approved names and symbols for human genes (Wain et al., 2002). Most of this data can be searched on the Web as text files. There are numerous public databases specifying gene and gene products that are associated with multiple organisms as well as with specific model organisms.

3.2 Knowledge Management

Artificial intelligence techniques have been used in knowledge management in biomedicine as early as the 1970s, when the *MYCIN* program was developed to support consultation and decision making (Shortliffe, 1976). In MYCIN, the knowledge obtained from experts was represented as a set of IF-THEN production rules. Systems of this type would be later known as *expert systems* and become very popular in the 1980s. Expert systems relied on expert knowledge that was *engineered* into it, which was a time-consuming and labor-intensive process.

The performance of MYCIN was encouraging and it even outperformed human experts in some cases (Yu et al., 1979). Despite its early success, it was never used in actual clinical settings. Other medical diagnostic systems were also seldom used clinically. The reasons were two-fold. First, people were skeptical about computer technologies and system performances. Computers were not popular at that time, and many physicians did not believe that computers could perform better than humans. Second, computers were big, expensive machines in the 1970s. It was not feasible to support complex programs like MYCIN on an affordable computer to provide fast responses (Shortliffe, 1987). However, with the improved performance and lower cost of modern computers and medical knowledge-based systems, we believe there is a great opportunity for adopting selected knowledge management systems and technologies in the biomedical context, in particularly, not as a human replacement (i.e., expert systems) but as a biomedical decision making aide.

3.2.1 Knowledge Management in Health Care

It has been generally recognized that patient record management systems is highly desired in clinical settings (Heathfield and Louw, 1999; Jackson, 2000; Abidi, 2001). The major reasons include physicians' significant information needs (Dawes and Sampson, 2003) and clinical information overload. Hersh (1996) classified textual health information into two main categories: patient-specific clinical information and knowledge-based information, which includes research reported in academic journals, books, technical reports, and other sources. Both types of information are growing at an overwhelming pace.

Although early clinical systems were mostly simple data storage systems, knowledge management capabilities have been incorporated in many of them since the 1980s. For example, the *HELP* system, developed at the Latter Day Saints Hospital in Utah, provides a monitoring program on top of a traditional medical record system. Decision logic was stored in the system to allow it to respond to new data entered (Kuperman et al., 1991). The *SAPHIRE* system performs automatic indexing of radiology reports by utilizing the UMLS Metathesaurus (Hersh et al., 2002). The clinical data repository at Columbia-Presbyterian Medical Center (Friedman et al., 1990) is another example of a database that is used for decision support (Hripcsak, 1993) as well as well as physician review. The clinical data repository at the University of Virginia Health System is another example (Schubart and Einbinder, 2000). In their data warehouse system, clinical, administrative, and other patient data are available to users through a Web browser. Case-based reasoning also has been proposed to allow physicians to access both operative knowledge and medical literature based on their medical information needs (Montani and Bellazzi, 2002). Janetzki et al. (2004) use a natural language processing approach to link electronic health records to online information resources. Other advanced text mining techniques also have been applied to knowledge management in health care and will be discussed in more detail later in the chapter.

3.2.2 Knowledge Management for Biomedical Literature

Besides clinical information, knowledge management has been applied to research articles and reports, mostly via selected information retrieval and digital library techniques. The National Library of Medicine (NLM) offers the PubMed service, which includes over 13 million citations for biomedical articles from MEDLINE and other relevant journals. Many search systems have been built to help users retrieve relevant biomedical research papers and reports in database systems and over the Web. Automatic indexing and retrieval techniques are often applied. For example, the *Telemakus* system offers researchers a framework for information retrieval, visualization, and knowledge discovery (Fuller et al., 2002; Fuller et al., 2004; Revere et al., 2004). Using information extraction and visualization techniques, the system allows researchers to search the database of research articles for a statistically significant finding. The *HelpfulMed* system allows users to search for biomedical documents from several databases including MEDLINE, CancerLit, PDQ, and other evidence-based medicine databases (Chen et al., 2003). The HelpfulMed database includes high-quality health care-related Web pages collected from reputable sites using a neural-network-based spreading activation algorithm (Chau and Chen, 2003). The system also provides a term-suggestion tool called *Concept Mapper*, which allows users to consult a system-generated thesaurus and the NLM's UMLS to refine their search queries (Houston et al., 1999; Leroy and Chen, 2001).

MARVIN is an example of medical information retrieval systems that applied selected machine learning techniques (Baujard et al., 1998). Built on a multi-agent architecture, the system filters relevant documents from a set of Web pages and follows links to retrieve new documents. While MARVIN's filtering was based on simple document similarity metrics, other algorithms such as maximum-distance, artificial neural networks, and support vector machines have been applied to filtering medical Web pages (Palakal et al., 2001; Chau and Chen, 2004). A Bayesian model based on term strength analysis also has been used in biomedical document retrieval (Wilbur and Yang, 1996). Shatkay et al. (2000; 2002) use a probabilistic similarity-based search to retrieve biomedical documents that share similar themes.

Other text mining techniques also have been used to facilitate the management and understanding of biomedical literature. For example, natural language processing and noun phrasing techniques have been applied to extract noun phrases from medical documents (Tolle and Chen, 2000). Noun phrases often convey more precise meanings than single terms and are often more useful in further analysis. Named-entity extraction also has been widely applied to automatically identify from text documents the names of entities of interest (Chau et al., 2002). While mostly tested on general entities such as people names, locations, organizations, dates, times, number expressions, and email addresses (Chinchor, 1998), named-entity extraction has been used to extract specific biomedical entities such as gene names, protein names, diseases, and symptoms with promising results (Fukuda et al., 1998; Leroy et al., 2003). The extracted entities and relations are useful for information retrieval and knowledge management purposes. Both entity and relation extraction techniques will be discussed in more detail in our review of text mining later in the article.

3.2.3 Accessing Heterogeneous Databases

In the post-genome era, biomedical data are now being generated at a speed much faster than researchers can handle using traditional methods (National Research Council, 2000). The abundance of genomic and biomedical data has created great potential for research and applications in biomedicine, but the data are often distributed in diverse databases. As biological phenomena are often complex, researchers are faced with the challenge of information integration from heterogeneous data sources (Barrera et al., 2004). Many techniques have been proposed to allow researchers and the general public to share their data more effectively. For example, Sujansky (2001) proposes a framework to integrate heterogeneous databases in biomedicine by providing a uniform conceptual schema and using selected query-translation techniques. The *BLAST* programs are widely used to search protein and DNA databases for sequence similarities (Altschul et al., 1997). The MedBlast system, making use of BLAST, allows researchers to search for articles related to a given sequence (Tu et al., 2004). Sun (2004) uses automated algorithms to identify equivalent concepts available in different databases in order to support information retrieval. A software agent architecture also has been proposed to help users retrieve data from distributed databases (Karasavvas et al., 2004).

3.2.4 Information Visualization and Multimedia Information Access

Information (and knowledge) visualization for biomedical informatics is critical for understanding and sharing knowledge. With the rapid increase in computer speed and reduction in cost, graphical visualization has become increasingly popular in biomedical applications. Visualization techniques support display of more meaningful information and facilitate user understanding. Maps, trees, and networks are among some of the most popular information visualization representations. In the HelpfulMed system discussed earlier, documents retrieved from different databases are clustered using a self-organizing map algorithm (Kohonen, 1995) and a two-dimensional map is generated to display the document clusters (Chen et al., 2003). Bodenreider and McCray (2003) apply radial diagrams and correspondence analysis techniques to visualize semantic groups in the UMLS semantic network. Han and Byun (2004) use a three-dimensional display to visualize protein interaction networks. Virtual reality also has been applied in visualizing metabolic networks (Rojdestvenski, 2003).

Three-dimensional displays, interactive visualization, multimedia displays, and other advanced visualization techniques have been applied successfully in many biomedical applications. The most prominent example is the NLM's Visible Human Project (Ackerman, 1991), which produces three-dimensional representations of the normal male and female human bodies by obtaining transverse CT, MR, and cryosection images of representative male and female cadavers. The data is complete and anatomically detailed as the male was sectioned at one millimeter intervals and the female at one-third of a millimeter intervals. The data provides a good testbed for medical imaging and multimedia processing algorithms and has been applied to various diagnostic, educational, and research uses.

Because text processing algorithms cannot be applied to multimedia data directly, image processing and indexing techniques are often needed for selected biomedical applications. These techniques enable users to visualize, retrieve, and manage multimedia data such as X-ray and CAT-scan images more effectively and efficiently. For example, Yoo and Chen (1994) developed a system to provide a natural navigation of patient data using three-dimensional images and surface rendering techniques. Antani et al. (2004) study different shape representation methods to measure the similarity between X-ray images in order to enable users to manage and organize these images. Their system allows users to retrieve vertebra shapes significant to the pathology indicated in the query. Due to the increasing popularity and maturity of medical imaging systems, we foresee a pressing need for advanced multimedia processing and knowledge management capabilities in biomedicine.

3.3 Data Mining and Text Mining

Data mining techniques have been widely used to find new patterns and knowledge from biomedical data. While Bayesian models were widely used in the early days, more advanced machine learning methods, such as artificial neural networks and support vector machines, have been applied in recent years. These techniques are used in different areas of biomedicine, including genomics, proteomics, and medical diagnosis, among others. In the following, we review some of the major applications of data mining and knowledge discovery techniques in the field.

3.3.1 Data Mining for Health care

Because of their predictive power, data mining techniques have been widely used in diagnostic and health care applications. Data mining algorithms can learn from past examples in clinical data and model the oftentimes non-linear relationships between the independent and dependent variables. The resulting model represents formalized knowledge, which can often provide a good diagnostic opinion.

Classification is the most widely used technique in medical data mining. Dreiseitl et al. (2001) compare five classification algorithms for the diagnosis of pigmented skin lesions. Their results show that logistic regression, artificial neural networks, and support vector machines performed comparably, while *k*-nearest neighbors and decision trees performed worse. This is more or less consistent with the performances of these classification algorithms in other applications (e.g., Yang and Liu, 1999). Classification techniques are also applied to analyze various *signals* and their relationships with particular diseases or symptoms. For example, Acir and Guzelis (2004) apply support vector machines in automatic spike signal detection in ElectroEncephaloGrams (EEG), which can be used in diagnosing neurological disorders related to epilepsy. Kandaswamy et al. (2004) use artificial neural network to classify lung sound signals into six different categories (e.g., *normal*, *wheeze*, and *rhonchus*) to assist diagnosis.

Data mining is also used to extract rules from health care data. For example, it has been used to extract diagnostic rules from breast cancer data (Kovalerchuk et al., 2001). The rules generated are similar to those created manually in expert systems and therefore can be easily validated by domain experts. Data mining has also been applied to clinical databases to identify new medical knowledge (Prather et al., 1997; Hripcsak et al., 2002).

3.3.2 Data Mining for Molecular Biology

New sequencing technologies and low computation cost have resulted in an overwhelming abundance of biological data that can be accessed easily by researchers. It is not feasible to analyze these data manually, and the gap between the amount of submitted sequence data and related annotations, structures, or expression profiles is rapidly growing.

Data mining has begun to play an important role in addressing this problem. Clustering is probably the most widely used data mining technique for biological data. For example, clustering analysis is often applied to microarray gene expression data to identify groups of genes sharing similar expression profiles. Eisen et al. (1998) applied hierarchical clustering on the *Saccharomyces cerevisiae* gene expression data and achieved promising results. Various other clustering algorithms also have been tested on gene expression data, including *k*-means clustering (Herwig et al., 1999), backpropagation neural networks (Sawa and Ohno-Machado, 2003), self-organizing maps (Tamayo et al., 1999; Herrero et al., 2001), fuzzy clustering (Belacel et al., 2004), expectation maximization (Qu and Xu, 2004), and support vector machines (Brown et al., 2000). Qin et al. (2003) used the idea of kernel (as in support vector machines) and combined it with hierarchical clustering. Gene expression analysis also has been applied in cancer class discovery and prediction (Golub et al., 1999; Hsu et al., 2003).

Besides clustering, other predictive data mining techniques also have been applied to biomedical data. For example, neural network models have been widely used in predicting protein secondary structure (Qian and Sejnowski, 1988; Hirst and Sternberg, 1992). Increasingly, data mining algorithms also have been used for prediction in various biomedical applications including protein backbone angle prediction (Kuang et al., 2004), protein domains (Nagarajan and Yona, 2004), biological effects (Krishnan and Westhead, 2004), and DNA binding (Ahmad et al., 2004). These predictive methods are often based on classification (supervised learning) algorithms such as neural networks or support vector machines.

3.3.3 Text Mining for Literature and Clinical Records

Text mining has been widely used to analyze biomedical literature. Because of the large amount of research articles in public databases and the diversity of biomedical research, it is not uncommon that researchers encounter some sequences or new genes that they have no knowledge about. It is quite likely that some important relationships between biological entities remain unnoticed because relevant data are scattered and no researcher has linked them together (Swanson, 1986; Smalheiser and Swanson, 1998). Given the large amount of published literature and that many researchers only specialize in a small sub-domain (e.g., several particular genes), text mining techniques could be invaluable in discovering new knowledge patterns or hypotheses from the large amount of existing and new literature in biomedicine (Yandell and Majoros, 2002).

Text mining for biomedical literature often involves two major steps. First, it must identify biomedical entities and concepts of interests from free text using natural language processing techniques. For instance, if we want to study the relationship between a gene (e.g., *p53*) and a disease (e.g., *brain tumors*), the names of both entities need to be correctly identified from the relevant textual documents. Many text mining algorithms have been applied to this problem. For example, Fukuda et al. (1998) use simple morphological clues to recognize the names of proteins and other materials with high accuracy. Support vector machines have been used in entity extraction by classifying words into the 24 entity classes in the *GENIA* corpus (Kazama et al., 2002). Tanabe and Wilbur (2002) use part-of-speech tagging and a Bayesian model to identify genes and proteins in text. Hatzivassiloglou et al. (2001) compared three machine learning techniques, namely Naïve Bayesian model, decision trees, and inductive rule learning, to resolve the classification of a biological entity (e.g., protein, gene, and RNA) after it was identified. Their results showed that the three learning models had comparable performance. Other studies have investigated the mapping between abbreviations and full names such that these names will not be considered by the system as different entities (Yu et al., 2002).

After the entity names have been identified, further analyses are performed to see whether these entities have any relationships, such as gene regulations, metabolic pathways, or protein-protein interactions (Blaschke et al., 1999; Dickerson et al., 2003). *Shallow parsing* is often used to focus on specific parts of the text to analyze predefined words such as verbs and nouns (Leroy et al., 2003). Sekimizu et al. (1998) identified the set of most frequently used verbs in a collection of abstracts and developed a set of rules to identify the subjects and objects of the verbs. Pustejovsky et al. (2002) used relational parsing and finite state automata to identify *inhibit* relationships from biomedical text. The *GENIES* system, based on the *MedLEE* parser (Friedman and Hripcsak, 1998), also has been used to extract molecular pathways from texts (Friedman et al., 2001). The *Telemakus* system extracts information by analyzing the headings and surrounding texts of tables and figures (Fuller et al., 2002; Revere et al., 2004). The *Genescene* system utilizes an ontology-based approach to relation extraction by integrating the *Gene Ontology*, the *Human Genome Nomenclature*, and the *UMLS* (Leroy and Chen, forthcoming). The system combines natural language processing and co-occurrence analysis techniques to identify terms and gene pathway relations from biomedical abstracts. The *EDGAR* system extracts drugs, genes, and relationships from text (Rindflesch et al., 2000). Wren et al. (2004) developed a system that uses a random network model to rank the relationships identified from text. Machine learning techniques also have been used to automate the process of annotation. For instance, Kretschmann et al. (2001) used a *C4.5* algorithm to generate rules for keyword annotation in the *SWISS-PROT* database.

Text mining also has been applied to patient records and other clinical documents to facilitate knowledge management. It adopts a process similar to that of text mining from literature. For example, the system reported by Harris et al. (2003) extracts terms from clinical texts. Using natural language processing techniques, the *MedLEE* system (Friedman and Hripcsak, 1998) has been applied to free-text patient records. It extracts useful entities in order to identify patients having tuberculosis or breast cancer based on their admission chest radiographs and mammogram reports, respectively (Knirsch et al., 1999; Jain and Friedman, 1997). Chapman et al. (2004) use a similar text mining approach for automated fever detection from clinical records to detect possible infectious disease outbreaks.

3.4 Ethical and Legal Issues for Data Mining

Medical records and biological data generated from human subjects contain private and confidential information. Patients' and human subjects' data must be handled with great caution in order to protect their privacy and confidentiality. Researchers do not automatically acquire the rights to use patient or subject data for data mining purposes unless they obtain the patients' or subjects' consent (Berman, 2002). In the US, the 1996 Health Insurance Portability and Accountability Act (HIPAA) set the standards for using and handling patient data in electronic format. The "Common Rule" also specifies how to protect human subjects in federally-funded research. In Europe, the EU Data Protection Directive specifies rules on handling and processing any information about individuals. Violations of these standards could result in legal responsibilities and penalties including fine and imprisonment. Data mining results that are relevant to patients and subjects need to be interpreted in the proper medical context and with the help of the biomedical professionals.

In biomedical data mining, under most conditions patient data should not be *individually identifiable*, i.e., no record should provide sufficient data to identify the individual related to the record. These include anonymous data (data collected without patient-identification information), anonymized data (data collected with patient-identification information which is removed later), or de-identified data (data with patient-identification information encoded or encrypted) (Cios and Moore, 2002).

4. SUMMARY

In this chapter we provide a broad overview of selected knowledge management, data mining, and text mining techniques and their use in various emerging biomedical applications. However powerful they may be, these techniques need to be used with great care in the biomedical applications. One concern, as discussed earlier, is that medical data are often sensitive and involve private and confidential information. It is important that patients' confidentiality and privacy are not compromised due to the introduction of advanced knowledge management, data mining, and text mining technologies. Another caveat is that findings generated from selected machine learning techniques need to be interpreted carefully. Knowledge and patterns discovered by computers need to be experimentally or clinically validated in order to be considered rigorous, just like any knowledge generated by human. Errors and incorrect associations could propagate quickly through electronic media, especially when large databases and powerful computational techniques are involved.

Nonetheless, these new knowledge management, data mining, and text mining techniques are changing the way new knowledge is discovered, organized, applied, and disseminated. With the increasing speed of computers, the connectivity of the Internet, the abundance of biomedical data, and the advances in medical informatics research, we believe we will continue to generate, manage, and harvest biomedical knowledge effectively and efficiently, allowing us to better understand the complex biological processes of life and assist in addressing the well-being of human kind.

REFERENCES

Abidi, S. S. R. (2001). "Knowledge Management in Healthcare: Towards 'Knowledge-driven' Decision-support Services," *International Journal of Medical Informatics*, 63, 5-18.

Acir, N. and Guzelis, C. (2004). "Automatic Spike Detection in EEG by a Two-stage Procedure Based on Support Vector Machines," *Computers in Biology and Medicine*, 34(7), 561-575.

Ackerman, M. J. (1991). "The Visible Human Project," *Journal of Biocommunication*, 18(2), 14.

Ahmad, S., Gromiha, M. M., and Sarai, A. (2004). "Analysis and Prediction of DNA-binding Proteins and Their Binding Residues Based on Composition, Sequence, and Structural Information," *Bioinformatics*, 20(4), 477-486.

Altschul, S. F., Madden, T. L., Schaffer, A. A., Zhang, J., Zhang, Z., Miller, W., and Lipman, D. J. (1997). "Gapped BLAST and PSI-BLAST: A New Generation of Protein Database Search Programs," *Nucleic Acids Research*, 25(17), 3389-3402.

Antani, S., Lee, D. J., Long, L. R., and Thoma, G. R. (2004). "Evaluation of Shape Similarity Measurement Methods for Spine X-ray Images," *Journal of Visual Communication and Image Representation*, 15, 285-302.

Baclawski, K., Cigna, J., Kokar, M. W., Mager, P., and Indurkhya, B. (2000). "Knowledge Representation and Indexing Using the Unified Medical Language System," in *Proceedings of the Pacific Symposium on Biocomputing*, 493-504.

Barrera, J., Cesar-Jr, R. M., Ferreira, J. E., and Gubitoso, M. D. (2004). "An Environment for Knowledge Discovery in Biology," *Computers in Biology and Medicine*, 34, 427-447.

Baujard, O., Baujard, V., Aurel, S., Boyer, C., and Appel, R. D. (1998). "Trends in Medical Information Retrieval on the Internet," *Computers in Biology and Medicine*, 28, 589-601.

Belacel, B., Cuperlovic-Culf, M., Laflamme, M., and Ouellette, R. (2004). "Fuzzy J-Means and VNS Methods for Clustering Genes from Microarray Data," *Bioinformatics*, 20(11), 1690-1701.

Belew, R. K. (1989). "Adaptive Information Retrieval: Using a Connectionist representation to Retrieve and Learn about Documents," in *Proceedings of the 12th ACM-SIGIR Conference*, Cambridge, MA, June 1989.

Berman, J. J. (2002). "Confidentiality Issues for Medical Data Miners," *Artificial Intelligence in Medicine*, 26(1-2), 25-36.

Blaschke, C., Andrade, M. A., Ouzounis, C. and Valencia, A. (1999). "Automatic Extraction of Biological Information from Scientific Text: Protein-Protein Interactions," in *Proceedings of the International Conference on Intelligent Systems for Molecular Biology*, 60-67.

Bodenreider, O. and McCray, A. T. (2003). "Exploring Semantic Groups through Visual Approaches," *Journal of Biomedical Informatics*, 36, 414-432.

Breiman, L. and Spector, P. (1992). "Submodel Selection and Evaluation in Regression: The X-random Case," *International Statistical Review*, 60(3), 291-319.

Brown, M. P. S., Grundy, W. N., Lin, D., Cristianini, N., Sugnet, C. W., Furey, T. S., Ares, M., and Haussler, D. (2000). "Knowledge-based Analysis of Microarray Gene Expression Data by Using Support Vector Machines," in *Proceedings of the National Academy of Sciences*, 97, 262-267.

Campbell, K. E., Oliver, D. E., and Shortliffe, E. H. (1998). "The Unified Medical Language System: Toward a Collaborative Approach for Solving Terminologic Problems," *Journal of the American Medical Informatics Association*, 5(1), 12-16.

Carbonell, J. G. Michalski, R. S., Mitchell, T. M. (1983). "An Overview of Machine Learning," in R. S. Michalski, J. G. Carbonell, and T. M. Mitchell (Eds.), *Machine Learning, An Artificial Intelligence Approach*, Palo Alto, CA: Tioga.

Cavideds, J. E. and Cimino, J. J. (2004). "Towards the Development of a Conceptual Distance Metric for the UMLS," *Journal of Biomedical Informatics*, 37, 77-85.

Chapman, W. W., Dowling, J. N., and Wagner, M. M. (2004). "Fever Detection from Free-text Clinical Records for Biosurveillance," *Journal of Biomedical Informatics*, 37, 120-127.

Chau, M. and Chen, H. (2003). "Comparison of Three Vertical Search Spiders," *IEEE Computer*, 36(5), 56-62.

Chau, M. and Chen, H. (2004). "Using Content-based and Link-based Analysis in Building Vertical Search Engines," in *Proceedings of the International Conference on Asian Digital Libraries*, Shanghai, China, December 13-17, 2004.

Chau, M., Xu, J. J., and Chen, H. (2002). "Extracting Meaningful Entities from Police Narrative Reports," in *Proceedings of the National Conference for Digital Government Research*, Los Angeles, California, USA, May 19-22, 2002, 271-275.

Chen, H. (2001). *Knowledge Management Systems: A Text Mining Perspective*, Tucson, AZ: The University of Arizona.

Chen, H. and Chau, M. (2004). "Web Mining: Machine Learning for Web Applications," *Annual Review of Information Science and Technology*, 38, 289-329.

Chen, H. and Kim, J. (1995). "GANNET: A Machine Learning Approach to Document Retrieval," *Journal of Management Information Systems*, 11(3), 9-43.

Chen, H., Lally, A. M., Zhu, B., and Chau, M. (2003). "HelpfulMed: Intelligent Searching for Medical Information over the Internet," *Journal of the American Society for Information Science and Technology*, 54(7), 683-694, 2003.

Chen, H. and Ng, T. (1995). "An Algorithmic Approach to Concept Exploration in a Large Knowledge Network (Automatic Thesaurus Consultation): Symbolic Branch and Bound Search vs. Connectionist Hopfield Net Activation," *Journal of the American Society for Information Science*, 46(5), pp. 348-369.

Chinchor, N. A. (1998). "Overview of MUC-7/MET-2," in *Proceedings of the Seventh Message Understanding Conference (MUC-7)*, Virginia, USA, April 29 - May 1, 1998.

Cimino, J. J., Min, H., and Perl, Y. (2003) "Consistency across the Hierarchies of the UMLS Semantic Network and Metathesaurus," *Journal of Biomedical Informatics*, 36, 450-461.

Cios, K. J. and Moore, G. W. (2002). "Uniqueness of Medical Data Mining," *Artificial Intelligence in Medicine*, 26(1-2), 25-36.

Cohen, P. R. and Feigenbaum, E. A. (1982). *The Handbook of Artificial Intelligence: Volume III*, Reading, MA: Addison-Wesley.

Dawes, M. and Sampson, U. (2003). "Knowledge Management in Clinical Practice: A Systematic Review of Information Seeking Behavior in Physicians," *International Journal of Medical Informatics*, 71, 9-15.

Dickerson, J. A., Berleant, D., Cox, Z., Fulmer, A. W., and Wurtele, E. (2003). "Creating and Modeling Metabolic and Regulatory Networks Using Text Mining and Fuzzy Expert Systems," in J. T. L. Wang, C. H. Wu, and P. P. Wang (Eds.), *Computational Biology and Genome Informatics*, World Scientific.

Dreiseitl, S., Ohno-Machado, L., Kittler, H., Vinterbo, S., Billhardt, H., Binder, M. (2001). "A Comparison of Machine Learning Methods for the Diagnosis of Pigmented Skin Lesions," *Journal of Biomedical Informatics*, 34, 28-36.

Duda, R. O. and Hart, P. E. (1973). *Pattern Classification and Scene Analysis*, New York: John Wiley and Sons.

Dunham, M. H. (2002). *Data Mining: Introductory and Advanced Topics*, New Jersey, USA: Prentice Hall.

Efron, B. (1983). "Estimating the Error Rate of a Prediction Rule: Improvement on Cross-Validation," *Journal of the American Statistical Association*, 78(382), 316-330.

Efron, B. and Tibshirani, R. (1993). *An Introduction to the Bootstrap*, Chapman and Hall.

Eisen, M., Spellman, P., Brown, P., and Botstein, D. (1998). "Cluster Analysis and Display of Genome-wide Expression Patterns," in *Proceedings of the National Academy of Sciences*, 95, 14863-14868.

Fayyad, U. M., Piatetsky-Shapiro, G., and Smyth, P. (1996). "From Data Mining to Knowledge Discovery in Databases," *AI Magazine*, 17(3), 37-54.

Fisher, D. H. (1987). "Knowledge Acquisition via Incremental Conceptual Clustering," *Machine Learning*, 2, 139-172.

Fogel, D. B. (1994). "An Introduction to Simulated Evolutionary Optimization," *IEEE Transactions on Neural Networks*, 5, 3-14.

Friedman, C. Hripcsak, G., Johnson, S. B., Cimino, J. J., Clayton, P. D. (1990). "A Generalized Relational Schema for an Integrated Clinical Patient Database," in *Proceedings of the 14th Annual Symposium on Computer Applications in Medical Care*, 335-339.

Friedman, C. and Hripcsak, G. (1998). "Evaluating Natural Language Processors in the Clinical Domain," *Methods of Information in Medicine*, 37, 334-344.

Friedman, C., Kra, P., Yu, H., Krauthammer, M., and Rzhetsky, A. (2001). "GENIES: A Natural-language Processing System for the Extraction of Molecular Pathways from Journal Articles," *Bioinformatics*, 17(Supp. 1), S74-S82.

Fukuda K., Tamura A., Tsunoda T., and Takagi T. (1998). "Toward Information Extraction: Identifying Protein Names from Biological Papers," in *Proceedings of the Pacific Symposium on Biocomputing*, 707-718.

Fuller, S., Revere, D., Soderland, S., Bugni, P., Kadiyska, Y., Reber, L., Fuller, H., and Martin, G. (2002). "Modeling a Concept-Based Information System to Promote Scientific Discovery: The Telemakus System," in *Proceedings of the AMIA 2002 Annual Symposium*, 1023.

Fuller, S., Revere, D., Bugni, P., Fuller, H., and Martin, G. (2004). "A Knowledgebase System to Enhance Scientific Discovery: Telemakus," *Biomedical Digital Libraries,* 1(2-15).

Goldberg, D. E. (1989). *Genetic Algorithms in Search, Optimization, and Machine Learning,* Reading, MA: Addison-Wesley.

Golub T. R., Slonim D. K., Tamayo P., Huard C., Gaasenbeek M., Mesirov J. P., Coller H., Loh M. L., Downing J. R., Caligiuri M. A., Bloomfield C. D., Lander E. S. (1999). "Molecular Classification of Cancer: Class Discovery and Class Prediction by Gene Expression Monitoring," *Science*, 286(5439), 531-537.

Gruninger, M. and Lee, J. (2002). "Ontology: Applications and Design," *Communications of the ACM*, 45(2), 39-41

Han, K., and Byun, Y. (2004). "Three-dimensional Visualization of Protein Interaction Networks," *Computers in Biology and Medicine*, 34, 127-139.

Harris, M. R., Savova, G. K., Johnson, T. M., and Chute, C. G. (2003). "A Term Extraction Tool for Expanding Content in the Domain of Functioning, Disability, and Health: Proof of Concept," *Journal of Biomedical Informatics*, 36, 250-259.

Hatzivassiloglou, V., Duboue, P. A., and Rzhetsky, A. (2001). "Disambiguating Proteins, Genes, and RNA in Text: A Machine Learning Approach," *Bioinformatics*, 17(Supp. 1), S96-S106.

Hayes-Roth, F. and Jacobstein, N. (1994). "The State of Knowledge-based Systems," *Communications of the ACM*, 37, 27-39.

Hearst, M. A. (1999). "Untangling Text Data Mining," in *Proceedings of ACL'99: the 37th Annual Meeting of the Association for Computational Linguistics*, Maryland, June 20-26.

Heathfield, H. and Louw, G. (1999). "New Challenges for Clinical Informatics: Knowledge Management Tools," *Health Informatics Journal*, 5(2), 67-73.

Herrero, J., Valencia, A., and Dopazo, J. (2001). "A Hierarchical Unsupervised Growing Neural Network for Clustering Gene Expression Patterns," *Bioinformatics*, 17, 126-136.

Hersh, W. (1996). *Information Retrieval: A Health Care Perspective*. Berlin, Germany: Springer-Verlag.

Hersh, W., Mailhot, M., Arnott-Smith, C., and Lowe, H. (2002). "Selective Automated Indexing of Findings and Diagnoses in Radiology Reports," *Journal of Biomedical Informatics*, 34, 262-273.

Herwig, R., Poustka, A., Müller, C., Bull, C., Lehrach, H., and O'Brien, J. (1999). "Large-scale Clustering of cDNA Fingerprinting Data," *Genome Research*, 9, 1093-1105.

Hirst, J. D. and Sternberg, M. J. E. (1992). "Prediction of Structural and Functional Features of Protein and Nucleic Acid Sequences by Artificial Neural Networks," *Biochemistry*, 31, 7211-7218.

Holland, J. H. (1975). *Adaptation in Natural and Artificial Systems*, Ann Arbor, MI: University of Michigan Press.

Hopfield, J. J. (1982). "Neural Network and Physical Systems with Collective Computational Abilities," in *Proceedings of the National Academy of Science,* USA, 1982, 79(4), pp. 2554-2558.

Houston, A. L., Chen, H., Hubbard, S. M., Schatz, B. R., Ng, T. D., Sewell, R. R. and Tolle, K. M. (1999). "Medical Data Mining on the Internet: Research on a Cancer Information System," *Artificial Intelligence Review*, 13, 437-466.

Hsu, A. L., Tang, S., and Halgamuge, S. K. (2003). "An Unsupervised Hierarchical Dynamic Self-organizing Approach to Cancer Class Discovery and Market Gene Identification in Microarray Data," *Bioinformatics*, 19(16), 2131-2140.

Hripcsak, G. (1993). "Monitoring the Monitor: Automated Statistical Tracking of a Clinical Event Monitor," *Computers and Biomedical Research, 26(5), 449-466.*

Hripcsak, G., Austin, J. H., Alderson, P. O., and Friedman, C. (2002). "Use of Natural Language Processing to Translate Clinical Information from a Database of 889,921 Chest Radiographic Reports," *Radiology*, 224(1), 157-163.

Humphreys, B. L., Lindberg, D. A. B., and McCray, A. (1993). "The Unified Medical Language System," *Methods of Information in Medicine*, 32(4), 281.

Humphreys, B. L., Lindberg, D. A. B., Schoolman, H. M., and Barnett, G. O. (1998). "The Unified Medical Language System: An Informatics Research Collaboration," *Journal of the American Medical Informatics Association*, 5(1), 1-11.

Jackson, J. R. (2000). "The Urgent Call for Knowledge Management in Medicine," *The Physician Executive*, 26(1), 28-31.

Jain, A. K., Dubes, R. C. and Chen, C. (1987). "Bootstrap Techniques for Error Estimation," *IEEE Transactions on Pattern Analysis and Machine Learning*, 9(5), 628-633.

Knirsch, C.A., Jain, N. L., Pablos-Mendez, A., Friedman, C., and Hripcsak, G. (1996). "Respiratory Isolation of Tuberculosis Patients Using Clinical Guidelines and an Automated Clinical Decision Support System," *Infection Control and Hospital Epidemiology*, 19(2), 94-100.

Jain, N. L. and Friedman, C. (1997). "Identification of Findings Suspicious for Breast Cancer Based on Natural Language Processing of Mammogram Reports." in *Proceedings of the Fall 1997 AMIA Conference*, Philadelphia, USA, 829-833.

Janetzki, V., Allen, M., and Cimino, J. J. (2004). "Using Natural Language Processing to Link from Medical Text to On-line Information Resources," *Proceedings of Medinfo*, 2004, 1665.

Joachims, T. (1998). "Text Categorization with Support Vector Machines: Learning with Many Relevant Features," in *Proceedings of the European Conference on Machine Learning*, Berlin, 1998, pp. 137-142.

Kandaswamy, A., Kumar, C. S., Ramanathan, R. P. Jayaraman, R., and Malmurugan, N. (2004). "Neural Classification of Lung Sounds Using Wavelet Coefficients," *Computers in Biology and Medicine*, 34, 523-537.

Karasavvas, K. A., Baldock, R., and Burger, A. (2004). "Bioinformatics Integration and Agent Technology," *Journal of Biomedical Informatics*, 37, 205-219.

Kazama, J., Maino, T., Ohta, Y., and Tsujii, J. (2002). "Tuning Support Vector Machines for Biomedical Named Entity Recognition," in *Proceedings of the Workshop on Natural Language Processing in the Biomedical Domain*, Philadelphia, USA, July 2002, 1-8.

Kohavi, R. (1995). "A Study of Cross-validation and Bootstrap for Accuracy Estimation and Model Selection," in *Proceedings of the 14th International Joint Conference on Artificial Intelligence*, San Francisco, CA, 1995, Morgan Kaufmann, pp. 1137-1143.

Kohonen, T. (1995). *Self-organizing Maps*, Springer-Verlag, Berlin.

Kononenko, I. (1993). "Inductive and Bayesian Learning in Medical Diagnosis," *Applied Artificial Intelligence*, 7, 317-337, 1993.

Kovalerchuk, B., Vityaev, E., and Ruiz, J. F. (2001). "Consistent and Complete Data and 'Expert' Mining in Medicine," in Cios, K. J. (Ed.), *Medical Data Mining and Knowledge Discovery*, New York, USA: Physica Verlag.

Kretschmann, E., Fleischmann, W., and Apweiler, R. (2001). "Automatic Rule Generation for Protein Annotation with the C4.5 Data Mining Algorithm Applied on SWISS-PROT," *Bioinformatics*, 17(10), 920-926.

Krishnan, V. G. and Westhead, D. R. (2003). "A Comparative Study of Machine-Learning Methods to Predict the Effects of Single Nucleotide Polymorphisms on Protein Function," *Bioinformatics*, 19(17), 2199-2209.

Kuperman, G. J., Gardner, R.M., Pryor, T.A. (1991). *The HELP System*, New York: Springer-Verlag.

Kwok, K. L. (1989). "A Neural Network for Probabilistic Information Retrieval," in *Proceedings of the 12th ACM-SIGIR Conference on Research and Development in Information Retrieval*, Cambridge, Massachusetts, June 1989, pp.21-30.

Langley, P. and Simon, H. (1995). "Applications of Machine Learning and Rule Induction," *Communications of the ACM*, 38(11), 55-64.

Leroy, G. and Chen, H. (2001). "Meeting Medical Terminology Needs – The Ontology-Enhanced Medical Concept Mapper," *IEEE Transactions on Information Technology in Biomedicine*, 5(4), 261-270.

Leroy, G. and Chen, H. (forthcoming). "Genescene: An Ontology-enhanced Integration of Linguistic and Co-occurrence-based Relations in Biomedical Texts" *Journal of the American Society for Information Science and Technology*, forthcoming.

Leroy, G., Chen, H., and Martinez, J. D. (2003). "A Shallow Parser Based on Closed-class Words to Capture Relations in Biomedical Text," *Journal of Biomedical Informatics*, 36, 145-158.

Lippmann, R. P. (1987). An Introduction to Computing with Neural Networks, *IEEE Acoustics Speech and Signal Processing Magazine*, 4, 4-22.

Maniezzo V. (1994). "Genetic Evolution of the Topology and Weight Distribution of Neural Networks," *IEEE Transactions on Neural Networks*, 5(1), 39-53.

Mendes, R. R. F., Voznika, F. B., Freitas, A. A. and Nievola, J. C. (2001). "Discovering Fuzzy Classification Rules with Genetic Programming and Co-evolution," *Principles of Data Mining and Knowledge Discovery, Lecture Notes in Artificial Intelligence*, 2168, pp. 314-325. Springer-Verlag, 2001.

Michalewicz, Z. (1992). Genetic Algorithms + Data Structures =Evolution Programs. Berlin: Springer-Verlag.

Mitchell, T. (1997). *Machine Learning*, McGraw Hill, 1997.

Montani, S. and Bellazzi, R. (2002). "Supporting Decisions in Medical Applications: The Knowledge Management Perspective," *International Journal of Medical Informatics*, 68, 79-90.

Nagarajan, N. and Yona, G. (2004). "Automatic Prediction of Protein Domains from Sequence Information Using a Hybrid Learning System," *Bioinformatics*, 20(9), 1335-1360.

National Research Council (2000). *Bioinformatics: Converting Data to Knowledge: Workshop Summary*, Washington, D.C.: National Academies Press.

Paass, G. (1990), "Probabilistic Reasoning and Probabilistic Neural Networks," in *Proceedings of the 3rd International Conference on Information Processing and Management of Uncertainty*, pp.6-8.

Palakal, M., Mukhopadhyay, S., Mostafa, J., Raje, R., N'Cho, M., and Mishra, S. (2001). "An Intelligent Biological Information Management System," *Bioinformatics*, 18(10), 1283-1288.

Prather, J. C., Lobach, D. F., Goodwin, L. K., Hales, J. W., Hage, M. L., and Hammond, W. E. (1997). "Medical Data Mining: Knowledge Discovery in a Clinical Data Warehouse," in *Proceedings of the AMIA Annual Symposium Fall 1997*, 101-105.

Perl, Y. and Geller, J. (2003). "Research on Structural Issues of the UMLS – Past, Present, and Future," *Journal of Biomedical Informatics*, 36, 409-413.

Pustejovsky J., Castano J., Zhang J., Kotecki M., and Cochran B. (2002). "Robust Relational Parsing over Biomedical Literature: Extracting Inhibit Relations," *Pacific Symposium on Biocomputing*, 362-373.

Qian, N. and Sejnowski, T. J. (1988). "Predicting the Secondary Structure of Globular Proteins Using Neural Network Models," *Journal of Molecular Biology*, 202, 865-884.

Qin, J., Lewis, D. P., and Noble, W. S. (2003). "Kernel Hierarchical Gene Clustering from Microarray Expression Data," *Bioinformatics*, 19(16), 2097-2104.

Qu Y. and Xu., S. (2004). "Supervised Cluster Analysis for Microarray Data Based on Multivariate Gaussian Mixture," *Bioinformatics*, 20(12), 1905-1913.

Quinlan, J. R. (1983) "Learning Efficient Classification Procedures and Their Application to Chess End Games," in R. S. Michalski, J. G. Carbonell, and T. M. Mitchell (Eds.), *Machine Learning: An Artificial Intelligence Approach*, Palo Alto, CA: Tioga.

Quinlan, J. R. (1993). *C4.5: Programs for Machine Learning*, Los Altos, CA: Morgan Kaufmann.

Revere, D., Fuller, S. S, Bugni, P. F., and Martin, G. M. (2004). "An Information Extraction and Representation System for Rapid Review of the Biomedical Literature," in *Proceedings of Medinfo*, 2004.

Rindflesch, T. C., Tanabe, L., and Weinstein, J. N., and Hunter, L. (2000). "EDGAR: Extraction of Drugs, Genes and Relations from the Biomedical Literature," in *Proceedings of the Pacific Symposium on Biocomputing* , 514-525.

Rindflesch, T. C. and Fiszman, M. (2003) "The Interaction of Domain Knowledge and Linguistic Structure in Natural Language Processing: Interpreting Hypernymic Propositions in Biomedical Text," *Journal of Biomedical Informatics*, 36, 462-477.

Rojdestvenski, I. (2003). "VRML Metabolic Network Visualizer," *Computers in Biology and Medicine*, 33, 169-182.

Rosse, C. and Mejino, J. L. V. (2003). "A Reference Ontology for Biomedical Informatics: The Foundational Model of Anatomy," *Journal of Biomedical Informatics*, 36, 478-500.

Rumelhart, D. E., Hinton, G. E., and McClelland, J. L. (1986a). "A General Framework for Parallel Distributed Processing," in D. E. Rumelhart, J. L. McClelland, and the PDP Research Group (Eds.), *Parallel Distributed Processing*, pp. 45-76, Cambridge, MA: The MIT Press.

Rumelhart, D. E., Hinton, G. E., and Williams, R. J. (1986b). "Learning Internal Representations by Error Propagation," in D. E. Rumelhart, J. L. McClelland, and the PDP Research Group (Eds.), *Parallel Distributed Processing*, pp. 318-362, Cambridge, MA: The MIT Press.

Samuelson, C. and Rayner, M. (1991). "Quantitative Evaluation of Explanation-based Learning as an Optimization Tool for a Large-scale Natural Language System," in *Proceedings of the 12th International Joint Conference on Artificial Intelligence*, Sydney, Australia, 1991, pp. 609-615.

Sawa, T. and Ohno-Machado, L. (2003). "A Neural Network-based Similarity Index for Clustering DNA Microarray Data," *Computers in Biology and Medicine*, 33, 1-15.

Schubart, J. R. and Einbinder, J. S. (2000). "Evaluation of a Data Warehouse in an Academic Health Sciences Center," *International Journal of Medical Informatics*, 60, 319-333.

Sekimisu, T., Park, H. S., and Tsujii, J. (1998). "Identifying the Interaction between Genes and Gene Products Based on Frequently Seen Verbs in MEDLINE Abstracts," *Genome Informatics*, 9, 62-71.

Shatkay, H., Edwards, S., and Boguski, M. (2002). "Information Retrieval Meets Gene Analysis," *IEEE Intelligent Systems*, 17(2), 45-53.

Shatkay, H., Edwards, S., Wilbur, W. J., and Boguski, M. (2000). "Genes, Themes, and Microarrays: Using Information Retrieval for Large-scale Gene Analysis," in *Proceedings of the International Conference on Intelligent Systems for Molecular Biology*, 317-328.

Shortliffe, E. (1976). *Computer-based Medical Consultations: MYCIN*, New York: Elsevier/North Holland.

Shortliffe, E. (1987). "Computer Programs to Support Clinical Decision Making," *Journal of the American Medical Association*, 258, 61-66.

Simon, H. A. (1983). "Why Should Machines Learn?" In R. S. Michalski, J. Carbonell, and T. M. Mitchell (Eds.), *Machine Learning: An Artificial Intelligence Approach*. Palo Alto, CA: Tioga Press.

Smalheiser, N. R. and Swanson, D. R. (1998). "Using ARROWSMITH: A Computer-assisted Approach to Formulating and Assessing Scientific Hypotheses," *Computer Methods and Programs in Biomedicine*, 57, 149-153.

Stone, M. (1974). "Cross-validation Choices and Assessment of Statistical Predictions," *Journal of the Royal Statistical Society*, 36, 111-147.

Sujansky, W. (2001). "Heterogeneous Database Integration in Biomedicine," *Journal of Biomedical Informatics*, 34, 285-298.

Sun, Y. (2004). "Methods for Automated Concept Mapping between Medical Databases," *Journal of Biomedical Informatics*, 37, 162-178.

Swanson, D. R. (1986). "Fish Oil, Raynaud's Syndrome, and Undiscovered Public Knowledge," *Perspectives in Biology and Medicine*, 30(1), 7-18.

Tamayo, P., Slonim, D., Mesirov, J., Zhu, Q., Kitareewan, S., Dmitrovsky, E., Lander, E. S., and Golub T. R. (1999). "Interpreting Patterns of Gene Expression with Self-organizing Maps: Methods and Application to Hematopoietic Differentiation," in *Proceedings of the National Academy of Sciences*, 96, 2907-2912.

Tanabe, L. and Wilbur, W. J. (2002). "Tagging Gene and Protein Names in Biomedical Text," *Bioinformatics*, 18(8), 1124-1132.

The Gene Ontology Consortium (2000). "Gene Ontology: Tool for the Unification of Biology," *Nature Genetics*, 25(1), 25-29.

Tolle, K. and Chen, H. (2000) "Comparing Noun Phrasing Techniques for Use with Medical Digital Library Tools," *Journal of the American Society for Information Science*, 51(4), 352-370.

Tu, Q., Tang, H., and Ding, D. (2004). "MedBlast: Searching Articles Related to a Biological Sequence," *Bioinformatics*, 20(1), 75-77.

Vapnik, V. (1998). *Statistical Learning Theory*, Wiley, Chichester, GB, 1998.

Wain, H. M., Lush, M., Ducluzeau, F., Povey, S. (2002). "Genew: The Human Gene Nomenclature Database," *Nucleic Acids Research*, 30(1), 169-171.

Wilbur, W. J. and Yang, Y. (1996). "An Analysis of Statistical Term Strength and Its Use in the Indexing and Retrieval of Molecular Biology Texts," *Computers in Biology and Medicine*, 26(3), 209-222.

Yandell, M. D. and Majoros, W. H. (2002). "Genomics and Natural Language Processing," *Nature Reviews Genetics*, 3(8), 601-610.

Yang, Y. and Liu, X. (1999). "A Re-examination of Text Categorization Methods, in *Proceedings of the 22nd Annual International ACM Conference on Research and Development in Information Retrieval* (SIGIR'99), 1999, pp. 42-49.

Yoo, T. S., and Chen, D. T. (1994). "Interactive 3D Medical Visualization: A Parallel Approach to Surface Rendering 3D Medical Data," in *Proceedings of the Symposium for Computer Assisted Radiology*, North Carolina, USA, June 12-15, 1994, 100-105.

Yu, H., Hatzivassiloglou, V., Rzhetsky, A., and Wilbur, W. J. (2002). "Automatically Identifying Gene/Protein Terms in MEDLINE Abstracts," *Journal of Biomedical Informatics*, 35, 322-330.

Yu, V. L., Fagan, L. M., Wraith, S. M., Clancey, W. J., Scott, A. C., Hannigan, J. Blum, R. L., Buchanan, B. G., and Cohen, S. N. (1979). "Antimicrobial Selection by a Computer: A Blinded Evaluation by Infectious Disease Experts," *Journal of the American Medical Association*, 242(12), 1279-1282.

Zadeh, L. A. (1965). "Fuzzy sets," *Information and Control*, 8, 338-353.

Zhang, L., Perl, Y., Halper, M., and Geller, J. (2003). "Designing Metaschemas for the UMLS Enriched Semantic Network," *Journal of Biomedical Informatics*, 36, 433-449.

SUGGESTED READINGS

Shortliffe, E. H. and Perreault, L. E. (2002). Medical Informatics: Computer Applications in Health Care and Biomedicine, Springer.
This excellent introductory book provides a comprehensive overview of the applications of computer and information technologies in health care and biomedicine.

Baldi, P. and Brunak, S. (2000). *Bioinformatics: The Machine Learning Approach*, The MIT Press.
The book describes bioinformatics from a technical perspective and explains in detail the application of data mining algorithms for biomedical sequence and structure analysis.

Mitchell, T. (1997). *Machine Learning*, McGraw Hill, 1997.
This introductory book includes useful reviews of various machine learning techniques and their applications.

Chen, H., Lally, A. M., Zhu, B., and Chau, M. (2003). "HelpfulMed: Intelligent Searching for Medical Information over the Internet," *Journal of the American Society for Information Science and Technology*, 54(7), 683-694, 2003.
This article provides an overview of medical information retrieval techniques on the Internet, including Web crawling, co-occurrence analysis, and document visualization.

Eisen, M., Spellman, P., Brown, P., and Botstein, D. (1998). "Cluster Analysis and Display of Genome-wide Expression Patterns," in *Proceedings of the National Academy of Sciences*, 95, 14863-14868.
This article presents a study on performing clustering techniques on gene expression data.

Swanson, D. R. (1986). "Fish Oil, Raynaud's Syndrome, and Undiscovered Public Knowledge," *Perspectives in Biology and Medicine*, 30(1), 7-18.
This article describes the interesting story of how public knowledge could remain "undiscovered" as there were no researchers linking the literature in two separate fields, and how the computer was used to discover such knowledge.

Yandell, M. D. and Majoros, W. H. (2002). "Genomics and Natural Language Processing," *Nature Reviews Genetics*, 3(8), 601-610.
This article reviews research studies that apply natural language processing and text mining techniques in genomics.

ONLINE RESOURCES

National Center for Biotechnology Information (NCBI) http://www.ncbi.nlm.nih.gov/
NCBI, a division of the National Library of Medicine, provides access to many excellent molecular biology resources, including GenBank (an annotated collection of all publicly available DNA sequences), Entrez (a cross-database search engine), and BLAST (a sequence similarity search engine).

Unified Medical Language Systems (UMLS)
http://www.nlm.nih.gov/research/umls/

Developed by the Lister Hill Center of the NLM, UMLS provides a large-scale and widely-used medical ontology for information retrieval and text mining applications in biomedicine. The three major components include the Metathesaurus, the Semantic Network, and the Specialist Lexicon.

ExPASy Proteomics Server
http://us.expasy.org/
The ExPASy (Expert Protein Analysis System) proteomics server is hosted by the Swiss Institute of Bioinformatics (SIB). It focuses on the analysis of protein sequences and structures. It provides access to Swiss-PROP, TrEMBL, and other proteomics and sequence analysis tools and resources.

Protein Data Bank
http://www.rcsb.org/pdb/
The Protein Data Bank is the single worldwide repository for 3-D biological macromolecular structure data.

European Bioinformatics Institute (EBI)
http://www.ebi.ac.uk/
EBI is the European equivalent of NCBI and is part of the European Molecular Biology Laboratory (EMBL). It manages several biological databases including: nucleic acid, protein sequences, and macromolecular structures.

GenomeNet
http://www.genome.jp/
Developed in Japan, GenomeNet includes several databases for genome research and molecular and cellular biology. Its services include the Kyoto Encyclopedia of Genes and Genomes (KEGG) and the DBGET Integrated Database Retrieval System, among others.

GenomeWeb
http://www.rfcgr.mrc.ac.uk/GenomeWeb/
This site provides a comprehensive directory of genome-related Web sites and information.

Saccharomyces Genome Database (SGM)
http://www.yeastgenome.org
This database contains information about the molecular biology and genetics of the yeast Saccharomyces cerevisiae. Commonly known as the baker's or budding yeast, its genome has been widely studied in bioinformatics.

The Visible Human Project
http://www.nlm.nih.gov/research/visible/
This site includes a detailed description of NLM's Visible Human Project, instructions on how to obtain the data, and some other related resources and conference information.

The UCI Machine Learning Repository
http://www.ics.uci.edu/~mlearn/MLRepository.html
This repository at the University of California, Irvine, contains data in many different domains (including biomedicine) that have been widely used to test and compare machine learning techniques.

WEKA
http://www.cs.waikato.ac.nz/ml/weka/
Developed at the University of Waikato in New Zealand, WEKA is an open-source machine learning software written in Java, containing a wide range of useful algorithms.

QUESTIONS FOR DISCUSSION

1. What are the similarities and differences between bioinformatics and medical informatics? How can research in the two areas be beneficial to each other?

2. What is an intelligent system? Can an intelligent system be more intelligent than humans? What are the important characteristics of an intelligent system in biomedicine?

3. Discuss the characteristics of major machine learning paradigms and their applicability in biomedicine.

4. Explain what knowledge management is and why it is useful for medical informatics. What are some of the good examples of biomedical knowledge management systems? How can a knowledge management system be created and used in industry?

5. Please compare the knowledge discovery process by computers with that in humans. Do you think that data mining and text mining techniques have begun to change the way that research is done in biomedicine?

6. What are the social, ethical, and legal concerns for future biomedical knowledge management, data mining, and text mining applications?

Chapter 2
MAPPING MEDICAL INFORMATICS RESEARCH

Shauna Eggers[1], Zan Huang[1], Hsinchun Chen[1], Lijun Yan[1], Cathy Larson[1], Asraa Rashid[1], Michael Chau[2], and Chienting Lin[3]

[1]Artificial Intelligence Lab, Department of Management Information Systems, Eller College of Management, The University of Arizona, Tucson, Arizona 85721; [2]The University of Hong Kong, School of Business, Hong Kong; [3]Department of Information Systems, Pace University, New York, NY 10038

Chapter Overview
The ability to create a big picture of a knowledge domain is valuable to both experts and newcomers, who can use such a picture to orient themselves in the field's intellectual space, track the dynamics of the field, or discover potential new areas of research. In this chapter we present an overview of medical informatics research by applying domain visualization techniques to literature and author citation data from the years 1994-2003. The data was gathered from NLM's MEDLINE database and the ISI Science Citation Index, then analyzed using selected techniques including self-organizing maps and citation networks. The results of our survey reveal the emergence of dominant subtopics, prominent researchers, and the relationships among these researchers and subtopics over the ten-year period.

Keywords
information visualization; domain analysis; self-organizing map; citation networks

1. INTRODUCTION

The rapid evolution of medical informatics and its subdomains makes it crucial for researchers to stay abreast of current developments and emerging trends. This task is made difficult, however, not only by the large amounts of available information, but by the interdisciplinary nature of the field. Relevant information is spread across diverse disciplines, posing a particular challenge for identifying relevant literature, prominent researchers, and research topics (Sittig, 1996, Andrews, 2002, Vishwanatham, 1998). Any attempt to understand the intellectual structure and development of the field must furthermore consider all of the contributing disciplines; as Börner et al. (2003) point out, "researchers looking at the domain from a particular discipline cannot possibly have an adequate understanding of the whole." In this chapter we report the results of an analysis of the medical informatics domain within an integrated knowledge mapping framework. We provide a brief review of the literature on knowledge mapping, then describe in detail the analysis design and results of our medical informatics literature mapping with three types of analysis: basic analysis, content map analysis, and citation network analysis.

2. KNOWLEDGE MAPPING: LITERATURE REVIEW

Domain analysis is a subfield of information science that attempts to reveal the intellectual structure of a particular knowledge domain by synthesizing disparate information, such as literature and citation data, into a coherent model (White and McCain 1997, Small 1999). Such a model serves as an overview to newcomers to the field, and reveals the field's dynamics and knowledge transfer patterns to experts.

A significant portion of domain analysis research has been focused on citation analysis. Historically, a great deal of manual effort was needed to gather citation data for this type of analysis by combining different literature resources and tracing through the citations. A manual analysis approach, however, is inherently subjective, and is impractical for the vast amounts of time-sensitive information available for most domains today (Börner et al., 2003). Digital citation indexes such as ResearchIndex (formerly CiteSeer) developed by NEC Research Institute (Lawrence et al. 1999) and ISI's Science Citation Index (SCI) eliminate the need for manual data collection, but still lead to large amounts of citation data that are difficult to analyze using traditional techniques. Recent developments in the field of domain visualization attempt to alleviate this citation information overload problem by applying information visualization techniques to produce visual (and often interactive) representations of the underlying intellectual structure of the domain reflected in the large-scale citation data. A wide range of techniques have been applied to citation visualization, including clustering display based on co-citation (Small, 1999), the "Butterfly" display (Mackinlay et al., 1999), Pathfinder network scaling (Chen and Paul, 2001), and hyperbolic trees (Aureka, 2002).

Content, or "semantic," analysis is another important branch of domain analysis. This type of analysis relies on natural language processing techniques to analyze large corpora of literature text. Techniques ranging from simple lexical statistics to key phrase co-occurrence analysis to semantic and linguistic relation parsing are applied to reveal topic distribution and associations within the domain. To alleviate the similar information overload problem as for the citation data, many visualization techniques have been developed to produce content maps of large-scale text collections. Prominent examples include ThemeScape and Galaxies (Wise et al., 1995), the underlying techniques of which are multidimensional scaling and principle component analysis, and WebSOM (Honkela et al., 1997) and ET Map (Chen et al., 1996) which are based on the self-organizing map algorithm.

The application of visualization techniques to both citation and content analysis is consistent with the exploratory nature of domain analysis and forms the foundation of knowledge (domain) mapping. These visualization results provide valuable support for users' visual exploration of a scientific domain to identify visual patterns that may reflect influential researchers and studies, emerging topics, hidden associations, and other findings regarding the domain.

The effectiveness of domain analysis specifically in medical informatics is demonstrated by surveys by Sittig (1996) and Vishwanathan (1998), who used citation-based analyses to identify core medical informatics literature, and by Andrews (2002), who uses author co-citation analysis (ACA) to create multidimensional maps of the relationships between influential authors. We have also seen large-scale content mapping of the general medical literature (Chen et al., 2003), but not specifically of the medical informatics field.

In this study, we adopt the knowledge mapping framework proposed by Huang et al. (2003) that leverages large-scale visualization tools for knowledge mapping in fast-evolving scientific domains. Under this framework we perform three types of analysis -- basic analysis, content map analysis, and citation network analysis -- to provide a multifaceted mapping of the medical informatics literature. Through analyzing documents and citation information we identify influential researchers in the field and the nature of their contributions, track knowledge transfer among the researchers, and identify domain subtopics and their trends of development. The results of our study present a comprehensive picture of medical informatics over the past ten years.

3. RESEARCH DESIGN

The Huang et al. (2003) framework proposes a generic set of analytical units, three analysis types, and various visualization technologies for representing the results of patent analysis. The analytical units include geographical regions, industries/research fields, sectors, institutions, individuals, and cross-units. Our medical informatics analysis focuses on individuals (authors), and research fields (subtopics) as units of analysis. We rely on two visualization techniques: self-organizing maps (SOMs) for revealing semantic grouping of topics, authors, and development trends; and citation networks for exploring knowledge transfer patterns. The details of our application of the Huang et al. three-pronged analysis are outlined below.

3.1 Basic Analysis

This first type of analysis provides "performance evaluation," namely, a measure of the level of an analytical unit's contribution to the field. Two types of measures are used for the contribution analysis, the productivity (or quantity) measures and impact (or quality) measures. We perform basic analysis at the author level to identify major researchers in medical informatics. The most prolific authors are determined by the number of publications attributed to them in our data set, with the highest-ranking authors deemed the most productive. A simple and commonly-used author impact measure is the number times an author is cited by others. The idea is that citation implies an acknowledgement of authority on the part of the citing author to the cited one, and that an author's citation level reflects the community's perceived value of their contribution to the field. This idea is supported by a substantial amount of academic literature on citation indexing. Garfield's 1955 vision of an interdisciplinary science citation index introduced the concept of citation as an impact factor indicator, and the concept has since been applied by the ResearchIndex in its citation context tool (Lawrence et al. 1999), Liu et al. (2004) in their AuthorRank indicator, and several domain analysis surveys (Andrews, 2002, Vishwanatham, 1998, Sittig, 1996, White and McCain, 1997, Chen et al., 2001, Noyons et al., 1999).

We expand on simple citation count by assigning authors an Authority score based on the HITS algorithm (Kleinberg, 1998), which was intended for identifying important web pages based on hyperlink citation structure. Following the formulation of the original HITS algorithm, two types of scores are defined for each author in our author citation analysis: an Authority score and a Hub score. An author with a high Authority score has a significant impact/influence on other authors, meaning his/her work has been extensively cited (directly and indirectly) by other authors. A high Hub score, on the other hand, indicates that an author's work has cited many influential studies. The Authority and Hub scores mutually reinforce each other: authors citing influential authors (with high Authority scores) tend to have high Hub scores; authors cited by authors who have cited influential authors (with high Hub scores) tend to be influential (with high Authority scores). With an author citation data set, we initialize the Authority scores as the number of times the authors are cited by others and the Hub scores as the number of times the authors cite others. The two scores are then computed following an iterative updating procedure:

$$\text{Authority Score}(p) = \sum_{q \text{ has cited } p} \text{Hub Score}(q)$$

$$\text{Hub Score}(q) = \sum_{q \text{ has cited } p} \text{Authority Score}(p)$$

The Authority score we use for our study is obtained with three iterations of score updating. It essentially incorporates the number of citations received by an author, the authors citing him/her, authors citing those citing authors, and so on.

3.2 Content Map Analysis

Content analysis is used in the Huang et al. framework to identify and track dominating themes in a field. Analyzing the content of the work produced by a specific analytical unit also provides valuable information on what subdisciplines that unit contributes to, and how the contribution changes over time. This approach augments traditional citation-based performance indicators (such as author co-citation) by operating directly on literature content, instead of inferring content from relationships between analytical units.

We use the self-organizing map (SOM) algorithm to perform content mapping of the medical informatics literature. Initially proposed by Kohonen (1990), the SOM algorithm analyzes similarities of entities with a large number of attributes and produces a map of the entities, in which the geographical distances correspond to the attribute-based similarities. In our study, we perform content mapping of papers and authors.

To generate the content maps, the text of each paper (a combination of titles and abstracts, in our study) is analyzed using the Arizona Noun Phraser, which identifies the key noun phrases based primarily on linguistic patterns (Tolle and Chen, 2000). These noun phrases, representing key concepts, are then used to represent the content of a paper by forming a binary vector, each element of which represents the occurrence of a particular noun phrase. The self-organizing map algorithm (SOM) typically produces a two-dimensional map to represent the content distribution of a set of documents. Each location in the map, that is, a node in a two-dimensional grid, is also assigned a key phrase vector, like the papers. These map node vectors are typically real-valued (for example, between 0 and 1) and initialized with random values. For each input paper, the SOM algorithm identifies a winning node that has the largest vector similarity measure to the input paper. The vector values of this winning node and its close neighbors are then updated to be more similar to the input paper vector. With all input papers used to perform the node vector updating process, the final configuration of the map, that is, the vector values of all map nodes, presents a content distribution of the input papers. The papers then obtain their locations in the map by finding the map nodes with the largest vector similarity measures. A map of authors is similarly generated by forming a key phrase vector for each author. The key phrase vector is created by combining the vectors for an author's papers, then used as input to the SOM algorithm in the same way as paper vectors.

We applied the multilayer SOM algorithms developed by Chen et al. (1995) to produce topic maps by adding a hierarchical topic region layer on top of a map of papers. We also perform longitudinal mapping, that is, a series of chronically sequential SOMs, to reveal the evolution of medical informatics subdisciplines. From the maps, a researcher can observe what disciplines exist at different points in time, when particular disciplines emerge, and their rate of growth and decline. A domain expert can potentially use such longitudinal maps to forecast emerging trends (Börner et al., 2003).

We also created an author map using the SOM algorithm. Based on the positions of the authors in the map, we identify groups of authors that had papers with similar contents.

3.3 Citation Analysis

Visualizing citation data as a network is a classic method for intuitively displaying knowledge transfer patterns among analytical units. Citation networks consist of nodes representing the analytical units, with directional links representing citations between them. When the analytical unit is an author, such networks can be used to quickly identify strong communication channels in the domain, and the structure of those channels. Since citation between authors implies a human judgment that a work by the cited author is relevant to one by the citing author, frequently-occurring citations can indicate that two authors work in a similar field. Hence, citation networks can be used to identify communities of researchers. For this study, we gathered citation information from ISI's Science Citation Index for the years 1994-2003 for a core group of researchers identified by the basic analysis. We then use the freely-available graphing program NetDraw (http://www.analytictech.com/netdraw.htm) to visualize the result.

4. DATA DESCRIPTION

Andrews (2002) points out that an author co-citation analysis is only as good as the analyst's choice of authors. The same can be said for domain analysis in general. We used a number of measures to collect as comprehensive a data set for our survey as possible. First, we used NLM's expansive MEDLINE database of biomedical literature to provide source documents for our analysis. We then used four criteria to locate documents in MEDLINE relevant to medical informatics. For an article to be included in our collection, at least one of the following had to be true:

1. The article was published in one of 22 prominent journals in the medical informatics domain. These journals consist of the 18 identified by Andrews (2002) and additionally two journals and two conference proceedings that are frequently cited in (Shortliffe et al., 2000). The complete list of journal titles is given in Table 2-1.
2. The article abstract or title contains one of the selected medical informatics keywords listed in Table 2-2.
3. The article is indexed by MEDLINE under the MeSH term "Medical Informatics." MeSH is widely acknowledged to be an authoritative indexing system.
4. The article was authored by a fellow of the American College of Medical Informatics (ACMI), a group of scholars who are determined by their peers to have made "significant and sustained contributions to the field" (http://www.amia.org/acmi/acmi.html).

The use of ACMI fellows as a test set on which to perform domain analysis is supported by Andrews (2002), who also cites the use of ACMI by Greenes and Siegel (1987).

Using the above criteria, we identified 24,495 medical informatics articles in MEDLINE, as of August 2004. Restricting our data set to articles published during our ten-year test bed, 1994-2003, yielded 16, 964 articles.

Table 2-1. Prominent medical informatics journals included in our study.

Journal Name
Artificial Intelligence in Medicine
Biomedizinische Technik (Biomedical Engineering)
Computer Methods and Programs in Biomedicine
Computers, Informatics, Nursing: CIN
IEEE Engineering in Medicine and Biology Magazine
IEEE Transactions on Information Technology in Biomedicine
International Journal of Medical Informatics
International Journal of Technology Assessment in Health Care
Journal of Biomedical Informatics
Journal of Cancer Education: The Official Journal of the American Association for Cancer Education
Journal of Evaluation in Clinical Practice
Journal of the American Medical Informatics Association (JAMIA)
M.D. Computing: Computers in Medical Practice
Medical and Biological Engineering and Computing
Medical Informatics and the Internet in Medicine
Medical Decision Making
Methods of Information in Medicine
Proceedings of the American Medical Informatics Association (AMIA) Annual Fall Symposium
Proceedings of the Annual Symposium on Computer Applications in Medical Care
Statistical Methods in Medical Research
Statistics in Medicine

Table 2-2. Keywords used to identify MEDLINE documents relevant to medical informatics.

Keyword
Medical informatics
Clinical informatics
Nursing informatics
Health informatics
Bioinformatics
Biomedical informatics

As White and McCain (1997) state, "we wished to let 'the field' dictate its top authors rather than choosing them ourselves." This means that in addition to using ACMI fellows for our analysis, we allowed our document set to determine the remainder of our author set: anyone identified as an author of an article in the medical informatics collection was included in our collection of authors. A count of the most frequently-occurring names in the collection determined the most prolific authors in the field, as listed in Table

2-3. These authors comprise the "core" set used to gather citation data from the Science Citation Index (SCI). As of this study, SCI is only searchable through the online Web of Science. A "citation search" was manually performed in the Web of Science for each author in our core set, to gather information on who has cited them, and who they cite. This search yielded some commonly-cited names that are not included in our core set, which can be seen in Tables 2-4 and 2-5. Together the core set and frequently-cited names list some of the most recognizable and influential researchers in the field, and citation information for all of these authors was used for our citation analysis.

5. RESULTS

5.1 Basic Analysis

Our basic analysis focused on authors as the analytical unit, with the results presented in Tables 2-3, 2-4, and 2-5. These tables offer different perspectives - productivity and impact factor, respectively - on the most highly contributing researchers in the domain. Table 2-3 lists the 96 most prolific authors, that is, those with the most publications attributed to them in our data set. James J. Cimino at Columbia University tops the list with 62 publications, followed closely by Arie Hasman at the University of Maastricht in the Netherlands, Robert A. Greenes of Harvard Medical School, and Perry L. Miller at Yale University. The citation search described in Section 4 above yielded some frequently cited authors that do not appear in the core set shown in Table 2-3. Citation counts were gathered for these authors in addition to those in the core set, and the most frequently cited of the combined list are shown in Table 2-4. Some authors of note in the list that do not appear among the core authors in Table 2-3 are Lucian L. Leape at the Harvard School of Public Health, Mor Peleg at Stanford University, and Suzanne Bakken at Columbia University.

Table 2-5 ranks the authors in the combined list by their citation-based Authority scores. James Cimino is again among the five highest scoring in this table, along with Mark A. Musen at Stanford University, Edward H. Shortliffe at Columbia University (formerly at Stanford), George Hripcsak at Columbia, and Paul D. Clayton, who was at Columbia until 1998 and is currently Chief Medical Informatics Officer at Intermountain Health Care in Salt Lake City. The latter four authors are shown in Table 2-3 to have approximately half the number of publications as the most prolific author, yet their Authority scores indicate the significant impact of their publications.

Table 2-3. Publication counts for prolific authors.

Author name	Number of publications in collection	Author name	Number of publications in collection
Cimino, James J.	62	Van der Lei, J.	22
Hasman, A.	52	Kahn, Michael G.	22
Greenes, Robert A.	45	Friedman, Carol	22
Miller, Perry L.	44	Rector, Alan L.	22
Haux, Reinhold	42	Whitehead, J.	21
Musen, Mark	39	Cerutti, S.	21
Patel, Vimla L.	38	Tierney, William M.	21
Safran, Charles	37	Warner, Homer R.	21
Barnett, Octo G.	35	Habbema, J. D.	20
Stefanelli, Mario	35	Friedman, Charles P.	20
Miller, Randolph A.	31	Beck, J. Robert	20
Shortliffe, Edward	31	Royston, P.	19
Van Bemmel, J. H.	30	Zhou, X. H.	19
Haug, Peter	29	McDonald, Clement	19
Hripcsak, George	29	Wigton, Robert S.	19
Fagan, Larry	29	Shahar, Y.	18
Kohane, Issac	28	Fieschi, M	18
Weinstein, M. C.	27	Lui, K. J.	18
Degoulet, Patrice	27	Haynes, R. Brian	18
Bates, David W.	27	Brinkley, James	18
Lenert, Leslie A.	27	Brennan, Patricia F.	18
Durand, L. G.	26	Kuperman, Gilad J.	18
Timpka, T.	26	Stead, William W.	18
Chute, Christopher	26	Tuttle, Mark S.	18
Clayton, Paul D.	26	Pinciroli, F.	17
Johnson, Stephen B.	26	Bolz, A.	17
Sittig, Dean F.	26	Spiegelhalter, D. J.	17
Greenland, S.	25	Simon, R.	17
Pfurtscheller, G.	25	Mitchell, Joyce A.	17
Hersh, William R.	25	Ohno-Machado, Lucila	17
Donner, A.	24	Tang, Paul C.	17
Thompson, S. G.	24	Tu, Samson W.	17
Huff, Standley M.	24	Van Ginneken, A.M.	16
Gardner, Reed M.	24	Dössel, O.	16
Dudeck, Joachim	24	Freedman, L. S.	16
Nadkarni, Prakash	24	Groth, T.	16
Teich, Jonathan M.	24	Meinzer, H. P.	16
Bellazzi, R.	23	Altman, Russ B.	16
Cooper, Greg	23	Reggia, James A.	16
Scherrer, Jean-Raoul	23	Slack, Warner V.	16
Wigertz, Ove	23		

Table 2-4. Citation counts for frequently cited authors.

Author name	Times cited by authors in medical informatics collection	Author name	Times cited by authors in medical informatics collection
Bates, D. W.	989	Greenes, R. A.	142
Cimino, J. J.	691	Lui, K. J.	137
McDonald, C. J.	359	Giuse, D. A.	135
Patel, V. L.	356	Neuper, C.	134
Hripcsak, G.	331	McCray, A. T.	131
Pfurtscheller, G.	306	Hersh, W. R.	129
Friedman, C.	301	Rind, D. M.	128
Miller, R. A.	289	Riva, A.	127
Musen, M. A.	287	Montani, S.	123
Greenland, S.	280	Huff, S. M.	123
Bellazzi, R.	243	Kuhn, K. A.	123
Overhage, J. M.	225	Johannesson, M.	122
Leape, L. L.	219	Kaplan, B.	120
Peleg, M.	215	Baud, R. H.	119
Hasman, A.	206	Lenert, L. A.	119
Bakken, S.	196	Combi, C.	117
Campbell, K. E.	188	Fox, J.	117
Chute, C. G.	183	Zeng, Q.	114
Shahar, Y.	180	Das, A. K.	114
Haux, R.	175	Degoulet, P.	113
Kushniruk, A. W.	167	Perl, Y.	113
Elkin, P. L.	167	Spackman, K. A.	112
Zhou, X. H.	164	Johnston, M. E.	112
Kuperman, G. J.	162	Safran, C.	112
Boxwala, A. A.	157	Owens, D. K.	111
Simon, R.	155	Andreassen, S.	111
Evans, R. S.	152	Friedman, C. P.	111

Table 2-5. Authority score ranking for frequently cited authors.

Author name	Authority score	Author name	Authority score
Clayton, P. D.	4.06	Tierney, W. M.	1.93
Cimino, J. J.	4.00	Tuttle, M. S.	1.89
Hripcsak, G.	3.86	Johnston, M. E.	1.84
Musen, M. A.	3.66	Hasman, A.	1.80
Shortliffe, E. H.	3.58	Brennan, P. F.	1.77
Safran, C.	3.54	McDonald, C. J.	1.63
Barnett, G. O.	3.33	Miller, P. L.	1.58
Greenes, R. A.	3.31	Shea, S.	1.57
Campbell, K. E.	3.01	Stefanelli, M.	1.56
Hersh, W. R.	2.95	Overhage, J. M.	1.49
Stead, W. W.	2.90	Ohnomachado, L.	1.42
Gardner, R. M.	2.90	Haynes, R. B.	1.37
Bates, D. W.	2.87	Friedman, C.	1.36

continued

Author name	Authority score	Author name	Authority score
Chute, C. G.	2.82	Lobach, D. F.	1.38
Kuperman, G. J.	2.76	Humphreys, B. L.	1.34
Friedman, C. P.	2.73	Haux, R.	1.33
Rector, A. L.	2.68	Rind, D. M.	1.29
Teich, J. M.	2.67	Evans, R. S.	1.25
Sittig, D. F.	2.64	Zielstorff, R. D.	1.21
Shahar, Y.	2.47	Peleg, M.	1.20
Warner, H. R.	2.45	McCray, A. T.	1.18
Slack, W. V.	2.41	Kohane, I. S.	1.16
Haug, P. J.	2.23	Dolin, R. H.	1.11
Tang, P. C.	2.19	Leape, L. L.	1.10
Patel, V. L.	2.12	Tu, S. W.	1.09
Miller, R. A.	2.09	Owens, D. K.	1.02
Shiffman, R. N.	2.00	Spackman, K. A.	1.02
Huff, S. M.	1.98	Van Bemmel, J. H.	1.01

5.2 Content Map Analysis

5.2.1 Topic Map Analysis

The content map analysis uses time-series topic maps to present development trends in medical informatics over the ten years. For this temporal analysis we created topic maps of three periods, 1994-1997, 1998-2000, and 2001-2003. By breaking the medical informatics papers published over the past decade into three periods, we hope to glean the recent evolution and topic changes of the field. To generate the maps, the abstracts and titles of 5,837 papers in our collection were processed for 1994-1997, 5,755 for 1998-2000, and 5,375 for 2001-2003.

In these topic maps clusters of papers are represented by shaded regions and labeled by representative noun phrases appearing in those papers. The medical noun phrases were extracted using the Arizona Noun Phraser as described previously. These noun phrases were extracted from the original text and the capitalization varies. However, phrases with capitalization variations were treated as the same phrases for the phrase vector representation. Numbers of papers within each cluster are presented in parentheses after the topic labels. As described previously in Section 3.2, neighboring topic regions have high content similarities. Users can click on the map regions to browse the papers.

The first topic map (Figure 2-1) displays an assortment of dominating themes for the first time period. There are many prominent but general medical information topics that occupy large regions, including: "Electronic Medical Records," "Computer-Based Patient Record," "Health Care,"

"Information Technologies," "Computer Programs," "Medical Students," etc. A few specific medical informatics applications also occupy large regions, including: "Hospital Information Systems" and "Clinical Information Systems." In addition, we also notice several small but distinct topic regions that are related to data analysis and mining, e.g., "Decision Support Systems," "Statistical Analysis," "Regression Models," "Artificial Neural Networks," and "Neural Networks." It appears that data mining and knowledge discovery research had already begun to emerge in 1994-1997, the first era of our analysis.

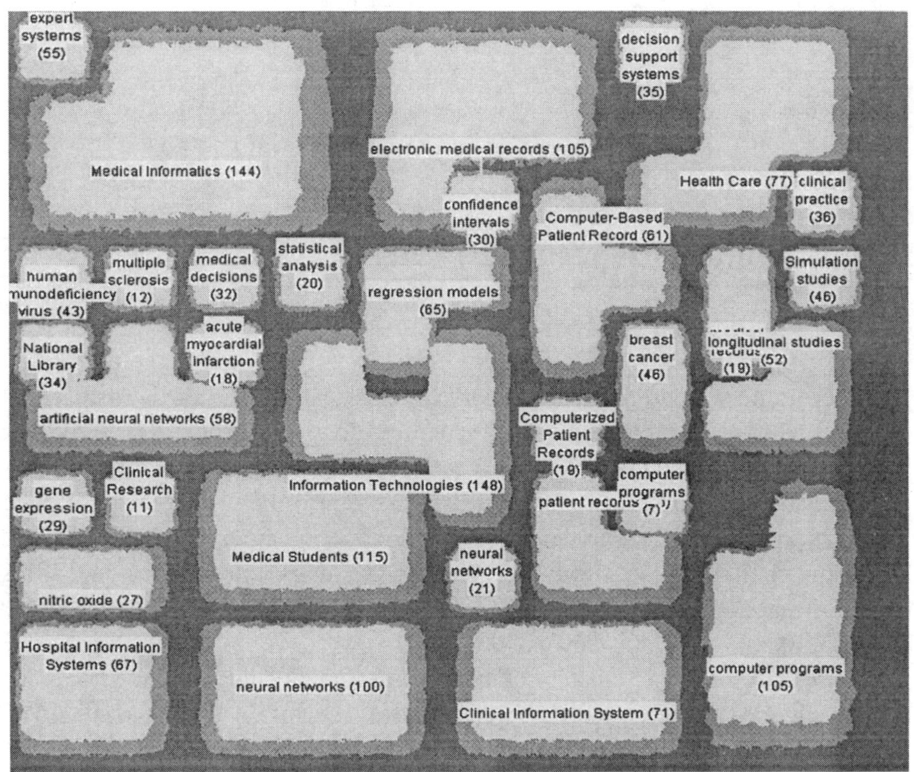

Figure 2-1. Top level content map for 1994-1997.

The topic regions in the second and third time periods were colored to reflect the growth rate of the topic compared with the previous time period (not shown here due to production reasons), which is computed as the ratio between the number of papers in the region for the current time period, and the number of papers in the region of the same topic label in the previous time period. The color legend of the growth rate is presented as well below these two content maps. In Figure 2-2, regions such as "Human Genome" and "Medical Imaging" correspond to the right end of the color legend, which represents newly emerged topic regions, while regions with lighter colors such as "Hospital Information System" corresponds to color legends close to the left end, which represent topic regions that had a slow or average growth rate.

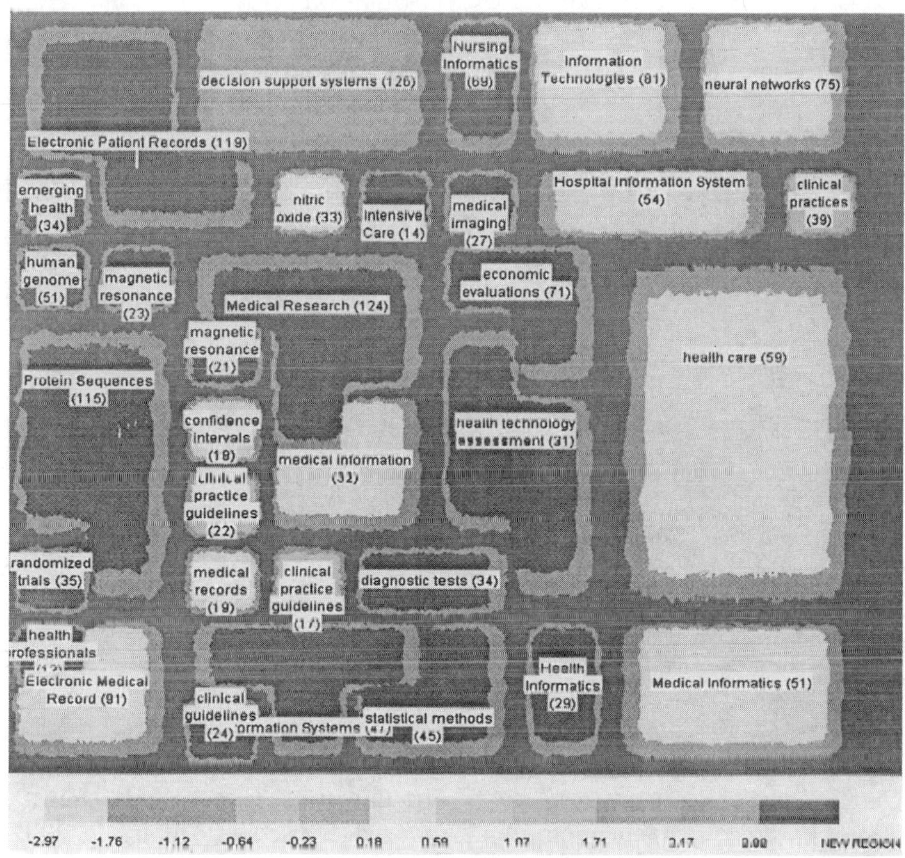

Figure 2-2. Top level content map for 1998-2000.

In the second map (Figure 2-2), we see the continued presence of several important, but general medical informatics topic regions, including: "Health Care," "Information Technologies," "Electronic Medical Records," "Hospital Information Systems," etc. Several data analysis and mining topics began to occupy larger regions than in 1994-1997, e.g., "Decision Support Systems" and "Neural Networks." In addition, "Protein Sequence" and "Human Genome" topics emerged the first time, increasing the scope of biomedical data. There is also an increased diversity of applications and methodologies such as: "Nursing Informatics," Medical Imaging," "Economic Evaluation," and "Health Technology Assessment."

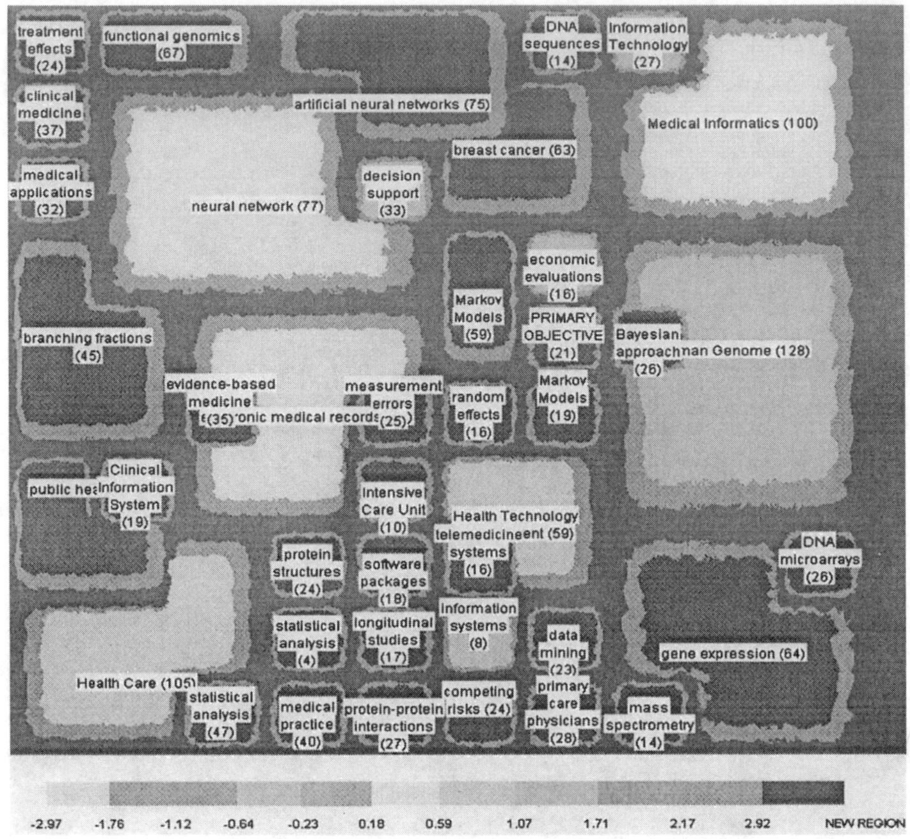

Figure 2-3. Top level content map for 2001-2003.

In addition to some of the general medical informatics topics ("Health Care," "Medical Informatics," etc.), the third map (Figure 2-3) shows a strong presence of data mining and knowledge discovery topics in 2001-2003 including: "Neural Networks," "Artificial Neural Networks," "Bayesian Approach," "Data Mining," "Markov Models," etc. Most interestingly, we see an explosion of biological and genomic data types and applications, including: "DNA Microarrays," "DNA Sequences," "Gene Expression," "Mass Spectrometry," "Protein-Protein Interactions," "Functional Genomics," etc.

The pattern of mixed topics observed between maps is consistent with the observation that medical informatics is a fast-growing, multidisciplinary field (Andrews, 2002). Sittig (1996) and Greenes and Siegel (1987) recount the difficulty of defining the boundaries of the medical informatics domain, and the resulting diversity of subfields attributed to it. Despite such challenges, we observed a consistent focus on health care, electronic medical records, and information technologies topics in general in the three eras of analysis. In addition, we also see overwhelming evidence of the presence of many emerging and exciting data mining and knowledge discovery research applications, especially those which leverage the opportunities presented by a wide spectrum of new, diverse, and large-scale biological and genomic data and problems.

5.2.2 Author Map Analysis

The author map in Figure 2-4 attempts to group individual researchers in the domain space, based on their common research interests. For this analysis we used the core author set from Table 2-3 as the input data. The result presents five major clusters of authors who had papers with similar contents. Each resulting cluster has been assigned a label indicating the common concept(s) that the cluster represents. The labels were manually selected from the keywords extracted by the SOM algorithm, a process which requires human judgment, but as Andrews (2002) points out, consistent with other cluster analysis methods. The keywords used to determine each label are listed in Table 2-6, and the individual groups are shown in detail in Figures 2-5 through 2-8 (with the exception of Group 3, which was decided not to be dense enough to require a zoomed in view).

Table 2-6. Top keywords generated from authors' texts and used to label author map groups.

Group 1	Group 2	Group 3
Decision support system	Clinical trials	Clinical applications
Decision support	Breast cancer	Clinical information system
Expert system	Risk factors	
Knowledge-based system	Cardiovascular disease	
	Coronary heart disease	
Group 4	**Group 5**	
Patient care	Clinical trials	
Medical record	Cohort study	
Electronic medical record	Confidence intervals	
Unified medical language system	Multivariate analysis	

The largest group in the center of the author map, Group 1, is labeled "Decision support and knowledge-based systems." This group contains 37 of the 96 authors, including W.R. Hersh, C.G. Chute, and M.A. Musen. Author proximity on the map indicates a degree of similarity between the research interests. Group 2, "Clinical trials for diseases," contains 15 authors, including R.A. Miller, Y. Shahar, and M. Stefanelli. Group 3, "Clinical applications and information systems," contains 6 authors, among them D.W. Bates and P.D. Clayton. Group 4, labeled "Patient care and electronic medical records," is comprised of such prolific authors as J.J. Cimino, D.F. Sittig, R.A. Greenes, C. Friedman, and E.H. Shortliffe.

Finally, Group 5, "Clinical trials and analysis," contains 8 authors, among them A. Donner and K.J. Lui. Authors in our original 96 that are not included in a group can be seen in the overall map in Figure 2-4.

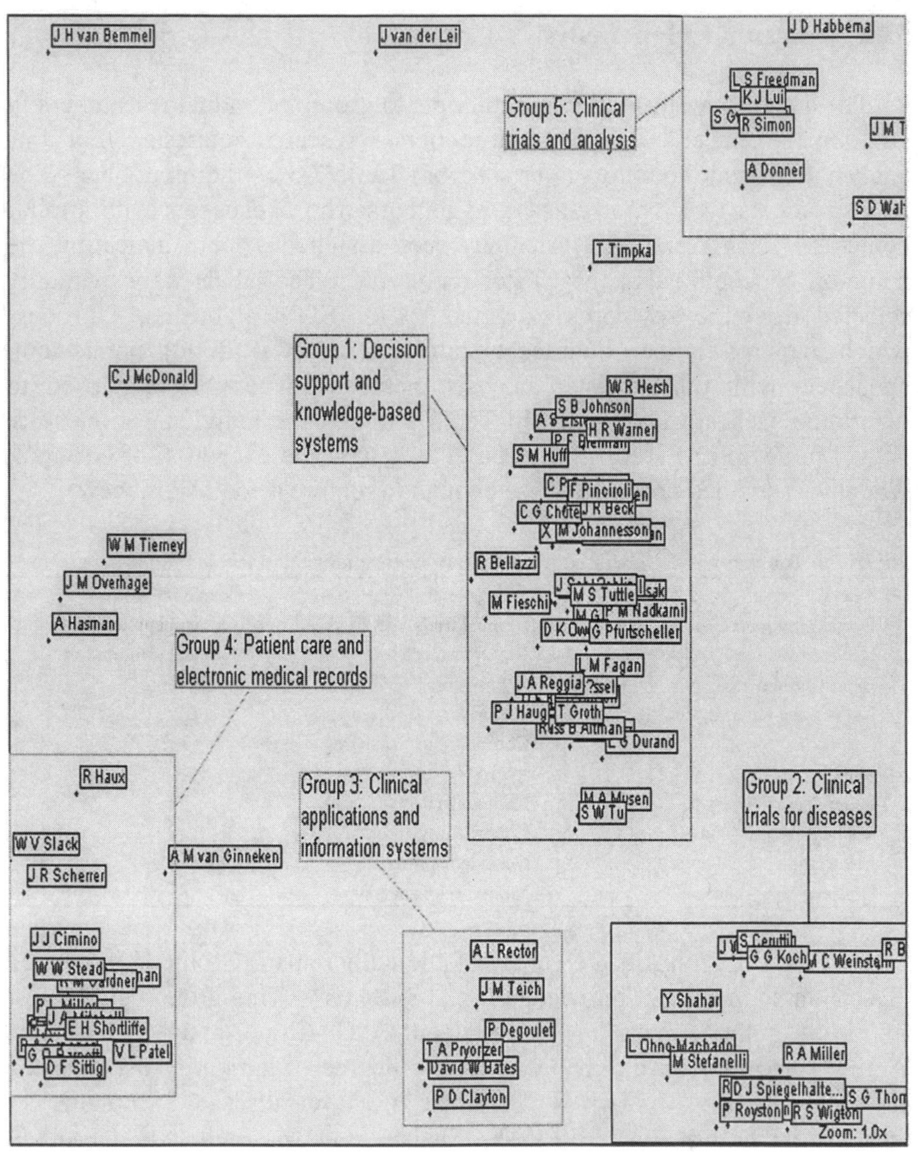

Figure 2-4. Overall author similarity map.

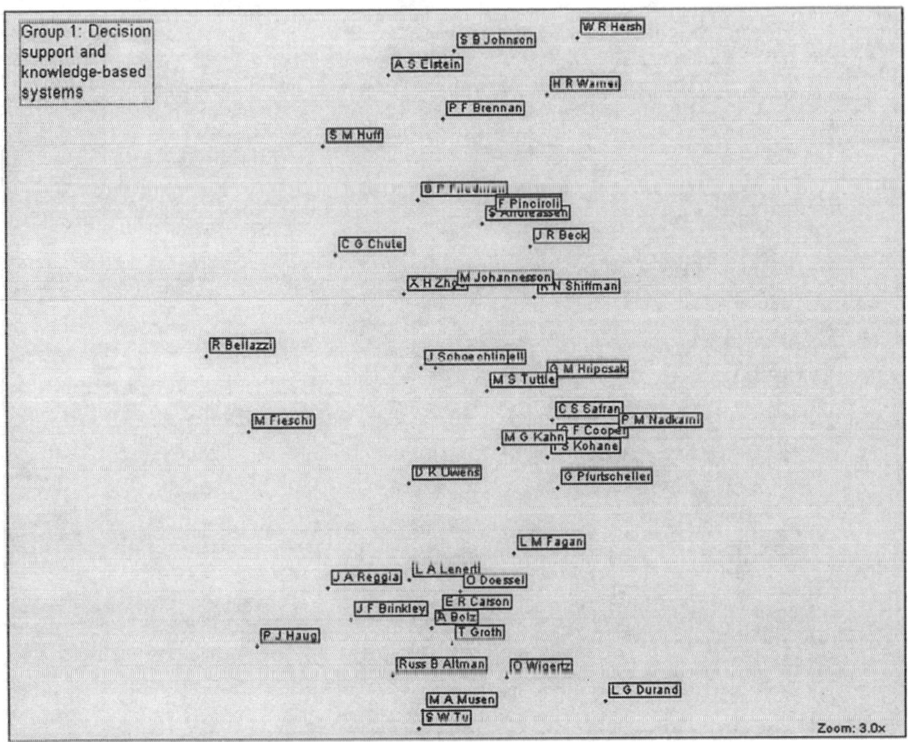

Figure 2-5. Author map - Group 1.

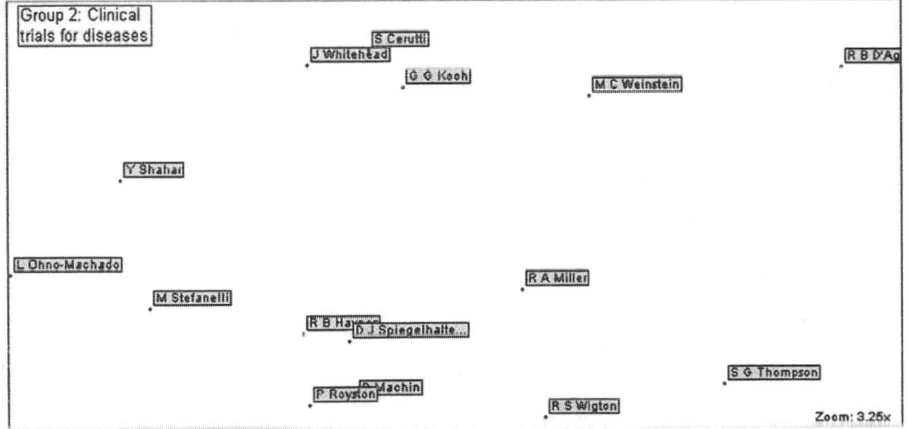

Figure 2-6. Author map - Group 2.

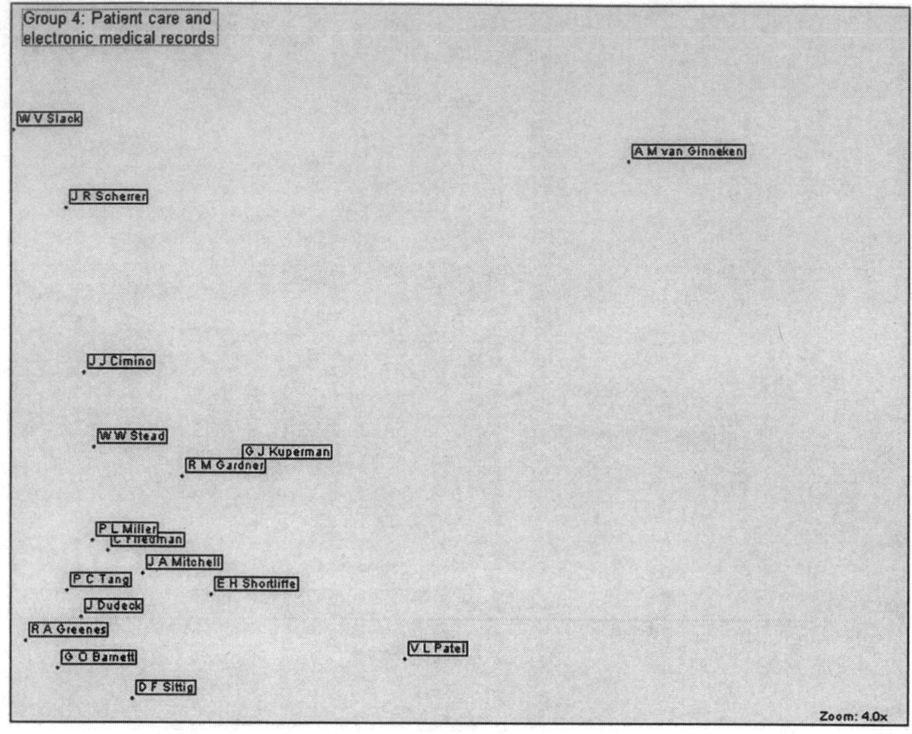

Figure 2-7. Author map - Group 4.

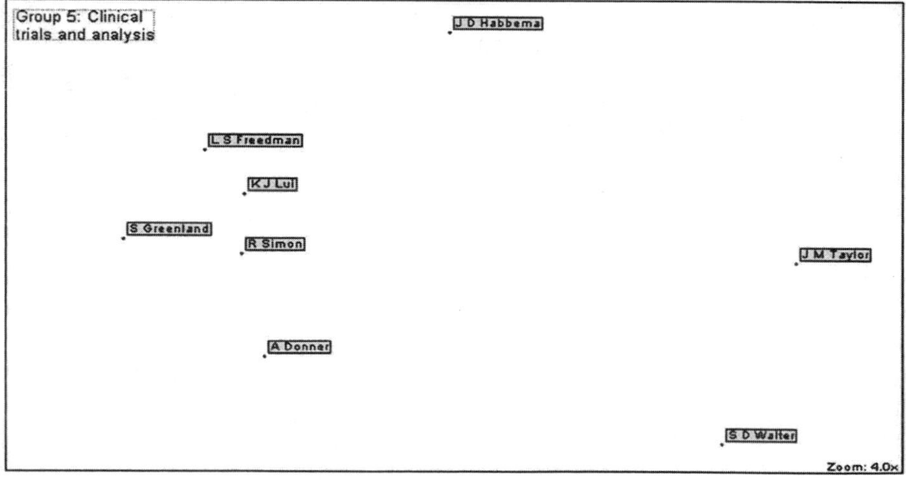

Figure 2-8. Author map - Group 5.

5.3 Citation Network Analysis

Using the data gathered from SCI, we created two citation networks of the most prominent researchers in medical informatics, as identified by our basic analysis. Both networks present views of the same data with different levels of filtering. A link from author A to author B indicates that A frequently cites B. In the visualization results, triangles indicate "core" authors (presented in Table 2-3) and circles represent "non-core" authors. In order to reveal only the strongest communication patterns, links associated with a small number of citations are filtered from the networks. Figure 2-9 is filtered by a link threshold of 10, that is, only links associated with 10 or more citations are shown. The result is a rather dense cluster, but hubs can still be observed around the major players from our basic analysis results: Edward H. Shortliffe, Paul D. Clayton, George Hripcsak, David W. Bates, James J. Cimino, and William R. Hersh, to name a few. These authors are not only frequently published and cited, they are cited repeatedly by consistent sets of other authors. Figure 2-10 is a view of the same citation data, filtered by a threshold of 20. In this view, clearer subgroups of citations emerge. One distinct subgroup of eight authors is disconnected from the larger graph. This group appears in the upper right-hand part and consists of four "core" authors from Table 2-3, and four "non-core" authors from Table 2-4. In the larger graph itself, hubs from Figure 2-9 begin to pull apart into subgroups. The most distinct group clusters around David Bates and William M. Tierney, and includes high-ranking authors from the basic analysis, such as Dean F. Sitting and Jonathan M. Teich. Other subgroups of the larger graph can be observed but are much less distinct. Obvious hubs are James Cimino, George Hripcsak, and Edward Shortliffe. Tightly connecting these are Carol Friedman, Vimla L. Patel, and Robert A. Greenes.

It should be noted that as a result of filtering by link strength, the citation networks do not reflect an overall qualitative performance measure of the authors, but rather the nature of their communication channels. That is, the graphs do not show who is the most cited, but who most frequently cites whom. It can be observed, for example, that there are no links to William Hersh in the 20-threshold network; however, our basic analysis indicates that Hersh is highly influential in the field, and is cited by numerous other authors. According to Figure 2-10, he is simply not cited more than 19 times by the same author. In contrast, there are two incoming links to Christopher G. Chute (from James Cimino and Peter L. Elkin). Chute is only slightly below Hersh in Authority ranking, but frequently cites and is cited by two specific authors, so is connected to the main graph.

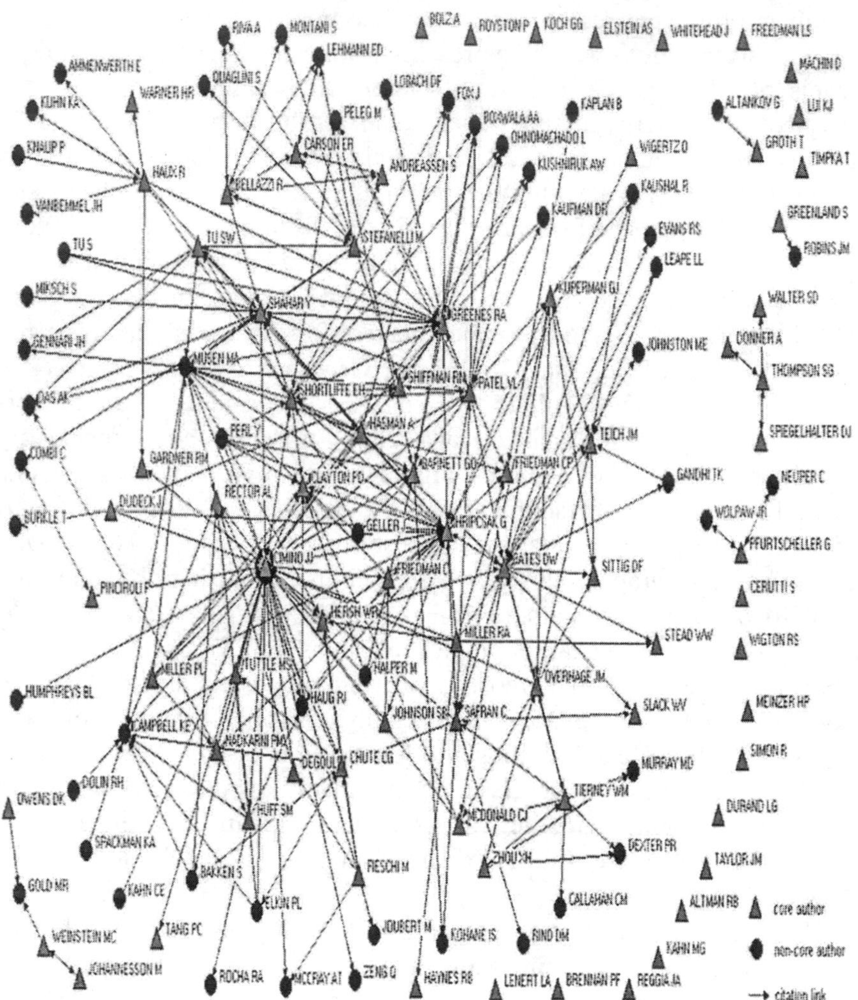

Figure 2-9. Author citation network (minimum cites per link: 10).

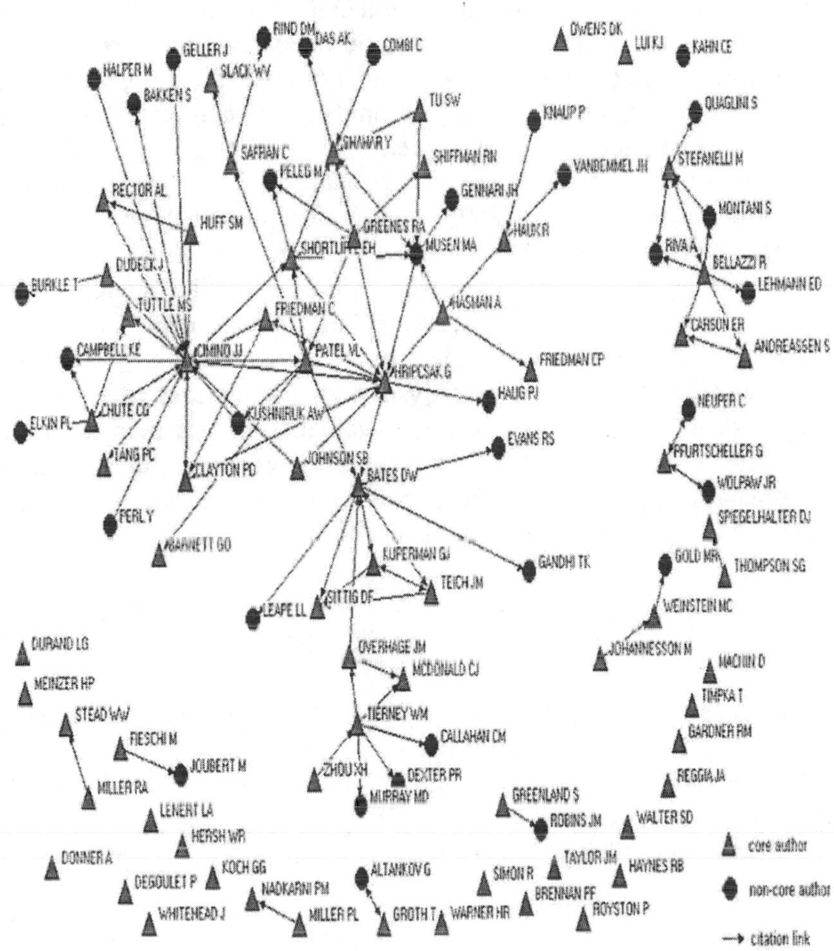

Figure 2-10. Author citation network (minimum cites per link: 20).

6. CONCLUSION AND DISCUSSION

For a fast-growing, interdisciplinary knowledge domain such as medical informatics, it is valuable to be able to create a picture of the state of the research from a variety of perspectives. Such a picture helps organize the vast amounts of information available in order to determine past and current (and possibly future) directions of the field, as well as prominent researchers, their relationships to each other, and the parts of the domain to which they contribute. Automatic information visualization techniques can perform these knowledge tasks efficiently and systematically. In this study we augment classic domain analysis techniques with visualization tools to create a variety of views of medical informatics over the past ten years. The results of our study present development trends of subtopics of the field, a performance evaluation of the prominent researchers, and graphs of knowledge transfer among researchers.

This study was designed in the context of the analysis framework developed by Huang et al. (2003), and implements the three types of analysis presented in that work: basic analysis, content maps, and citation networks. Based on the data set extracted from widely-used data sources such as the MEDLINE database and SCI, we believe our analysis helps reveal the coverage and evolution of the field. It would be interesting to compare the particular findings from our analysis with the pictures of the field in the minds of the domain experts. Such evaluation would help determine how accurate our analysis results are and reveal interesting discrepancies between automatic analysis results and expert knowledge that might enhance our understanding of the state of the field.

7. ACKNOWLEDGEMENTS

This research was supported by the following grants: (1) NIH/NLM, 1 R33 LM07299-01, 2002-2005, "Genescene: A Toolkit for Gene Pathway Analysis" and (2) NSF, IIS-9817473, 1999-2002, "DLI - Phase 2: High Performance Digital Library Classification Systems: From Information Retrieval to Knowledge Management."

REFERENCES

Andrews, J. (2002). "An Author Co-citation of Medical Informatics," *Journal of the Medical Library Association,* 91(1), 47-56.

Borgatti, S. and Chase, R. (n.d.) NetDraw Network Visualization Tool. Version 1.39, Retrieved Aug. 30, 2004, http://www.analytictech.com/netdraw.htm

Börner, K., Chen, C., and Boyack, K. (2003). "Visualizing Knowledge Domains," in Blaise Cronin (Ed.), *Annual Review of Information Science and Technology*, 37: 179-255. Information Today, Inc. and American Society for Information Science and Technology, 2003.

Boyak, K. W. and Börner, K. (2003). "Indicator-assisted Evaluation and Funding of Research: Visualizing the Influence of Grants on the Number and Citation Counts of Research Papers," *Journal of the American Society for Information Science and Technology*, 54(5), 447 - 461.

Chen, C. and Paul, R. J. (2001). "Visualizing a Knowledge Domain's Intellectual Structure," *IEEE Computer,* 34(3), 65-71.

Chen, C., Paul, R. J., and O'Keefe, B. (2001). "Fitting the Jigsaw of Citation: Information Visualization in Domain Analysis," *Journal of the American Society for Information Science and Technology*, 52(4), 315-330.

Chen, H., Houston, A. L., Sewell, R. R., and Schatz, B. R. (1998). "Internet Browsing and Searching: User Evaluation of Category Map and Concept Space Techniques," *Journal of the American Society for Information Science*, 49(7), 582-603.

Chen, H., Lally, A., Zhu, B., and Chau, M. (2003). "HelpfulMed: Intelligent Searching for Medical Information over the internet," *Journal of the American Society for Information Science and Technology (JASIST)*, 54(7), 683-694.

Chen, H., Schuffels, C., and Orwig, R. (1996). "Internet Categorization and Search: A Self-organizing Approach," *Journal of Visual Communication and Image Representation*, 7(1), 88-102.

Garfield, E. (1979). *Citation Indexing: Its Theory and Application in Science, Technology and Humanities*. New York, NY: John Wiley.

Garfield, E. (1995). "Citation Indexes for Science: a New Dimension in Documentation Through Association of Ideas," *Science*, 122, 108-111.

Greenes, R.A. and Siegel, E.R. (1987). "Characterization of An Emerging Field: Approaches to Defining the Literature and Disciplinary Boundaries of Medical Informaticcs," in *Eleventh Annual Symposium on Computer Applications in Medical Care*. 411–415.

Honkela, T., Kaski, S., Lagus, K., and Kohonen, T. (1997). "WebSom - Self-Organizing Maps of Document Collections," in *Proceedings of the Workshop on Self-Organizing Maps*, 310-315.

Huang, Z., Chen, H., Yip, A., Ng T. G., Guo, F., Chen, Z. K., and Roco, M. C. (2003). "Longitudinal Patent Analysis for Nanoscale Science and Engineering: Country, institution and Technology Field," *Journal of Nanoparticle Research*, 5, 333-386.

Kleinberg, J. (1998). "Authoritative Sources in a Hyperlinked Environment," in *Proceedings of the ACM-SIAM Symposium on Discrete Algorithms*, 668-677

Kohonen, T. (1990) "The Self-Organizing Map," in *Proceedings of the IEEE*, 78(9), 1464-1480.

Lawrence, S., Giles, C. L., and Bollacker, K. (1999). "Digital Libraries and Autonomous Citation Indexing," *IEEE Computer*, 32(6), 67-71.

Lin, C., Chen, H., and Nunamaker, J. F. (2000). "Verifying the Proximity Hypothesis for Self-organizing Maps," *Journal of Management Information Systems*, 16(3), 57-70.

Liu, X., Bollen J., Nelson M. L., and Van De Sompel, H. (2004). "All in the Family? A Co-authorship Analysis of JCDL Conferences (1994 - 2003)," http://lib-www.lanl.gov/~xliu/trend.pdf

Mackinlay, J. D., Rao, R., and Card, S. K. (1999). "An Organic User Interface for Searching Citation Links," in *Proceedings of the CHI'95, ACM Conference on Human Factors in Computing Systems*, 67-73.

Noyons, E. C. M., Moed, H. F., and Luwel, M. (1999). "Combining Mapping and Citation Analysis for Evaluative Bibliometric Purposes: A Bibliometric Study," *Journal of American Society for Information Science*, 50(2), 115-131.

Shortliffe, E. H, Fagan, L., Perreault, L. E., and Wiederhold, G. (Eds.) (2000). *Medical Informatics: Computer Applications in Health Care and Biomedicine* (2nd Edition). New York, NY: Springer Verlag.

Sittig, D. F. (1996). "Identifying a Core Set of Medical Informatics Serials: An Analysis Using the MEDLINE Database," *Bulletin of the Medical Library Association*, 84(2), 200-204

Small, H. (1999). "Visualizing Science by Citation Mapping," *Journal of the American Society for Information Science*, 50(9), 799-812.

Tolle, K. and Chen, H. (2000). "Comparing Noun Phrasing Techniques for Use with Medical Digital Library Tools," *Journal of the American Society for Information Science, Special Issue on Digital Libraries*, 51(4), 518-22.

Vishwanatham, R. (1998). "Citation Analysis in Journal Rankings: Medical Informatics in the Libary and Information Science Literature," *Bulletin of the Medical Library Association,* 86(4), 518-22.

White, H. D. and McCain, K. (1998). "Visualizing a Discipline: An Author Co-citation Analysis of Information Science, 1972 - 1995," *Journal of the American Society for Information Science,* 49(4), 327 - 355.

Wise, J. A., Thomas, J. J., Pennock, K., Lantrip, D., Pottier, M., Schur, A., and Crow, V. (1995). "Visualizing the Non-Visual: Spatial Analysis and Interaction with Information from Text Documents," in *Proceedings of the IEEE Information Visualization 95 (InfoViz'95),* 51-58.

SUGGESTED READINGS

Andrews, J. (2002). "An author co-citation of medical informatics," *Journal of the Medical Library Association,* 91(1), 47-56.
Andrews applies multivariate analyses and visualization techniques to map relationships between the fifty most-cited ACMI fellows for the years 1994 to 1998.

Cronin, B. (Ed). (2003). *Annual Review of Information Science and Technology*, Vol 37. Medford, NJ: Information Today, Inc./American Society for Information Science and Technology.
Number 37 in a series that offers a comprehensive overview of information science and technology. This volume contains chapters on indexing and retrieval for the web, and visualizing knowledge domains in general.

Chen, C. (2003). *Mapping Scientific Frontiers: The Quest for Knowledge Visualization*. Secaucus, NJ: Springer-Verlag.
A thorough investigation of the effectiveness of using visualization tools to reveal shifts in scientific paradigms, and of the need for interdisciplinary research in information visualization and information science.

Chen, C., Paul, R. J. (2001). Visualizing a knowledge domain's intellectual structure. *IEEE Computer*. 34(3), 65-71.
Introduces Pathfinder network scaling to produce a 3D knowledge landscape from science citation patterns. The authors propose a four-step approach to "extends and transform" traditional author citation and co-citation analysis.

Garfield, E. (1979). Citation Indexing: Its theory and application in science, technology and humanities. John Wiley, New York.
Garfield's influential review of the creation and usefulness of citation indexes for understanding knowledge domains, especially since his seminal 1955 paper on the subject (*Science*, 122, 108-111).

Honkela, T., Kaski, S., Lagus, K., Kohonen, T. (1997). WebSom - Self-Organizing Maps of Document Collections. *Proceedings of the Workshop on Self-Organizing Maps*. 310-315.
Introduces WEBSOM, a well-known application of the SOM algorithm to organize high dimensional text documents according to similarity, and to present the results in an intuitive user interface.

Kohonen, T. (1990) The Self-Organizing Map, *Proceedings of the IEEE.* 78(9), 1464-1480.
Influential review and demonstration of various applications of the SOM algorithm.

Small, H. (1999). Visualizing science by citation mapping. *Journal of the American Society for Information Science.* 50(9), 799-812.
Demonstrates the use of associative trails and virtual reality software to create and navigate spatial representations of a sample of multidisciplinary science citation data. The author also provides a nice overview discussion and justification for applying information visualization techniques to science.

White, H. D., McCain, K. (1998). Visualizing a discipline: An author co-citation analysis of information science, 1972 - 1995. *Journal of the American Society for Information Science.* 49(4), 327 - 355.
The authors use author co-citation data to map the field of information science.

ONLINE RESOURCES

ISI Science Citation Index, through the Web of Science
ISI Journal Citation Reports
http://isi6.isiknowledge.com/portal.cgi

ResearchIndex (also known as CiteSeer)
http://citeseer.ist.psu.edu/
http://www.neci.nec.com/~lawrence/researchindex.html

Entrez PubMed, from NLM
Access to NCBI's MeSH, MEDLINE, and journal databases:
http://www.ncbi.nlm.nih.gov/entrez/query.fcgi

American College of Medical Informatics
http://www.amia.org/acmi/facmi.html

NetDraw, network visualization tool
http://www.analytictech.com/netdraw.htm

Information analysis and visualization demos
SOM and GIS: http://ai.bpa.arizona.edu/go/viz/index.html
SOM: http://www.cis.hut.fi/research/som_pak/
CiteSpace: http://www.pages.drexel.edu/~cc345/citespace/
SPIRE and Themescape: http://nd.loopback.org/hyperd/zb/spire/spire.html

QUESTIONS FOR DISCUSSION

1. What analytical units in addition to authors and documents can be used to examine the state of medical informatics research? What kind of perspectives on the field would these analytical units provide?

2. What is the relationship between citation data and the topology of a knowledge domain? What is the motivation for using such data for domain analysis?

3. What are the advantages of using content analysis over citation analysis for identifying domain subtopics? What are the advantages of using citation analysis over content analysis?

4. How effective are the results of visualization technologies (such as citation networks and self-organizing maps) at presenting domain knowledge in an intuitive way? Are the results informative, easy to understand?

Chapter 3

BIOINFORMATICS CHALLENGES AND OPPORTUNITIES

Peter Tarczy-Hornoch[1] and Mark Minie[2]

[1]Department of Medical Education and Biomedical Informatics; [2]Health Sciences Libraries, University of Washington, Seattle, WA 98195

Chapter Overview

As biomedical research and healthcare continue to progress in the genomic/post genomic era a number of important challenges and opportunities exist in the broad area of biomedical informatics. In the context of this chapter we define bioinformatics as the field that focuses on information, data, and knowledge in the context of biological and biomedical research. The key challenges to bioinformatics essentially all relate to the current flood of raw data, aggregate information, and evolving knowledge arising from the study of the genome and its manifestation. In this chapter we first briefly review the source of this data. We then provide some informatics frameworks for organizing and thinking about challenges and opportunities in bioinformatics. We use then use one informatics framework to illustrate specific challenges from the informatics perspective. As a contrast we provide also an alternate perspective of the challenges and opportunitics from thc biological point of view. Both perspectives are then illustrated with case studies related to identifying and addressing challenges for bioinformatics in the real world.

Keywords

bioinformatics; computational biology; biomedical informatics; Human Genome Project; applied informatics; foundations of informatics; information access; sociotechnical dimensions; evaluation

1. INTRODUCTION

As biomedical research and healthcare continue to progress in the genomic/post genomic era, a number of important challenges and opportunities exist in the broad area of biomedical informatics. Biomedical informatics can be defined "as the scientific field that deals with biomedical information, data, and knowledge – their storage, retrieval, and optimal use for problem-solving and decision making" (Shortliffe et al.., 2001). To understand the challenges and opportunities for informatics within the field of bioinformatics (defined most broadly as informatics in the domains of biology and biomedical research) it helps to understand the broader context in which they exist.

In the broader context, the key challenges to bioinformatics essentially all relate to the current flood of raw data, aggregate information, and evolving knowledge arising from the study of the genome and its manifestation. The genome can be thought of as the machine code or raw instructions for creation and operation of biological organisms (its manifestation). The information encoded in DNA results in the creation of proteins which serve as the key building blocks for biological function (a protein on the surface of one cell (neuron) in the brain can recognize a chemical signal sent by a neighboring neuron). Proteins physically aggregate to create more complex units of biological function termed protein complexes (the protein that recognizes the signal from a neuron might be part of a protein complex that translates that signal into an action such as turning on another protein that was in "standby mode"). Proteins and protein complexes interact with one another in networks or pathways to carry out higher level biological processes (such as the neuronal signaling pathway). These pathways include regulatory mechanisms whereby the function of the pathway overall is controlled by relevant input parameters (such as frequency and intensity of input from the part of the nervous system related to sensing pain). This regulation is complex and can include feedback and interaction among the proteins and protein complexes of the pathway, as well as regulation and interaction of other pathways. Interestingly, mechanisms include also the regulation of the conversion (translation) of the raw information encoded in the DNA into the intermediate messages (mRNA) and regulation of the conversion of the mRNA into proteins, as well as modification of the proteins themselves. The pathways in turn are assembled into more complex systems of multiple interacting pathways (pathways involved in evasive response to painful stimuli). In multi-cellular animals these complex systems in turn interact to control the function of their basic building blocks, namely the cells (for example, a brain cell or neuron). The cells in turn interact with one another and form higher order structures termed organs (the brain, for example). These organs interact with one another to form systems (such as the nervous system, which includes the brain as well as the input from sensory organs and the output to muscles and other organs). These systems interact to carry out higher order functions such as seeking out food sources (thus for example the nervous system guides the organism to seek food, the digestive system breaks down food, the metabolic system helps control the conversion of food to sugars, and the circulatory system helps deliver this energy to cells). Expanding beyond this level one can think of organisms interacting to form ecosystems in turn resulting in the Earth's biosphere. This hierarchical progression is illustrated in Figure 3-1. This cursory overview of the modern view of biological systems begins to shed light on the challenges faced by the fields of modern biology and biomedical research and the roles that bioinformatics might play.

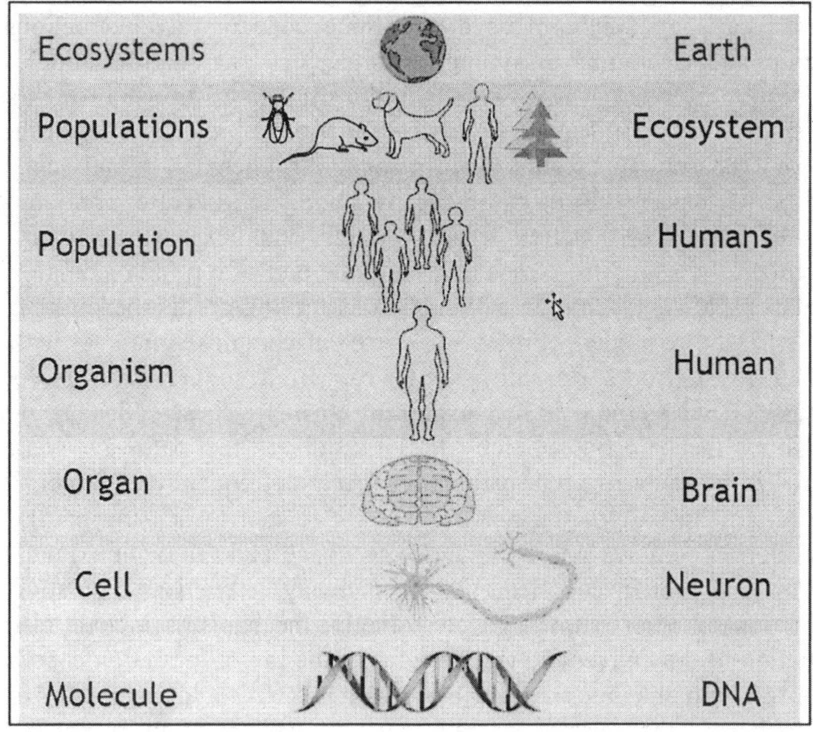

Figure 3-1. Hierarchy of biological systems.

In the broader context, to understand the opportunities for both biomedical research and bioinformatics, it helps to understand the genesis of this flood of information and more importantly the vision of how this information might be used. The roots of both the large quantity of information and the guiding vision can be traced to the start of the modern era of biomedical research, which is felt to be the discovery by Watson and Crick in 1953 of DNA as the information storage mechanism for cells. Research into the genome continued at a relatively linear pace until the establishment in 1989 of the National Center for Human Genome Research (NCHGR) to carry out the role of the National Institutes of Health (NIH) in the International Human Genome Project (HGP: see Online Resources). The HGP served to accelerate the pace of data generation from a linear to an exponential growth pattern as shown in Figure 3-2.

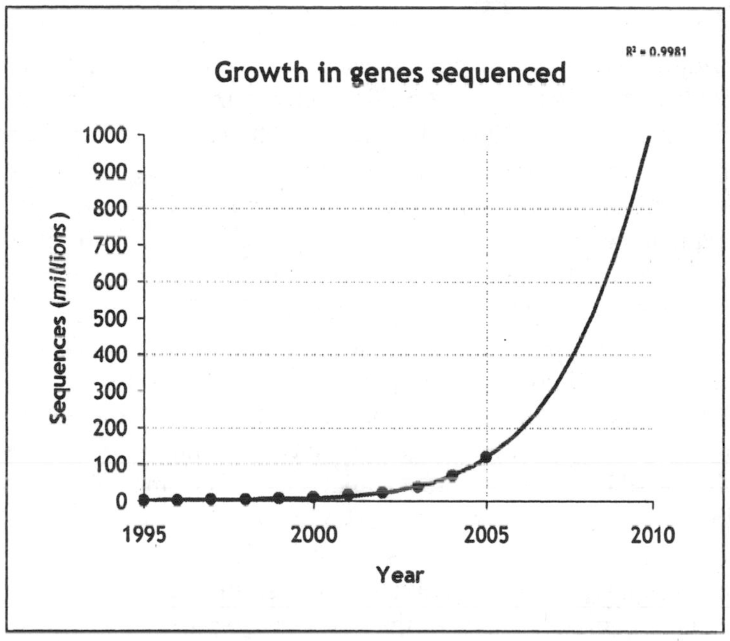

Figure 3-2. Growth in genes sequenced.

The seed of the vision for the HGP and the investment that has been made can be found in the mission of the National Institutes of Health (NIH) which is "science in pursuit of fundamental knowledge about the nature and behavior of living systems and the application of that knowledge to extend healthy life and reduce the burdens of illness and disability." The relationship of this mission to the grand vision of the HGP was published in 1990 as part of the first five year plan for the HGP: "The information generated by the human genome project is expected to be the source book for biomedical science in the 21st century and will be of immense benefit to the field of medicine. It will help us to understand and eventually treat many of the more than 4000 genetic diseases that afflict mankind, as well as the many multifactorial diseases in which genetic predisposition plays an important role." (See Online Resources). The flood of data, information, and knowledge we face today in biology and biomedical research can be traced directly to the coordinated international investment of large amounts of funding to sequence the human genome as a first step in arriving at a deeper understanding of the basis of human health and disease (Collins and McKusick, 2001). Research into the genomics and basic biology of diverse other organisms was galvanized by this effort as well and has been proceeding in parallel over the last decade and a half. With the completion of the sequencing of the DNA of humans and other organisms we have however only begun to explore the hierarchy discussed above and shown in Figure 3-1.

A guiding vision for the next phases of the HGP was articulated in a paper published in *Nature* on the 50th anniversary of Watson and Crick's discovery (Collins et al., 2003). This paper outlines fifteen grand challenges clustered into three broad areas: Genomics to Biology (improving our understanding of complex biological systems), Genomics to Health (developing and applying our understanding of the genomic basis for health and disease), and the sometimes underappreciated Genomics to Society (broadly, the ethical, legal, and social implications of our understanding).

These challenges, of course, present opportunities as well. As an example of a grand challenge presenting opportunities for biologists and informatics researchers in the Genomics to Biology area, consider, "Grand Challenge I-2: Elucidate the organization of genetic networks and protein pathways and establish how they contribute to cellular and organismal phenotypes." An example from the Genomics to Health area is, "Grand Challenge II-3: Develop genome-based approaches to prediction of disease susceptibility and drug response, early detection of illness, and molecular taxonomy of disease states." In response to the challenges posed by a post-genome sequencing era of biomedical research the NIH has identified the intersection of the computing and biological and biomedical fields as a key opportunity for future research based on the challenges and potentials outlined above. A critical articulation of this was provided by the report that led to the creation of the National Institutes of Health Biomedical Information Science and Technology Initiative (BISTI). (See Online Resources for the URL.) This introduction provides a high level overview of the opportunities and challenges for the field of bioinformatics. In the following sections we outline from an informatics perspective some more specific challenges and illustrate this with case studies/examples.

2. OVERVIEW OF THE FIELD

2.1 Definition of Bioinformatics

The definition of bioinformatics used in this chapter is the broadest possible definition of the field, namely *all informatics research and application in support of the biological research endeavor*. In the context of the definition of biomedical informatics given in the introduction "as the scientific field that deals with biomedical information, data, and knowledge – their storage, retrieval, and optimal use for problem-solving and decision making" (Shortliffe et al., 2001), we define bioinformatics as the subset of the field that focuses on information, data, and knowledge in the context of biological and biomedical research. By our definition the culture and environment (context) in which bioinformatics is studied and applied are that of the researcher in the laboratory seeking new knowledge. This includes a broad range of research ranging from a) basic molecular and cellular level research seeking to understand the way cancer results in unregulated growth of cells to, b) whole animal applied research looking at ways to block the spread of cancers, to c) clinical research involving patients looking at genetic factors influencing susceptibility to cancer. It is distinct from clinical informatics which focuses on the culture and environment of clinical care involving patients and healthcare providers in settings ranging from one's own home, to outpatient (clinic) and inpatient (hospital) care. This definition is similar to the one used by the BISTI website: "Research, development, or application of computational tools and approaches for expanding the use of biological, medical, behavioral or health data, including those to acquire, store, organize, archive, analyze, or visualize such data" (see Online Resources).

There are a number of other definitions of the term "bioinformatics" and in reading the literature it is important to be sure one is clear on the meaning being used. For some, the term is fairly narrow and refers primarily to developing and validating and applying algorithms for processing and analyzing sequences of DNA (the phrase "computational biology" is also being used for this area). Others expand the definition of bioinformatics to include any algorithmic or statistical approach to the analysis of biological data. Some make a distinction between mathematical modeling in biology and bioinformatics, whereas others view the former a subset of the later. For some, bioinformatics refers to the basic research in the area, whereas the applied side of deploying systems is termed biocomputational infrastructure. For others, bioinformatics refers to the set of computational tools used by biologists to carry out their research. A very interesting alternate broad definition is, "The study of how information is represented and transmitted in biological systems, starting at the molecular level" (Bergeron, 2002). For the remainder of this chapter, we will use this last broader, more inclusive definition of the term.

2.2 Opportunities and Challenges – Informatics Perspective

2.2.1 Frameworks for Describing Informatics Research

The field of biomedical informatics is relatively young and there are a number of ways to organize important research questions and areas (and in turn to discuss challenges and opportunities).

The American Medical Informatics Association developed the following framework, shown in Tables 3-1 and 3-2, categorizing research papers in the discipline submitted for review at the 2003 annual meeting (Scientific Program Committee Chair: Mark A. Musen, Foundations Track Chair: Charles P. Friedman, Applications Track Chair: Jonathan M. Teich, see http://www.amia.org/meetings/archive/f03/call.html#categorizing).

The Foundations Track, shown in Table 3-1, focuses on theories, models, and methods relevant to biomedical informatics broadly (applicable to clinical informatics, bioinformatics, and public health informatics). Bold faced categories are foundational approaches often referred to in publications in the bioinformatics arena. Each of these represents ongoing areas of inquiry and thus potential challenges and opportunities for bioinformatics, both in terms of research and in terms of application. As will be discussed later in this chapter, some foundational areas are not currently active areas of research in bioinformatics and may represent important opportunities for future research (in particular many of the areas in C).

Table 3-1. Categories of Informatics Research*

I. Foundations of Informatics Building Models and Methods for Biomedical Information Systems
A. Modeling Data, Ontologies, and Knowledge
1. Controlled terminologies and vocabularies, ontologies, and knowledge bases
2. Data models and knowledge representations
3. Knowledge acquisition and knowledge management
B. Methods for Information and Knowledge Processing
1. Information retrieval
2. Natural-language processing, information extraction, and text generation
3. Methods of simulation of complex systems
4. Computational organization theory and computational economics
5. Uncertain reasoning and decision theory
6. Statistical data analysis
7. Automated learning, discovery, and data mining methods

continued

I. Foundations of Informatics Building Models and Methods for Biomedical Information Systems
B. Methods for Information and Knowledge Processing *(continued)*
8. Software agents, distributed systems
9. Cryptography, database security, and anonymization
10. Image representation, processing, and analysis
11. Advanced algorithms, languages, and computational methods
C. Human Information Processing and Organizational Behavior
1. Cognitive models of reasoning and problem solving
2. Visualization of data and knowledge
3. Models for social and organizational behavior and change
4. Legal issues, policy issues, history, ethics

**Used with permission from the American Medical Informatics Association*

The Applications Track, shown in Table 3-2, focuses on real world systems: their design, implementation, deployment, and evaluation. Bold faced categories are applications often referred to in publications in the bioinformatics arena. Each category represents ongoing areas of inquiry and thus potential challenges and opportunities for bioinformatics, both in terms of research and in terms of application. As will be discussed later in this chapter, some application areas, similar to the theoretical track, are not currently active and may represent important opportunities for future research: for example, B, or the intersection of bioinformatics with C1.

Table 3-2. Categories of Informatics Research*

II. Applied Informatics - Real World Solutions for Real World Problems
A. Advanced Technology and Application Infrastructure
1. Data standards and enterprise data exchange
2. System security and assurance of privacy
3. Human factors, usability, and human-computer interaction
4. Wireless applications and handheld devices
5. High-performance and large-scale computing
6. Applications of new devices and emerging hardware technologies
B. Evaluation, Outcomes, and Management Issues
1. Organizational issues and enterprise integration
2. System implementation and management issues
3. Health services research: health care outcomes and quality
C. Information, Systems and Knowledge Resources for Defined Application Areas
1. Care of the patient
a. Electronic medical records
b. Computer-based order entry
c. Clinical decision support, reference information, decision rules, and guidelines
d. Workflow and process improvement systems
e. Nursing care systems

continued

II. Applied Informatics - Real World Solutions for Real World Problems

C. Information, Systems and Knowledge Resources for Defined Application Areas
- 1. Care of the patient *(continued)*
  - f. Ambulatory care and emergency medicine
  - g. Telemedicine and clinical communication
  - h. Patient self-care, and patient-provider interaction
  - i. Disease management
- 2. Care of populations
  - a. Disease surveillance
  - b. Regional databases and registries
  - c. Bioterrorism surveillance and emergency response
  - d. Data warehouses and enterprise databases
- 3. Enhancements for education and science
  - a. Consumer health information
  - b. Education, research, and administrative support systems
  - c. Library applications
- **4. Bioinformatics and Computational Biology**
  - **a. Genomics**
  - **b. Proteomics**
  - **c. Studies linking the genotype and phenotype**
  - **d. Determination of biomolecular structure**
  - **e. Biological structure and morphology**
  - **f. Neuroinformatics**
  - **g. Simulation of biological systems**

**Used with permission from the American Medical Informatics Association*

The University of Washington Biomedical and Health Informatics Graduate Program has taken a less granular approach to categorizing the broad field of biomedical informatics with three application domains and four foundational areas. The three application domains are: a) Biomedical Research, b) Clinical Care, and c) Public Health. The four foundational areas are: a) Biomedical Data and Knowledge, b) Biomedical Information Access and Retrieval, c) Biomedical Decision Making, and d) Socio-Technical Dimensions of Biomedical Systems. In addition to the application domains and the foundational areas the University of Washington requires grounding in methodologies including programming, statistics, research design and evaluation. The need for evaluation methodologies is especially important as is discussed below. The next sections will use these foundational areas to illustrate challenges and opportunities in the bioinformatics domain.

2.2.2 Opportunities and Challenges – Biomedical Data and Knowledge

The volume and diversity of biomedical data is growing rapidly, presenting a number of challenges and opportunities ranging from data capture, data management, data analysis, and data mining. The analysis of this data is generating new knowledge that needs to be captured. As the volume of this knowledge grows, so does the need to develop formal ways of representing this knowledge. Knowledge bases and formal approaches including ontologies are potential solutions. This particular area of biomedical data and knowledge will be explored in more depth than the other areas given the emphasis of this book.

Analysis of gene expression (microarray) experiments illustrates diverse aspects of the problem with modern biological data. In a gene expression experiment the biologist measures the level of expression of all genes in a particular tissue under a given condition, and then frequently compares expression levels to those in the same tissue under a different condition (a process known as differential gene expression). Thus, for example, one might measure the level of gene expression (the degree to which certain genes are turned on or off) by comparing cancer cells that have received a cancer drug to ones that have not.

The first challenge is management of the experimental data since a single gene expression measurement results in thousands of data points. In turn typically one repeats each experimental condition and control condition multiple times. Frequently, the measurements are repeated at multiple time points (for example, before treatment with a drug, one hour after, four hours after, eight hours after, twenty-four hours after). A number of open source and commercial packages help researchers collect and manage gene expression data.

The next challenge is data analysis and data mining. There are a number of commercial expression array analysis packages but they often do not implement the latest algorithms and methods for data analysis. Important open source collaborations aim to develop tools to assist researchers in developing and using new tools for array analysis. This collaboration is the BioConductor project (http://www.bioconductor.org) and is built on top of the R programming environment (Ihaka and Gentleman, 1996).

Finally, there is the need to mine large data sets of gene expression data. A number of studies have been published using a variety of data mining techniques from computer science and this is still a rapidly evolving area. An example of this class of problems is trying to predict the outcome of cancer patients based on analyses of the gene expression in their cancerous tissue (e.g. gene expression in a piece of breast cancer removed by the surgeon). A classic study used DNA microanalysis and a supervised classifier to predict outcome of breast cancer far better than any other classifiers (van 't Veer et al., 2002).

The data capture and data management problem is compounded by the fact that modern biological experiments frequently involve diverse types of data ranging from analysis of mutations (changes in the DNA sequence) to gene expression to protein expression to biochemical measurements to measurements of other properties of organisms (frequently termed phenotype). In order to make sense out of these diverse experimental results and to incorporate data, information, and knowledge from public domain databases (such as databases of protein function) data integration is needed. A number of data integration systems for biomedical data have been developed. These data integration approaches are reviewed in a number of articles (Sujansky, 2001). The BioMediator system (formerly GeneSeek) (Donelson et al., 2004; Mork et al., 2001; Mork et al., 2002; Shaker et al., 2002; and Shaker et al., 2004) is one such system for data integration. It is designed to allow biologists to develop their own views of the way in which diverse private (experimental data) and public databases and knowledge bases relate to one another and to map this view (the mediated schema) onto the specific sources they are interested in querying. The interfaces, or wrappers, to these diverse sources are written in a general purpose fashion to permit the same wrappers to be reused by diverse biologists. The custom views (mediated schemata) are captured in a frames based knowledge base (implemented in Protégé) (Stanford, 2002). The system architecture permits in a single environment both the integration of data from diverse sources and the analysis of this data (Mei et al., 2003). The system works well but an important set of challenges surrounds the need to develop tools that permit the biologists to manipulate the mediated schema in a more intuitive fashion. Another challenge is to incorporate such systems into the workflow of the typical biological lab.

Ultimately all this data generates new knowledge which needs to be captured and shared. The volume of this knowledge is growing only linearly as shown in Figure 3-3 in contrast to the growth of the data.

An important challenge to knowledge creation is developing ways to increase the rate of knowledge generation to keep up with the rapid growth of data. Even with the linear growth of knowledge the volume of it is such that it is becoming difficult for one person to keep up with it all systematically. In order to access and use this knowledge it is becoming more and more important that the knowledge be captured in computable form using formalisms from the computer science community such as ontologies. These topics are discussed in more detail in other chapters.

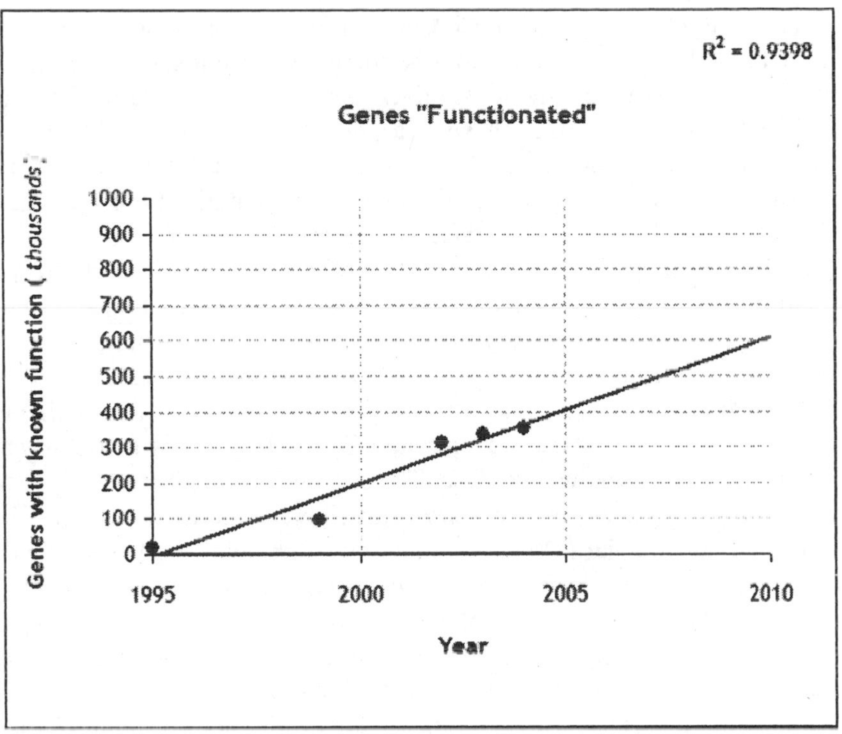

Figure 3-3. Growth rate of Genes with known function

The power and the challenges of these approaches can be illustrated by three important bioinformatics related knowledge bases. The first is the Foundational Model of Anatomy (FMA) (Rosse and Mejino, 2003) which is a centrally curated knowledge base capturing anatomic knowledge about the body from levels of granularity ranging from the whole body down to cells, sub cellular compartments and molecules (proteins). The FMA is becoming widely adopted as a reference standard for describing a variety of biologic processes in terms of where they occur and what they impact (serving as the anatomic component of the Unified Medical Language System (Tuttle, 1994; Bodenreider et al., 2002, among others). Some important challenges remain, though, in that: a) the FMA describes only human anatomy yet much work is being done on other species; b) the centralized curation process ensures internal consistency and quality control yet does not scale well to match the expansion of the FMA; c) the FMA describes normal physical structures but needs to be extended to describe abnormal (or disease related) structures; and d) the FMA needs to be extended to describe processes and functions of the physical structures.

The Gene Ontology (GO) Consortium (Gene Ontology Consortium, 2000) takes a different approach to describing the current state of knowledge about proteins and their functions. Given the evolving nature of the field a centralized top down approach such as that taken by the FMA was not possible. The GO is thus created and curated in a distributed fashion by a consortium of experts in molecular biology. The strength of this approach is that it scales well and adapts well to the rapidly changing state of our knowledge. Challenges to the GO approach include a) difficulty in maintaining internal consistency of the knowledge base; b) capturing in computable form from biologists subtle aspects of function; and c) maintaining referential integrity as the knowledge base evolves.

The third example of a bioinformatics knowledge base is the PharmGKB project (Klein et al., 2001, and PharmGKB, n.d.*)* which is a sophisticated pharmacogenomics knowledge base. The strength of this knowledge base is that it was centrally designed with distributed input to capture in a computable form a large amount of knowledge relevant to the field of pharmacogenomics - the interaction between an individual's genes, the medicines taken and the variability in response to these medicines. The challenges with this approach, however, are: a) it is dependent up on human curation (this is a shared challenge with FMA and GO as well); and b) extending the knowledge base to other areas of biology will be a challenge since unlike GO and FMA, the scope of PharmGKB was designed to be deep and narrow (pharmacogenomics) rather than broad and comprehensive (anatomy or molecular function).

2.2.3 Opportunities and Challenges – Biomedical Information Access and Retrieval

As the volume of data and knowledge grows it is becoming critical to biologists that they be able to access and retrieve the relevant pieces when they need it. The older paradigm of keeping up with the contents of the handful of top journals relevant to one's biological research area no longer works.

There are three key factors contributing to this. The first factor is that the sheer volume of new information is such that systematically keeping up is no longer a viable option. The second and related factor is that with the growth in new information has come a growth in the number of places in which information is published. Related to the dispersion of information across diverse sources is the fact that interdisciplinary and interprofessional research is becoming the norm, thus important research findings are published in a wider range of journals. The third factor is that information is becoming more and more available in electronic form and no longer just in condensed form in journals, resulting in a proliferation of biological databases, knowledge bases, and tools.

The University of Washington BioResearcher Toolkit (see Online Resources) illustrates the opportunity and the challenge this presents for biologists and for bioinformatics researchers and developers. Simply trying to find the right resource for a particular task from among hundreds is a challenge to say nothing of finding the right information within that resource. Given the volume of data and the fact that it exists as a combination of data in databases and free text, an important part of information access and retrieval has been both data integration and data mining, as discussed above. Intelligent parsing of queries, frequently involving natural language processing of both queries and sources, is becoming a key component of information access and retrieval. The challenges, opportunities, and state of the art of information retrieval (and data mining) in bioinformatics is covered in more depth in other chapters.

2.2.4 Opportunities and Challenges – Biomedical Decision Making

Thus far the field of bioinformatics has done little explicit research into the area of decision making. Within clinical or medical informatics there is a rich history of research into systems designed to help care providers and patients (healthcare consumers) make optimal decisions surrounding diagnosis (what disease or illness is it that a patient has) and management (which of the options for treatment are best factoring in details of the circumstances and the values of the patient). Approaches and methods used have included Bayesian belief networks, decision analytic models, and rule-based expert systems, among others. An important area on the clinical side for decision support systems has been genetic testing which has obvious ties (though one step removed) to bioinformatics research. Though this area of decision making is outside the primary scope of this book, it is worth noting that there appears to be a great potential opportunity to explore the development of tools for biologists to explicitly assist them in their decision making processes. The challenge is the paucity of literature and study in this potential arena. The first steps likely would be needs assessments and development and validation of models of decision making for biologists to see if in fact there is a niche for decision making tools in biomedical research.

2.2.5 Opportunities and Challenges – Evaluation and Socio Technical Dimensions of Biomedical Systems

The bioinformatics literature has a large number of papers published on theoretical frameworks for bioinformatics systems and a large number of papers on specific bioinformatics applications. There is, however, a relative lack of formal evaluations of bioinformatics systems and models. There is also a relatively sparse literature that formally and systematically examines the needs of biologists for specific tools (for example, Yarfitz and Ketchell, 2000). In part this is due to the relatively young nature of the field. A related factor is that to date much tool development has been driven by experienced biologists solving recurring problems they face through computational tools and sharing these tools with others. Though evaluation per se is outside the scope of this book it is important to learn from the experience of the clinical (medical) informatics community. Careful assessment and evaluation of the needs of users of the system is an important factor in guiding future development both on the theoretical (foundational) front as well as the applied front. Equally importantly formal evaluations and comparisons of alternate solutions (both applied and theoretical) are needed in order to guide development as well. An excellent resource on evaluation of systems in the clinical (medical) informatics arena is *Evaluation Methods in Medical Informatics* (Friedman and Wyatt, 1997); to date there is no similar book for bioinformatics evaluation.

The socio-technical environment in which informatics research and application development occur is becoming increasingly important on the clinical (medical) informatics front. It appears likely this will be true on the bioinformatics front as well. There are a number of ways of looking at this contextualization of informatics. The AMIA community has coalesced interests and activities in this area around the "People and Organizational Issues" working group. Their mission as quoted from their website is "a) To apply the knowledge of human behaviors toward the use of information technology within a health care environment; b) To effectively describe the benefits and impacts of information technology before paradigm shifts fully occur; c) To incorporate organizational change management and human concerns into information technology projects; and d) To distinguish between the human and technology issues when system successes or failures occur." As the field of bioinformatics grows and matures many of these challenges and opportunities will arise and need to be addressed. Already there are anecdotal reports of the purchase and deployment of complex expensive bioinformatics software packages that are unused despite apparent demand - a finding not unlike what has been seen with the development and deployment of unsuccessful clinical information systems.

Another perspective is provided by the description of the core graduate program courses at the University of Washington, "Sociotechnical Issues in Biomedical Informatics"; quoting from the course description: "Essentially all informatics work - whether purely theoretical or purely applied - is conceived, designed, built, tested, and implemented in organizations. Organizations are comprised of individuals and individuals are human beings, complete with philosophies, ideas, biases, hopes and fears. To build effective and valued informatics systems, the informaticist must understand how and why people behave as individuals, in groups, in organizations, and in society, and then build tools and systems that consider these human factors. The premise of this course is that the thoughtful consideration and application of the management sciences offers the opportunity to mitigate these risks." As bioinformatics projects are smaller in scope, these issues have not risen to the forefront, but as larger scale bioinformatics endeavors are undertaken it is almost certain they will.

2.3 Opportunities and Challenges – Biological Perspective

The exponential growth in basic biological data and the incorporation of that raw information into highly integrated databases on the Internet, along with the relatively linear but nonetheless rapid changes in our understanding of biological systems present several opportunities and challenges. These challenges faced by biologists and biomedical researchers present a complementary view to the perspective of the bioinformatics researcher. As noted in the section on Socio-technical Dimensions, understanding and addressing the challenges of the biologists in the trenches are critical to successful deployment of bioinformatics applications. We now discuss some of the challenges and opportunities viewed from the biological perspective.

2.3.1 Data Storage, Standardization, Interoperability and Retrieval

The huge growth in biological information being acquired at every level of the biological organization, from simple DNA sequences on up to the global ecosystem, has created serious challenges in data storage, retrieval and display. These challenges are being met by new developments in nanotechnology, search algorithms, and virtual/augmented reality tools as well as more conventional approaches.

2.3.2 Data Publication and Knowledge Sharing

NIH now requires all data generated by research it funds to be published in easily accessible and sharable electronic format, creating overwhelming challenges for current approaches such as journals and websites. New technologies such as wikis (see http://wiki.org/ and http://en.wikipedia.org/) and bibliomics tools (such as Telemakus: http://www.telemakus.net/ and PubGene: http://www.pubgene.org/) will need to be applied to these challenges in publication. The very meaning of "publication" has already started to evolve, and libraries in particular are becoming directly involved in providing for the distribution and archiving of raw data from scientific experiments (see DSpace: http://www.dspace.org/). Additionally, increased use of "telepresence" tools such as the Access Grid (http://www.accessgrid.org/) and online collaboration/knowledge sharing tools such as AskMe (http://www.askmecorp.com/) provide new and novel infrastructure in support of the basic biology research effort.

2.3.3 Analysis/annotation Tool Development and Distribution/access

The intense development of Open Source bioinformatics tools within different departments/groups at Universities and other institutions has created a need to develop the means of making these "home brew" tools available to the general bioresearch community. At present there is no integrated package analogous to Microsoft Office or an electronic medical record for biomedical researchers. The BioResearcher Toolkit (http://healthlinks.washington.edu/bioresearcher) provides a mechanism for the dissemination and sharing of such tools via its "UW HSL Bioinformatics Tools" section. There, tools developed by national biomedical researchers as well as local biomedical researcher (such as the web based protein structure prediction tool developed by Dr. Robert Baker of the UW Biochemistry Department, Robetta (http://robetta.bakerlab.org/), are made available to users. Other networked software tools, such as Vector NTI and PubGene are also available through the BioResearcher Toolkit site.

2.3.4 Hardware Development and Availability

Many bioinformatics applications require tremendous computational power. This challenge is being met by the availability of clusters constructed from readily available desktop computers (http://www.bio-itworld.com/news/083004_report5927.html) as well as specially constructed supercomputing devices such as IBM's BlueGene (http://www.research.ibm.com/bluegene/). Furthermore, the evolution of a new class of "BioIT" specialists such as "The BioTeam" (http://www.bioteam.net/) has increased the availability and utility of hardware needed to meet developments in bioinformatics. Though this may not per se be a challenge for bioinformatics researchers, it does present a challenge to biomedical researchers seeking to use powerful tools; thus, it is a challenge for the discipline of bioinformatics.

2.3.5 Training and Education

The constantly changing nature of bioinformatics tools and the rapid growth in biological information has created a need for the development of better and more effective training and education programs in bioinformation data retrieval and analysis. The EDUCOLLAB Group at the National Center for Biotechnology Information (NCBI) has developed a series of introductory and advanced training programs for bioinformatics tool use, and the University of Washington Health Sciences Library has developed a 3-Day intensive training program to train students, faculty and staff in the use of NCBI online resources, commercial software and new developments in biology such as RNAi. These training sessions have been successfully given using telepresence tools such as the Access Grid. Additionally, commercial training companies such as OpenHelix (http://www.openhelix.com/) are now developing to meet the challenge and opportunity presented by the need for such training and education. There has also been a growing realization that a new type of profession, that of "bioinformationist", may be necessary to contend with the vast amount of data and analysis requirements resulting from what is essentially the digital imaging of Earth's biosphere (Lyon et al., 2004; and Florance et al., 2002].

2.3.6 Networking and Communications Tools

The highly dispersed nature of the modern biological research enterprise has from its inception required a very high degree of networking and communications among individual researchers and organizations—the Human Genome Project itself would not have been possible without the use of the Internet to promote and facilitate the distributed approach to sequencing and annotating the human genome. This had led to more extensive use of telecommunications tools such as WebEx and also to the development of so called "virtual" organizations such as VirtualGenomics.org (http://www.virtualgenomics.org/). NIH Director Elias A. Zerhouni has specifically described the need for the development of research teams spread out over large distances and many disciplines as a critical part of the NIH Roadmap, and the particular challenge provides the opportunity to develop new organizational structures and networking and communications tools. The Cornell University Life Sciences Initiative VIVO website (http://vivo.library.cornell.edu) provides a prototype for such a tool in a University context, while the Community of Science (COS-http://www.cos.com/) is a commercial enterprise tool for promoting collaborative research.

2.3.7 Publication/comprehension of Biological Information

Novel means of publication of data—wikis with their potential for rapid and constant peer review, data posting on websites such as the Gene Expression Omnibus (GEO: http://www.ncbi.nlm.nih.gov/geo/), modeling efforts such as the e-cell Project (http://www.e-cell.org/) and virtual disease models such as the Entelos Diabetes virtual patients (http://www.entelos.com/) and computer generated animations (http://www.wehi.edu.au/education/wehi-tv/dna/index.html) to help understand biological systems—are becoming essential to making efficient use of digital biological information for both clinicians and basic biology researchers. Additionally, new paradigms such as Systems Biology are providing new and important intellectual frameworks for comprehending biological information.

2.3.8 Physical Infrastructure and Culture

Conferencing facilities at university libraries for virtual meetings, computer laboratories for training, and architectural designs to promote contact among researchers can further promote collaboration and sharing of data, knowledge and expertise. Bio-X (http://biox.stanford.edu/) at Stanford University is an example of one such effort.

2.3.9 Research Center Coordination

Many of the resources for biological research are extremely expensive and mechanisms for sharing such resources must be developed. One example of the use of high speed Internet systems to allow the sharing and operation for advanced tools remotely is the Telescience Portal at the University of California, San Diego (https://telescience.ucsd.edu/), which provides for a collaborative environment for telemicroscopy and remote science. As high-speed connectivity and real-time videoconferencing tools become the norm, "Portals" allowing the use of complex and expensive scientific instruments such as high voltage electron microscopes remotely will allow researchers all over the world to perform experiments remotely and to form collaborative research teams driven by research needs rather than location.

2.3.10 Public Outreach

As the stem cell research issue and sometimes emotional debates concerning biodefense, genetics, nanotechnology, and robotics (GNR) developments show (Joy, 2000) it is critical to educate the public as to the science behind such fields as bioinformatics. Public understanding of the Human Genome Project, for example, will greatly enhance decision making as to how the results of that project will be used in the delivery of genomics based health care and technologies. High School Education projects such as the Seattle Biomedical Research Institute's BioQuest (http://www.sbri.org/sci-ed/index.asp) as well as direct connection with public media such as the Sci-Fi Channel (which has recently elected to produce science fiction classics such as the "Andromeda Strain" and Greg Bear's "Darwin's Radio and Darwin's Children") and other organizations with influence in the public understanding of science and its roles and effects on society are critically important.

3. CASE STUDY

3.1 Informatics Perspective – The BIOINFOMED Study and Genomic Medicine

The BIOINFOMED study funded by the European Commission (Martin-Sanchez et al., 2004) is an excellent case study at multiple levels. First it is a study focusing on formally developing a list of challenges and opportunities within bioinformatics and thus provides yet another perspective on opportunities and challenges. Secondly, it explicitly identifies these challenges in a particular sociotechnical context providing a first hand example of the issues identified under evaluation and sociotechnical dimensions. Finally, it articulates the fact that in order to achieve the promise of the Human Genome Project it is critical that work be done at the intersection of bioinformatics and clinical (medical) informatics.

The broad context of the BIOINFOMED study is that of the promise of the Human Genome Project as articulated in the beginning of this chapter. The specific focus is captured by the title of the resulting paper, "Synergy between medical informatics and bioinformatics: facilitating genomic medicine for future health care." The methods used were a prospective study of the relationships and potential synergies between bioinformatics and medical informatics. The starting point for the study was a written survey developed by the lead institute (Institute of Health Carlos III in Spain) that addresses a number of questions related to research directions and the future of both bioinformatics and medical informatics with an emphasis on opportunities to exchange knowledge across the two subdisciplines. A group of thirty professionals with expertise in medical informatics, bioinformatics, genomics, public health, clinical medicine and bioengineering met twice to analyze and synthesize the results of the survey.

The sociotechnical perspective was the articulation of the various stakeholders' interests and the resultant opportunities and challenges. For the focus of their paper (informatics in support of genomic medicine) they identified the following stakeholders: a) scientists/researchers; b) those executing clinical trials; c) health care professionals; d) health care consumers; e) systems providing healthcare; f) policy decision makers; g) industry; and h) society at large. For each stakeholder they identified different challenges and opportunities for biomedical informatics overall. From an evaluative point of view the study identified a number of gaps and synergies between the fields of bioinformatics and medical informatics.

The result of the study was a list of research priorities proposed by the BIOINFOMED study. Each item on the list included a description of the barrier(s) (e.g. the challenges), a proposed solution (e.g. the opportunities), a priority rating and a risk rating. The prioritization was High vs. Medium. The risk was defined as the probability that focusing on the research priority would fail to deliver results and given a rating of High, Medium, or Low risk. The items were grouped into four areas. The first area was enabling technologies. An example of one item is, "Barrier: Need to expand current interoperability standards for new genetic data infrastructure, Proposed Solution: Data Communication Standards, Priority: High, Risk: Medium." The second area was medical informatics in support of functional genomics. An example of one item is, "Barrier: Patient care data have not been systematically used in genomic research, Proposed Solution: phenotype databases suitable for genomic research, Priority: High, Risk: Low." The third area was bioinformatics in support of individualized healthcare. An example of one item is, "Barrier: Unavailability of models for including genetic data into electronic health records, Proposed Solution - Genetics data model for the EHR, Priority: Medium, Risk: Medium." The fourth area was the unified field of biomedical informatics in support of genomic medicine. An example of one item is, "Barrier: Linking environmental and lifestyle information to genetic and clinical data, Proposed Solution: Population based repositories, Priority: High, Risk: Low."

3.2 Biological Perspective – The BioResearch Liaison Program at the University of Washington

The University of Washington Health Sciences Library BioResearcher Liaison (http://healthlinks.washington.edu/hsl/liaisons/minie/) provides direct access to bioinformation consulting tools and training, and is a model program for contending with the issues discussed in Section 2.3. The BioResearcher Liaison program evolved out of an earlier effort called the BioCommons, and has been fully integrated into the Library's "informationists" infrastructure. The most visible part of this program is the BioResearcher Toolkit (http://healthlinks.washington.edu/bioresearcher) as shown in Figure 3-4, which provides a "portal" to biological information links, laboratory services, bioinformatics tools and consulting through the Library's Liaisons program (http://healthlinks.washington.edu/hsl/liaisons/). The contrast between the BIOINFOMED study and the BioResearcher toolkit is that the former lays out a research agenda for the future at the intersection of bioinformatics and medical informatics whereas the later is designed to address problems here and now. It is informative to compare and contrast the two case studies looking at the difference between grand challenges and on the ground realities.

The BioResearcher Toolkit is the second most visited part of the HealthLink's website (http://healthlinks.washington.edu/) after the more clinically oriented Care Provider Toolkit (see website at: http://healthlinks.washington.edu/care_provider/) with over 3,000 unique hits per month.

Since the consolidation of the BioCommons into the Library in 2002, the networked software and webware offerings have been the most used part of the BioResearcher Toolkit part of the website, with over 800 registered users of the various software packages available from the site and over 1,200 downloads over the past two years. These users are from that total pool of faculty, staff and students at the University of Washington, and come from large variety of departments as shown in Figure 3-5.

In addition to the BioResearcher Toolkit, the BioResearcher Liaison also provides a 3-day course given every quarter, the BioResearcher Tune-Up. The BioResearcher Tune-Up is a 3-Day intensive class with three modules—Module I: NCBI Online, Module II: Bioinformatics Software Workshop and Module III: Advanced Topics. Module I is a highly interactive tutorial which is taught in a computer lab using a web based PowerPoint template that allows students to directly follow the trainer through a tutorial on how to use NCBI databases using a single biologically relevant example: Huntington's disease.

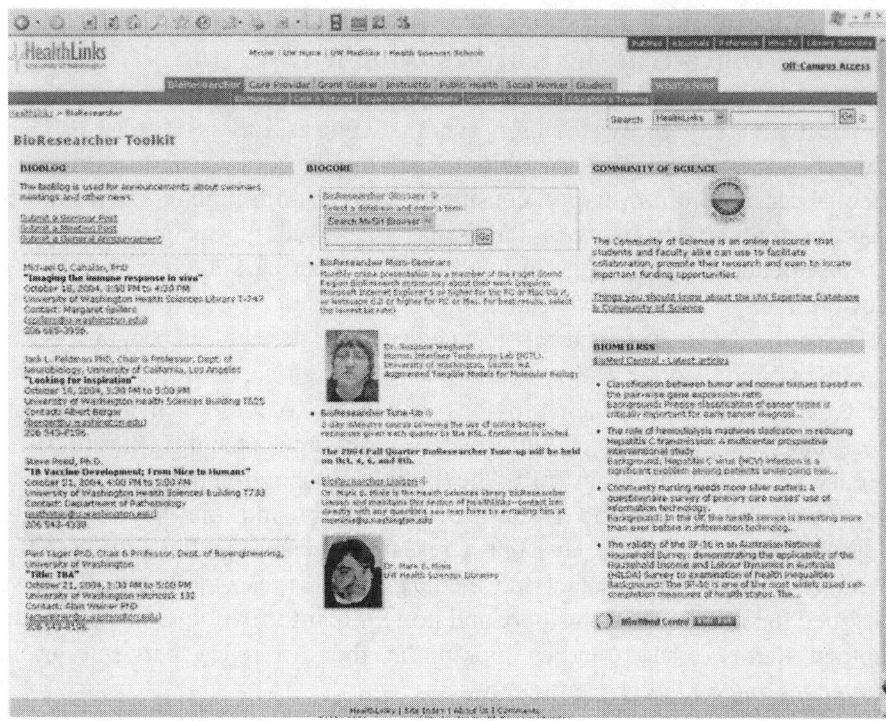

Figure 3-4. The BioResearcher Toolkit.

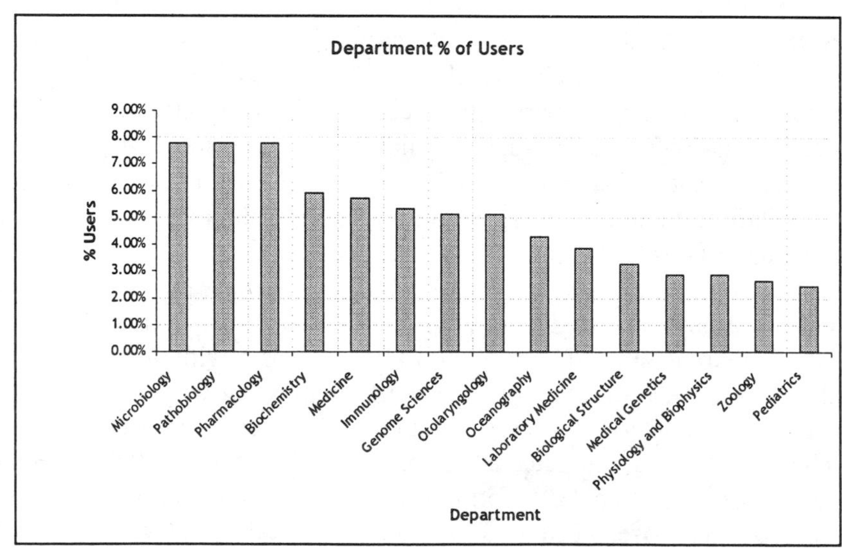

Figure 3-5. Users of the BioResearcher Toolkit

The Huntington Disease theme allows exploration of every database on Entrez, Blast and LocusLink (as well as Entrez Gene) and thus allows for a simple way to follow a story that touches on all aspects of the disease from the molecular level on up. Additionally, the module allows for the demonstration of discovery through digital data mining—a heretofore little known relationship between Huntington's disease and Type I diabetes is revealed while exploring the expression resources on GEO and SAGE Genie. The course is highly popular, with more applicants than there is room each quarter, and each attendee evaluates each module using an online form identical to that used by NCBI to evaluate similar modules taught by the EDUCOLLAB Group. Typically, this Module is rated as "Very Good" to "Excellent."

Module II is usually run by a guest vendor, who typically presents a tutorial workshop on a bioinformatics tool available from the BioResearcher Toolkit Computing and Laboratory section. For example, a Workshop/Tutorial on GeneSifter (www.genesifter.net), a web-based microarray analysis tool for analyzing gene expression data has been given as part of the BioResearcher Tune-Up with students participating both in the computer lab onsite and offsite via WebEx. The use of WebEx in particular is highly interactive, and has allowed more students, faculty and staff to attend than would be otherwise possible. Note also that one of the services provided by GeneSifter is the archiving of raw data, and the means to release that data in a highly interoperable format to the general scientific public in compliance with NIH's new rules on this issue.

Module III, the "Advanced Topics" part of the Tune-Up, covers a wide variety of relevant research oriented topics, ranging from seminars on DNA based nanotechnology to eukaryotic gene regulatory mechanisms as shown in Figure 3-6.

The BioResearch Liaison program also provides for one-on-one consulting with basic biology researchers at the University of Washington. For example, a client recently requested assistance in identifying a simple bioinformatics tool that would process molecular sequence data into graphical maps of alternative splice products for the gene studied. This led to a recommendation for the freely available NIH/NCBI tool "SPIDEY." A web-based open source program that was readily adapted to the clients needs. All consult encounters are followed-up with online evaluation forms to track the BioResearch Liaison's effectiveness, and the results are usually "Very Good" to "Excellent."

Recently, the Health Sciences Library BioResearch Liaison provided a version of the BioResearcher Tune-Up as an online training session to Alaska, Hawaii, Montana, Nevada, Utah, and Wyoming using the Access Grid videoconferencing technology shown in Figure 3-7. This provided a training session focused on the use of the sequence alignment tool BLAST (http://www.ncbi.nlm.nih.gov/BLAST/), and was a success from both the technical and teaching perspectives—the conferencing technology worked without glitches and online evaluations of the course by attending students gave it a "Very Good" to "Excellent" rating.

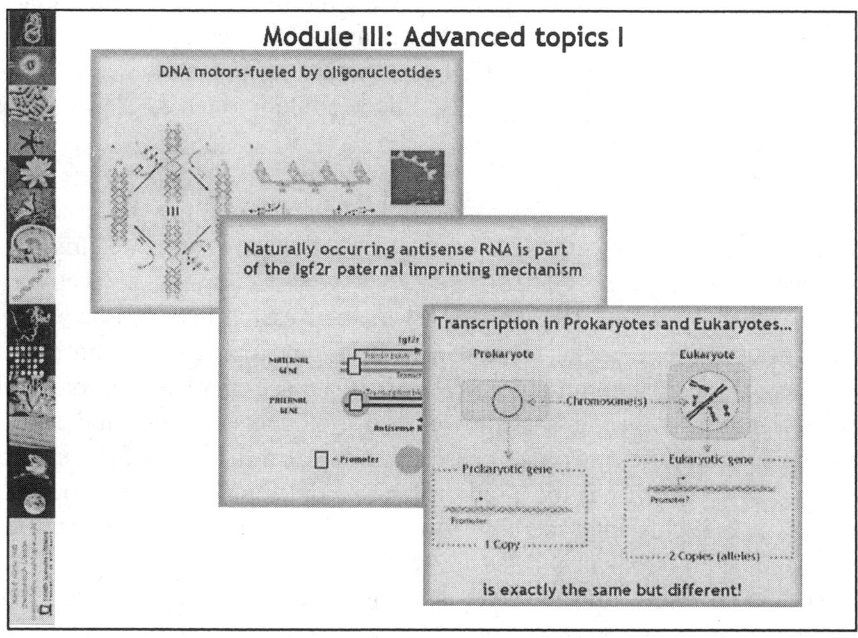

Figure 3-6. Advanced Topics section of the Tune-Up

Finally, the BioResearch Liaison program has had a number of notable successes in Public Outreach, with an important one involving providing a presentation to a large audience of science fiction writers and their publishers on how to access genomics information online as part of the "Science Friday" part of the recent 2004 Nebula Awards Conference in Seattle. One end result: an offer to publish a scientifically factual review of molecular biology and genomics in a prominent science fiction magazine widely read by the general public.

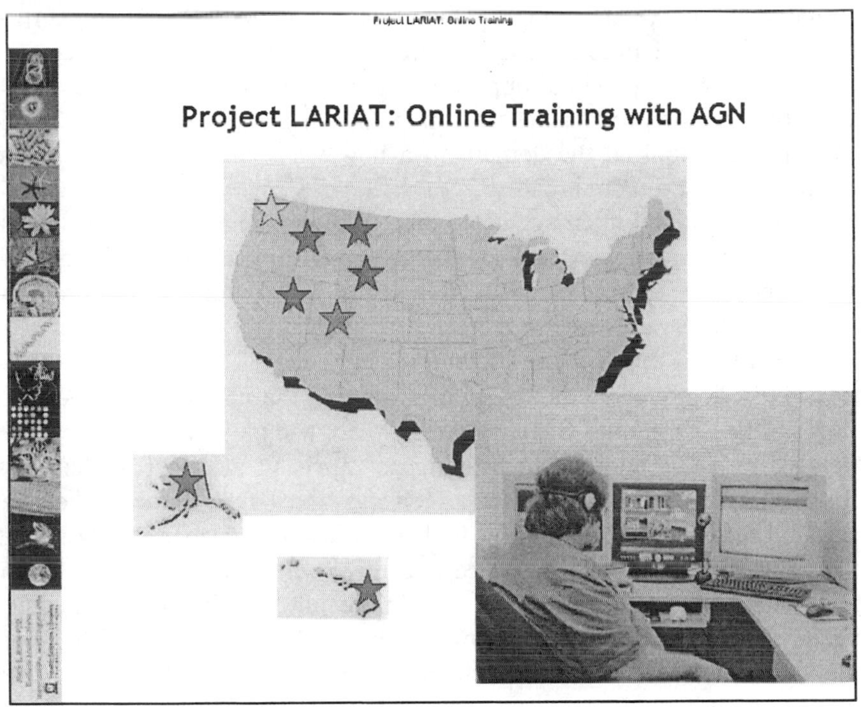

Figure 3-7. An Online BioResearcher Tune-Up

4. CONCLUSIONS AND DISCUSSION

The field of bioinformatics (defined both as foundational research and applied development of systems in support of biomedical research) presents a number of exciting challenges and opportunities for biologists, computer scientists, information scientists and bioinformaticists. These challenges sit at the intersection of biology and information. Ideally, larger scale work in this broad area involves a partnership between those with expertise in relevant foundational domains (e.g. computer scientists) and application domains (e.g. biologists) as well as bioinformaticists to serve as a bridge. The potential benefits of addressing some of these challenges are great both in terms of improving our understanding in general of how biological systems work and in terms of applying a better understanding of how the human biological system works in order to help improve health and treat disease.

Though many definitions of bioinformatics exist we have chosen to focus on the more inclusive definition to provide a richer picture of the opportunities and challenges. Indeed, it is possible that from the new perspectives of this more broadly defined bioinformatics the very informational nature of living systems may lead to a paradigm shift in biology. Our illustrations using specific examples nevertheless represent only a subset of the potential opportunities, and the inclusion of the broader framework for categorizing research papers will perhaps stimulate a reader of this book to look at the domain in a new way leading to unanticipated benefits to the field more broadly.

An important and often neglected area in biomedical informatics broadly (and bioinformatics by extension) is the human dimension captured in the socio-technical aspects of biomedical systems. In this context it is important to note two observations from the field of clinical (medical) informatics: a) the majority of applications developed in the lab have failed to be successfully deployed in the real world, and b) the majority of time, these failures relate to human factors rather than technical factors. It is also noteworthy that in addition to training scientists in the field in the use of online bioinformatics resources such as NCBI's Entrez, the very accessibility of these tools on the Internet allow for the possibility that the general public may directly use and possibly even participate in the further development of a "digital biology." Unlike other major developments in science in the 20th Century, the inherently "webified" nature of genomics information makes it relatively accessible to all - amateur scientists really can "try this at home."

We have presented two closely related but contrasting perspectives (the biological and the informatics perspectives) on the opportunities and challenges for bioinformatics. We have done so to a) illustrate some subtle but important distinctions, and b) demonstrate the value of having diverse perspectives as one explores the field of bioinformatics.

Finally, we have illustrated again the potential benefits of further work in this field through two case examples which also illustrate how researchers are going about trying to realize this potential. Here again the differences in the informatics and biological perspectives are worth noting. Particularly intriguing are both the emerging realization from both perspectives that biological systems are inherently digital, and the emerging parallel "Digital Biosphere" deriving from bioinformatics research activities. A true theoretical biology is at last emerging, where it may eventually be possible to understand complex biological systems by modeling them *in silico*. Significant progress in this direction has already taken place, with the publication of a detailed computer model of the regulatory network responsible for the control of flagellar biosynthesis in E.coli based on quantitative gene expression data (Herrgard and Palsson, 2004). This model is now being tested against the well defined genetic system of E. coli, and has already provided a system for developing new insights into this biological process. Particularly intriguing and revealing is the ready exchange of information between the *in vivo* and *in silico* systems.

5. ACKNOWLEDGEMENTS

Peter Tarczy-Hornoch: NIH NHGRI and NLM. "Interdisciplinary Center for Structural Informatics," #P20 LM007714; "BioMediator: Biologic Data Integration and Analysis System," # R01 HG02288; "Biomedical and Health Informatics Training Program," #T15 LM07442.

REFERENCES

Bergeron, B. (2002). *Bioinformatics Computing*, Upper Saddle River NJ: Prentice Hall.

Bodenreider, O., Mitchell, J., and McCray, A. (2002), "Evaluation of the UMLS as a Terminology and Knowledge Source for Biomedical Informatics," in *Proceedings of the AMIA Symposium*, p. 504-8.

Collins, F.S. and McKusick, V.A. (2001). "Implications of the Human Genome Project for Medical Science," *JAMA*, 285(5), 540-544.

Collins, F.S., et al. (2003). "A Vision for the Future of Genomics Research," *Nature*, 422(6934), 835-847.

Donelson, L., Tarczy-Hornoch, P., and Mork, P. (2004). "The BioMediator System as a Data Integration Tool to Answer Diverse Biologic Queries," in *Proceedings of MedInfo*, IMIA, an Francisco, CA.

Florance, V., Guise, N., and Ketchell, D. (2002). "Information in Context: Integrating Information Specialists into Practice Settings," *Journal of the Medical Library Association*, 90(1), 49-58.

Friedman, C. and Wyatt, J. (1997). *Evaluation Methods in Medical Informatics*, New York: Springer.

Gene Ontology Consortium, (2000). "Gene Ontology: Tool for the Unification of Biology," *Nature Genet.* 25, 25-29.

Herrgard, M. and Palsson, B. (2004). "Flagellar Biosynthesis in Silico: Building Quantitative Mdels of Regulatory Networks," *Cell*, 117(6), 689-90.

Ihaka, R. and Gentleman, R. (1996). R. "A Language for Data Analysis and Graphics," *Journal of Computational and Graphical Statistics*, 5, 299-314.

Joy, B. (2000). "Why the Future Doesn't Need Us," *Wired*, 8(4).

Klein, T.E., et al. (2001). "Integrating Genotype and Phenotype Information: An Overview of the PharmGKB Project," *The Pharmacogenomics Journal*, 1, 167-170.

Lyon, J., et al. (2004). "A Model for Training the New Bioinformationist," *Journal of the Medical Library Association*, 92(2), 188-195.

Martin-Sanchez, F., et al. (2004). "Synergy between Medical Informatics and Bioinformatics: Facilitating Genomic Medicine for Future Health Care," *Journal of Biomedical Informatics*, 37(1), 30-42.

Mei, H., et al. (2003). "Expression Array Annotation Using the BioMediator Biological Data Integration Systems and the BioConductor Analytic Platform," in *Proceedings of the American Medical Informatics Association (AMIA) Annual Symposium*. Washington, DC: American Medical Informatics Association.

Mork, P., A. Halevy, and Tarczy-Hornoch,P. (2001). "A Model for Data Integration Systems of Biomedical Data Applied to Online Genetic Databases," in *Journal of the American Medical Informatics Association, Fall Symposium Supplement*, p. 473-477.

Mork, P., et al., (2002). "PQL: A Declarative Query Language over Dynamic Biological Schemata," in *Proceedings of the American Medical Informatics Association Fall Symposium*, p. 533-537.

PharmGKB, *http://www.pharmgkb.org*.

Rosse, C. and Mejino, J. (2003). "A Reference Ontology for Bioinformatics: The Foundational Model of Anatomy," *Journal of Biomedical Informatics*, 36, 478-500.

Shaker, R., et al. (2004). "The BioMediator System as a Tool for Integrating Biologic Databases on the Web," in *Proceedings of the Workshop on Information Integration on the Web, held in conjunction with VLDB 2004.*

Shaker, R., et al. (2002). "A Rule Driven Bi-directional Translation System Remapping Queries and Result Sets between a Mediated Schema and Heterogeneous Data Sources," in *Journal of the American Medical Informatics Association, Fall Symposium.*

Shortliffe, E., et al. (2001). *Medical Informatics: Computer Applications in Health Care and Biomedicine.* Second ed. New York, Springer-Verlag.

Stanford (2002). Protegé Home Page, *http://protege.stanford.edu/*

Sujansky, W. (2001), "Heterogeneous Database Integration in Biomedicine," *Journal of Biomedical Informatics,* 34(4), 285-98.

Tuttle, M., Nelson, J. (1994). "The Role of the UMLS in 'Storing' and 'Sharing' across Systems," *Internet Journal of BioMedical Computing*, 34(1-4), 207-37.

van 't Veer, L., et al. (2002). "Gene Expression Profiling Predicts Clinical Outcome of Breast Cancer," *Nature*, 415(6871), p. 530-6.

Yarfitz, S. and D. Ketchell, "A Library-based Bioinformatics Services Program," *Bulletin of the Medical Library Association*, 88(1), 36-48.

SUGGESTED READINGS

Dan E. Krane. Michael L. Raymer. *Fundamental Concepts of Bioinformatics*. Publisher: Benjamin Cummings, 2003.
Provides a good overview of Molecular Biology and Biological Chemistry for the non-biologist then addresses important problems in bioinformatics for both the biologist and informaticist (sequence alignment, substitution, phylogenetics, gene identification, structure prediction, and proteomics).

Andreas D. Baxevanis. B.F. Ouellette, eds. *Bioinformatics: A Practical Guide to the Analysis of Genes and Proteins.* 2nd Edition. Publisher: Wiley-Interscience. 2001.
Provides an overview of internet accessible tools for the biologist with an emphasis on NCBI resources aimed at a mixed audience of biologists and developers. Each section is a blend of the underlying biology, the computing principles involved, and some practical hands on advice and tips.

Stanley I. Leovsky, ed. *Bioinformatics Databases and Systems.* Publisher: Kluwer Academic Publishers, 1999.
Provides an excellent overview of biological databases and computing systems aimed more at the developer than the biologist but useful to both. A series of deployed bioinformatics databases are described in some detail by the developers of the databases (e.g. NCBI, KEGG, FlyBase). Then a series of deployed tools are described by their developers in a 2nd section (e.g. BioKleisli, SRS, ACDEB).

Z. Lacroix, and T. Critchlow, eds. *Bioinformatics: Managing Scientific Data.* Publisher: Morgan Kaufmann, 2003.

Provides an excellent overview from more of a computer science standpoint of data management issues in biological research with an emphasis on data integration using selected examples from both academics and industry.

ONLINE RESOURCES

NIH Biomedical Information Science and Technology Initiative (BISTI)
http://www.bisti.nih.gov/

Report of the Working Group on Biomedical Computing, Advisory Committee to the Director, National Institutes of Health, Co-Chairs: David Botstein and Larry Smarr "The Biomedical Information Science and Technology Initiative"
http://www.nih.gov/about/director/060399.htm

NIH All About the Human Genome
http://www.genome.gov/10001772/

The University of Washington BioResearcher Toolkit
http://healthlinks.washington.edu/bioresearcher/

The NCBI website; the entry point for such search and analysis tools as Entrez and BLAST etc. Also a source for online tutorials on the use of PubMed, CN3D (free structure viewing tool), BLAST etc.
http://www.ncbi.nlm.nih.gov/

University of Pittsburg biolibrary website
http://www.hsls.pitt.edu/guides/genetics/

The Cornell University Life Sciences Library website, VIVO
http://vivo.library.cornell.edu/

QUESTIONS FOR DISCUSSION

1. Define interesting challenges in knowledge management and data mining in biomedical informatics based on the primary bioinformatics literature.

2. Define interesting challenges in knowledge management and data mining in biomedical informatics based on the primary biomedical research literature or based on interviews with biomedical researchers.

3. Define key unmet needs related to bioinformatics tools based on the primary bioinformatics literature and the primary biomedical research literature. Compare and contrast the unmet needs from these two perspectives.

4. What are the implications of the different perspectives from biology, biomedical research, and bioinformatics research?

5. What are the possibilities for theoretical biology based on knowledge mining of biological databases?

6. Discuss the implications for medical systems of virtual patients and disease modeling.

7. Given the informational nature of biological systems—what are the implications for our definition and understanding of life?

8. Point-of-Care Diagnostics and biomedical informatics—what might be the implications for future medical care and costs?

Chapter 4

MANAGING INFORMATION SECURITY AND PRIVACY IN HEALTHCARE DATA MINING

State of the Art

Ted Cooper[1] and Jeff Collman[2]

[1]Department of Ophthalmology, Stanford University Medical School, Palo Alto, California, 94304; [2]ISIS Center Georgetown University School of Medicine; Department of Radiology; Georgetown University Medical Center, Washington D.C., 20057

Chapter Overview

This chapter explores issues in managing privacy and security of healthcare information used to mine data by reviewing their fundamentals, components and principles as well as relevant laws and regulations. It also presents a literature review on technical issues in privacy assurance and a case study illustrating some potential pitfalls in data mining of individually identifiable information. The chapter closes with recommendations for privacy and security good practices for medical data miners.

Keywords

information security; privacy; confidentiality; integrity; availability; HIPAA; anonymity; risk management; privacy appliance; human subjects protection

1. INTRODUCTION

As the health care delivery system adopts information technology, vast quantities of health care data become available to mine for valuable knowledge. Health care organizations generally adopt information technology to reduce costs as well as improve efficiency and quality. Medical researchers hope to exploit clinical data to discover knowledge lying implicitly in individual patient health records. These new uses of clinical data potentially affect healthcare because the patient-physician relationship depends on very high levels of trust. To operate effectively physicians need complete and accurate information about the patient. However, if patients do not trust the physician or the organization to protect the confidentiality of their health care information, they will likely withhold or ask the physician not to record sensitive information (California HealthCare Foundation, 1999). This puts the patient at risk for receiving less than optimum care, the organization at risk of having incomplete information for clinical outcome and operational efficiency analysis, and may deprive researchers of important data. Numerous examples exist of inappropriate disclosure of individually identifiable data that has resulted in harm to the individual (Health Privacy Project, 2003). Concerns about such harm have resulted in laws and regulations such as the privacy rules of the Health Insurance Portability and Accountability Act (HIPAA) of 1996 directly governing the use of such information by most health care providers, health plans, payors, clearinghouses, and researchers. These laws and regulations may also indirectly govern the use of this data by the business partners of these entities. None of these laws forbid research or using technologies such as data mining. All require medical investigators, whether conducting biomedical research or quality assurance reviews, to take sound precautions to respect and protect the privacy and security of information about the subjects in their studies.

Data mining especially when it draws information from multiple sources poses special problems. For example, hospitals and physicians are commonly required to report certain information for a variety of purposes from census to public health to finance. This often includes patient number, ZIP code, race, date of birth, gender, service date, diagnoses codes (ICD9), procedure codes (CPT), as well as physician identification number, physician ZIP code, and total charges. Compilations of this data have been released to industry and researchers. Because such compilations do not contain the patient name, address, telephone number, or social security number, they qualify as de-identified and, therefore, appear to pose little risk to patient privacy. But by cross linking this data with other publicly available databases, processes such as data mining may associate an individual with specific diagnoses. Sweeney (1997) demonstrates how to re-identify such data by linking certain conditions with the voting list for Cambridge, Massachusetts which contains demographic data on over 50 thousand voters. Birth date alone can uniquely identify the name and address of up to 12% of people on such compilations with birth date and gender up to 29%, birth date and 5-digit ZIP code up to 69%, and full postal code and birth date up to 97%.

Recent work has demonstrated ways to determine the identity of individuals from the trail of information they leave behind as they use the World Wide Web (Malin and Sweeney, 2001) (Malin et al., 2003). IP addresses from online consumers have been linked with publicly available hospital data that correlates to DNA sequences for disease. Data collected as individuals use the Internet to obtain health information, services and products also pose hazards to privacy but much less law and regulation governs the use and disclosure of this type of information (Goldman and Hudson, 2000).

In this chapter we explore issues in managing privacy and security of healthcare information used to mine data by reviewing their fundamentals, components and principles as well as relevant laws and regulations. We also present a literature review on technical issues in privacy assurance and a case study illustrating some potential pitfalls in data mining of individually identifiable information. We close the chapter with recommendations for privacy and security good practices for medical data miners.

2. OVERVIEW OF HEALTH INFORMATION PRIVACY AND SECURITY

Often voluminous, heterogeneous, unstructured, lacking standardized or canonical form, and incomplete, as well as surrounded by ethical considerations and legal constraints, the characteristics of patient health care records make them "messy." Because they originate primarily as a consequence of direct patient care with the presumption of benefit for the patient, their use for research or administrative purposes must happen with care to ensure no harm to the patient. Inappropriate disclosure, loss of data integrity, or unavailability may each cause harm (Cios and Moore, 2002). Recent laws and regulations such as HIPAA provide patients with legal rights regarding their personally identifiable healthcare information and establish obligations for healthcare organizations to protect and restrict its use or disclosure. Data miners should have a basic understanding of healthcare information privacy and security in order to reduce risk of harm to individuals, their organization or themselves.

2.1 Privacy and Healthcare Information

The term "privacy" bears many meanings depending on the context of use. Common meanings include being able to control release of information about one's self to others and being free from intrusion or disturbance in one's personal life. To receive healthcare one must reveal information that is very personal and often sensitive. We control the privacy of our healthcare information by what we reveal to our physicians and others in the healthcare delivery system. Once we share personal information with our caregivers, we no longer have control over its privacy. In this sense, the term "privacy" overlaps with "confidentiality" or the requirement to protect information received from patients from unauthorized access and disclosure. For example, the HIPAA Privacy Standard (Department of Health and Human Services, 2002) requires healthcare providers, health plans and health plan clearinghouses to establish appropriate administrative, technical, and physical safeguards to protect the use and disclosure of individually identifiable health information. HIPAA draws on ethical standards long developed in the health care disciplines that identify protecting the confidentiality of patient information as a core component of the doctor-patient relationship and central to protecting patient autonomy. Thus, ethics, laws and regulations provide patients with certain rights and impose obligations on the healthcare industry that should keep patient health information from being disclosed to those who are not authorized to see it.

2.2 Security and Healthcare Information

Use of the Internet has resulted in recognition that information technology security is of major importance to our society. This concern seems relatively new in healthcare, but information technology security is a well established domain. A large body of knowledge exists that can be applied to protect healthcare information. A general understanding of security can be obtained by understanding:

1. Security Components
2. Security Principles
3. Threats, Vulnerabilities, Control Measures and Information Assurance
4. Achieving Information Security: Administrative, Physical, Technical Safeguards

2.2.1 Security Components

Security is achieved by addressing its components: confidentiality, integrity, availability and accountability.

1. Confidentiality is the property that data or information is not made available or disclosed to unauthorized persons or processes.
2. Integrity is the property that data or information have not been altered or destroyed in an unauthorized manner.
3. Availability is the property that data or information is accessible and useable upon demand by an authorized person.
4. Accountability is the ability to audit the actions of all parties and processes which interact with the information and to determine if the actions are appropriate.

2.2.2 Security Principles

In 1997 the International Information Security Foundation published the latest update to this set of generally-accepted system security principles (International Security Foundation, 1997):

1. Accountability Principle

The responsibilities and accountability of owners, providers and users of information systems and other parties concerned with the security of information systems should be explicit.

2. Awareness Principle

In order to foster confidence in information systems, owners, providers and users of information systems and other parties should readily be able, consistent with maintaining security, to gain appropriate knowledge of and be informed about the existence and general extent of measures, practices and procedures for the security of information systems.

3. Ethics Principle

Information systems and the security of information systems should be provided and used in such a manner that the rights and legitimate interests of others are respected.

4. Multidisciplinary Principle

Measures, practices and procedures for the security of information systems should take account of and address all relevant considerations and viewpoints, including technical, administrative, organizational, operational, commercial, educational and legal.

5. Proportionality Principle
Security levels, costs, measures, practices and procedures should be appropriate and proportionate to the value of and degree of reliance on the information systems and to the severity, probability and extent of potential harm, as the requirements for security vary depending upon the particular information systems.

6. Integration Principle
Measures, practices and procedures for the security of information systems should be coordinated and integrated with each other and with other measures, practices and procedures of the organization so as to create a coherent system of security.

7. Timeliness Principle
Public and private parties, at both national and international levels, should act in a timely coordinated manner to prevent and to respond to breaches of security of information systems.

8. Reassessment Principle
The security of information systems should be reassessed periodically, as information systems and the requirements for their security vary over time.

9. Equity Principle
The security of information systems should be compatible with the legitimate use and flow of data and information in a democratic society.

2.2.3 Threats, Vulnerabilities, Control Measures and Information Assurance

Numerous threats exist to computer systems and the information they contain originating from within and outside organizations. Some common threats include malicious code such as viruses, Trojan horses, or worms. Malicious code often takes advantage of vulnerabilities in operating system software but depends, too, upon organizational weaknesses such as the failure to deploy, update or train workers in the use of antivirus software. Malicious code may enable denial of service attacks, impersonation, information theft and other intrusions. Attacks by famous malicious code such as the Melissa or Lovebug viruses highlight the threat of "hackers", outsiders with intent to harm specific organizations or network operations in general. Insiders with privileged access to network operations and a grudge against their employer actually wreak the most harm to say nothing of ill-trained workers unintentionally making mistakes.

For individuals with responsibility for protecting the security of computerized information assets, the important point to remember is that each computer system with its host organization has its own security weaknesses or vulnerabilities. To minimize the likelihood of harm from threats, organizations must perform an information security risk assessment which serves as the foundation for an information assurance plan. Because computer security is relative, i.e. absolute security does not exist, an information assurance plan seeks to apply cost-effective control measures to reduce to acceptable levels the likelihood of loss to an organization from likely threats. In other words, the information assurance plan is designed to manage risk. Control measures include policies, procedures, and technology. Risk assessments should be repeated periodically because both threats and vulnerabilities change over time and used to update information assurance plans.

The HIPAA Security Standard reflects good practice in the information security industry and, thus, provides guidance to medical dataminers about how to proceed. Thanks to HIPAA many resources have emerged in the last several years to help, including *The CPRI Toolkit: managing information security in healthcare* (see http://www.himss.org/resource) and *Managing Information Security Risks: The OCTAVE^sm Approach* (Alberts and Dorofee, 2003). The website of the National Institute of Standards and Technology contains a wealth of guidance on computer information security in general as well as specific topics (see http://crst.nist.gov, particularly the Special Publications section).

2.2.4 Achieving Information Security: Administrative, Physical, and Technical Safeguards

The measures to control threats and vulnerabilities can be organized into three categorizes of safeguards: administrative, physical and technical. The HIPAA Security Standard describes "Administrative Safeguards" as administrative actions, policies and procedures "to manage the selection, development, implementation, and maintenance of security measures to protect electronic protected health information and to manage the conduct of the covered entity's workforce in relation to the protection of that information" (Department of Health and Human Services, 2003, pg. 261). Administrative safeguards include policies and procedures such as risk assessment and management, assigning responsibility for information security, developing rules and procedures for assigning access to information, sanctioning misbehavior, responding to security incidents and implementing a security training and awareness program. Physical safeguards include policies, procedures and measures to control physical access to information assets such as computer sites, servers, networks, and buildings. HIPAA focuses special attention on workstations processing patient information requiring hospitals to identify their uses as well as controls on physical access. Technical controls include the various devices typically associated with "information security" such as passwords, firewalls, and encryption as well as technical measures for assuring health information integrity. Virtual private networks, tokens for user access, audit logs and public/private key infrastructure (PKI) are examples of technical safeguards.

2.2.5 Laws and Regulations

The following regulations found in the Code of Federal Regulations (CFR) are likely to apply to the use of health care data in data mining in the United States:

1. Standards for Privacy of Individually Identifiable Health Information; Final Rule Title 45 CFR Parts160 and 164, known as the HIPAA Privacy Standard (Department of Health and Human Services, 2002),
2. Security Standards Final Rule Title 45 CFR Parts 160, 162, and 164, known as the HIPAA Security Rule, (Department of Health and Human Services, 2003), and
3. Department of Health and Human Services (HHS) or the Food and Drug Administration (FDA) Protection of Human Subjects Regulations, known as the Common Rule Title 45 CFR part 46 (Department of Health and Human Services, 2001) or Title 21 CFR parts 50 and 56, respectively (Food and Drug Administration, 2002).

The European Union, Canada, and Australia have instituted their own laws and regulations in this area (see list of websites cited in Section 9. resources).

A full explanation of the HIPAA regulations is beyond the scope of this chapter, however understanding some of its basic requirements are essential for those engaging in healthcare data mining. These regulations set the national floor for the use and disclosure of most personally identifiable health information in the health care delivery system in the United States. While they supersede contrary state laws, they do not supersede state laws and regulations that are more stringent. Many states have more stringent laws and regulations. A discussion of common questions follows (see also http://www.hhs.gov/ocr/hipaa/)

Who must comply with HIPAA?

Most healthcare providers, health plans and healthcare clearinghouses must comply with HIPAA. Excluded from HIPAA are healthcare providers that do not transmit electronic information, records covered by the Family Educational Rights and Privacy Act and employment health records held by a covered entity in its role as employer. Healthcare information collected by entities not covered by HIPAA is not subject to these regulations.

What information is protected by HIPAA?

The HIPAA Privacy Standard applies to individually identifiable health information including oral, written and electronic information used by covered entities to make decisions. For providers this includes medical records and billing records. For health plans this includes enrollment, payment, claims adjudication, and case or medical management record systems records. This protected health information is known as PHI. The HIPAA Security Standard only applies to electronic information and does not cover oral or paper information.

What rights does HIPAA grant patients?

Patients have a right to:

1. a notice of information practices from providers and health plans that states how PHI is used and protected,
2. obtain copies of their healthcare records,
3. amend their healthcare records, and
4. an accounting of disclosures made for purposes other than treatment, payment and healthcare operations.

What must entities covered by HIPAA do?

Covered entities must:

1. provide a notice of information practices and abide by the notice,
2. designate an individual to be responsible for privacy,
3. provide appropriate administrative, physical and technical safeguards for PHI.
4. only use and disclose PHI in accordance with the HIPAA Privacy Standard, and
5. have written agreements with business associates with whom they share PHI requiring the business associate to protect the PHI.

What are the key rules for use and disclosure of PHI?

1. Except for specific exclusions, an authorization from the patient is required for covered entities to use or disclose PHI for purposes other than treatment, payment and healthcare operations.
2. Only the minimum necessary amount of PHI may be used or disclosed to satisfy the purpose of the use or disclosure, with the exception that physicians may disclose the entire record to other providers for treatment purposes.

What is meant by healthcare operations?

Healthcare operations are the usual business operations of healthcare providers and health plans. Specifically included are: quality assessment and improvement activities, outcomes evaluation, development of clinical guidelines, population-based activities relating to improving health or reducing health care costs, protocol development, case management and care coordination, contacting of health care providers and patients with information about treatment alternatives; and related functions that do not include treatment, reviewing the competence or qualifications of health care professionals, evaluating practitioner, provider performance and health plan performance, conducting training programs in which students, trainees, or practitioners in areas of health care learn under supervision to practice or improve their skills as health care providers, training of non-health care professionals, accreditation, certification, licensing, or credentialing activities, underwriting, premium rating, and other activities relating to the creation, renewal or replacement of contracts, conducting or arranging for medical review, legal services, and auditing functions, including fraud and abuse detection and compliance programs, business planning and development, such as conducting cost-management and planning-related analyses for managing and operating the entity, including formulary development and administration, development or improvement of methods of payment or coverage policies; customer service, resolution of internal grievances, sale, transfer, merger, or consolidation.

What is the difference between health care operations and research?

For HIPAA, research means a systematic investigation, including research development, testing, and evaluation, that is designed to develop or contribute to generalizable knowledge. If the same query is used on the same data in one case to improve efficiency, and in the second case to contribute generalizable knowledge, it is not research in the first case but is in the second case. Additional protections must be in place for research.

What does HIPAA require for research?

HIPAA research requirements only apply to HIPAA covered entities (providers, health plans and health plan clearinghouses).

PHI used or disclosed for research must be authorized by the patient unless:

1. A waiver has been granted by a privacy review board or institutional review board (IRB):
   - The review board or IRB must document that they believe the PHI used or disclosed involves no more than minimal risk to the privacy of individuals based on:
     - An adequate plan to protect PHI identifiers from improper use and disclosure;
     - An adequate plan to destroy those identifiers at the earliest legal and practical opportunity consistent with the research, and
     - Adequate written assurances that the PHI will not be reused or disclosed to any other person or entity except as required by law, for authorized oversight of the research study, or for other research for which the use or disclosure of the PHI is permitted by the Privacy Rule.
   - The research could not practicably be conducted without the requested waiver.
   - The research could not practicably be conducted without access to and use of the PHI.
2. For reviews preparatory to research and with the researcher making the following written assertions:
   - The use or disclosure is sought solely to review PHI as necessary to prepare the research protocol or other similar preparatory purposes;
   - No PHI will be removed from the covered entity during the review; and
   - The PHI that the researcher seeks to use or access is necessary for the research purposes.
3. For research on decedent's information, the covered entity is assured by the researcher that the use or disclosure is solely for research on the PHI, and is necessary for research purposes.
4. If the PHI has been de-identified in accordance with the standards of the Privacy Rule and therefore is no longer PHI. HIPAA describes two approaches for de-identification, including 1) a person with appropriate knowledge and experience applies and documents generally accepted statistical and scientific methods for de-identifying

information, or 2) remove 18 specific identifiers listed in section 164.514 of the rule.

5. If the information is released as a limited data set as prescribed by the Privacy Standard (with 18 specific identifiers removed) and a data usage agreement with the researchers stating that they will not attempt to re-identify the information.

What are the exceptions to the requirement for authorizations prior to disclosure of PHI?

HIPAA permits disclosures without authorizations as required by law for public health, health oversight activities, victims of abuse, neglect or domestic violence, judicial and administrative proceeding, law enforcement, for specialized government functions (military and veteran activities, national security and intelligence, medical suitability, correctional institutions, public benefit programs) and research. It should be noted that for each of these exceptions there are additional provisions that govern the details of the disclosures. It should also be noted that HIPAA permits but does not require any disclosures.

In addition to the HIPAA Privacy Rule researchers must comply with the HHS and FDA Protection of Human Subjects Regulations.

What is the relationship of these regulations?

"There are two main differences. First, the HHS and FDA Protection of Human Subjects Regulations are concerned with the risks associated with participation in research. These may include, but are not limited to, the risks associated with investigational products and the risks of experimental procedures or procedures performed for research purposes, and the confidentiality risks associated with the research. The Privacy Rule is concerned with the risk to the subject's privacy associated with the use and disclosure of the subject's PHI.

Second, the scope of the HHS and FDA Protection of Human Subjects Regulations differs from that of the Privacy Rule. The FDA regulations apply only to research over which the FDA has jurisdiction, primarily research involving investigational products. The HHS Protection of Human Subjects Regulations apply only to research that is conducted or supported by HHS, or conducted under an applicable Office for Human Research Protections (OHRP)-approved assurance where a research institution, through their Multiple Project Assurance (MPA) or Federal-Wide Assurance (FWA), has agreed voluntarily to follow the HHS Protection of Human

Subjects Regulations for all human subjects research conducted by that institution regardless of the source of support. By contrast, the Privacy Rule applies to a covered entity's use or disclosure of PHI, including for any research purposes, regardless of funding or whether the research is regulated by the FDA" (National Institutes of Health, pg. 5, February 5, 2004).

What are the differences between the HIPAA Privacy Rule's requirements for authorization and the Common Rule's requirements for informed consent?

"Under the Privacy Rule, a patient's authorization is for the use and disclosure of protected health information for research purposes. In contrast, an individual's informed consent, as required by the Common Rule and the Food and Drug Administration's (FDA) human subjects regulations, is a consent to participate in the research study as a whole, not simply a consent for the research use or disclosure of protected health information. For this reason, there are important differences between the Privacy Rule's requirements for individual authorization, and the Common Rule's and FDA's requirements for informed consent. However, the Privacy Rule's authorization elements are compatible with the Common Rule's informed consent elements. Thus, both sets of requirements can be met by use of a single, combined form, which is permitted by the Privacy Rule. For example, the Privacy Rule allows the research authorization to state that the authorization will be valid until the conclusion of the research study, or to state that the authorization will not have an expiration date or event. This is compatible with the Common Rule's requirement for an explanation of the expected duration of the research subject's participation in the study. It should be noted that where the Privacy Rule, the Common Rule, and/or FDA's human subjects regulations are applicable, each of the applicable regulations will need to be followed (National Institutes of Health, pg. 10, February 5, 2004)

Under the Common Rule, when may individually identifiable information be used for research without authorization or consent?

"Research involving the collection or study of existing data, documents, records, pathological specimens, or diagnostic specimens, if these sources are publicly available or if the information is recorded by the investigator in such a manner that subjects cannot be identified, directly or through identifiers linked to the subjects" (Department of Health and Human Services, pg 5, August 10, 2004).

3. REVIEW OF THE LITERATURE: DATA MINING AND PRIVACY AND SECURITY

In the previous sections managing health information privacy and security has been described as required by organizations involved in the industry of delivering healthcare; e.g. healthcare providers, health plans, payors, and clearinghouses. In this section we will explore the additional issues that large scale data mining presents for managing health information privacy and security. Data mining offers many possible benefits to the medical community, including administrators as well as researchers. One example of the value that can be derived from large data collections is demonstrated by Kaiser Permanente's Northern California Region reduction of the risk of their members dying from cardiovascular causes so that it is no longer their number one cause of death. According to the 2002 Annual Report of the National Committee for Quality Assurance (2002, pg. 23), "Since 1996, appropriate cholesterol control (as defined by HEDIS, an LDL level of less than 130) among the CAD population has improved from 22 percent to 81 percent. Among eligible patients discharged after a heart attack, 97 percent were on beta-blockers. The mortality rate from heart attacks at KPNC hospitals are up to 50 percent lower than at similar hospitals across the state." This was made possible by the development of a clinical data repository to support real-time direct healthcare delivery to its membership (over three million individuals), evidence-based medical knowledge and use of this data to guide their healthcare delivery processes (Levin et al., 2001) (Pheatt et al., (2003).

As information technology has become commonly used to support the core processes of healthcare, enormous volumes of data have been produced. Numerous organizations desire access to this data to apply techniques of knowledge discovery. Privacy concerns exist for information disclosed without illegal intrusion or theft. A person's identity can be derived from what appears to be innocent information by linking it to other available data. Concerns also exist that such information may be used in ways other than promised at the time of collection. Ways to share person specific data while providing anonymity of the individual are needed. Stated another way, controls are needed to manage the inferences about individual identity that can be made from shared person specific data. The Federal Office of Management and Budget (1994) has developed an approach to limit disclosure from government data so that the risk that the information could be used to identify an individual, either by itself or in combination with other information, is very small. This Report on Statistical Disclosure Limitation Methodology, Statistical Policy, discusses both tables and microdata. The report includes a tutorial, guidelines, and recommendations for good practice; recommendations for further research; and an annotated bibliography. Techniques, rules and procedures for tables (magnitude versus frequency, counts, suppression, random versus controlled rounding, confidentiality editing) and microdata (sampling, removing identifiers, demographic detail, high visibility variables, adding random noise, rank swapping, blank and imputation for randomly selected records and blurring) are documented.

3.1 General Approaches to Assuring Appropriate Use

Past experience has shown three approaches to be common for using personal data in research and secondary analysis: using personal data only with the subject's consent, using personal data without explicit consent with a public interest mandate, and making the data anonymous before use (Lowrance, 2002). The discussion above of HIPAA and the Common Rule address the first two of these techniques, obtaining subject consent and authorized public interest use such as public health. Developing methods for assuring data anonymity offers promise for the future.

> As Sweeney (2003, pg. 15) says: "The goal of pioneering work in data anonymity is to construct technology such that person-specific information can be shared for many useful purposes with scientific assurances that the subjects of the data cannot be re-identified."

A discussion of specific methods for making data anonymous follows below. Each specific approach embodies one or more of four general approaches to the problem of assuring against disclosure of confidential information when querying statistical databases containing individually identifiable information including: conceptual, query restriction, data perturbation, and output perturbation approaches (Adam and Wortmann, 1989). Unfortunately none of these approaches offers a completely satisfactory solution. The conceptual model has not been implemented in an on-line environment and the others involve considerable complexity and cost and may obscure medical knowledge.

3.1.1 Conceptual approach

In the conceptual model, a user can access only the properties of population (i.e. a collection of entities that have common attributes and its statistics such as patients of certain ages, genders and ZIP codes) and tables that aggregate information. The user thus knows the attributes of the population and its origin, but may not manipulate the data or launch queries that merge and intersect subpopulations from the collection. The user may only see statistical tables that contain either zero or at least two individuals never information on a single individual. With no access to the data and tables with only data on more than one individual, disclosure of information about a single individual is prevented. Tables 4-1 and 4-2 illustrate acceptable and unacceptable data tables using the conceptual approach.

Table 4-1 displays attributes only about types of persons by age, sex and ZIP and includes cells with numbers larger than one. Table 4-2 displays the same types of data but includes one cell with information about only a single individual (Female, Age 31-40), a presentation not permitted in the conceptual approach. While this model is thought to provide anonymity, it has never been implemented at a practical level with a production software system.

Table 4-1. Population Attributes Acceptable Table

Attributes	# Male	# Female	ZIP
Age 10-20	2031	2301	94027
Age 21-30	231	243	94027
Age 31-40	24	27	94027

Table 4-2. Population Attributes Unacceptable Table

Attributes	Male	Female	ZIP
Age 10-20	231	241	94027
Age 21-30	23	24	94027
Age 31-40	2	1	94027

3.1.2 Query restriction approach

Five methods have been developed to restrict queries:

1. query-set-size control - a method that returns a result only if its size is sufficient to reduce the chances of identification,
2. query-set-overlap control - a method that limits the number of overlapping entities among successive queries of a given user,
3. auditing – a method that creates up-to-date logs of all queries made by each user and constantly checks for possible compromise when a new query is issued,
4. cell suppression – a method that suppresses cells that might cause confidential information to be disclosed from being released, and
5. partitioning – a method that clusters individual entities into a number of mutually exclusive subsets thus preventing any subset from containing precisely one individual.

3.1.3 Data perturbation

This approach alters the data before permitting access to users. For example the source data is replaced with data having the same probability distribution (Islan and Brankovic, 2004). In other words, noise is inserted in the data that seeks to achieve anonymity and at the same time not change the statistical significance of query results. Users do not have access to the original data.

3.1.4 Output perturbation

This approach permits use of the original data, but modifies or renders the output incomplete. Techniques of output perturbation include processing only a random sample of the data in the query, adding or subtracting a random value that will not alter the statistical difference from the result, and rounding up or down to the nearest multiple of a certain base. Murphy and Chueh have published a successful implementation of query output perturbation to determine if a research database contains a set of patients with specific characteristics of sufficient size for statistical significance (Murphy and Chueh, 2002). In this example, the query result alters the number of patients with the specific characteristics by adding or subtracting a small random number. In addition, for values nearing zero, a result of less than three is presented.

3.2 Specific Approaches to Achieving Data Anonymity

Rendering data anonymous assures freedom from identification, surveillance or intrusion for the subjects of medical research or secondary data analysis while allowing data to be shared freely among investigators (Meany, 2001). Achieving complete data anonymity poses a considerable challenge. For example, 87% of individuals in the United States can be uniquely identified by their date of birth, gender and 5-digit ZIP code (Sweeney, 2002). True anonymity also poses ethical problems of its own, including loss of the possibility of benefit to the individual patient from knowledge discovered from the data, greatly increased complexity of maintaining an up-to-date database, and elimination of some checks against scientific fraud (Behlen and Johnson, 1999).

A number of techniques exemplifying or combining the general approaches described above have been advocated to help address this issue, including:

1. Data aggregation

2. Data de-identification
3. Binning
4. Pseudonymisation
5. Mediated access

3.2.1 Data Aggregation (an example of the Conceptual Approach)

Providing access only to aggregate data while prohibiting access to records containing data on an individual constitutes one approach commonly advocated to reduce risks to privacy. Although this approach does protect privacy, it critically limits medical research. Clinical research requires prospectively capturing and analyzing data elements associated with individual patients. Outliers are often a major focus of interest. Aggregate data does not support such efforts.

3.2.2 Data de-identification (an example of the Data Perturbation Approach)

The HIPAA Privacy Standard excludes de-identified health information from protected health information. De-identified health information may be used and disclosed without authorization. The HIPAA Privacy Standard considers information to have been de-identified by the use of either a statistical verification of de-identification or by removing 18 explicit elements of data. Such data may be used or disclosed without restriction. The details of these approaches are described in the pamphlet, *Protecting Personal Health Information in Research: Understanding the HIPAA Privacy Rule* (Department of Health and Human Services, July 13, 2004).

While these approaches to de-identification provide compliance with the HIPAA Privacy Standard, they do not guarantee anonymity. A number of reports have appeared recently that criticize these approaches for being too complicated and posing a threat to clinical research and care (Melton, 1997) (Galanddiuk, 2004). A variety of approaches for this issue have been published including successful implementation of policies, procedures, techniques and toolkits that meet academic medical center needs and comply with the Privacy Standard (UCLA) (Sweeney 1996) (Moore et al, 2000) (Ruch et al, 2000) (Moore et al 2001) (Thomas et al, 2002), (Lin et al, 2004) (Oiliveira and Zaïane, 2003) (Saul, 2004).

Goodwin and Prather performed a study that de-identified the data in the Duke TMR perinatal database in accordance with the HIPAA Privacy Standard and assigned a coded identifier to each patient to permit re-identification of patients under controlled circumstances. The database contained data on 19,970 patients with approximately 4,000 potential variables per patient (Goodwin and Prather, 2002). They noted several issues:

1. To meet the requirement for removing all elements of date except year required the conversion of dates to days since conception to permit the data to be useful for pregnancy studies.
2. Clinician users were still able to identify one patient by her extremely young age.
3. The process was tedious, time-consuming and expensive.

They concluded that it is imperative to maintain the public's trust by doing everything possible to protect patient privacy in clinical research, and privacy protection will require careful stewardship of patient data.

3.2.3 Binning (Another example of the Data Perturbation Approach)

Binning deploys a technique for generalizing records in a database by grouping like records into a category and eliminating their unique characteristics (for example, grouping patients by age rather than date of birth). Elegant work has been done using this approach. One approach permits the level of anonymity to be controlled and matched to a user profile indicating the likelihood that data external to the database would be used permitting re-identification (Sweeny, 1997). Another report provides a measure of the information loss due to binning (Lin et al, 2002).

3.2.4 Pseudonymisation (Another example of the Data Perturbation Approach)

This technique involves replacing the true identities of the individuals and organizations while retaining a linkage for the data acquired over time that permits re-identification under controlled circumstances (Quatin et al., 1998). A trusted third party and process is involved. The trusted third party and process must be strictly independent, adhere to a code of conduct with principles of openness and transparency, have project-specific privacy and security policies and maintain documentation of operating, reporting and auditing systems (Claerhout et al., 2003).

3.2.5 Mediated access (A combination of Query Restriction and Output Perturbation Approaches)

Mediated access puts policy, procedure and technology between the user and the data and, thus, illustrates a general point that all medical investigators should bear in mind: sound health information privacy and security programs include a range of controls (Wiederhold and Bilello, 1998). "The system is best visualized as residing on a distinct workstation, operated by the security officer. Within the workstation is a rule-based system which investigates queries coming in and responses to be transmitted out. Any query and any response which cannot be vetted by the rule system is displayed to the security officer for manual handling. The security officer decides to approve, edit, or reject the information. An associated logging subsystem provides both an audit trail for all information that enters or leaves the domain, and provides input to the security officer to aid in evolving the rule set, and increasing the effectiveness of the system." (Wiederhold et al, 1996). The workstation, nonetheless, depends on and functions as a component of a broader security architecture that provides layered protection against unauthorized access by deploying sound practices such as encryption of transmissions, intrusion prevention with firewalls and a public/private key infrastructure. When functioning as a whole, the workstation and technical infrastructure provide several security controls, including:

1. authentication of users (optionally more extensive for external users),
2. authorization (determination of approved role),
3. processing of requests for data using policy-based rules,
4. initiating interaction of security officer oversight for requests that conflict with rules,
5. communication of requests that meet the rules to the internal databases,
6. communication from the internal databases of unfiltered results,
7. processing the unfiltered results to ensure that policy rules are met,
8. initiating interaction with security officer oversight when results do not meet the rules,
9. writing origin identification, query, action and results to a log file, and
10. transmission of data meeting rules to requestor (Wiederhold, 2002).

Ferris and colleagues report an approach that adds de-identification and re-identification to other security controls and supports the HIPAA requirement for accounting for disclosures (Ferris et al, 2002). There are two modules in this approach:

1. Key Escrow Module

This module consists of a privacy manager that uses key escrow to support de-identification and re-identification, user authentication, logging of user sessions, generation and storage of query-specific public/private keys and manages role-based access.

2. Biomedical Database Module

This module associates the research database with the database manager and audit database. It is used for accessing the research data, generating an audit trail and de-identifying results when required.

3.3 Other Issues in Emerging "Privacy Technology"

Two kinds of privacy issues for computer science research have been identified: those inherent in applications of developing technology and those related to information practices needed in the development of technology. New efforts in "privacy technology" attempt to protect individual privacy while permitting the collection, sharing and uses of person-specific information. This research addresses two major concerns: disclosure of individually identifiable sensitive information by the linkage of information with other publicly available databases, and the use of information obtained for one purpose for another purpose. Threats to Homeland Security have made considerable funding available to investigate this topic in order to support bio-terrorism surveillance and protect individual privacy.

For example, Sweeney and colleagues at Carnegie-Mellon University have built "CertBox" to provide privacy protection in biosurveillance.

> "Emergency room visits and other healthcare encounters will be reported daily to the state's public health department under the authority of public health law. Collected health information will be filtered in real-time by a self-contained machine called a CertBox, which automatically edits combinations of fields (often demographics) so that released information relates to many people ambiguously. Settings are preset for a specific population and set of data fields and then sealed to prohibit tampering. CertBox technology de-identifies health information in accordance to the scientific standard of de-identification allowed under HIPAA. The resulting de-identified data is then shared with bio-terrorism surveillance systems. CertBox technology (more generally termed a "privacy appliance" by DARPA) allows us to certify that resulting data are properly de-identified and to warranty that resulting data remain practically useful for anomaly detection algorithms in bioterrorism surveillance" (Sweeney, L., 2003, p.15).

Other aspects of privacy technology include detecting and removing or replacing identifying information from information in text (e.g. medical reports, letters, notes, email) (Sweeney, L, 1996) (Ruch, et al., 2000) as well as facial images (Newton, et al., 2003). Techniques have been reported for embedding encrypted digital watermarking and patient identifiers in medical images (Tzelepi, 2002) to protect privacy during use and transmission.

Data mining investigators have begun encouraging their colleagues to take a research interest in issues related to protecting the privacy and security of personal information. For example, Berman argues that:

> "Human subjects issues are a legitimate area of research for the medical data miners. Novel protocols for achieving confidentiality and security while performing increasingly ambitious studies (distributed network queries across disparate databases, extending the patient's record to collect rich data from an expanding electronic medical record, linking patient records to the records of relatives or probands, peer-to-peer exchange of medical data) will be urgently needed by the data mining community" (Berman, 2002).

The techniques of data mining have been used to address the issue of auditing access and use of data as well as for testing devices for intrusion detection and access control. Commercial products exist that automatically correlate and compare suspicious information gathered from different points in computer systems, draw conclusions, and act on potential attacks and security violations (Dicker, 2003).

Berman's suggestion illustrates a general point: research into privacy and security technology necessarily entails the study of values and their embodiment in technological artifacts. Instead of assuming that ensuring privacy necessarily requires sacrificing research efficiency and efficacy, Berman's suggestion pushes researchers toward considering their relationship in specific instances and developing new approaches to both privacy and research design. In this respect, Berman echoes core concerns of a major body of research in the field of Human-Computer Interaction, known as "Value Sensitive Design" (Friedman, Kahn, and Borning, draft June 2003; Taipale, 2003).

3.4 "Value Sensitive Design": A Synthetic Approach to Technological Development

"Value Sensitive Design" attempts to incorporate relevant important considerations (values) into new technology throughout the entire lifecycle of design, development, deployment and retirement. Deriving inspiration from related ideas in computer ethics, social informatics, computer supported cooperative work and participatory design, value sensitive design implements the basic proposition that all technology relates in design and use to important values and, therefore, cannot fundamentally emerge as "value-neutral" (Friedman, Kahn, and Borning, draft June 2003). Value sensitive design enables incorporating any value into the design process but places high priority on human values "with ethical import" including privacy as well as related values in health care such as human welfare, trust, autonomy, accountability, identity, and informed consent.

In practice, value sensitive design includes conceptual, empirical and technical investigations. In conceptual investigations, designers consider key questions about the context of technological implementation such as "Who are the direct and indirect stakeholders and how are they affected?" "What values are implicated and how do the designers evaluate their relative importance and priority?" Empirical investigations entail study of the actual human contexts of technological implementation. Technical investigations evaluate the relative support particular technological designs provide for the interests of specific stakeholders and realization of specific values. These investigations often identify conflicts among stakeholders and values that must be addressed in the design process. For example, in designing data mining technologies for medical investigations, stakeholders include investigators, study subjects and patients with the disease. Values include assuring integrity of research data as well as enhancing the welfare of patients and protecting subject privacy. The properties of specific technical designs may provide greater support for the interests and values of one group of stakeholders (for example, the subjects and their privacy) than for others. All design methodologies inevitably make choices of these kinds. Value sensitive design has developed means for making explicit the choices and their rationale (Friedman, Kahn, and Borning, draft June 2003).

In a spirited defense of data mining in bioterrorism surveillance, Taipale invokes the principles of value sensitive design in justifying privacy protections slated for development under the Terrorist Information Awareness (TIA) program (Tiapale, 2003) (see case study below for detailed review of TIA). TIA included programs for developing privacy appliances incorporating what Taipale calls rule-based processing, selective revelation, and strong credentialing and auditing. Rule-based processing entails research on intelligent query agents that negotiate access to specific data bases depending on the inquirer's authorization and meta-data labels about specific data items. Selective revelation technologies employ

> "an iterative, layered structure that reveals personal data partially and incrementally in order to maintain subject anonymity. Initial revelation would be based on statistical or categorical analysis .... This analysis would be applied to data that was sanitized or filtered

in a way so that it did not reveal personally identifying information. Based on initial results, subsequent revelations may or may not be justified. At each step, legal and technical procedures can be built in to support particular privacy policies (or other policies, such as security clearances, etc.) (Taipale, 2003 pg. 79)".

Strong, encrypted tamper-proof auditing mechanisms that log access to distributed databases help protect against insider and outsider threats. Sweeney's "CertBox" constitutes an example of such a privacy appliance.

3.5 Responsibility of Medical Investigators

In addition to the usual security risks, medical research may add potential loss of life or health and requires special emphasis on privacy, confidentiality and data integrity (Berman, 2002). The Common Rule provides for subject safety and, with the HIPAA Privacy Standard, specifies accountability for the use and disclosure of protected health information. The medical data miner must conduct only valid research and in a way that protects human subjects. A privacy review board or institution review board is required to provide oversight for each project. Careful attention must be given to assure that there is proper justification and documentation of the process especially for waivers for individual consent or use of research conducted under the exemption of the common rule or the HIPAA Privacy Standard (Berman, 2002).

The Utah Resource for Genetic and Epidemiologic Research (RGE) (http://www.research.utah.edu/rge/) is an example of a well established medical data mining implementation where responsibilities have been made explicit (Wylie J.E., and Mineau, G.P., 2003). It has been functioning since 1982 when it was established on executive order of the Governor of Utah to be a data resource for the collection, storage, study and dissemination of medical and related information for the purpose of reducing morbidity or mortality. It does not perform research but maintains and improves data for research projects. The RGE contains over six million records. It includes genealogies on the founders of Utah and their descendants, cancer records, birth and death certificates, driver's license records, census records and follow-up information from the Health Care Financing Administration.

It has the following policies and procedures:

1. All data received by the RGE comes from contributors who have contracts that specify the conditions for use of the data and requires the data contributors to approve projects that use their data.

2. A committee composed of contributors and others familiar with the issues of medical data and research review all requests for access.
3. IRB approval is required for each project.
4. Access is project specific and may not be used for other purposes.
5. Data must be destroyed or returned at the end of the project.
6. Projects must justify reasons for access to information that identifies individuals.
7. Identifying information is removed from records and stored in separate tables in a relational database and requires the RGE staff to recombine the data for record linking and data management.
8. If projects wish to contact individuals, they must arrange for data contributors or their designees to contact the individual about interest in the proposed project. Identifying information is provided to the project only for those individuals wishing to participate. Other information or biospecimens are only collected after informed consent is obtained.
9. For-profit organizations may not have direct access to RGE data. They may participate with university or other non-profit entities. Commercial project sponsors may participate directly only in research activities that involve no individually identifying information.

Sweeney (2003, pg. 13) makes a plea to fellow computer scientists whose point applies as well to medical investigators:

> "Most computer scientists can no longer afford to do their work in an ivory tower and rely on the social scientists and lawyers to make decisions about limits of its use. First, policy makers and lawyers may not fully understand the technology. Second, decisions will often be made as a reaction to biased or sensationalized public opinion. Third, policy decisions are often crude and sub-optimal, and tend to legislate over simple technical remedies. Finally, there is a horrible temporal mismatch -- policy can be a function of years but new technology is a function of months, so policy enacted on today's technology may be totally inappropriate for tomorrow's and policy supporting technology today can prohibit it tomorrow. Computer scientists can and must insulate their creations from such risk."

In other words, medical investigators must proactively take responsibility for assuring adequate privacy controls in their projects. To ignore this responsibility risks potentially ceding control or completely losing their projects. The case study of the Terrorist Information Awareness Program illustrates just such a scenario.

4. CASE STUDY: THE TERRORIST INFORMATION AWARENESS PROGRAM (TIA)

4.1 The Relevance of TIA to Data Mining in Medical Research

This case study examines the controversy surrounding termination of the Terrorist Information Awareness (TIA) Program, a very large counterterrorism effort organized by the Defense Advanced Research Projects Agency (DARPA). At first glance, the TIA program would appear to have no relevance to data mining in medical data analysis and research because of its focus on crime prevention and law enforcement. This difference in mission, however, should not distract analysis from some core commonalities with respect to privacy and security as well as functionality. Data mining in counterterrorism and medical data analysis face the problem of developing legal, ethically appropriate and secure methods of managing individually identifiable information. Whether beginning with identified possible subjects for investigation and medical research, or discovering possible subjects in the course of analysis, organizations conducting terrorist investigations and medical data analysis must obey relevant privacy laws, establish appropriate policies and procedures, train their workforce, and implement risk-based administrative, physical and technical privacy and security safeguards.

As will be explained below, TIA lost funding and faced censure from journalists, the Inspector General of the Department of Defense, and Congress partially because DARPA paid insufficient attention to some of these core controls. Medical investigators similarly ignore the guidance of HIPAA at their peril and may take TIA's experience as an object lesson in what to avoid. As medical researchers study the TIA program, they will also find some of its proposed data mining capabilities very attractive, particularly the powerful data aggregation, analysis and linking tools as well as the virtual collaboration tools described below. Quite rightfully wanting to apply such capabilities to their research, medical investigators confront another shared characteristic with TIA, having to balance the rights and welfare of individuals with possible benefits of their work for society as a whole. Patients and terrorists potentially face entirely different sets of consequences if identified by investigators as fit subjects for analysis. Patients may benefit or contribute to knowledge that benefits others in their situation. Terrorists potentially face punishment. But, in both cases, the investigators using data mining techniques of various kinds potentially gather data about unrelated persons risking invasion of privacy and potentially broader harms. The American public basically accepts as legitimate the aims of medical research and counterterrorism. As the fate of TIA demonstrates, however, individual programs must carefully assess, judiciously weigh and clearly explain the trade-offs between individual and societal welfare in specific instances, particularly in these times of struggle.

4.2 Understanding TIA

DARPA's "Report to Congress regarding the Terrorism Information Awareness Program" (DARPA, 2003) outlined the intended goals, responsible organizational structure, system components, expected efficacy, and relevant federal laws and regulations as well as privacy concerns of the program known as the Terrorism Information Awareness Program or TIA. Under the original name of the "Total Information Awareness Program", DARPA planned to create a system of systems "to integrate technologies into a prototype to provide tools to better detect, classify, and identify potential foreign terrorists." (DARPA, 2003 p. 3) The target technologies existed in a range of states of development from desirable but not yet acquired or developed to transitioning to operational use (DARPA, 2002 p 15.) Multiple universities, small and large defense contractors, and Federally-funded Research and Development Organizations worked on the various component technologies, many of which were funded as part of DARPA's ongoing research in counterterrorist technologies. DARPA created the Information Awareness Office (IAO) in January 2002 "to integrate advanced technologies and accelerate their transition to operational users." (DARPA, 2003, p. 1) The TIA prototype network included a main node in the Information Operations Center of the US Army Intelligence and Security Command (INSCOM) with additional nodes at subordinate INSCOM commands as well as other defense and intelligence agencies such as the National Security Agency, the Central Intelligence Agency and the Defense Intelligence Agency. In the post-September 11, 2001 world, TIA was supposed to help accelerate development and deployment of core tools in the fight against the "asymmetric threat" posed by terrorists "in full compliance with relevant policies, laws, and regulations, including those governing information about US persons." (DARPA, 2003, pg. 4)

TIA drew its component systems from three other categories of work coordinated by DARPA's IAO, including the Data Search, Pattern Recognition and Privacy Protection Programs, the Advanced Collaborative and Decision Support Programs, and the Language Translation Programs. The Data Search, Pattern Recognition and Privacy Protection Program coordinated development of technologies with the intention of seeking, analyzing and making available to decision-makers information about individually identifiable human beings potentially associated with terrorism. To find this data, TIA-sponsored data mining technologies would search disparate federal databases for transactions by persons – transactions such as applications for passports, visas, and airline ticket purchases – and attempt to link them with other events such as arrests or suspicious activities that, taken together, might indicate a terrorist act in the making. The Genisys program sponsored technologies to virtually aggregate data in support of effective analysis across heterogeneous databases and public sources. The Evidence Extraction and Link Discovery program enabled "connecting the dots" of suspicious activities beginning with a particular object such as a person, place or thing. Scalable Social Network Analysis sponsored technologies for separating terrorist groups from other groups using techniques of advanced social network analysis. (Taipale, 2003; DARPA, 2002; DARPA, 2003). In order to protect against abuses of the privacy of individually identified persons as well as protect sensitive data sources, the Genisys Privacy Protection Program intended to develop a "privacy appliance" for "providing critical data to analysts while controlling access to unauthorized information, enforcing laws and policies through software mechanisms, and ensuring that any misuse of data can be quickly detected and addressed." (DARPA, 2003, p. 6) The proposed privacy technology would also have sought to improve identity protection by limiting inference from aggregate sources. (DARPA, 2003) In other words, while data mining enables rapid identification of terrorist suspects and activities, the privacy technologies help prevent abuse of American law and regulations about individual citizen's privacy thus achieving security with privacy. (Taipale, 2003)

TIA intended to integrate the advanced data mining technologies with advanced collaboration tools sponsored under the Advanced Collaborative and Decision Support Programs and with advanced language translation and analysis tools sponsored under the Language Translation Programs. The advanced collaboration tools included sophisticated war-gaming and simulation capabilities as well as data sharing technology. The language translation tools were planned to enable rapid translation and preliminary analysis of foreign language materials from open and restricted source materials (DARPA, 2003). According to DARPA (2003), the integration of these powerful, computerized tools under TIA would support a series of steps among distributed, collaborating experts attempting to discover the plans and intentions of potential terrorists, including:

1. Develop terrorist attack scenarios;
2. Initiate automated searches of data bases using terrorist attack scenarios and other intelligence information as starting points;
3. Identify individuals suspected of involvement in terrorist activities;
4. Identify associations among suspect individuals;
5. Link such associations with associations of other individuals;
6. Develop competing hypotheses about the plans of suspect associates in conjunction with other types of intelligence data;
7. Introduce the behavior and activities of suspect associates into models of known patterns of behavior and activity indicative of terrorist attack;
8. Generate range of plausible outcomes with options for action;
9. Analyze risks associated with each option for action;
10. Present complete analysis to decision-maker; and,
11. Record all steps of process in a corporate knowledge base for futurc review and use.

DARPA (2003, pg. 4) intended this entire process to yield four major benefits to the fight against counterterrorism, including:

1. Increase by an order-of-magnitude the information available for analysis by expanding access to and sharing of data;
2. Provide focused warnings within an hour after occurrence of a triggering event or passing of an articulated threshold;
3. Automatically cue analysts based on partial matches with patterns in a database containing at least 90 per cent of all known foreign terrorist attacks;
4. Support collaboration, analytical reasoning, and information sharing among analysts to hypothesize, test and propose theories and mitigating strategies about possible futures; and, thus
5. Enabling decision-makers effectively to evaluate the impact of current or future policies.

4.3 Controversy

Beginning in November 2002, columnists from major newspapers and magazines including *The Washington Post, The New York Times,* and *The National Review* as well as scholarly organizations publicly criticized the TIA program on multiple grounds. Concerns about governmental abuse of personal information that sacrifices the privacy rights of individual American citizens in the name of national security in the post-9/11 world constituted the heart of their criticism. This specific controversy occurred in the context of an ongoing Congressional review of the privacy implications of multiple new security programs such as TIA and the Transportation Security Agency's Computer Assisted Passenger Prescreening System II (CAPPS II), to say nothing of HIPAA. Although other columnists and commentators attempted to defend TIA (Taipale, 2003; Taylor, 2002), the controversy produced an audit and an unfavorable report about the TIA program by the Inspector General of the Department of Defense. Congress ultimately withdrew funds for the TIA program (Office of the Inspector General, 2003). The controversy sheds important light on issues that any data mining project manipulating individually identifiable information must necessarily confront and manage well, especially projects that potentially harm or do not directly benefit the persons under surveillance.

Commentators argued that TIA posed multiple threats to the privacy of individual Americans, including:

1. Violates the Fourth Amendment of the Constitution by searching a data base containing detailed transaction information about all aspects of the lives of all Americans (Safire, 2002; Washington Post, 2002; Crews, 2002; Stanley and Steinhardt, 2003)
2. Undermines existing privacy controls embodied in the Code of Fair Information Practices, such as improper reuse of personal data collected for a specific purpose (Simons and Spafford, 2003; Safire, 2002; Crews, 2002)
3. Overcomes "privacy by obscurity" including inappropriate coordination of commercial and government surveillance (Safire, 2002; Washington Post, 2002; Stanley and Steinhardt, 2003)
4. Increases the risk of falsely identifying innocent people as terrorists (Crews, 2002; Simons and Spafford, 2003; Stanley and Steinhardt, 2003)
5. Increases the risk and cost of identity theft by collecting comprehensive archives of individually identifiable information in large, hard-to-protect archives (Simons and Spafford, 2003)
6. Accelerates development of the total surveillance society (Safire, 2002; Washington, 2002; Crews, 2002; Stanley and Steinhardt, 2003)

Other undesirable consequences in addition to invasion of privacy potentially flowed from TIA, including:

1. Undermining the trust necessary for the successful development of the information economy and electronic commerce (Crews, 2002; Simons and Spafford, 2003)
2. Undesirably altering the ordinary behavior of the American population including quelling healthy civil disobedience, "normalizing" terrorist behavior, and inhibiting lawful behavior (Crews, 2002; Simons and Spafford, 2003)
3. Creating new, rich targets for cyberterrorism and other forms of individual malicious abuse of computerized personal information (Crews, 2002; Simons and Spafford, 2003)

Some commentators also argued that TIA demonstrated important organizational shortcomings, including:

1. Poor choice of leadership with Admiral John Poindexter of Iran-Contra fame as program director (Safire, New York Times, November 14, 2002; Washington Post Editorial November 16, 2002)
2. Insufficient oversight (Safire, New York Times, November 14, 2002; Washington Post Editorial November 16, 2002; Simons and Spafford, January 2003)
3. Low likelihood of achieving its goal of "countering terrorism through prevention" (Crews, National Review November 26, 2002; Simons and Spafford, January 2003)

On December 12, 2003, the DOD Inspector General (DOD IG) issued a report on its audit of the TIA program entitled, "Terrorism Information Awareness Program (D-2004-033) (Office of the Inspector General, December 12, 2003). The DOD IG conducted the audit in response to questions from Senator Charles E. Grassley, Ranking Member of the Senate Finance Committee with supporting letters and questions from Senator Chuck Hagel and Senator Bill Nelson. The audit objectives included assessing "whether DARPA included the proper controls in developmental contracts for the TIA program that would ensure that the technology, when placed in operational environment, is properly managed and controlled." (DOD, Office of the Inspector General, 2003, p. 3). The audit focused particularly on DARPA's appreciation for the importance of protecting the privacy of individuals potentially subject to TIA surveillance. The DOD IG (2003, p. 4) summarized its conclusions as follows:

> "Although the DARPA development of TIA-type technologies could prove valuable in combating terrorism, DARPA could have better addressed the sensitivity of the technology to minimize the possibility for Governmental abuse of power and to help insure the successful transition of the technology into the operational environment."

While acknowledging the application of TIA-type technologies in foreign intelligence, the DOD IG expressed strong reservations about DARPA's inattention to the implications of TIA for potential governmental abuse in domestic intelligence and law enforcement purposes. The DOD IG particularly faulted DARPA program management for not having consulted experts in policy, privacy and legal matters to ensure successful transition to the operational environment. Four factors contributed to DARPA's inattention to these issues (DOD, Office of the Inspector General, 2003, pg. 4), including:

1. DARPA did not implement the best business practice of performing a privacy impact assessment (PIA);
2. Under Secretary of Defense for Acquisition, Technology and Logistics initially provided oversight of the TIA development and did not ensure that DARPA included in the effort the appropriate DOD policy, privacy and legal experts;
3. DARPA efforts historically focused on development of new technology rather than on the policies, procedures and legal implications associated with the operational use of technology; and,
4. The DARPA position was that planning for privacy in the operational environment was not its responsibility because TIA research and experiments used synthetic artificial data or information obtained through normal intelligence channels.

To have exercised due care, safeguarded taxpayers' money, and protected its program, DARPA should have taken several precautions, including:

1. Employed governmental best privacy practice by executing a PIA. In the words of the DOD IG (2003 p. 7), a PIA "consists of privacy training, gathering data on privacy issues, identifying and resolving the privacy risks, and approval by (the agency) privacy advocate";
2. Ensured adequate oversight by a responsible agency with experts in policy, privacy and legal matters;
3. Developed in advance policies and procedures as well as technology for protecting privacy; and,
4. Considered the ultimate use of the information in the operational environment not just the source of data used in research experiments.

Taking these precautions would have integrated privacy concerns into TIA's entire developmental and acquisition lifecycle instead of relegating that responsibility to end-users. DARPA could thus have avoided causing unnecessary alarm among the members of Congress and the American public and, had the program continued, avoided wasting taxpayer's money on expensive retrofits or redesign of the TIA applications.

By the time the DOD IG released its report, Congress had terminated all funding for TIA and most of its component applications. Nonetheless, the DOD IG made two additional recommendations to guide development of future TIA-type programs. Before resuming TIA-type research, DARPA should take specific steps to integrate privacy management into its research and development management process, including:

1. Conduct Privacy Impact Assessments on potential research and development projects using models such as the PIA of the Internal Revenue Service, endorsed by the Federal Chief Information Officer's Council, as a best practice for evaluating privacy risks in information systems; and
2. Appoint a Privacy Ombudsman to oversee PIAs and thoroughly scrutinize TIA-type applications from a privacy perspective.

4.4 Lessons Learned from TIA's Experience for Medical Investigators Using "Datamining" Technologies

The TIA program imagined integrating many innovative technologies into an effort to preempt terrorist attacks by identifying and sharing information about suspicious activities among relevant Federal agencies. "Data mining" technologies of various types with the purpose of examining individually identifiable information in Federal and commercial databases constituted the program's core functionality. While not as comprehensive as TIA, medical data analysis and research employing datamining of patient records invites comparison as well as contrast with DARPA's counterterrorism research and development program. In particular, medical investigators should not take for granted the good will of their patients, their institutions or their funding agencies. Unlike the program management of DARPA and TIA, principle investigators must take personal responsibility for assuring proper identification and implementation of privacy controls, thorough training of their staff in privacy responsibilities and communication of their efforts to all relevant audiences. TIA teaches medical researchers some specific lessons when translated into the environment of healthcare research and data analysis:

1. Medical researchers should take full advantage of the privacy functions of the Institutional Review Board (IRB). From the perspective of the DOD IG's report on TIA, the IRB represents an oversight board that is fully equipped to advise and monitor

researchers on privacy policies, procedures and practices. In most academic research institutions, HIPAA has strengthened the IRB's awareness and competence to manage privacy issues.

2. Medical researchers should devote great care in preparing the privacy and security portions of their IRB forms, particularly the informed consent form. The IRB review forms can function for the individual research project like the Privacy Impact Assessment in Federal agencies in helping to identify and propose mitigation plans for project privacy risks. The informed consent form provides an ideal vehicle for explaining to patient-subjects a project's privacy protections.
3. Medical investigators should cultivate an effective relationship with the medical center's HIPAA Privacy and Security Officers. Like the privacy ombudsman in Federal agencies, the HIPAA Privacy and Security Officers function as points of articulation and communication when necessary between the researcher, the patient-subject, the institution, and external agencies such as the Office of Civil Rights, Department of Health and Human Services.
4. Medical investigators should consider the advisability of a project external advisory board when conducting research or using datamining methods that might provoke special privacy concerns. If properly composed and chartered, an external advisory board can provide useful expertise in policy, privacy and legal matters external to a medical researcher's own institution and lend extra credibility to a project's good faith efforts in the event of controversy.
5. Medical investigators should formally develop and document in writing privacy and security policies and procedures for the research project or its parent unit. As HIPAA and the DOD IG report emphasize, these policies and procedures must include administrative and physical as well as technical privacy and security controls. These written policies and procedures should inform the information about privacy protections included in the IRB and informed consent forms.

5. CONCLUSIONS AND DISCUSSION

A formal approach to managing the use and disclosure of personal health information is in the best interests of patients, individual researchers, organizations and society. The risks to those who do not adhere to good security and privacy practices are considerable. Future laws and regulations are likely to increase penalties for inappropriate use or disclosure. While much attention has been given to research, organizations should implement the same general processes to support analyses done for the purpose of healthcare operations as for research.

"Researchers have no automatic right to review patient data. Besides developing strategies for minimizing patient risk, as described herein, investigators should take simple steps to characterized their compliance with human subjects requirements" (Berman, pg. 33, 2002).

A recent publication recommends:

"First, sensitive raw data like identifiers, names, addresses and the like, should be modified or trimmed out from the original database, in order for the recipient of the data not to be able to compromise another person's privacy. Second, sensitive knowledge which can be mined from a database by using data mining algorithms, should also be excluded, because such a knowledge can equally well compromise data privacy, as we will indicate. The main objective in privacy preserving data mining is to develop algorithms for modifying the original data in some way, so that the private data and private knowledge remain private even after the mining process. The problem that arises when confidential information can be derived from released data by unauthorized users is also commonly called the "database inference" problem." (Vcrykios et al, pg. 1, 2004).

While these are good recommendations, they are insufficient for medical data mining. As long as the original data is available, there is risk to confidentiality, integrity and availability of the data. Thus, an effective privacy program depends upon implementing robust security controls. Medical dataminers should be sure to employ several important security practices, including:

1. Mandatory oversight by a privacy board or institutional review board with approval for each project should be established.
2. The methods sections of research proposals and publication submissions should include a description of steps to minimize patient risks and that IRB approval has been obtained.
3. Good access control and authorization should be used for each session and query.
4. Where possible, the common identifiers (e.g. names, addresses) of the data subjects should be removed or hidden from the data user.
5. Robust audit practices should be instituted.
6. Training for all principle investigators that reinforces their responsibilities should be required.
7. Sanctions should be applied for violations of policy and/or procedures.
8. Trends in breaches and sanctions should be tracked and trended over time and used in the process of security awareness and training.

6. ACKNOWLEDGEMENTS: FUNDING SOURCES OR RESEARCH PARTNERS

Dr. Cooper and Dr. Collmann thank Kim Schwarz, Adam Robinson, Georganne Higgins and Jim Wilson for their assistance in various aspects of this paper. National Library of Medicine contract NO1-LM-3-3506, "Applications of advanced network infrastructure in health and disaster management: Project Sentinel Collaboratory" supported Dr. Collmann's work on this chapter. Opinions, interpretations, conclusions and recommendations are those of the authors and are not necessarily endorsed by the National Library of Medicine.

REFERENCES

Adam, N.R., Wortmann, J.C. (1989). "Security-control Methods for Statistical Databases: A Comparative Study," *ACM Computing Surveys (CSUR)* 21(4) 515 - 556.

Alberts C, Doroffe A. (2003). *Managing Information Security Risks: The OCTAVEsm Approach.* Boston, MA, Addison-Wesley.

Behlen, F.M., Johnson, S.B. (1999). "Multicenter Patient Records Research: Security Policies and Tools," *J Am Med Inform Assoc.* 6(6) 435-43.

Berman, J.J. (2002). "Confidentiality Issues for Medical Data Miners," *Artif Intell Med.* 26(1-2):25-36.

California HealthCare Foundation (1999). Medical Privacy and Confidentiality Survey Summary and Overview, http://www.chcf.org/documents/ihealth/survey.pdf.

Cios, K.J., Moore, G.W. (2002). "Uniqueness of Medical Data Mining," *Artif Intell Med.* 26(1-2), 1-24.

Claerhout, B., De Moor, G.J., De Meyer, F. (2003). "Secure Communication and Management of Clinical and Genomic Data: The Use of Pseudonymisation as Privacy Enhancing Technique," *Stud Health Technol Inform.* 95:170-5.

Crews, Jr., C.W., November 26, 2002). "The Pentagon's Total Information Awareness Project: Americans Under the Microscope?", *Techknowledge*, Issue #45, originally in *National Review Online*, November 25, 2002.

Defense Advanced Research Project Agency (July 19, 2002). "Total Information Awareness Program (TIA) System Description Document (SDD)," Version 1.1.

Defense Advanced Research Project Agency (May 20, 2003). Information Awareness Office, "Report to Congress regarding the Terrorist Information Awareness Program: In response to Consolidated Appropriations Resolution, Pub.L. No. 108-7, Division M, § 111(b)", Detailed Information.

Department of Defense (December 12, 2003). Office of the Inspector General, Information Technology Management, "Terrorist Information Awareness Program" (D-2004-033).

Department of Health and Human Services (August 10, 2004). Office for Human Research Protections Guidance on Research Involving Coded Private Information or Biological Specimens, http:// www.hhs.gov/ohrp/humansubjects/guidance/cdebiol.pdf.

Department of Health and Human Services (July 13, 2004). Protecting Personal Health Information in Research: Understanding the HIPAA Privacy Rule, (NIH Publication Number 03–5388), http://privacyruleandresearch.nih.gov/pr_02.asp.

Department of Health and Human Services (2002). Final Privacy Standard, Title 45 CFR Parts 160 and 164, http://www.hhs.gov/ocr/hipaa/privrulepd.

Department of Health and Human Services (2003). Final Security Standard, Title 45 CFR Parts 160, 162, and 164, www.cms.hhs.gov/hipaa/hipaa2/regulations/security/03-3877.pdf.

Department of Health and Human Services (2001). Human Subjects Regulations Common Rule Title 45 part 46, http://www.hhs.gov/ohrp/humansubjects/guidance/45cfr46.htm.

Department of Health and Human Services (2001). Office for Human Research Protections, Code of Federal Regulations, Title 45, Part 46, Subpart A, 46.101 (b) (4); http://www.hhs.gov/ohrp/humansubjects/guidance/45cfr46.htm#subparta.

Department of Health and Human Services (2004). Protecting Personal Health Information in Research: Understanding the HIPAA Privacy Rule, (NIH Publication Number 03–5388), http://privacyruleandresearch.nih.gov/pr_02.asp .

Department of Health and Human Services (August 14, 2002). Office of the Secretary. 45 CFR Part 160, 162, and 164, Standards for Privacy of Individually Identifiable Health Information: Final Rule, Federal Register, Vol. 67, No. 157, 53181-53273.

Department of Health and Human Services (February 20, 2003). Office of the Secretary. 45 CFR Part 160, 162, and 164, Security Standards: Final Rule. Federal Register, Vol. 68, No. 34, 8333-8381.

Dicker, K.M. (2003). "The Evolution of Data Mining and Related Security Correlation Technology," *SANS Institute*, http://www.giac.org/practical/GSEC/Keith_Dickter_GSEC.pdf.

Federal Office of Management and Budget (1994). Statistical Policy Working Paper 22, *Report on Statistical Disclosure Limitation Methodology*, http://www.fcsm.gov/working-papers/wp22.html.

Ferris, T.A., Garrison, G.M., Lowe, H.J. (2002). "A Proposed Key Escrow System for Secure Patient Information Disclosure in Biomedical Research Databases," in *Proc AMIA Symp*. 245-9.

Food and Drug Administration (2002). Protection of Human Subjects Regulations Title 21 CFR parts 50 and 56, http://vm.cfsan.fda.gov/~lrd/cfr50.html.

Friedman, B., Kahn, JR., P.H. and Borning, A., et al. (Draft of June 2003). Value Sensitive Design: Theory and Methods, http://www.ischool.washington.edu/vsd/vsd-theory-methods-draft-june2003.pdf.

Galandiuk, S. (2004). Legislative Threat to Clinical Science: The Obfuscation and De-identification of Protected Health Information," *Br J Surg*. 91(3) 259-61

Goldman, J. and Hudson, Z. (2000). "Perspective Virtually Exposed: Privacy and E-Health," *Health Affairs,* 19(6), 140-8.

Goodwin, L.K. and Prather, J.C. (2002). "Protecting Patient Privacy in Clinical Data Mining," *J Healthc Inf Manag,* 16(4):62-7.

Health Privacy Project (2003). *Medical Privacy Stories*, http://www.healthprivacy.org/usr_doc/Privacy_storiesupd.pdf

International Information Security Foundation (1997). Generally-Accepted System Security Principles, http://web.mit.edu/security/www/GASSP/gassp021.html

Islan, M.Z., and Brankovic, L., A. (2004). "Framework for Privacy Preserving Classification in Data Mining, School of Electrical Engineering and Computer Science," *Australasian Computer Science Week.*

Levin, E.G., Arango, J., Steimle, A.E., Lee, P.C., Fireman, B. (2001). "Innovative Approach to Guidelines Implementation Is Associated with Declining Cardiovascular Mortality in a Population of Three Million [abstract]," in *American Heart Association's Scientific Sessions*, Anaheim, California.

Lin, Z,. Hewett, M., Altman, R.B. (2002). "Using Binning to Maintain Confidentiality of

Medical Data," in *Proc AMIA Symp*. 454-8.

Lin, Z., Owen, A.B., Altman, R.B. (2004). "Genetics. Genomic Research and Human Subject Privacy," *Science*, 9:305(5681):183.

Lowrance, W. (2002). "Learning from Experience: Privacy and the Secondary Use of Data in Health Research," The Nuffield Trust, www.nuffield trust.org.uk

Malin B., Sweeney L. (2001). "Re-identification of DNA through an Automated Linkage Process," in *Proc AMIA Symp*. 423-7.

Malin, B., Sweeny, L., and Newton, E. (2003). "Trail Re-identification: Learning Who You Are from Where You Have Been," Carnegie Mellon University, School of Computer Science Data Privacy Laboratory, *Technical Report*, LIDAP-WP12 (Pittsburgh).

Meany, M.E. (2001). "Data Mining, Dataveillance, and Medical Information Privacy," in *Privacy in Health Care.* J, Humber, ed., Humana Press, pp. 145-164.

Melton, L.J. (1997). "The Threat to Medical-Records Research," *N Engl J Med.*, 13;337(20) 1466-70.

Moore, G.W., Brown, L.A., Miller, R.E. (2001). "Gödelization of a Pathology Database: Re-Identification by Inference," *Johns Hopkins Autopsy Resource*, http://www.netautopsy.org

Moore, G.W., Brown, L.A., Miller, R.E. (2000). "Set Theory Definition and Algorithm for Medical De-identification," *Johns Hopkins Autopsy Resource*, http://www.netautopsy.org

Murphy, S.N., Chueh, H.C. (2002). "A Security Architecture for Query Tools Used to Access Large Biomedical Databases," in *Proc AMIA Symp*. 552-6.

National Committee for Quality Assurance (2002). *Annual Report.*

National Institute of Health (2004). HIPAA Privacy Rule, Frequently Asked Questions # 17; http://privacyruleandresearch.nih.gov/faq.asp#17

National Institute of Health (2004). HIPAA Privacy Rule, Clinical Research and the HIPAA Privacy Rule, http://privacyruleandresearch.nih.gov/clin_research.asp

Newton, E., Sweeney, L. and Malin, B. (2003). *Preserving Privacy by De-identifying Facial Images*, Carnegie Mellon University, School of Computer Science, *Technical Report*, CMU-CS-03-119 (Pittsburgh).

Oliveira, S.R.M., Zaïane, O.R. (2003). "Protecting Sensitive Knowledge by Data Sanitization," in *Proceedings of the Third IEEE International Conference on Data Mining,* Melbourne, Florida, USA, 613-616.

Pheatt, N., Brindis, R., Levin, E. (2003). "Putting Heart Disease Guidelines into Practice: Kaiser Permanente Leads the Way," *The Permanente Journal*, 7(1) 18-23, http://xnet.kp.org/permanentejournal/winter03/guides.html

Quantin, C., Bouzelat, H., Allaert, F.A., Benhamiche, A.M., Faivre, J., Dusserre, L. (1998). "Automatic Record Hash Coding and Linkage for Epidemiological Follow-up Data Confidentiality," *Methods Inf Med*, 37(3) 271-7.

Ruch, P., Baud, R. H ., Rassinoux A., Bouillon, P., Robert, G. (2000). "Medical Document Anonymization with a Semantic Lexicon," in *Proc AMIA Symp* 729-733.

Safire, W. (November 14, 2002). "You are a Suspect," *New York Times.*

Saul, M. (2004). "De-Identification Tool for Patient Records Used in Clinical Research," *Health Services Library System*, 9(3). http://www.hsls.pitt.edu/about/news/hslsupdate/2004/june/iim_de_id/

Simons, B. Spafford, E.H. (2003). Co-chairs, US ACM Policy Committee, Association for Computing Machinery, Letter to Honorable John Warner, Chairman, Senate Committee on Armed Forces.

Stanley, J., Steinhardt, B., (January 2003). *Bigger Monster, Weaker Chains: The Growth of an American Surveillance Society*, American Civil Liberties Union, Technology and Liberty Program.

Sweeney, L. (1997). "Weaving Technology and Policy Together to Maintain Confidentiality,"

J Law Med Ethics, 25(2-3):98-110, 82.

Sweeney, L. (1997). "Guaranteeing Anonymity When Sharing Medical Data, The Datafly System," in *Proc AMIA Symp* 51-55.

Sweeney, L. (2002). "K-anonymity: A Model for Protecting Privacy," *International Journal on Uncertainty, Fuzziness, and Knowledge-based Systems,* 10(7) 557-570.

Sweeney, L. (2003). "Navigating Computer Science Research through Waves of Privacy Concerns: Discussions among Computer Scientists at Carnegie Mellon University," *ACM Computers and Society*, 34(1):1-18.

Sweeney, L. (1996). "Replacing Personally-Identifying Information in Medical Records, The Scrub System," in *Proc. AMIA,* 333-337.

Taipale, K.A. (2003). "Data Mining and Domestic Security: Connecting the Dots to Make Sense of Data," *The Columbia Science and Technology Law Review*, Vol. V, 5-83, http://www.stlr.org/cite.cgi?volume=5&article=2

Taylor, S., (December 2002). "Big Brother and Another Overblown Privacy Scare," *Atlantic Online*

Thomas, S.M., Mamlin, B., Schadow, G., McDonald, C. (2002). "A Successful Technique for Removing Names in Pathology Reports Using an Augmented Search and Replace Method," in *Proc AMIA Symp*. 777-81.

Tzelepi, S., Pangalos, G. and Nikolacopoulou, G. (2002). "Security of Medical Multimedia," *Med. Inform,*. 27(3):169–184.

UCLA DataServer - An open sourcc xml data gateway, UCLA medical imaging informatics, http://www.mii.ucla.edu/dataserver/docs/features/deidentification.html

Verykios, V.S., et al. (2004). "State-of-the-art in Privacy Preserving Data Mining," *SIGMOD Record,* 33(1):1-8.

Washington Post (November 16, 2002). "Total Information Awareness," Saturday.

Wiederhold, G., Bilello, M. (1998). "Protecting Inappropriate Release of Data from Realistic Databases," in DEXA *'98 Workshop on Security and Integrity of Data Intensive Applications*, http://www-db.stanford.edu/pub/gio/TIHI/DEXAgio.html

Wiederhold, G., Bilello, M., Sarathy, V., Qian, X. (1996). "A Security Mediator for Health Care Information," in *Proc AMIA Symp*. 120-4.

Wiederhold, G. (2002). "Future of Security and Privacy in Medical Information," *Stud Health Technol Inform*, 80:213-29.

Wylie J.E., and Mineau, G.P. (2003). "Biomedical Databases: Protecting Privacy and Promoting Research," Trends Biotechnol, 21(3):113-6.

SUGGESTED READINGS

Department of Defense, Office of the Inspector General, Information Technology Management, "Terrorist Information Awareness Program (D-2004-033), December 12, 2003.

The DOD IG's report describes core institutional issues in protecting privacy in data mining.

Department of Health and Human Services, *Protecting Personal Health Information in Research: Understanding the HIPAA Privacy Rule*, (NIH Publication Number 03–5388). The NIH Guide to privacy considerations in human subject research provides a good introduction to these issues.

Berman, J.J., "Confidentiality Issues for Medical Data Miners." *Artif Intell Med.*, 26(1-2):25-

36 (2002).
This article describes some of the innovative computational remedies that will permit researchers to conduct research and share their data without risk to patient or institution.

Sweeney, L., "Navigating Computer Science Research Through Waves of Privacy Concerns: Discussions among Computer Scientists at Carnegie Mellon University." *ACM Computers and Society*. 34 (1) (2003).
This article introduces the nature of privacy concerns in computer science research and explains the potential benefits and risks.

ONLINE RESOURCES

Asia

Asia-Pacific Privacy Charter Initiative
http://www.bakercyberlawcentre.org/appcc/announce.htm

Australia

Federal Privacy Law
http://www.privacy.gov.au/act/

Data Matching
http://www.privacy.gov.au/act/datamatching/index.html

Complaint Case Notes and Determinations
http://www.privacy.gov.au/act/casenotes/index.html

Canada

Canada's Health Information Infostructure
http://www.privcom.gc.ca/information/02_03_02_e.asp#002

Personal Information Protection and Electronic Documents Act
http://www.medicalpost.com/mpcontent/article.jsp?content=20031211_144135_3716

Department of Health and Human Services

Office of Civil Rights - HIPAA Security and Privacy
http://www.hhs.gov/ocr/hipaa/

HIPAA Privacy Rule and Public Health Guidance from CDC and the U.S. Department of Health and Human Services
http://www.cdc.gov/mmwr/preview/mmwrhtml/m2e411a1.htm

Common rule: Title 45 Part 46 Protection of Human Subjects
http://www.hhs.gov/ohrp/humansubjects/guidance/45cfr46.htm

HIPAA and research
http://privacyruleandresearch.nih.gov/

European Union

Directive 95/46/EC of the European Parliament and of the Council of 24 October 1995 on the protection of individuals with regard to the processing of personal data and on the free movement of such data.
http://www.cdt.org/privacy/eudirective/EU_Directive_.html

Briefing Materials on the European Union Directive on Data Protection
http://www.cdt.org/privacy/eudirective/

US Department of Commerce – Safe Harbor
http://www.export.gov/safeharbor/

Health Information and Systems Society
CPRI Toolkit: Managing Information Security in Healthcare, http://www.himss.org/asp/cpritoolkit_toolkit.asp

Institute of Medicine

For the Record: Protecting Electronic Health Information
http://books.nap.edu/html/for/

Health services research: Protecting Data Privacy in Health Services Research
http://www.nap.edu/books/0309071879/html/

Institutional Review Boards and Health Services Research Data Privacy: A Workshop Summary
http://books.nap.edu/books/NI000228/html/

National Institutes of Health
Research Repositories, Databases, and the HIPAA Privacy Rule
http://privacyruleandresearch.nih.gov/research_repositories.asp

Protecting Personal Health Information in Research: Understanding the HIPAA Privacy Rule
http://privacyruleandresearch.nih.gov/pr_02.asp

Clinical Research and the HIPAA Privacy Rule
http://privacyruleandresearch.nih.gov/clin_research.asp

Institutional Review Boards and the HIPAA Privacy Rule
http://privacyruleandresearch.nih.gov/irb_default.asp

Privacy Boards and the HIPAA Privacy Rule
http://privacyruleandresearch.nih.gov/privacy_boards_hipaa_privacy_rule.asp

Security Research Centers
Carnegie Mellon University, Privacy Technology Center
http://center.privacy.cs.cmu.edu/index.html

Johns Hopkins Autopsy Resource
Http://www.netautopsy.org

Purdue University, Center for Education and Research in Information Assurance and Security
http://www.cerias.purdue.edu/

Stanford University Security Laboratory
http://theory.stanford.edu/seclab/index.html

AT&T Labs- Research
http://www.research.att.com

IBM Privacy Research Institute
http://www.research.ibm.com/privacy

Microsoft Research
http://research.microsoft.com

QUESTIONS FOR DISCUSSION

1. How should medical investigators address the components of security (confidentiality, integrity, availability and accountability) for a new project?

2. What types of oversight should medical investigators establish when planning data mining projects on patient data.

3. What are the trade-offs that should be considered in developing a risk management plan for a medical data mining project?

4. Describe what a medical investigator must do to respect the rights granted to patients by the HIPAA Privacy Standard and the requirements of the Common Rule.

5. How is data mining used to enhance security and brainstorm potential avenues of research in this area?

Chapter 5

ETHICAL AND SOCIAL CHALLENGES OF ELECTRONIC HEALTH INFORMATION

Peter S. Winkelstein

Department of Pediatrics and School of Informatics, University at Buffalo, State University of New York, Buffalo, NY 14222

Chapter Overview

The development of modern bioethics has been strongly influenced by technology. Important ethical questions surround the use of electronic health records, clinical decision support systems, internet-based consumer health information, outcome measurement, and data mining. Electronic health records are changing the way health information is managed, but implementation is a difficult task in which social and cultural issues must be addressed. Advice produced by decision support systems must be understood and acted upon in the context of the overall goals and values of health care. Empowering health care consumers through readily-available health information is a valuable use of the internet, but the nature of the internet environment raises the spectre of abuse of vulnerable patients. Outcome studies have inherent value judgments that may be hidden. Data mining may impact confidentiality or lead to discrimination by identifying subgroups. All of these issues, and others, require careful examination as more and more health information is captured electronically.

Keywords

ethics; bioethics; Internet; electronic health records; decision support; consumer health informatics; outcomes; data mining

1. INTRODUCTION

Modern bioethics and health informatics are intimately connected. It is no coincidence that the two fields emerged at the same time. Modern bioethics originated and had evolved in response to increasingly perplexing questions about the goals of health care. These questions arise largely as a result of the advent of modern technology. Prior to the development of the mechanical ventilator and the dialysis machine, medicine was limited in its ability to extend the life span of patients with incurable conditions. With the introduction of these new devices, it suddenly became possible to imagine machine-dependent patients and, more importantly, to imagine patients who would prefer death to such dependency. Hence the beginnings of modern bioethical thought.

Computers and other informatics technologies and techniques have only made matters more complex. Difficult ethical questions surround the use of electronic health records, clinical decision support and prognostic systems, internet-based consumer health information, outcome measurement, and data mining. Some of these questions are new, raised by new technologic capabilities; some are old, but recast in new forms by the use of informatics.

A fundamental question is: what is the proper role of technology in health care? What health care decisions should be entrusted to computers? Are there decisions that computers should not make or roles they should not play? In attempting to answer these questions, it is important to recognize that computers are merely tools. Tools do not, in themselves, change the underlying goals of health care. They are only properly used to advance the goals of the underlying endeavor. Over 20 years ago, Moor (1979) concluded that the one task that should never be assigned to computers is the choice of the goals of medicine itself.

Although it is clear that the use of health informatics tools should be judged in the same way as the use of any medical tool, i.e. by their ability to advance the goals of medicine, that judgment can be very complex because the tools themselves are complex. In addition, the extraordinary usefulness of computer-based technology creates a digital imperative: a strong incentive to adjust all of health care so as to be compatible with computers. It is much less expensive, for example, to move digital information than to move actual patients or laboratory samples. While the use of informatics can have enormous benefits (e.g. telemedicine for geographically isolated patients) the quality of care can also be degraded in the rush to digitization.

There have been attempts to use bioinformatics tools to work on some of the most difficult bioethics problems that face health care, including cost-effectiveness and futility of treatment, especially near the end-of-life. For example, computer-based prognostic systems have been developed for some critical care settings (Knaus et al., 1991). Their output, however, has not provided answers to questions such as "when should we stop treatment" or "what treatments should be tried" because the answers to those questions are not entirely quantitative and scientific. Families and caregivers often have very personal values and views about what constitutes appropriate health care. For some, extending life by a week through aggressive interventions may be a great blessing; for others, the same treatment may be considered torture. It is not possible to separate personal values from these decisions.

Rather than prescribe particular consequences for particular actions, modern bioethical thought tends to focus on fundamental truths, laws, or motive forces. The language of bioethics often invokes "principles," which are compact statements of fundamental import. The four key bioethical principles developed by Beauchamp and Childress (1994), for example, are autonomy (the right of an individual to determine his or her own health care), beneficence (the duty of health care workers to improve the welfare of their patients), nonmaleficence (the duty of health care workers to avoid doing intentional harm), and justice (both to individuals and society at large). Unfortunately, the term "principles" has become diluted and is often used to describe simple lists of issues.

The development of modern bioethics has been strongly influenced by technology. But technology itself does not determine the ethics of medicine. Technological advances need to be seen in, and judged by, the light of health care goals. The digital imperative must be resisted unless a clear benefit of computerization can be demonstrated. Technology cannot give us answers to questions that require personal and social value judgments.

2. OVERVIEW OF THE FIELD

2.1 Electronic Health Records

Electronic health records (EHRs), also known as electronic medical records or computerized patient records, are found in an increasing number of physician offices and hospitals. One estimate puts current penetration at 10%, increasing to 25% in 3 years (private communication, 2004). EHRs represent more than a simple computerization of the traditional paper chart. They provide the ability to manage health information using modern information techniques that are impossible to apply to paper record keeping. The use of these techniques has the potential to dramatically change how both individuals and society view health care. Dramatic changes in health care have always been accompanied by equally important ethical challenges, and adoption of EHRs is no exception.

Before enumerating the ethical issues raised by EHRs, it is reasonable to ask why we need EHRs at all. Fundamentally, the answer is that we cannot expect to provide quality medical care without optimal information management. Proper medical treatment depends on timely access to accurate patient medical histories, laboratory results, and many other pieces of data. Problems such as missing or misplaced charts, paper-based laboratory reporting, and illegible handwriting are common roadblocks to care. We are in an era where health care providers can generate enormous amounts of information about a patient but have only antiquated and inadequate methods of managing that information. This discrepancy will only get worse as health care skills get better, especially with the advent of readily-available genomic data. In addition, health care providers cannot begin to empower patients with access to their own medical information if the providers can't manage that information themselves. EHRs provide the tools that can be used to begin to solve these problems.

In addition to improving the quality of care, adoption of EHRs holds out other promises, including improvements in the efficiency with which health care is provided, increased patient satisfaction, opportunities for research, quality improvement and, especially, reduction of errors. Although the EHR has been touted as the key tool for reducing medical errors, the evidence of success is still scant. While recognizing that EHRs are a key component of error reduction in health care the Institute of Medicine (2000) cautions "ALL technology introduces new errors, even when its sole purpose is to prevent errors."

Another promise for which there is as yet no empirical support is return on investment (ROI). EHRs are expensive and complex to install. Savings from reduced file room staffing and square footage, along with savings from reduced dictation expenses, may not fully cover the cost of the EHR. As a result, one can imagine a "have" and "have not" condition, where some practices (perhaps specialty practices in affluent areas) can afford EHRs, but other practices (perhaps inner-city primary care) cannot. That disparity could jeopardize the quality of care for the patients of the "have not" practices. It could also interfere with public health issues because data from the "have not" patients would be much less readily available and therefore underrepresented in public health databases.

The implementation of an EHR system is no easy task (Ash 2003). The conversion from a paper chart to an EHR system puts a great deal of stress on the complex social systems that exist within health care institutions. It requires reconceptualizing the medical record and medical communication, including organizational-level changes in workflow. Resistance to even minor changes is a normal response, especially in complex environments, and a change of the magnitude represented by the EHR engenders resistance of the same order.

Although EHRs are an information technology (IT) product, the decision to implement an EHR and the selection of an appropriate vendor are not solely within the IT realm. The end-users of these systems must be included in the decision process. Clinicians, especially physicians, are the *de facto* arbiters of EHR acceptance in any health care institution. EHR implementation therefore requires a strong, committed physician champion with the time to devote to the project. It is also critical to manage expectations. Clinicians often wish to believe that an EHR will immediately and completely eliminate all perceived barriers to access to clinical data, and they become frustrated when they find that it does not.

Clinicians are also most concerned about clinical data entry, which is the component least improved by the EHR. Most EHR benefits initially accrue to back-office staff at the perceived cost of clinician's time. In addition, clinicians may perceive the EHR as a barrier to provider-patient communication and family-centered care. Clinicians are also focused on the content of an EHR, such as clinical documentation templates, alerting capabilities, and patient lists. Unfortunately, EHRs are far from turn-key at this time. They come with very limited content and require a great deal of customization in order to function in a particular clinical environment, customization that must largely be done by the end-user. This is a significant barrier to EHR adoption. EHR vendors are beginning to understand that and are devoting more resources to content development.

Another barrier to adoption is that current EHRs are largely stand-alone systems. They typically interface with billing systems and, in a hospital setting, may have connections to laboratory and other systems. But EHRs do not typically communicate with each other. There are currently no well-accepted standards for EHR interoperability. In order to apply the tools of modern information management to health data, especially for population-based studies, there must be a way to aggregate data from many EHRs. The National Health Information Infrastructure project is beginning to address some of these issues (Yasnoff 2004).

The need for EHR interoperability, along with the expense of EHR systems, is likely to drive fundamental changes in how medical records are stored. Centralized third-party medical record keeping, in the form of data "banks," may supplant the current model of record keeping by individual practices. Centralized record keeping would enable health care workers, and patients themselves, to access medical records where and when needed. It would also, of course, require strong security measures.

No discussion of EHRs can ignore the concerns of privacy, confidentiality, and security. Privacy is the ability of a patient to control the information about him or herself. Confidentiality is the commitment of another person or organization to the patient to control information about the patient. Security measures are safeguards against inadvertent or malicious breaches of confidentiality. Security measures also include protections against loss of information. It is generally accepted that privacy of medical data is an important right of the individual. Privacy may be viewed either as a utilitarian concept (i.e. patients will not honestly and completely discuss their medical problems without assurances of confidentiality) or as a right in and of itself. Privacy is also essential to the exercise of autonomy in medical decision making, just as a secret ballot is fundamental to the exercise of democracy.

Maintaining privacy and confidentiality through appropriate security is one of the key challenges of EHRs. It has long been recognized for related uses of electronic media, such as email (Kane 1998). Aside from technical issues, there are a number of factors that contribute to the challenge. Determining the proper security measures for medical records must be done in the context of the goals for the records. For instance, an important goal of EHRs is to improve access to medical records, for both providers and patients (Delbanco 2004). A perfectly secure EHR would be one to which no access was allowed, so a balance between security and access must be struck. However, the answer to the question of what is the "correct" balance is not a technical or scientific one but rather a social and political one. The answer depends on the values of the participants. These values vary widely. Some people see the benefits of access and are perfectly comfortable with their medical data recorded on computers while others are concerned about of breaches of confidentiality and resist such record-keeping.

EHRs are not the only systems where this balance must be struck. Electronic toll badges, for example, allow for the convenient payment of tolls without actually stopping at the tollbooth via electronic identification. Many people use such devices without concern. Others refuse to use them, fearing the use of the data to track their movements. One feature of the electronic toll badge is that a driver can opt-out of its use at any time simply by leaving it at home and paying the toll in cash. The ability to opt-out of technology on an as-desired basis is important to the acceptance of the technology. At the moment, it is difficult for patients to opt-out of having at least medical billing information entered into a computer. As EHRs become more ubiquitous, opting-out may become impossible.

The issues of privacy, confidentiality, and security have attracted the attention of government regulators. In 2002, the Health Insurance Portability and Accountability Act (HIPAA) privacy regulations went into effect (c.f. http://www.hhs.gov/ocr/hipaa/). The HIPAA regulations are designed to restrict the inappropriate flow of medical information without disrupting medical care. In particular, the regulations target health data belonging to specific, identifiable patients (PHI). They regulate data flow by dividing medical information use into categories. PHI that is used for payment and other health care operations is subject to the "minimum necessary" restriction, which simply means that only the minimum amount of information necessary (as determined by a reasonable person) to accomplish the task should be used. PHI that is used for medical treatment of the patient is not subject to the "minimum necessary" restriction. Use of PHI for any other purpose requires explicit authorization from the patient.

The structure of the HIPAA regulations puts the burden of determining the "minimum necessary" amount of information and of detecting inappropriate disclosures on the health care provider. This structure is consistent with the current model of record keeping, namely that medical records are largely held by health care providers. With a shift to centralized record-keeping, it becomes possible to give patients more control over and responsibility for the confidentiality of their records. Patients could receive periodic or on-demand reports of the audit trail of accesses to their records. They would then be responsible for detecting and reporting inappropriate uses of their records in the same way that consumers are responsible for reviewing their credit card statements for fraudulent uses of their credit.

The HIPAA regulations require technological, policy, and educational interventions. They affect all PHIs, whether electronic or paper. They also affect how research is conducted, including data mining of medical records, which is considered by the HIPAA regulations to be a form of human-subjects research. The HIPAA regulations essentially extend the concept of harm for research participants to include breaches of confidentiality. Specific procedures for obtaining approval for such research are outlined in the regulations.

Electronic health records are changing the way health information is managed. Especially with interoperability, and possibly centralization, EHRs will allow the application of modern information management tools to health care data. Implementation of EHRs, however, is a difficult task. When implementing an EHR, social and cultural issues must be addressed, and expectation management is critical. Acceptance of an EHR is dependent on acceptance of the underlying goals of the implementation.

2.2 Clinical Alerts and Decision Support

One of the promises of EHRs is that the information they contain can be used to provide automatic alerts such as drug-drug interactions and suggestions for treatment or diagnosis. This naturally raises the question of who is in charge of making medical decisions, the clinician or the computer?

The "standard view" (Miller, 1990) is that human clinicians should retain the ultimate authority to make decisions and that computers should provide advice only. There are two reasons for this standard view. One is simply that computer decision support systems have so far not been shown to be clinically useful, especially in general diagnostic situations (Berner 1994). This does not mean, however, that such systems will never be useful, only that the construction of useful systems is complex. Whether computers will ever provide powerful enough decision support to supplant human clinicians in at least some situations is an empirical question (Moor 1979). The second reason for the standard view is that medical decisions are more than the simple "mapping from patient data to a nosology of disease states" (Mazoué 1990). In other words, many medical decisions cannot be made on entirely scientific grounds. Rather, they require the careful consideration of the underlying goals and values of health care in the context of the individual patient and society at large. This sort of judgment can only be made by those who understand these values and have the skills required to make decisions based on them—namely, humans.

Even if computers remain in an advice-only mode, however, there may still be powerful reasons for following that advice. For example, EHRs and other prescription-writing and dosing programs (e.g. http://www2.epocrates.com, http://www.pdr.net) routinely perform medication interaction checking. If a clinician ignores a warning provided by one of these programs, it is clear that he runs the risk of providing inferior medical care, not to mention of being subject to legal action. Even though the computer has provided "only" advice, the clinician ignores it at his peril. It is easy to imagine that it would be even more difficult to ignore diagnosis or treatment advice.

Because computer alerts and decision support systems can have such power, it is essential to be sure that these systems are properly designed, evaluated, and maintained (Anderson 1994). "Properly" in this context means that clinical decision support systems must adhere to the underlying goals of medicine, which may be different from the underlying goals of a commercial systems designer. Commercial systems must adhere to such health care goals as standard of care, primacy of the best interest of the patient, and informed consent (Goodman 2001). They cannot operate under the usual free-market ethic of *caveat emptor*.

Computer decision support systems that are designed to provide prognosis information are particularly problematic. Obtaining an accurate prognosis can be a difficult task for clinicians, but it is an important one because many treatment decisions are based on prognosis. In addition, the prognosis is a critical piece of information for patients, particularly in cases of life-threatening illness. Prognostic scoring systems can potentially be used for several purposes (Sasse 1993): quality assessment, resource management (including triage and rationing), and individual patient care decisions.

Quality assessment and improvement is an important goal for any health care institution. Prognostic scoring systems could be used, for example, to compare actual to expected outcomes. This is a reasonable use of scoring systems, assuming they have been properly evaluated, in that it is aligned with, and furthers, the goals of health care. Whether such use will actually improve the quality of care has yet to be demonstrated.

Resource management is a much more difficult problem. We currently do not have a societal consensus on how to manage our health care resources. It is generally accepted that we are no longer able to do everything for everyone but, nonetheless, the health care system continues to function on that premise. Clinicians often find themselves in ethically problematic positions where their traditional role of patient advocate is in conflict with their duty to manage society's health care resources. No computer system can solve this problem. It is possible, however, that accurate information provided by computer systems may assist in the process of making these kinds of difficult decisions.

The most problematic use of prognostic scoring systems is in making individual care decisions. As noted above, decisions based on prognosis are not entirely scientific but are also value-driven (Knaus 1993), thus putting them outside the realm of computers. In addition, prognostic systems are by their very nature based on, and provide, a statistical score. Applying population statistics to individuals is fraught with problems (Thomasma 1988). At the same time prognostic scores can have an aura of certainty and objectivity that they do not warrant. They can also be self-fulfilling: if care is withdrawn due to a poor prognostic score, the patient will certainly die, thus apparently confirming the score. On the other hand, good prognostic statistics are a key to good medical decision making. What is critical is that the data provided by a prognostic scoring system be properly interpreted and applied, which means, in turn, that the users of these systems must be properly trained and qualified (Goodman 2001).

The development of computer-based diagnostic programs has received a great deal of attention in the field of medical informatics. Diagnosis programs are designed to process information about a patient and produce a differential diagnosis list, usually rank-ordered by probability. Most of these programs (c.f. http://www.lcs.mgh.harvard.edu/dxplain.htm) use the Bayes theorem (c.f. Fletcher 1996) to calculate the probability of a diagnosis based on the probabilities that the input signs and symptoms are associated with the diagnosis (there are some notable exceptions to this approach, for example, http://www.isabel.org.uk/). However, these programs have not so far shown a great deal of promise, especially for general diagnosis (Berner

1994). One problem is that Bayesian calculations are strongly dependent on the underlying population statistics. For example, there is a high likelihood that a child with a fever has a viral syndrome, but that is not a useful piece of information for a computer to communicate to a physician. Computer-based diagnostic programs are needed most to remind clinicians about rare or unusual diagnoses, not to determine common diseases. The nature of a Bayesian calculation does not lend itself to detection of rare events. There may be other knowledge discovery tools that are more sensitive to unusual events and that clearly warrant further development in this area.

A similar problem occurs with some processes designed to help with medical decision making. Decision trees (Detsky 1997) can be constructed for some clinical situations, with branches representing outcomes and intermediate states with their associated likelihoods. By simple Bayesian calculation, the likelihood of an outcome can be determined from these trees. By giving each outcome a value ("utility," typically on a scale of 0 to 1, where 0 represents death and 1 represents healthy life), a patient or clinician can get some indication of the most desirable course of action (i.e. the optimal combination of probability and value). Unfortunately, this process has at least three major pitfalls. First, if any of the branch points depend on population statistics (i.e. likelihood of a disease), then the Bayesian calculation is generally overwhelmed by that point, making the rest of the tree irrelevant. Second, a utility scale of 0 to 1 does not capture the full range of possible values. In particular, it is certainly possible to imagine states worse than death (i.e. with negative utilities) (Patrick 1994). Lastly, utilities may be very individual (for example, palliative chemotherapy may be intolerable to some, worthwhile to others). Individual utilities are at least burdensome, and perhaps impossible, to determine accurately. Substituting population-based utilities (averages of utilities chosen by many people given the same situation) (Bell 2001) erases any ability of a patient or clinician to adjust the decisions produced by the tree to reflect personal values.

Clinical alerting and decision making systems can, without question, improve the quality of health care, but they must be implemented properly. Users need training and education about the abilities and limitations of the systems. Systems must be evaluated and maintained. System designers and vendors must understand that their systems will be held to the high standards of medical care. Most importantly, the advice produced by these systems must be understood and acted upon in the context of the overall goals and values of health care.

2.3 Internet-based Consumer Health Information

The dramatic increase in accessibility of information provided by the internet, and especially the World Wide Web protocols, has of course extended to the field of health care. This development is fundamentally good for health care because information is the lifeblood of evidence-based medicine (see Section 2.4 below). Medical information designed for patients can also strengthen patients' ability to make informed judgments about their own care. But the open nature of the web also brings with it the danger of inaccurate or misleading information, both by omission and by comission.

The fundamental basis of the doctor-patient relationship (or the relationship between any reputable health care provider and their client) is the primacy of the patient's best interests. When a provider suggests a course of treatment, the patient can reasonably expect that the provider is suggesting what in the provider's judgment is best for the patient. Sometimes there are several reasonable courses of action that may be appropriate in a given situation and the provider will assist the patient in making choices through the process of informed consent. The concepts of best interest and informed consent derive directly from the principles of autonomy, beneficence, and nonmaleficence.

The relationship between a salesman and a customer is quite different. There, the suggestions by the salesman of choices that might be made by the customer are largely based on the self-interest of the salesman, not the best interest of the customer. A good salesman has the ability to make the customer believe, however, that he has the customer's interests in mind. Experienced consumers are well aware of this and understand the nature of the relationship, namely *caveat emptor*.

The concepts of informed consent and *caveat emptor* may come into direct conflict on the World Wide Web. Because authorship and, more importantly, the author's intent can be difficult to determine on web sites, it is difficult for patients to know what ethical construct to apply to a particular site. Some sites provide authoritative medical information that is designed to enhance the ability of patients to understand their conditions and make appropriate health care choices. Other sites are designed to sell something to those same patients while masquerading as sources of information. Patients may be unable to distinguish between the two types of sites, either because the latter sites have purposely been made to appear like the former or simply because patients may not be aware of the purposes of commercial medical web sites. To confuse matters further, there may be a mix of commercial information and authoritative information on the same site, either well or poorly distinguished.

Several certification programs purport to assist consumers in determining the quality of a medical web site (e.g. http://www.hiethics.org, http://www.truste.org, http://www.hon.ch). Such programs are only as good as the awareness they generate among consumers and the quality of their underlying requirements for certification. The effectiveness of certification programs has yet to be demonstrated. Requirements for certification vary widely, but there are several themes that appear consistently. These include:

- Clear mission and appropriate use statements
- Clear attribution and dating of medical material and claims
- Clear indication of advertising material
- Commitment to use of scientifically supported medical information
- Clear contact and complaint resolution information
- Appropriate security for PHI
- Ability to opt-out or opt-in to sharing of PHI
- Ability to amend PHI
- Agreement to bind business partners to policies of site
- Clear and timely notifications of any changes to policies
- Extensive disclosures, including
  - privacy practices
    - data sharing with third parties
    - aggregation and re-identification
    - use of tracking technology
  - financial
    - ownership
    - sponsors
    - third party revenues from data sharing

Many of the requirements are disclosure-dependent, meaning that the commitment is to inform the consumer about the site practices, rather than to eschew certain practices, such as preferentially including information from a financial sponsor, altogether. As a result, the burden of determining the quality of a site's information rests squarely on the shoulders of the consumer, even though the site bears a logo of certification.

Empowering health care consumers through readily-available health information is a valuable use of the internet. The nature of the internet environment, however, raises the spectre of abuse of vulnerable patients. Reputable web sites with health information must be careful to inform users about the nature of the site's information, both through extensive disclosures and avoidance of deceptive marketing practices. Patients must be educated to approach commercial web sites with *caveat emptor* firmly in mind.

2.4 Evidence-based Medicine, Outcome Measures, and Practice Guidelines

Modern medicine is defined in part by its use of therapies demonstrated to work by scientific evidence. Outcome studies and practice guidelines purport to provide such evidence. Outcome studies data must therefore be accepted unless convincing reasons to discard them exist. Outcomes research, however, faces a number of practical and philosophical problems which raise important ethical questions about the proper use of their results.

For this discussion, I define outcomes research as the statistical examination of outcomes as a function of diagnostic or treatment strategies using large numbers of subjects. Outcomes research often utilizes multiple studies combined via meta-analysis. Results from outcome studies are usually descriptive and use the language of statistics and probability. Practice guidelines typically combine outcomes research results with "expert" or "consensus" panels to produce prescriptive recommendations for clinical practice, often in the form of algorithms.

Outcomes research would not be possible without computational power. It is through data mining, knowledge discovery, and meta-analysis that results are obtained and all of these endeavors are impractical without computers. The internet is also critical in that it serves as the primary medium through which the results of outcomes studies and practice guidelines are distributed (c.f. http://www.cochrane.org, http://www.guideline.gov).

Outcomes studies have significant practical problems. Potential methodological flaws include the inherent difficulties of meta-analysis, inaccurate description of variables, and potentially inadequate sample sizes to detect small effects. Results may generalize poorly due to limited study populations (for example, males only or outdated therapeutic regimens) (Lagasse 1996, Gifford 1996). Most significantly, the design of outcomes research contains inherent value judgments which may not be apparent in the reporting of the results.

These value judgments are evident in various aspects of study design. For example, the outcome measure of "cost effectiveness" is value-driven because the answer to the question of what constitutes a reasonable expenditure of health care dollars is not scientific, but rather social and political. Also, complex systems do not lend themselves to simple measures, so a choice of measure must be made. This choice often involves the values of the researcher or the funding agency. Some important outcomes (for example pain, quality of life, reassurance, or justice) may be difficult, or even impossible, to measure and may therefore be inappropriately ignored

(Kerridge 1998). Fundamentally, the choice of outcome measure and its use is value-driven.

If the results of outcome studies are linked to resource allocation, then several additional problems occur. First, it is often impossible to compare studies of the outcomes of treatments for different conditions, but such comparisons must be made if treatments for different conditions are to compete for health care resources. Second, even if we use outcome studies only to compare treatments for a single condition, we must know which treatment is better and also by how much in order to understand how to manage expenditures (Shiell 1997). Third, the agenda of treatments studied may itself be driven by considerations of cost rather than health (Tanenbaum 1994). It is all too tempting for those who manage health care resources to assume that a lack of evidence for the efficacy of a treatment implies that the treatment has no efficacy and should therefore receive little support.

The purported ability of outcomes research to improve health care rests on the single assumption that data about probabilities of outcomes is valuable in making optimal decisions about health care delivery. That assumption is open to criticism on several fronts. First, as in the case of prognostic scoring systems, it can be very difficult to apply probabilistic, population-based data to individual patients, especially if the study population poorly matches the individual's background. Second, as was demonstrated by the poor performance of general diagnostic systems (Berner 1994), the technique of probability-based decision making itself is of limited value, due in part to the sensitivity of the Bayes theorem to population statistics. Finally, the use of probability-based rather than causal-based reasoning as the primary method of health care decision making is a significant departure from historical precedent. Probability-based reasoning implies an acceptance of a utilitarian philosophy which is not consistent with much of the moral philosophy of medicine. In addition, the use of probability-based reasoning implies an acceptance of induction (the expectation that one event will follow another from past experience of such sequences) as a reasoning model, as opposed to the more familiar use of causal models such as pathophysiology (Goodman 1996). Given that computer diagnostic systems do not demonstrate the efficacy of such reasoning, and given the inherently value-driven nature of outcome studies, it is clear that the results of outcome studies must be used with great caution.

2.5 Data Mining

The use of computers in health care has engendered an explosion of the quantity of electronically encoded data. A central theme of medical informatics is the use of this data for knowledge discovery, a process commonly termed "data mining." The purpose of data mining is to identify significant data patterns that would otherwise go undetected. Data mining is used in outcomes research, epidemiology, drug and genome discovery, biomedical literature searching, and many other areas. Data mining can also be used to detect unusual data patterns which might be indicative of disease outbreaks or fraudulent activities.

One of the great promises of the EHR is that clinical data will become available for data mining. As increasing amounts of health data become computerized, it is easy to imagine that data mining of EHRs will become the primary form of clinical research. Electronic records containing PHI need proper protections. The HIPAA regulations incorporate procedures for such research, recognizing that a breach of privacy is a form of harm.

In addition to regulating research on databases containing PHI, HIPAA also provides a mechanism to de-identify data. Once de-identified, data is free from regulation under HIPAA. Two methods of de-identification are allowed. One requires a statistical determination of the level of de-identification necessary to make re-identification unlikely. The other prescribes the removal of a specific set of identifiers (the "safe harbor" method) (http://privacyruleandresearch.nih.gov/research_repositories.asp). The safe harbor method is much simpler and is likely to be the method of choice for most situations. However, the safe harbor method removes a great deal of information that might be critical for answering relevant questions. For example, date elements must be removed except for the year, making determination of age to the accuracy necessary in the pediatric population essentially impossible. There is very little known about how useful or useless data de-identified by the safe harbor method will be.

The goal of de-identification is to make it statistically unlikely that the PHI of an individual patient can be reconstructed from a de-identified data set. Whether the safe harbor method accomplishes that goal has not been verified. There is evidence that some information that can be included in de-identified data may in fact be unique to a particular patient. For example, the ICD-9 diagnosis code, especially when combined with other data such as medications, may map closely to patient identifiers (Clause 2004). It is the clear duty of a researcher using de-identified data to avoid re-identification, but once such data sets become public it will be impossible to limit re-identification activities. Worse, because so much personal information is publicly available, it may be possible to use external sources of data in combination with de-identified medical data to construct fairly complete PHI information that could be used in ethically inappropriate ways (denial of health care, insurance, employment, etc.).

When selecting methods for rule-creation from data mining results, it is important to know what the rules will be used for. For example, we have already seen that using Bayesian rules are not particularly effective for general diagnosis. A system for general diagnosis would be much more valuable if it accurately detected rare events. This is not because of anything in the nature of medical diagnosis itself, but because in a practical sense computers are not needed to diagnose common illnesses. It is important that the strengths of a particular data mining technique match the intended use of the results.

Similarly, it is important that database design take into account the range of possible queries that might be made of it. At the most basic level, one cannot examine data that is not included in the database at all. On a more complex level, it may well be possible to detect patterns in data that cannot then be adequately explained using that data. For example, some outcome measures such as lung function of patients with cystic fibrosis could be constructed and calculated for a number of different institutions. Given the results, one would then naturally ask what the best institution was doing right and what the worst institution was doing wrong. The answers to those questions could well be impossible to obtain from the original data set. Without those answers, the information about institutional outcomes may be useless, and perhaps damaging (Donaldson 1994). The problem of identifying the information to be measured and recorded is as old as epidemiology itself and has been made more acute by the computational power available today.

A particular concern regarding data mining arises when those results identify new patterns in population subgroups. This can happen when doing population-based or genomic research. It can even occur in research on de-identified data sets. Invidious discrimination requires differentiation between groups. We have experienced the ongoing evil of discrimination along the familiar lines of race, sex, age, and others. It is therefore of concern if data mining creates new subgroups that could then be the target of discrimination. It is easy to imagine discrimination along genetic lines (slow vs. fast drug metabolizers, for example). But any subgroup could be affected. Research where there is the possibility of new subgroup identification should be carried out with great caution, carefully weighing the potential medical benefits against the risks of harm from discrimination.

Data mining will undoubtedly provide important information for epidemiology, clinical decision support, and the practice of evidence-based medicine. It is important to realize, however, that there are ethical and social concerns about the use of the results. As with any health informatics technique, data mining must be used with a clear understanding of, and to further, the underlying goals of health care.

REFERENCES

Anderson, J.G. and Ayden, C.E. (1994). "Evaluating Medical Information Systems: Social Contexts and Ethical Challenges, in K.W. Goodman, Ed., *Ethics, Computing and Medicine: Informatics and the Transformation of Health Care,* Cambridge: Cambridge University Press, 57-74.

Ash, J.S., Stavri, P. Z., and Kuperman, G. J. (2003). "A Consensus Statement on Considerations for a Successful CPOE Implementation," *J Am Med Inform Assoc.,* 10:229-234.

Beauchamp, T.L., and Childress, J.F. (1994). *Principles of Biomedical Ethics,* 4th Ed., New York: Oxford University Press.

Bell, C.M., and Chapman, R.H., et al., "An Off-the-shelf Help List: A Comprehensive Catalog of Preference Scores from Published Cost-utility Analyses," *Med Decis Making,* 21(4): 288-294.

Berner, E.S., and Webster, G.D., Et Al. (1994). "Performance of Four Computer-based Diagnostic Systems," *N. Engl J Med.,* 330:1792-1796.

Clause, S.L., and Triller, D.M., Et Al. (2004). "Conforming to HIPAA Regulations and Compilation of Research Data," *Am J Health Syst Pharm.,* 61:1025-1031.

Delbanco, T., and Sands, D.Z., "Electrons in Flight—E-mail Between Doctors and Patients," *N Engl J Med.,* 350:1705-1707.

Detsky, A.S., and Naglie, G., Et Al. (1997). "Primer on Medical Decision Analysis: Part 1-getting Started," *Med Decis Making,* 17(2):123-125.

Donaldson, M.S., Lohr, K.N., and Bulger, R.J. (1994). "Health Data in the Information Age: Use, Disclosure, and Privacy—part 1," *JAMA,* 271(17):1308.

Fletcher, R.H., Fletcher, S.W., and Wagner, E.H. (1996). *Clinical Epidemiology: The Essentials*, 3rd Ed., Baltimore: Williams & Wilkins.

Gifford, F. (1996). "Outcomes Research and Practice Guidelines: Upstream Issues For Downstream Users," *Hastings Cent Rep.*, 26(2):38-44.

Goodman, K.W. (1996). "Outcomes, Confidentiality, and Appropriate Use," *Crit Care Clin.,* 12 (1):109-122.

Goodman, K.W., and Miller, R.A. (2001). "Ethics and Health Informatics: Users, Standards and Outcomes," in Shortliffe, E.H, and Perreault, L.E., Eds., *Medical Informatics: Computer Applications in Health Care and Biomedicine*, New York: Springer, 257-281.

Institute of Medicine (2000). *To Err Is Human*, Washington, D.C., National Academy Press.

Kane, B., and Sands, D.Z. (1998). "Guidelines for the Clinical Use of Electronic Mail with Patients," *J Am Med Inform. Assoc.,* 5:104-111.

Knaus W.A., and Wagner D.P., Et Al. (1991). "The APACHE III Prognostic System. Risk Prediction of Hospital Mortality for Critically Ill Hospitalized Adults," *Chest,* 100(6):1619-1636.

Knaus, W.A. (1993). "Ethics Implications of Risk Stratification in the Acute Care Setting," *Camb Q Healthc Ethics,* 2:193-196.

Kerridge, I., Lowe, M., and Henry, D. (1998). " Ethics and Evidence Based Medicine," *BMJ,* 316:1151-1153.

Lagasse, R.S. (1996). "Monitoring and Analysis of Outcome Studies," *Int Anesthesiol Clin.,* 34(3):263-277.

Mazoué, J.G. (1990). "Diagnosis Without Doctors," *J Med Philos.,* 15:559-579.

Miller, R.A. (1990). "Why the Standard View Is Standard: People, Not Machines, Understand Patients' Problems," *J. Med Philos.*, 15:581-591.

Moor, J.H., 1979). "Are There Decisions Computers Should Never Make?" *Nat Syst*, 1:217-229.

Patrick, D.L., and Starks, H.E., Et Al. (1994). "Measuring Preferences for Health States Worse Than Death," *Med Decis Making*, 14:9-18.

Sasse, K. (1993). "Prognostic Scoring Systems: Facing Difficult Decisions with Objective Data," *Camb Q Healthc Ethics*, 2:185-91.

Shiell, A. (1997). "Health Outcomes Are about Choices and Values: An Economic Perspective on the Health Outcomes Movement," *Health Policy*, 39:5-15.

Tanenbaum, S.J. (1994). "Knowing and Acting in Medical Practice: The Epistemological Politics of Outcome Research," *J Health Polit Policy Law*, 19(1):27-44.

Thomasma, D.C. (1998). "Applying General Medical Knowledge to Individuals: A Philosophical Analysis," *Theor Med.*, 9: 187-200.

Yasnoff, W.A., and Humphreys, B.L., Et Al. (2004). "A Consensus Action Agenda for Achieving the National Health Information Infrastructure," *J Am Med Inform Assoc.*, 11:332-338.

SUGGESTED READINGS

Goodman, K.W. (ed.), *Ethics, Computing and Medicine: Informatics and the Transformation of Health Care*, Cambridge University Press, Cambridge.
This is the central text for the area of ethics in medical/health informatics.

Howell, J.H., and Sale, W.F. (eds.), *Life Choices: A Hastings Center Introduction to Bioethics*, 2nd ed., Georgetown University Press, Washington DC.
An introductory text to bioethics in general.

Institute of Medicine, *To Err is Human*, National Academy Press, Washington, D.C.
This report sets the agenda for the reduction of medical errors.

ONLINE RESOURCES

Certification organizations for medical web sites
http://www.hiethics.org
http://www.truste.org
http://www.hon.ch

Examples of evidence-based medicine and guideline distribution sites
http://www.cochrane.org
http://www.guideline.gov

Examples of web-based diagnostic systems
http://www.lcs.mgh.harvard.edu/dxplain.htm
http://www.isabel.org.uk

HIPAA Privacy Regulations overview and information on research
http://www.hhs.gov/ocr/hipaa/
http://privacyruleandresearch.nih.gov/research_repositories.asp

QUESTIONS FOR DISCUSSION

1. The mission statement for an effort by a medical organization to collect data from multiple local sources reads: "The mission is to transform what is now a disconnected set of data into a form that is complete for any given patient, no matter where they are seen. It should be available to different groups or health care professionals for different reasons. This will involve an assessment of the current data in terms of its location, accuracy, and accessibility to different parties, identification of these parties with an understanding of the kinds of data they may need, and then a matching of these needs to the restructuring of the database itself." Discuss the technical and political challenges of this mission. How does HIPAA impact this mission? Where is the primary focus of this project, on the patients or providers?

2. You are the CIO of a large hospital system. An eight-year old girl who is a patient at one of your hospitals is in need of a liver transplant. A suitable donor has not yet been found. Many people die while awaiting liver transplants because of a shortage of organs. The girl's parents wish to set up a web site describing their daughter's illness and prognosis, particularly the critical need for a liver, in the hope that this will help them find a donor. They ask you to make this part of your hospital system's public web site because they think that that location will give it more legitimacy and attract more internet traffic. Do you allow this? Why or why not? Discuss in terms of the HIPAA regulations and the principles of Beauchamp and Childress. Include any other factors or reasoning you consider important.

3. Describe three barriers to EHR adoption. What is meant by "minimum necessary" in the HIPAA regulations? Describe what technical and policy measures would be necessary in order to have third-party centralized EHRs. Discuss the advantages and disadvantages of centralized EHRs.

4. You wish to code into an EHR a clinical algorithm for the treatment of hyperbilirubinemia in the newborn (http://aappolicy.aappublications.org/cgi/content/full/pediatrics;114/1/297). "This algorithm depends on the age of the newborn in hours and the level of total serum bilirubin (a blood test). "Turn-around time for this result ranges from 30 minutes to two hours. Describe technical barriers to this project. How would you test the code? What mechanisms do you need to

maintain the code if the algorithm changes? How would you detect and handle missing or inaccurate data?

5. What are the characteristics necessary for a useful general diagnostic support system? What data mining techniques are available besides Bayesian algorithms? Are any of them suitable for a general diagnostic support system?

6. How would you measure the loss of data due to safe harbor de-identification? How would you determine ease of re-identification?

UNIT II

Information and Knowledge Management

Chapter 6
MEDICAL CONCEPT REPRESENTATION

Christopher G. Chute

Mayo Clinic College of Medicine, Rochester, MN 55905

Chapter Overview
The description of concepts in the biomedical domain spans levels of precision, complexity, implicit knowledge, and breadth of application that makes the knowledge representation problem more challenging than that in virtually any other domain. This chapter reviews some of this breadth in the form of use-cases, and highlights some of the challenges confronted, including variability among the properties of terminologies, classifications, and ontologies. Special challenges arise at the semantic boundary between information and terminology models, which are not resolvable on one side of either boundary. The problems of aggregation are considered, together with the requirement for rule-based logic when mapping information described using detailed terminologies to high-level classifications. Finally, the challenge of semantic interoperability, arguably the goal of all standards efforts, is explored with respect to medical concept representation.

Keywords
vocabulary; ontology; classification; biomedical concepts; terminology

'When I use a word,' Humpty Dumpty said in rather a scornful tone, 'it means just what I choose it to mean – neither more nor less.'
'The question is,' said Alice, 'whether you CAN make words mean so many different things.'
'The question is,' said Humpty Dumpty, 'which is to be master – that's all.'

Lewis Carroll, *Through the Looking Glass.* 1862

1. INTRODUCTION

Medical concepts, by their nature, are complex notions. Patient descriptions about diagnoses or procedures often invoke levels of detail and chained attributes that pose complex computer-science problems for data representation. Confounding this mechanical complexity is the sheer scale and scope of concepts that can figure into medical thought, ranging from molecular variance to sociologic environments, aptly illustrated by Blois (1988) two decades past. This breadth can be aggravated by invoking concepts and terms that can probe the depths of present knowledge, bordering unto arcane realms of science and clinical practice with limited understanding and less experience. Furthermore, a serious tension remains between making such expressions readable and understandable by humans while attempting to address the increasing need for machine-interpretable expressions that leverage computerized knowledge and decision support. Finally, had we perfect knowledge of medicine, our patients, or the spectrum of sciences medicine invokes, the task of consistently representing patient information would be hard enough. Sadly, oftentimes we struggle with incomplete information, partial understanding, and flawed models. Managing this morass to address efficiently and effectively the multiple uses of medical concepts is hard (Rector, 1999).

1.1 Use-cases

A purely abstract discussion of medical concept representation is unbounded. Enumeration of neurotransmitter molecules on a motor endplate receptor bears comparison to angels on pin-heads for clinical purposes. On the other hand, excessive reductionism may suggest a need for little more than rudimentary collections of medical terms served up as pick-lists, though this perspective typically has more to do with hiding complexity than eliminating it.

Use-case definitions serve as a practical means for defining beginning assumptions and scope as well as bounding problem spaces. Thus to frame the context of this chapter with respect to medical concept representation, some broad-based use-cases are outlined.

1.1.1 Information Capture

The most clinically familiar circumstance of representing medical concepts is documenting patient findings, conditions, interventions, and outcomes. This documentation ranges from the unstructured dictations of progress notes and summaries to the thoughtful management of fully encoded problem lists, flow sheets, and encounter codes. Care providers thus assimilate information about patients, make inferences from patterns of observation, and re-express observations and conclusions. These expressions comprise a kind of concept representation, though not always formalized.

Medical concept representation typically implies a formal or at least machinable manifestation of clinical information. However, for practical purposes the field covers the full spectrum of information capture, including natural language. The problem of mapping natural language expressions to controlled terminologies is a topic unto itself. However, many of the challenges outlined in this chapter pertain equally to natural language expressions and more formal manifestations.

1.1.2 Communication

Information about patients, specimens, and experiments often needs to be transferred among providers or within a health care enterprise. Transmission media range from the non-machinable "fax" of text images to a highly structured clinical message conforming to the HL7* Version 3 information model. Typically, medical concepts inherit the characteristics associated with their capture, with respect to detail, formalization, and structure. However, standard message protocols such as those developed by HL7 may impose degrees of formalization that require transformation of medical concepts into highly structured representations.

Communication of medical concepts occurs for a purpose and requires that the recipient can use the information. When the recipient is a human being, seeking to read an historical medical record, these concepts may require minimal information. However, for the machine-processable transfer of electronic medical records between a referring physician and a tertiary medical center, as envisioned by the National Health Information Infrastructure (Yasnoff et al., 2004), a higher degree of interoperability is needed. Similarly, for the systematic processing of drug orders to avoid medical errors within an enterprise, a machine-interpretable representation of concurrent medications and medical problems must be achieved. This latter communication begins to overlap the formalization requirements for decision support, as described here.

* http://www.hl7.org

1.1.3 Knowledge Organization

The organization of medical knowledge is among the oldest applications of classification, dating to Aristotle's efforts in biology and formal descriptions (Pellegrin, 1986). The subsequent history (Chute, 1998; Chute, 2000) of medical concepts tells the story of increasingly detailed classifications from the haphazard collection of causes of death in the 16th century London Bills of Mortality (Graunt, 1939) to the emergence of large description logic-based terminologies (Baader et al., 2002) such as SNOMED CT†.

The explosion of modern biomedical ontologies (Smith and Rosse, 2004) provides what Alan Rector has described as a "conceptual coat-rack" for medical knowledge that knowledge authors and users find irresistible. Furthermore, the boundaries between representing concepts in an ontology using acyclic graphs and complex relations begin to blur the distinction between knowledge representation and concept organization. This realization was articulated nearly 40 years ago (Lindberg et al., 1968; Bloise, Tuttle, and Sherertz, 1981).

1.1.4 Information Retrieval

Most information sources have an indexing infrastructure that facilitates rapid and accurate retrieval. The oldest biomedical database that supports indexed retrieval is Medline/PubMed, for which the MeSH (Medical Subject Heading) vocabulary (Nelson et al., 2004) was created and is maintained. Virtually every user of the medical literature has encountered MeSH concepts, if only indirectly. Most user interfaces to literature retrieval tools translate natural text entries into MeSH concepts and then retrieve medical journal articles that have these MeSH codes or their hierarchical children (concept explosion or recursive subsumption).

For clinical data, classifications such as ICD-9-CM‡ serve an indexing role roughly corresponding to MeSH. However, ICD codes in most countries are applied for billing purposes and may not accurately reflect the underlying clinical content (Chute et al., 1996).

Whether to use natural language or coded data is an old question (Cote, 1983), though most modern practitioners recognize that any subsequent inferencing on retrieved information, using statistical regressions or machine learning techniques, must ultimately categorize or "bin" the data. Taken to the limit, such categorization defines concept classification systems. The

† http://www.snomed.org

‡ http://www.cdc.gov/nchs/about/otheract/icd9/abticd9.htm

topic of medical information retrieval is addressed more completely by Hersh in this volume and elsewhere (Hersh, 2003).

1.1.5 Decision Support

Helping clinicians make better decisions all the time is arguably the ultimate goal of computer-assisted decision support systems. However, in order for such systems to work, the knowledge resources that drive decision rules must share the terms and concepts used by clinicians to describe the patient. For example, decision rules made to operate on sulfa drugs may not "fire" if they do not recognize drug trade names (e.g., Bactrim®) as equivalent. Failure to recognize semantic equivalence is a more serious challenge when confronting the myriad expressions and terms that can describe a disease. This equivalence can be daunting when a concept is fully represented using terminology composition in one setting but constitutes a combination of terms in specific fields where the information model or field semantics modify meanings in another setting. The classic example of this circumstance is "family history of heart disease" vs. "heart disease" in a field labeled "family history."

The Arden Syntax[§], a popular standard for expressing medical logic modules and decision support rules, suffers from an incomplete specification of rule triggers and vocabulary semantics. Often called the "curly braces problem" (Choi, Lussier, and Mendoca, 2003) after the typographical brackets used to contain trigger concepts and rule-logic terms, implementers of a decision rule published in the Arden Syntax were left to their own devices to interpret exactly what events and codes in their own organization best correspond to the concepts within the curly braces. This semantic challenge highlights the importance of shared concept representation among logic-rule authors, implementers, and users.

2. CONTEXT

The settings of use often define as much about concepts as any surface form or text string might convey. The famous linguistic example of contextual syntax is "Time flies like an arrow, but fruit flies like a banana." This example illustrates the profound changes of meaning that context can have on words, terms, and expressions. The biomedical domain, while often more structured than general language, does not escape the influence of context on the representation or interpretation of concepts.

[§] http://www.hl7.org/Special/committees/Arden/arden.htm

The definitively cited work on context, language, and concepts remains Ogden and Richards' 1923 opus, *The Meaning of Meaning* (Ogden et al., 1923). They describe the classic "semantic triangle" which distinguishes a purely abstract thought or human concept, a referent object in the real world, and language symbols we might culturally share to refer to this concept. Invoking the Shakespearian metaphor of a "'rose' by any other name...," Ogden outlines that the shared cultural context of a rose – merely a pretty flower or a symbol of love – dictates its interpretation. These shared cultural assumptions are little different in health care, though comprehensive medical concept representations in the guise of a fully-specified HL7 message leaves less context to assumption and more to explicit assertion.

2.1 Concept Characteristics

Disease descriptions exist along many axes of characteristics, defining continua of expression. These axes have implications for managing concept representation and interpreting concept instances.

- Certainty – Clinicians document medical concepts throughout a care episode, including periods when they are unsure of their own speculations. Clinical assertions range from differential diagnoses, which include broad possibilities, probable but uncertain observations, to final diagnoses (though these too are often revised). Hence, many concepts in patient records may comprise more noise than fact.
- Etiologic Precision – Diagnostic statements are fraught with vagueness, syndromic generalization, and final common pathway manifestations attributable to multiple causes. Consider the label "congestive heart failure," which exhibits myriad etiologies though shared clinical outcomes. Many medical concepts exhibit substantial clinical heterogeneity. Contrasting such vagueness is the emergence of an increasing number of clinical characterizations that correspond to precise molecular variations (Scriver, 2001; McKusick), such as hemoglobinopathies or specific tumors. Indeed, the entire genomic revolution will inevitably transform our understanding of disease and etiology in a manner analogous to the effect of the germ theory of disease.
- Granularity (specificity) – Disease hierarchies are not just the province of classifications, but find expression in clinical descriptions. There is a profound difference between a problem list entry of "cancer" and one that specifies "Stage IIb squamous cell carcinoma of the right upper lobe with metastatic extension to the liver..." Reference to a "granular" description implies a detailed expression, often as a composition. More

specific terms can be distinguished without composition (e.g., the granular "aortic insufficiency" contrasting with "heart disease").

- Completeness – specific use-cases often determine how completely clinical descriptions are expressed. Routine outpatient office visits may exhibit a limited amount of disease detail when compared with the detail provided through an elaborate clinical trial protocol. The boundary between completeness and granularity is often determined by how the information is represented between vocabulary expressions vs. information model structures.

2.2 Domains

The professional language or jargon of medicine differs markedly from general English (or any other natural human language). While health professionals doubtless share substantial biomedical sub-language elements, there is important sub-specialization by medical specialty. Neonatologists do not fully share the language of psychiatrists; similar contrasts could be drawn between the language of cardiologists and pathologists, radiologists and clinical pharmacologists, rehabilitation specialists and oncologists, and so on. These distinctions define concept domains, although domains are by no means limited to differences in clinical sub-specialties.

A palpable way to illustrate these distinctions is to examine how certain specialties might disambiguate simple and common abbreviations. The following table expands differently the abbreviation MS by some domain specialties. What is remarkable is that these expansions occur consistently *within* domains, but almost always inconsistently *among* domains. The exercise is equally repeatable with MI, MR, and countless other abbreviations.

Table 6-1. Domain-specific expansion of "MS"

Domain specialty	"MS" abbreviation expansion
Cardiology	mitral stenosis
Neurology	multiple sclerosis
Anesthesia	morphine sulfate
Obstetrics	magnesium sulfate
Research science	manuscript
Physics	millisecond
Education	Master of Science
U.S. Postal Service	Mississippi
Computer science	Microsoft
Correspondence	female name prefix

Domain-specific term disambiguation is not restricted to abbreviations. The NLM's UMLS** contains six meanings for "cold." One is an abbreviation expansion (chronic obstructive lung disease). However, each of these meanings carries a unique concept identifier (CUI) within the UMLS that can be invoked to represent a context-independent statement. Furthermore, concepts can be fully expressed in language to avoid ambiguity, although most human interfaces find fully disambiguated text expansions tedious at best and sometimes insufferable. Using widely understood shorthand expressions *within* a domain for human consumption is a practice not likely to languish anytime soon.

2.3 Structure

The meaning of a term is as much influenced by the company it keeps (structural context) as by who uses it (domain). However, the expression of structural context has a dual nature in medical concept representation, as illustrated in the figure below. Specifically, highly detailed, granular vocabulary expressions can be composed which express a complex notion illustrated by the vocabulary composition view. Semantically identical assertions can be expressed using shorter vocabulary elements within a specific information model that conveys the additional semantics – in this case, the qualification of "family history."

When one begins to deal with more complex information models and more expressive vocabulary spaces, the problem worsens. The following table is adapted from material suggested by David Markwell of the UK at the inaugural TermInfo meeting held at National Aeronautics and Space Administration (NASA) in Houston, TX during August 2004. This series of meetings was convened to examine the spectrum of concept modeling that can exist between terminology models (such as SNOMED or GO) and information models (such as HL7 reference information model (RIM) or caBIO), and in particular where these models generate a semantic overlap. The table highlights alternative ways of modeling the same information by using HL7 RIM and the SNOMED CT context model.

Table 6-2. HL7 RIM and SNOMED CT Context Model

HL7 RIM	SNOMED CT Attribute
targetSiteCode(Observation)	"finding site"
targetSiteCode(Procedure)	"procedure site"
methodCode(Observation & Procedure)	"method"
approachSiteCode(Procedure)	"approach," "access"
priorityCode(Act)	"priority"

** http://www.nlm.nih.gov/research/umls/

The conclusion, almost inescapably, is that there is no one correct way to represent complex medical concepts. Invoking higher-level information models such as the HL7 RIM or even just a "family history" box has equivalent validity and semantics to composition expressions built using vocabulary models and syntax. If both are valid, then what is the problem?

The resolution of complex, semantically equivalent expressions that differ in their allocation of meaning to an information model or compositional vocabulary expressions is difficult. Establishing semantic equivalence between such hybrid representations – or even their purely modeled archetypes of complete information model or vocabulary expression – is an under-developed research problem. Few solutions exist, and none scale to the scope of problems encountered in real-world clinical expressions. The practical implication is that virtually all use-cases that require communication or consistent recognition of content by a recipient (as in decision support) will fail, should care not be taken to negotiate the allocation of semantics between information and vocabulary models.

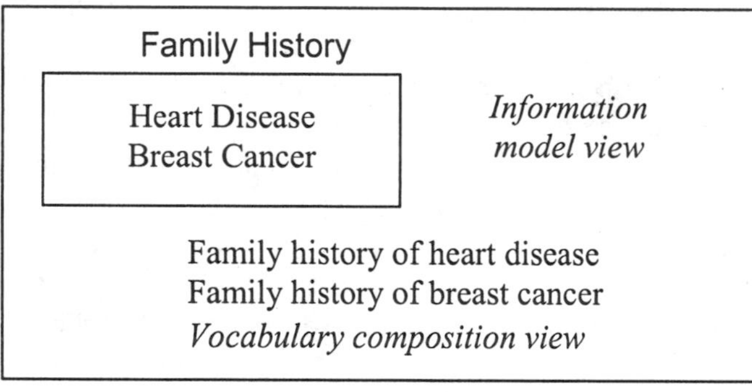

Figure 6-1. Information model view and vocabulary composition view.

3. BIOMEDICAL CONCEPT COLLECTIONS

3.1 Ontologies

Philosophers recoil at the pluralization of ontology. Originally, the term ontology referred to the consideration of what kinds of entities comprise reality. Computer scientists, in the era of artificial intelligence and knowledge representation, co-opted the term to mean an organization of concepts in domains, which might encompass medical concepts or enumerations and relations among Boeing 777 aircraft parts. Gradually, criteria discriminating formal ontologies from ordinary hierarchies of concepts included the requirement that ontologies exhibit internal consistency, acyclic polyhierarchies, and computable semantics. Within medicine, the pioneering work of Rector and the Galen project (Rector, Nowland, and the Galen Consortium, 1993) illustrated how medical concepts could be represented as a formal ontology and demonstrated applications where this formalized representation mattered (Rector, Rossi Mori, and Consorti, 1993).

Modern biomedical ontologies are becoming synonymous with concept collections assembled using description logics (Baader and Nutt, 2002). The venerable SNOMED has evolved to incorporate description logics which include role restrictions (Spackman et al., 2002). Concept collections pertaining to basic biology, such as the Gene Ontology, while criticized for lacking many formalizations now expected of terminology bearing the ontology banner (Smith, Williams, and Schulze-Kremer, 2003), are also evolving to become a semantically computable resource (Wroe et al., 2003).

The novel promise of ontologies is their ultimate connection as a distributed system of interlocking conceptual schema. For example, when a LOINC term invokes a drug within a drug-sensitivity evaluation, semantic interoperability is enhanced if there emerges agreement that the NDF-RT (Chute et al., 2003) (National Drug Formulary – Reference Terminology; a UMLS source vocabulary) would form the basis for drug references. Similarly, the designation of a common anatomy terminology that could span the spectrum of use-cases across biology and medicine would greatly enhance our ability to consistently create and interpret biomedical concepts; a leading candidate for a common anatomical ontology appears to be the Foundational Model of Anatomy developed by Smith and Rosse (2004).

The development and deployment of ontologies is being greatly accelerated by the emergence and adoption of the Protégé ontology editor developed by Musen and colleagues (Noy et al., 2003). Closely coupled with this tooling is agreement upon an expression syntax for description logics, published by the WC3. Called OWL (the Ontology Web Language), this standard (McGuinness and Harmelen, 2004) was approved only recently but has already penetrated the ontology authoring community almost completely. As of this writing, the OWL extensions to Protégé (Knublauch, 2004) provide the best authoring and editing environment for ontology development available anywhere; that this NIH-funded effort is now available[††] with an open-source license should further accelerate the quality and number of well-formed ontologies in biomedicine.

[††] http://protege.stanford.edu; funded in part by NIH grant P41-LM07885 to Mark Musen, MD PhD

3.2 Vocabularies and Terminologies

Cimino provides a detailed description of vocabularies and terminologies in this volume. For the purposes of this discussion, it is useful to distinguish vocabularies and terminologies from ontologies. Simplistically, vocabularies and terminologies are less formal than ontologies, uniformly lacking logical descriptions that serve to computationally define terms. As a practical matter, most large ontologies contain a sizable fraction, if not a majority, of "primitive" terms undefined by description logic formalisms – terminologies remain the major mode for biomedical terminologies, if not at some levels the exclusive mode for the present.

There is no commonly accepted distinction between a vocabulary and terminology, though many adherents in the field might suggest that terminologies have associated codes and hierarchies while a simple vocabulary may comprise little more than a bag of words. However, invocation of the moniker "controlled vocabulary," which may imply more formality than exists in a terminology, renders this tenuous differential inconsistent.

Vocabularies and terminologies are often described by intended role, though few adhere to these role expectations. The most common distinctions among terminology uses are:

- Entry Terminologies: specifically constructed to provide familiar and common terms and phrases readily recognized by humans. These term collections often sacrifice precision and rigor in favor of familiarity and jargon.
- Reference Terminologies: semi-formal representations of terms and concepts intended for machine interpretation.
- Administrative Terminologies: higher-level classifications which aggregate clinical findings for particular administrative purposes.

Common vocabularies and terminologies include LOINC, CPT (Current Procedural Terminology), HL7 Vocabulary Tables (over 100 of them), and NDC drug codes. People familiar with these systems can recognize wide variations in structure, quality, and consistency of these concept libraries. What makes a terminology good is beyond the scope of this work, though systematic evaluations of common terminologies exist (Chute et al., 1996; Campbell et al., 1997), as do generalized discussions of what makes a good terminology (Chute, Cohn, and Campbell, 1998; Cimino, 1998). The reality remains that most terminologies fail to adhere to good design principles, suffering from the recycling of abstract codes to inconsistent hierarchies to ambiguous groupings of concepts.

3.3 Aggregation and Classification

Historically, significant tension existed between the terminology and classification communities (Cote, 1983; Ingenerf and Giere, 1998), with each maintaining the advantages of their use-case. However, recent thinking has established their mutual advantages along a continuum of granularity or specificity (Chute, Cohn, and Campbell, 1998). Medical information, by its nature, is highly detailed. Hence a need for concept systems or terminologies that can reasonably capture highly detailed information will always exist. On the other hand, many use-cases work best with highly grouped data. Examples include public-health statistics, reimbursement categories, or administrative groupings of patients.

High-level aggregation systems, such as ICD-9-CM, have been unjustly criticized for not having enough granularity to function in decision support or clinical retrieval use-cases. The complaint is accurate but the criticism unjustified because high-level classifications such as ICD-9-CM were *never intended* to function as detailed terminologies. If criticism is to be made, it should be of Electronic Health Record (EHR) vendors and most implementing providers who insist on using ICD-9-CM for use-cases such as patient problem lists and clinical decision support triggers that more properly demand detailed terminologies, such as SNOMED-CT.

However, the specter of double-coding clinical findings, diagnoses, procedures, or adverse events, once in a detailed terminology and again in a required or mandated classification, reasonably discourages best-coding practice. Few providers have the resources to appropriately code cases for reimbursement and quality oversight, never mind code them again for clinical applications. Early drafts of the PITAC (President's Information Technology Advisory Committee) Report on Health Information Technology went so far as to suggest that providers code just once, in a detailed terminology, and that secondary re-use of clinical data be facilitated by appropriate mapping to requisite classifications, such as the newly drafted ICD-10-CM. The final version (U.S. President's Information Technology Advisory Committee, 2004) of this report, however, provided a much more balanced perspective on the important roles that high-level classifications can play, coupled with the many practical difficulties of accurately mapping detailed clinical terms to complex classifications.

Kent Spackman, editor of SNOMED-CT, proposed (personal communication) that mapping from detailed terminologies to complex classification would provide more reliable and consistent coding. However, he points out that to be done correctly, the coding rules of a classification, such as ICD-9-CM, must be made explicit and machineable. Most classifications rely on indentations, typographic conventions, index entries, and established professional coding lore as a basis for conveying the rules of coding. These rules can be quite elaborate, specifying complex inclusion and exclusion criteria for assignment to a specific code. As a simple example, pre-eclampsia is distinguished from ordinary hypertension in ICD-9-CM by obviously requiring female gender, pregnancy, and renal involvement.

These coding rules could define "Aggregation Logics," and should be published as machine-readable logic rules by developers of classifications. The analogy is often made to "Grouper" rules, by which collections of ICD codes are grouped into higher-level DRGs (Diagnostic Related Groups) by computer algorithms. Aggregation Logics would fill the gap between detailed clinical expressions and the intermediate classifications, such as ICD-9-CM or ICD-10-CM, when it becomes adopted. The point is to avoid duplicate coding by providers, and consistent with Spackman's assertion, likely provide more reliable and consistent coding into ICD-level classifications.

3.4 Thesauri and Mappings

3.4.1 The UMLS

No discussion of biomedical terminology and concept representation would be complete without mention of the Unified Medical Language System (*op. cit.*). Originally intended to serve as a Rosetta Stone to suggest translations among terminologies (Humphreys and Lindberg, 1989), it has taken a more practical role as the major semantic thesaurus of biomedical terms. The UMLS is comprised of over 100 separate terminology sources, including SNOMED CT, MeSH, and ICD-9-CM. However, it does not contain formal description logic assertions across terms from different vocabularies, though hierarchical assertions, broader/narrower relations, and "other" relationships are meticulously mapped and curated by human editors for the entire corpus.

The 2004 release of the UMLS Metathesaurus saw the most dramatic change in the file structures and formats of the UMLS since its original release in 1988. To accommodate the complex description logic assertions of SNOMED CT, the NLM introduced a Rich Release Format (RRF) (National Library of Medicine, 2003), which for the first time promised "source transparency." The intention was to permit users of the UMLS to extract terminologies from the Metathesaurus in a format that would transparently reflect the original content of a particular terminology. Previously, the UMLS formatting process resulted in a "lossy" information transfer. The modern vision of the UMLS, to become the definitive source and publication format for major biomedical terminologies, is thereby greatly advanced.

3.4.2 Word-Level Synonymy

An emerging requirement for natural language thesauri is not presently served by the UMLS, though it is approximated for general English by resources such as WordNet (Fellbaum, 1998). Consider the retrieval use-case for Renal Cancer when data may have been recorded as Kidney Cancer. The UMLS happens to include explicit English synonyms that map these two phrases, but this is not the case for all word-level synonyms and permutations that one might imagine in biomedical concepts.

The public sharing of word-level concept clusters has been widely proposed (Solbrig et al., 2000), and indeed some generalized methods for creating and evolving them have been explored (Pakhomov, Buntrock, and Chute, 2004). The broad creation, shared maintenance, and coordinated use of consensus-driven thesauri of common synonyms will be a great advance toward linking phrases entered by providers with elements of controlled terminologies. These resources, in a second generation of curation, could also include degrees of pleisionymy. Ultimately, these thesauri can be married with ontologies and terminologies to provide a horizontal (synonym) and vertical (terminologies) component to medical concept representation and retrieval (Chute, 2002).

4. STANDARDS AND SEMANTIC INTEROPERABILITY

Medical concepts, once expressed, must be understood by people or machines. The context of concept assertion can overlay additional semantics that must be understood. Fully specified messaging environments, such as HL7 or caBIO[‡‡], can carry sufficient information to explain context, but there is no replacement for agreed-upon content standards, to wit common vocabularies.

In the United States, the Federal eGov initiatives have spawned the Consolidated Health Informatics (CHI) set of standards[§§]. Working in concert with the NCVHS, the CHI working groups have proposed terminology and interchange standards that would be required for use among US Federal agencies. Intended to define a critical-mass tipping point for the

[‡‡] http://ncicb.nci.nih.gov/core/caBIO

[§§] http://www.whitehouse.gov/omb/egov/gtob/health_informatics.htm

general US health care economy, the explicit intention is that such Federal leadership would define de facto a common basis for content standards. While the CHI proposals are still new as of this writing, the intended effect appears to be taking place. Evidence for this can be seen in the NLM contract to HL7 (HHS N276 2004 43505C) to ensure that all HL7 vocabulary tables are CHI-compliant.

One may conclude that substantial progress and tangible resources have emerged in the past few years to support the consistent and comparable representation of medical concepts for a broad spectrum of use-cases. The rapid adoption of ontology languages such as OWL, their subsequent availability in high fidelity within the UMLS, and the active negotiation and specification of what contextual information belongs in an information model vs. a terminology model bring increasing problems of robust solutions. The common use of highly detailed and semantically coherent medical messages and retrievals is not yet realized, but progress has been dramatic in the past five years. The clichéd refrain that more work needs to be done certainly pertains, but that work is now more palpably satisfying and is vectoring toward consensus solutions and practical standards specifications.

5. ACKNOWLEDGEMENTS

I am grateful to Harold Solbrig, Mark Musen, and Alan Rector for fruitful discussions that contributed to my understanding of these issues. This work is funded in part by R01 LM007319.

REFERENCES

Baader, F., Calvanese, D., McGuinness, D. L. et al. (Eds.). (2002). *The Description Logic Handbook: Theory, Implementation and Applications*, Cambridge University Press.

Baader, F. D. and Nutt, F. W. (2002). "Basic Description Logics," in F. Baader et al. (Eds.), *The Description Logic Handbook*, Cambridge University Press, 47-100.

Blois, M., Tuttle, M., and Sherertz, D. (1981). "RECONSIDER: A Program for Generating Differential Diagnoses," in *Proceedings of the 5th Annual Symposium on Computer Applications in Medical Care*, Washington, D.C., Nov. 1-4, 1981, 263-268.

Blois, M. S. (1998). "Medicine and the Nature of Vertical Reasoning," *New England Journal of Medicine*, 318(13), 847-851.

Campbell, J. R., Carpenter, P. C., Sneiderman, C. et al. (1997). "Phase II Evaluation of Clinic Coding Schemes: Completeness, Taxonomy, Mapping, Definitions, and Clarity," *Journal of the American Medical Informatics Association*, 4(3), 238-251.

Choi, J., Lussier, Y. A., and Mendoca, E. A. (2003). "Adapting Current Arden Syntax Knowledge for an Object Oriented Event Monitor," in *AMIA Annual Symposium Proceedings*, 814.

Chute, C. G. (2002). "The Horizontal and Vertical Nature of Patient Phenotype Retrieval: New Directions for Clinical Text Processing," in *Proceedings of the AMIA Annual Fall Symposium*, 165-169.

Chute, C. G. (2000). "Clinical Classification and Terminology: Some History and Current Observations," *Journal of the American Medical Informatics Association*, 7(3), 298-303.

Chute, C. G. (1998). "The Copernican Era of Healthcare Terminology: A Re-centering of Health Information Systems," in *Proceedings of the AMIA Symposium*, 68-73.

Chute, C. G., Carter, J. S., Tuttle, M. S. et al. (2003). "Integrating Pharmacokinetics Knowledge into a Drug Ontology: As an Extension to Support Pharmacogenomics," in *Proceedings of the AMIA Symposium*, 170-174.

Chute, C. G., Cohn, S. P., and Campbell, J. R. (1998). "A Framework for Comprehensive Health Terminology Systems in the United States: Development Guidelines, Criteria for Selection, and Public Policy Implications. ANSI Healthcare Informatics Standards Board Vocabulary Working Group and the Computer-Based Patient Records Institute Working Group on Codes and Structures," *Journal of the American Medical Informatics Association*, 5(6), 503-10.

Chute, C. G., Cohn, S. P., Campbell, K. E. et al. (1996). "The Content Coverage of Clinical Classifications," *Journal of the American Medical Informatics Association*, 3(3), 224-233.

Cimino, J. J. (1998). "Desiderata for Controlled Medical Vocabularies in the Twenty-First Century," *Methods of Information in Medicine*, 37(4-5).

Cote, R. (1983). "Editorial: Ending the Classification Versus Nomenclature Controversy," *Medical Informatics*, 8(1), 1-4.

Fellbaum, C. (1998). "WordNet: An Electronic Lexical Database," in *Language, Speech, and Communication*, Cambridge, Mass: MIT Press.

Graunt, J. (1939). *Natural and Political Observations Made Upon the Bills of Mortality; London, 1662,* Baltimore, MD: The Johns Hopkins Press.

Hersh, W.R. (2003). *Information Retrieval: A Health and Biomedical Perspective*, New York: Springer.

Humphreys, B. L. and Lindberg, D. A. B. (1989). "Building the Unified Medical Language System," in *Symposium on Computer Applications in Medical Care*, 13, 475-480.

Ingenerf, J. and Giere, W. (1998). "Concept-oriented Standardization and Statistics-oriented Classification: Continuing the Classification Versus Nomenclature Controversy," *Methods of Information in Medicine*, 37(4-5).

Knublauch, H. (2004). "The Protégé OWL Plugin," in *7th International Protégé Conference*, Bethesda, MD, http://protege.stanford.edu/conference/2004/index.html

Lindberg, D. A. B., Rowland, L. R., Buck, C. R. et al. (1968). "CONSIDER: A Computer Program for Medical Instruction," in *Proceedings of the 9th IBM Med. Symposium*, White Plains, New York: IBM.

McGuinness, D. L. and Harmelen, Fv. (2004). "*OWL Web Ontology Language: Overview*," W3C, http://www.w3.org/TR/owl-features/

McKusick, V. A. *OMIM - Online Mendelian Inheritance in Man*, Bethesda: National Center for Biotechnology Information, NIH/NLM, http://www.ncbi.nlm.nih.gov/entrez/query.fcgi?db=OMIM

National Library of Medicine. (2003). *MLS Metathesaurus Rich Release (MR+) Format*, Bethesda, MD: National Institutes of Health, http://www.nlm.nih.gov/research/umls/white_paper.html

Nelson, S. J., Schopen, M., Savage, A. G. et al. (2004). "The MeSH Translation Maintenance System: Structure, Interface Design, and Implementation," in *Medinfo*, 67-9.

Noy, N. F., Crubezy, M., Fergerson, R. W. et al. (2003). "Protege-2000: An Open-source Ontology-Development and Knowledge-Acquisition Environment," in *Proceedings of the Annual AMIA Symposium*, 953.

Ogden, C. K., Richards, I. A., Malinowski, B. et al. (Eds.) (1923). *The Meaning of Meaning: A Study of the Influence of Language Upon Thought and of the Science of Symbolism*, London:Routledge & Kegan Paul.

Pakhomov, S. V., Buntrock, J. D., and Chute, C. G. (2004). "Using Compound Codes for Automatic Classification of Clinical Diagnoses," in *Medinfo*, 411-415.

Pellegrin, P. (1986). *Aristotle's Classification of Animals: Biology and the Conceptual Unity of the Aristotelian Corpus*, Berkeley: University of California Press.

Rector, A. L. (1999). "Clinical Terminology: Why is it So Hard?" *Methods of Information in Medicine*, 38(4-5), 239-252.

Rector, A. L., Nowland, W., and The Galen Consortium. (1993). "The GALEN Project," *Computer Methods and Programs in Biomedicine,* (45), 75-78.

Rector, A. L., Rossi Mori, A., Consorti, M. F. et al. (1998). "Practical Development of Re-usable Terminologies: GALEN-IN-USE and the GALEN Organization," *International Journal of Medical Informatics*, 48(1-3), 71-84.

Scriver, C. R. (2001). *The Metabolic and Molecular Bases of Inherited Disease*, 8th ed. New York: McGraw-Hill. 4 vols, (xlvii, 6338, I-140 p.).

Smith, B. and Rosse, C. (2004). "The Role of Foundational Relations in the Alignment of Biomedical Ontologies," in *Medinfo*, 444-448.

Smith, B., Williams, J., and Schulze-Kremer, S. (2003). "The Ontology of the Gene Ontology," in *AMIA Annual Symposium Proceedings*, 609-613.

Solbrig, H., Elkin, P., Ogren, P. et al. (2000). "A Formal Approach to Integrating Synonyms with a Reference Terminology," in *Journal of the American Medical Informatics Association Symposium Supplement.*

Spackman, K. A., Dionne, R., Mays, E. et al. (2002). "Role Grouping as an Extension to the Description Logic of Ontylog, Motivated by Concept Modeling in SNOMED," in *Proceedings of the AMIA Symposium*, 712-6.

United States. President's Information Technology Advisory Committee, and the National Coordination Office for Information Technology Research and Development. (2004). "Revolutionizing Health Care through Information Technology Report to the President," Arlington, VA: National Coordination Office for Information Technology Research and Development, http://www.itrd.gov/pitac/reports/20040721_hit_report.pdf

Wroe, C.J., Stevens, R., Goble, C. A. et al. (2003). "A Methodology to Migrate the Gene Ontology to a Description Logic Environment Using DAML+OIL," in *Pacific Symposium on Biocomputing*, 624-635.

Yasnoff, W. A., Humphreys, B.L., Overhage, J. M. et al. (2004). "A Consensus Action Agenda for Achieving the National Health Information Infrastructure," *Journal of the American Medical Informatics Association*, 11(4), 332-338.

SUGGESTED READINGS

Baader, F., Calvanese, D., McGuinness, D. L. et al. (Eds.) (2002) *The Description Logic Handbook: Theory, Implementation and Applications*, Cambridge University Press.

The definitive textbook which outlines the history and current state-of-the-art for Description Logics. Since Description Logics form the basis of modern ontologies, familiarity with this technology is increasingly required for mastery of concept representation.

Rector, A. L. (1999). "Clinical Terminology: Why is it so Hard?" *Methods of Information in Medicine,* 38(4-5), 239-252.
An outstandingly concise and complete exposition on the terminology problem in health care, effectively refuting commonly held expectations that health terminology should be trivial.

Chute, C. G. (2000). "Clinical Classification and Terminology: Some History and Current Observations," *Journal of the American Medical Informatics Association*, 7(3), 298-303.
A brief history of medical classification and description, providing background and context for the evolution of thinking and practice in health classifications through the last century.

ONLINE RESOURCES

http://umlsks.nlm.nih.gov/
The home side of the NLM's Unified Medical Language Systems

http://informatics.mayo.edu
The specification and open-source for the LexGrid project, terminology editor, and Common Terminology Services (from HL7).

http://protege.stanford.edu
The most widely used ontology editor, Protégé, and related resources.

http://www.co-odc.org/
The Collaborative Open Ontology Development Environment home page, including tutorials and resources

QUESTIONS FOR DISCUSSION

1. What are the relative roles of terminology models and information models in representing complex medical expressions?

2. What is the distinction between representing information and aggregating information? Specifically, what are the relative roles and relationships among terminologies and classifications?

3. How might a spectrum of secondary data uses, such as decision support, quality improvement, biomedical research, or administrative aggregation, impact information representation and display?

4. How might the retrieval of information be affected by differing ways of representing it? Specifically include discussion of granularity, detail, aggregations (lumping), or context?

Chapter 7

CHARACTERIZING BIOMEDICAL CONCEPT RELATIONSHIPS

Concept Relationships as a Pathway for Knowledge Creation and Discovery

Debra Revere and Sherrilynne S. Fuller

Department of Medical Education and Biomedical Informatics, University of Washington, Seattle, WA 98195

Chapter Overview

The importance of biomedical concept relationships and document concept interrelationships are discussed and some of the ways in which concept relationships have been used in information search and retrieval are reviewed. We look at examples of innovative approaches utilizing biomedical concept identification and relationships for improved document and information retrieval and analysis that support knowledge creation and management.

Keywords

concept relationships; document representation; information management; information search and retrieval; indexing systems; textual analysis

The process of tying two items together is the important thing.

Vannevar Bush, *As We May Think*, 1945

1. INTRODUCTION

Advances in the biomedical sciences have been accompanied by an overwhelming increase in the biomedical literature. It has become critically important to not only understand developments in one's own area of specialization, but to also be able to learn quickly about developments in related and occasionally unrelated subject areas. The ability to rapidly survey the literature and integrate information gathered by researchers from multiple fields of expertise constitutes the necessary first step toward enabling biomedical scientists and researchers to keep current in their field.

Interest in developing techniques and methods for processing documents and document collections, with the goal of providing the information most relevant to a user's need, pre-dates widespread use of computers. However, these efforts have accelerated as the wealth of scientific literature expands alongside the need to uncover information that is already present in large and unstructured bodies of text, commonly referred to as "non-interactive literatures" (Swanson and Smalheiser, 1997); i.e., literatures that do not cite each other but which, nevertheless, together present useful new information. In addition, the sequencing of the human genome has provided intensive impetus for developing effective tools to identify interrelated concepts and roles such as gene-disease connections and gene-drug interactions from the published literature, as well as from a variety of other types of databases.

This chapter will first explore the importance of biomedical concept relationships and document concept interrelationships, and some of the ways in which concept relationships have been used in information search and retrieval. We will then review a variety of approaches that have been used to represent biomedical concept relationships, beginning with the early concept identification systems developed in the 1950's. Finally, we will look at examples of innovative approaches utilizing biomedical concept identification and relationships for improved document and information retrieval and analysis that support knowledge creation and management.

Before continuing further, a few definitions are in order. A concept can be considered the atom or smallest unit of any knowledge domain or discipline. However, concepts do not exist in isolation; they occur in complex, multidimensional networks that represent "real world" relationships. For the purposes of this chapter, we are using the term "relationship" to denote a semantic association between two or more identified concepts. For example, some typical relationships include: *concept A* "is caused by" *concept B*; *concept A* "is associated with" *concept B*; and *concept A* "is a part of" *concept B*. We will explore the utility of biomedical concept relationships for improving document and information retrieval and analysis, both within individual documents and among document sets.

Although most of us have a common understanding of the term *relationship*, it is often difficult to explain what appears to be implicit in meaning, even though concept relationships are "...an integral part of the very foundation on which we build and organize our knowledge and understanding of the world in which we live. If concepts are seen as the basic building blocks of conceptual structure, then relationships are the mortar that holds it together" (Green et al., 2001).

The idea that concepts are related to one other is quite useless without knowing the meaning of the relationship. And before a relationship can be identified, "we must be able, first, to designate all the parties bound by the relationship and, second, to specify the nature of the relationship" by identifying the entities that participate in the relationship and the semantics and properties of the relationship (Green, 2001).

Yet, there is a need to be specific and precise when exploring what relationships are, how they are defined and how they can be represented. For example, in the field of mathematics, relational operators such as equals (=), less than (<) or greater than (>) express specific and precise meanings that are well understood by those who are familiar with numbers and mathematics. It would be ideal if our knowledge of relationships in other fields could be interpreted at the same level of precision.

Consider the following scenario: An Alzheimer Disease (AD) researcher is investigating the beneficial effects of caffeine ingestion to slow memory impairment. She knows that caffeine, like adenosine A(2A) receptor antagonists, blocks ß-amyloid–induced neurotoxicity in some rat models for AD. She also knows that caffeine has been shown to improve memory deficits in rat models for Parkinson's Disease (PD). The researcher wants to know if there is a relationship between the protective effects of caffeine consumption and adenosine A2A receptor antagonists for AD patients. She needs to know the level of caffeine dosage ingested over what time period, possible negative and positive associations of caffeine with other neurodegenerative diseases and association of caffeine with other conditions found in an elderly population, such as stroke, high blood pressure, etc. She also wants to know if treatment combining caffeine and adenosine A2A receptor blockers might further slow memory impairment. Using the PubMed®[1] search interface, she searches the MEDLINE® database. Maintained by the National Library of Medicine® (NLM®), MEDLINE

[1] *http://www.ncbi.nlm.nih.gov/entrez/query.fcgi*

contains over 14 million citations to biomedical articles with over 2000 citations added weekly.[2]

The researcher conducts numerous searches using the following terms: *Alzheimer's disease*, *memory*, *caffeine*, *neurodegenerative diseases* and *adenosine A2A receptor antagonists* (hereafter referred to as A2A blockers). Her first PubMed searches on "neurodegenerative diseases" and "Alzheimer's disease" respectively yield 119,829 and 40,785 potential documents. Searching on "A2A blockers" and "caffeine" each retrieve 1209 and 18,965 citations. The researcher then conducts searches combining terms. Combining the terms "memory" and "caffeine" retrieves 173 citations; combining "caffeine" and "A2A blockers" yields 94 citations. Next she tries "neurodegenerative disease" and "caffeine" (93 citations), "neurodegenerative disease" and " A2A blockers " (56 citations) and "Alzheimer's disease" and "caffeine" (21 citations). The researcher then combines "Alzheimer's disease," " A2A blockers " and "caffeine" which yields 9 citations—of which, after careful examination, only 3 appear to present actual answers to her questions.

This scenario includes all of the classic information science problems: precision (a measure of the number of relevant documents as a fraction of all the documents retrieved by the system); recall (a measure of the number of documents useful to the user as a fraction of all the relevant documents retrieved); "aboutness" (subject of the document); and vocabulary control. Traditional search engines and bibliographic database search-and-retrieval systems operate on retrieving a set of documents that reflect only one relationship: a similarity of content matching the keyword or subject terms in the user's query using a basic Boolean keyword retrieval (a query using the Boolean operators "and," "or" and "not"). The implicit interpretation is that there is an equivalent relationship between the user's concept and the document citations retrieved that represent the concept; i.e., the list of citations serves as a surrogate for the requested concept. While this relationship can be presumed and considered useful some of the time, it usually delivers a retrieval set that falls far short of the user's information need and often overwhelms the user with many irrelevant documents.

So the researcher's questions remain unanswered: How much and over what period of time must caffeine be consumed to slow memory loss? Will a combination of caffeine and adenosine A2A receptor antagonists shorten that period of time? Do the neurotoxicity-blocking effects of caffeine and adenosine A2A receptor blockers also come into play with other neurodegenerative diseases? What about possible negative associations with other conditions?

[2] For more information see *http://www.nlm.nih.gov/pubs/factsheets/pubmed.html*

When the researcher entered her terms or term pairs in PubMed, the searching algorithm used the standard keyword approach that counted the words in a query, looked for the presence of terms that matched the query in the database's bibliographic information (i.e., the abstract, title, keywords). While both terms may have been present in the same document, their inclusion in the retrieval set did not necessarily indicate that the two concepts had been studied *in relation* to one another. The term "caffeine" may be present in the abstract as part of an explanation of previous research. The indexer may have included "Alzheimer's disease" in the keywords because it was mentioned in the article introduction. Although the terms "caffeine" and "A2A receptor" are listed as keywords for a document, this does not guarantee that they were studied in relation to one another.

Most approaches to indexing and retrieval of documents to date have not exploited the structure of the document itself as a way of more precisely characterizing biomedical relationships. In addition, most biomedical literature mining has been performed on title and/or abstract words rather than all the words (full-text) in the documents. The researcher's primary information need is to know what was studied and the results of that study. This is data normally located in the Methods and Results sections of a biomedical research report—not necessarily evident in the citation, abstract, subject headings or keywords assigned to the document. Identifying the concept relationships that were studied and reported in the research document is a critical means of matching the biomedical literature user's questions with relevant literature, providing a way to rapidly review, integrate and inter-relate concepts of interest.

2. BACKGROUND AND OVERVIEW: THE USE OF CONCEPT RELATIONSHIPS FOR KNOWLEDGE CREATION

The thesaurus is a key tool developed by information science researchers for displaying the logical, semantic relationships among terms and rules for establishing compilations of terms to denote concepts and concept relationships. Comprised of the specialized vocabulary of a discipline or field of study, the thesaurus is a list of preferred terms to indicate two types of relationships between pairs of terms:

- synonyms, i.e., which of two or more equivalent terms can represent a concept, commonly denoted by Use or UseFor relationships; and
- hierarchical relationships, i.e., broader and narrower terms (parent/child relationships) and association (related terms, such as close siblings).

A more sophisticated type of thesaurus is the ontology, defined as an explicit, formal, systematic specification of all the categories of objects, concepts and other entities in a field or domain; the relations between these categories; and the properties and functions needed to define the objects and specify their actions (see Chapter 8, "Biomedical Ontologies," for a detailed description). Ontologies use rich semantic relationships among terms and strict rules about how to specify terms and relationships.

A key biomedical language resource is the Unified Medical Language System® (UMLS)®, developed by the NLM to overcome information retrieval (IR) problems caused by differences in biomedical terminology.[3] The UMLS consists of three multipurpose knowledge sources that together provide structured representation of concepts and relationships in the biomedical domain:

1. The UMLS Metathesaurus®, a large, multi-purpose, multi-lingual specialized vocabulary database that contains information about biomedical and health-related concepts, their various names and the semantic relationships between concepts.
2. The Semantic Network, a consistent categorization of all concepts represented in the UMLS Metathesaurus and to provide a set of useful relationships between these concepts.
3. The SPECIALIST Lexicon, a general English lexicon that includes biomedical vocabulary and a lexical entry that records each term's syntactic, morphological, and orthographic information. The lexical entry is of critical importance to natural language processing (NLP) systems.

One of the more thorough reviews of concept and document relationships occurred at the 1997 ACM/SIGIR workshop "Beyond Word Relations." Participants examined a number of relationship types as possibly significant for IR systems, beyond the traditional topic-matching relationship. The workshop proposed seven relationship types that could prove useful in IR systems:

1. Word-based relationships: documents that share the same vocabulary or word;
2. Attribute-based relationships: relationships based on shared characteristics (e.g., documents *A* and *B* share same author);
3. Document-document hierarchical relationships: situations in which one document is a sub-set or super-type of the other (e.g., document *A* is an appendix or sub-piece of document *B*);
4. Document-document topological relationships: a conceptual extension to the hierarchical relationship, this includes relationships that denote

[3] *http://www.nlm.nih.gov/pubs/factsheets/umls.html*

conceptual equivalence (e.g., document *A* is a translation of document *B*), commentary (e.g., document *A* updates document *B*), etc.

5. Document-to-document influence relationships: situations in which one document has affected the writing of another (e.g., document *A* builds on the work of document *B*);
6. Topic-based (or meta-topic based) relationships: this type includes the traditional topic-matching relationship, as well as situations in which documents are related but through less obvious topical resemblances (e.g., "non-interactive literatures" as mentioned earlier);
7. Usage-based relationships: documents that are related through the use of the documents, as in a user profile (Hetzler, 1997).

In addition to issues related to capturing a variety of types of concept relationships, the importance of document structure to indexing strategy, the difficulty of translating a user's information need into a query that can retrieve relevant and useful document representations (whether bibliographic citations or full-text documents) and the problem of how to represent biomedical concept relationships have been under investigation for many years. Our overview will cover various approaches to biomedical literature data mining that focus on utilizing concept relationships, including indexing and vocabulary strategies, information extraction (IE), NLP, text mining and literature-based discovery IR—however, much of the work has narrowly focused in the genome sciences domain. While we provide examples of methods and systems that represent each approach, much more work has been published in this area than can be referenced here.

2.1 Indexing Strategies and Vocabulary Systems

Traditionally, subject access to information has been provided in two ways: (1) classification, the process of describing the subject of an information object (its "aboutness") so it can be uniquely distinguished from all other items and (2) indexing, the process of assigning terms from a controlled vocabulary list (a collection of preferred terms that are used to assist in more precise retrieval of content).

There are two kinds of relationships between terms: semantic and syntactic. Semantic relationships are by definition permanent relationships; they exist independently of document content. For example, the concepts *animal* and *mammal* are related regardless of content. Syntactic relationships, however, consist of otherwise unrelated concepts that are brought together because of the document "space" they share. These relationships are not permanent.

An early innovation in information science research was the coordinate index—a list of terms that can be combined when indexing or searching a body of literature—which developed into subject-based terminology lists and, more recently, into thesauri and ontologies. In 1952, Taube pioneered the development of IR systems with his invention of the post-coordinate indexing system for subject retrieval. Post-coordinate indexing is the assigning of single concept terms from a controlled vocabulary to a record so that the user is able to "coordinate" or combine the terms using any combination of those concepts in any order when searching (Taube, 1953-57). Post-coordinate index can minimize the number of entries necessary to index all the concepts in a work.

An example of post-coordinate indexing is the MEDLINE database. Documents in the MEDLINE database are indexed using the Medical Subject Headings (MeSH) vocabulary, a post-coordinate indexing strategy. The user can select individual, indexer-assigned concepts from the MeSH controlled vocabulary and combine them with Boolean operators.

A more sophisticated indexing approach to capture biomedical relationships is pre-coordinate indexing, in which terms from the coordinate index are combined at the time of indexing into subject strings that capture concept relationships. Users do not have to coordinate these concepts themselves but can search on the pre-coordinated concepts, resulting in more precise retrieval. For example, MeSH terms can be assigned to a document with one or more of the possible 82 subheadings attached, such as "diagnosis" or "drug therapy." A search on the term "hypertension" with the attached subheading "drug therapy" will retrieve articles on the treatment of hypertension using drugs in a more precise manner than simply connecting the two concepts by an "and" operator, as in post-coordinate systems. The latter could result in retrieval of articles that are about hypertension with the drug therapy directed at another disease.

In coordinate indexing, syntactic relationships are displayed according to the syntax of a normal sentence, either through the syntax of the subject string (precoordinate indexing) or through devices such as facet indicators (postcoordinate indexing). Because of the absence of syntactic relationship indicators in postcoordinate systems, users are unable to distinguish between different contexts for the same term. This can result in retrieval of a set of documents that, while topically related, also contain "false drops" because there is not a mechanism for linking the terms to their respective composite subject or context (Foskett, 1982).

Farradane (1980) proposed Relational Indexing, a framework of nine relationship types, as a scheme for representing structures of syntactic relationships between terms in document descriptions with the goal of providing better retrieval of technical documents. In relational indexing, the meaning in information objects is denoted in the relationships between terms. This approach was not widely utilized, perhaps because the limited number of relationship types required manual indexing. However, Relational Indexing served as a precursor to some of the features of the UMLS and other thesaurus systems.

Similar to relational indexing, Craven (1978) proposed LIPHIS (Linked Phrase Indexing System), a system of computer-assisted subject indexing that used a network of terms in which arcs correspond to relationships denoted by prepositions. Like Farradane's scheme, the emphasis on concept relationships captured more of the content of an information object than individual term indexing alone.

In many vocabulary systems, conceptual relationships are characterized by generic relationships such as "broader than" and "related to." Other systems, including ontologies, utilize terminologic logic to describe a richer, more informative set of semantic relationships, such as "is_a," "connected_to" and "part_of." In biomedical literature, thesauri concept relationships conform to three general semantic classes of relationships that are used to express various dependencies and connections (Chowdury, 1999):

- Equivalence: which denotes the relationship between a preferred and non-preferred term and is shown through cross-references;
- Hierarchical: which represents pairs of terms in their superordinate (the whole) or subordinate (the part) status and is denoted by "Broader Than" and "Narrower Than" codes; and
- Associative: which describes the relationships between terms that are not in either the hierarchical or equivalence class and is shown by "Related" term codes.

In any natural language text, sequences of characters are combined into words, sequences of words are combined into sentences, sequences of sentences are combined into paragraphs and the sequences of paragraphs into texts. There is, thus, a hierarchy of levels of organization in text and there are corresponding levels of indexing to represent these textual levels. In the case of biomedical research reports, the structure is highly predictable (e.g., introduction, methods and design, research findings and conclusions sections). Approaches to relationship representation described so far have largely ignored the structure of the document in favor of representing isolated concepts. More recent work has focused on systems to extract biomedical relationships in the context of the document structure.

2.2 Integrating Document Structure in Systems

Much of the progress over the last several years in improving text understanding and retrieval has been due to systematic evaluations using complete, naturally-occurring texts as test data conducted at the Message

Understanding Conferences (MUC)[4] and Text REtrieval Conferences (TREC)[5]. MUC and TREC are currently sponsored by the U.S. Advanced Research Projects Agency (ARPA) and have enjoyed the participation of non-U.S. as well as U.S. organizations. MUC focuses on NLP while TREC is IR-focused. Both conferences provide the necessary infrastructure and large test corpora for large-scale, statistically valid performance figures and objective evaluation metrics of NLP and text retrieval methodologies.

Identification and classification of names of person, organization, location, etc., at accuracies exceeding 90% and successful extraction of binary relations among these entities at over 75% accuracy have been reported from these conferences (Aone et al., 1998). Also, as information extraction and retrieval systems have improved, attention has recently been turning to the potential contribution of document structure for text understanding and retrieval.

For example, Yeh et al. (2003) report the results of a Challenge Evaluation task created for the Knowledge Discovery and Data Mining (KDD) Challenge Cup to identify the set of genes discussed in a training corpus of 862 journal articles curated in FlyBase, a comprehensive database for information on the genetics and molecular biology of *Drosophila*. The common feature among the "winning" systems was use of document structure (i.e., concentrating on only certain sections of the document, for example, the "Results" or "Methods" sections and avoiding sections such as "References" in which citations will include names of genes not discussed in the paper) and/or linguistic structure (e.g., sections, paragraphs, sentences, and phrases), as well as table and figure captions, as a means of limiting where to look for features or patterns. The authors note that:

> This is in contrast to the information retrieval approach of treating a paper as just an unstructured set of words. We expect that systems will need to make more extensive use of linguistic and document structure to achieve better results and to accommodate more realistic tasks. (Yeh et al., 2003, pp. i338-9)

A similar approach is used by the PASTA (Protein Active Site Template Acquisition) Project system, which focuses on extracting information concerning the roles of particular amino acid residues in known three-dimensional protein structures. Text preprocessing includes a module that analyzes the text structure to determine which sections will proceed to continued processing. Since certain term classes may occur in only one particular section of text, by leveraging the standard structure of a scientific

[4] *http://www-nlpir.nist.gov/related_projects/muc/*

[5] *http://trec.nist.gov*

article, PASTA can exclude those portions of text that are not of interest. In addition, the PASTA system uses the document section to alter processing (Gaizauskas et al., 2000).

Another system that processes only specific parts of documents is FigSearch (Liu et al., 2004), a classification system that focuses specifically on a document's table and figure legends. The system ranks figures as likely to represent a certain type (e.g., protein interactions, signaling events) and allows users to search for these specialized subsets of figures from full-text.

Although incorporating the document's structure can help reduce the scope of material needing processing and potentially reduce inevitable "noise" in the results, this approach is not without its limitations. A major criticism of these systems is their specialization and consequent difficulty in porting to new domains or use in new applications. Also, the advantages of specialization (e.g., faster processing time) are achieved at the cost of limited terminology handling. For example, terminological issues of synonyms and term variants, expanding abbreviations, and lack of a mechanism for handling relations between terms continue to be problems encountered in systems that incorporate document structure.

2.3 Text Mining Approaches

Text mining refers to the process of extracting interesting and non-trivial patterns or knowledge from unstructured text documents. Text mining systems generically involve preprocessing document collections using text categorization and term extraction, storing the intermediate representations for analysis (e.g., distribution analysis, clustering, etc.) and visualizing the results. Association rules, which link pairs or larger groups of concepts and are usually assigned support and confidence values, are a dominant analysis method in text mining research. An association rule such as *concept A* → *concept B* indicates there may be a potentially interesting directional association from *A* to *B*. Typically these are discovered by exploiting the co-occurrence of concepts in the texts being mined (Hristovski et al., 2004).

Hirschman et al.,'s (2002) review of milestones in biomedical text mining research, notes that the field began by focusing on three approaches to processing text:

- Linguistic context of the text, such as the work of Fukuda et al. (1998) who pioneered identification of protein names;
- Pattern matching, as seen in the work of Ng and Wong (1999), who used templates that matched specific linguistic structures to recognize and extract protein interaction information from MEDLINE documents; and

- Word co-occurrence, such as Stapley and Benoit (2000) who extracted co-occurrences of gene names from MEDLINE documents and used them to predict their connections based on occurrence statistics.

As data mining technologies and NLP systems improve, more complex text can be processed and corpus-based approaches developed, as seen in the work of Pustejovsky et al. (2002), who used a corpus-based approach to develop rules specific to a class of predicates on a corpus of *inhibit*-relations; and Leroy and Chen (2005), whose Genescene system uses prepositions as entry points into phrases in the text, then fills in a set of templates of patterns of prepositions around verbs and nominalized verbs. NLP has also been used to capture specific relations in databases. For example, EDGAR is a system that extracts relationships between cancer-related drugs and genes from biomedical literature, incorporating a stochastic part of speech tagger, syntactic parser and semantic information from the UMLS (Rindflesch et al., 2000).

These systems have worked to overcome some of the limitations previously mentioned—such as decoding acronyms and abbreviations and detecting synonyms—using machine learning methods, NLP and incorporating ontologies (e.g., the Gene Ontology (GO)[6], a controlled vocabulary of genes and their products).

2.4 Literature-based Discovery IR Systems

As stated in the Introduction, the idea of discovering new relations from a bibliographic database was introduced by Swanson as "undiscovered public knowledge" that merit further investigation. Figure 7-1 illustrates Swanson's characterization of one of his first "mutually isolated literatures": Raynaud's disease, a peripheral circulatory disorder, and dietary fish oil. Although each of the two literatures were public knowledge, they were not bibliographically-related (i.e., did not cite one another), but were linked through intermediate literatures that had not been noticed before (Swanson and Smalheiser, 1997).

The premise of Swanson's approach is, given a body of literature reporting that *concept A* influences or is related to *concept B*; and given another body of literature reporting that *concept B* is related to or influences *concept C*; it may be inferred that *concept A* is linked to *concept C*, and if this relationship has not been experimentally tested, there is the potential to uncover previously "undiscovered public knowledge," form hypotheses, and investigate the relationship between *concept A* and *concept C*.

[6] *http://www.geneontology.org/*

In collaboration with Smalheiser over the past two decades, Swanson explored potential linkages via intermediate topics or specializations between bibliographically disconnected areas of specialization. Using this method, several concept relationships have been discovered and proposed for hypothesis testing, including the relationship between migraine and magnesium (Swanson and Smalheiser, 1997) and automatically identifying viruses that may be used as bioweapons (Swanson et al., 2001), among others. ARROWSMITH, an interactive discovery system based on Swanson and Smalheiser's methods, was created in 1991 and continues development today.[7]

Since the introduction of literature-based discovery, efforts to automate this approach and develop discovery algorithms that can be applied to a knowledge base or bibliographic database have resulted in several systems. One such system is BITOLA (Figure 7-2), which applies a general literature discovery algorithm to a knowledge base derived from the known relations between biomedical concepts (MeSH descriptors plus gene symbols in the document title and abstract fields) in the MEDLINE bibliographic database.

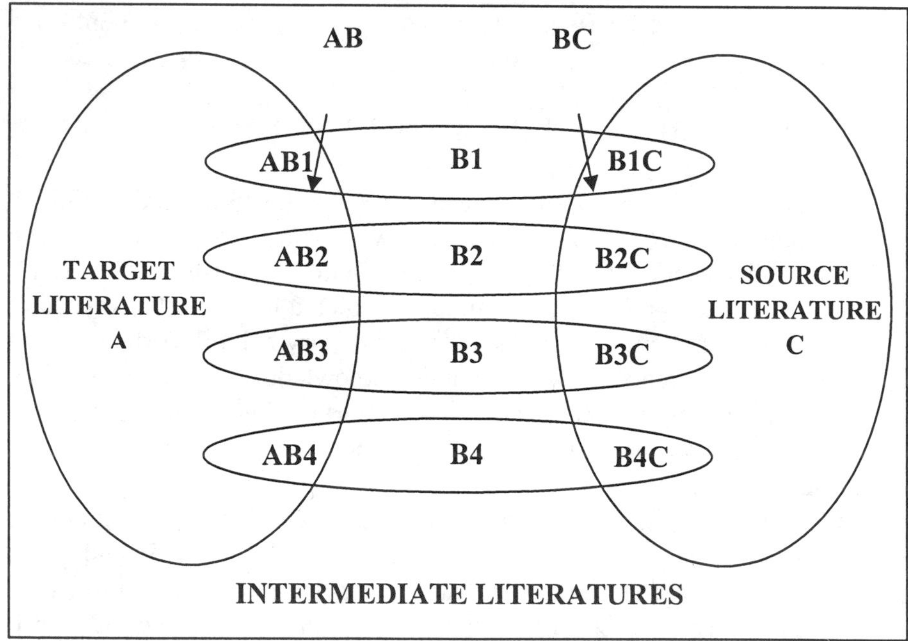

Figure 7-1. A Venn diagram that represents sets of articles or "literatures," *A* and *C*, that have no articles in common but which are linked through intermediate literatures *Bi* ($i = 1,2,\ldots$).

[7] There are two implementations available on the Web: *http://kiwi.uchicago.edu* and *http://arrowsmith.psych.uic.edu*

Such a structure may contain unnoticed useful information that can be inferred by combining pairs of intersections *ABi* and *BiC*. (From: Swanson and Smalheiser, 1997; reproduced with permission of the author.

BITOLA uses HUGO (Human Genome Organisation), the National Center for Biotechnology Information's (NCBI) LocusLink (a database of curated sequence and descriptive information about genetic loci) and OMIM (NCBI's Online Mendelian Inheritance in Man catalog of human genes and genetic disorders) as sources for gene symbols and names as well as gene locations. It also uses OMIM to obtain chromosomal locations. To decrease the number of candidate relations and make the system more suitable for disease candidate gene discovery, the system includes genetic knowledge about the chromosomal location of the starting disease as well as the chromosomal location of the candidate genes (Hristovski et al., 2004).

Similar to Swanson's procedure, BITOLA first finds all the *concepts Y* that are related to the starting *concept X* (e.g., if *X* is a disease then *Y* might be pathological functions). Then all the *concepts Z* related to *concepts Y* are found (e.g., if *Y* is a pathological function, *Z* might be a molecule related to the pathophysiology of *Y*). Finally, the medical literature is searched to check whether *concept X* and *concepts Z* appear together. If they do not appear together, there is the possibility that a new relationship between *concept X* and *concept Z* has been discovered. Figure 7-2 illustrates the BITOLA literature discovery system.

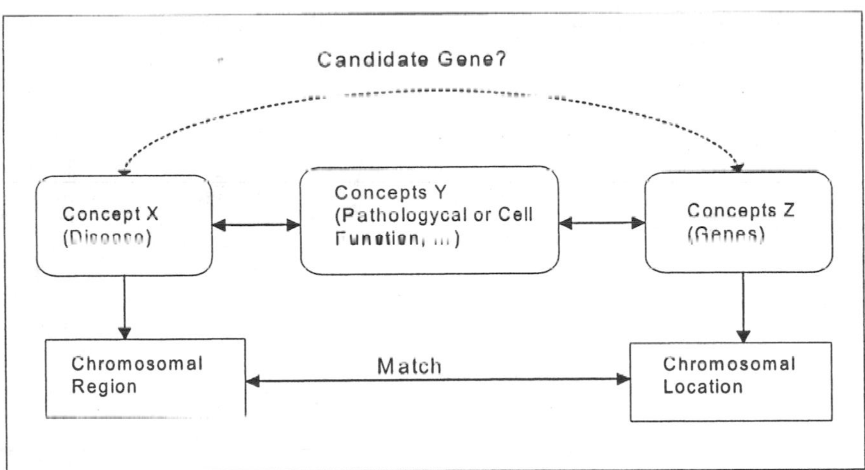

Figure 7-2. Discovery algorithm overview as applied to candidate gene discovery. For a starting disease *X*, we find the related concepts *Y* (disease characteristics) according to the literature (MEDLINE), then find the genes *Z* that are related to disease characteristics *Y*. (From: Hristovski et al., 2004; reproduced with permission of the author.)

As mentioned earlier, one criticism of literature-based discovery IR systems is the limitation imposed by utilizing only titles and/or abstracts. Another criticism, given the exponential growth in MEDLINE records, is their reliance on the use of words to "bridge" between unrelated domains; with new information and records added to MEDLINE on a daily basis, the scale of analysis required by these systems will continue to increase. In addition, other limitations cited include the absence of using synonyms to control vocabulary, the ambiguity produced when abbreviations are not automatically expanded and, particularly with gene symbols, the redundant use of a particular symbol that has differing meanings depending on the context (Wren et al., 2004).

2.5 Summary

Several of the systems already mentioned are, in fact, hybrid systems combining different text mining approaches (NLP, co-occurrence techniques, machine learning, etc.) with incorporation of different knowledge sources (UMLS Metathesaurus, GO, OMIM, etc.). It is obvious that there is not one approach that might be applied to the diverse and wide-ranging biomedical literature. Text formats vary from structured to unstructured and systems vary from free-text analysis to those that focus on document sections (titles and/or abstracts; specific document sections such as methods or table/figure captions). While most of the systems mentioned incorporate some controlled vocabulary component to reduce "noise," efforts to overcome terminologic issues—such as synonym, acronym and abbreviation ambiguity—vary widely.

It is notable, however, that most systems are genomic or proteomic-specific. The issue of scalability of these specialized systems will play an increasing role in their utility and future use as pathways for knowledge creation and discovery.

3. CASE EXAMPLES

Vannevar Bush, often referred to as one of the early pioneers of what later emerged into hypertext systems and the World Wide Web, suggested the use of associations as the main organizing mechanism when filing and retrieving records of information, and described an information space based on the use of associative "trails" to retrieve information (Bush, 1945). We have already mentioned several innovative approaches that are using biomedical concept relationships for improved document and information retrieval and analysis. In this section, we highlight two systems that include biomedical concept relationship extraction—Genescene and Telemakus—and embody the suggestions of Vannevar Bush in making the increasing body of recorded knowledge more easily accessible.

3.1 Genescene

Genescene,[8] which focuses on cellular processes, utilizes published MEDLINE abstracts and allows retrieval and visualization of biomedical relations extracted from the content of the abstracts. It uses a linguistic relation parser and Concept Space, an automatically-generated, corpus-based co-occurrence thesaurus of semantically related concepts. The system combines bottom-up and top-down approaches. The parser provides precise and semantically rich relations with a rule-based top-down algorithm. Concept Space captures the relations between semantic concepts from large collections of text using bottom-up techniques. The overall system offers a bottom-up view on the data in that the data is allowed to speak for itself, generating interesting patterns or associations that can be used to form new hypotheses. What follows is a summary of the system as described in Leroy and Chen (2005).

The process of creating the Concept Space begins with a download of MEDLINE abstracts in XML format. Abstract and title areas are selected. Using the AZ Noun Phraser (Tolle and Chen, 2000) optimized for biomedical language by using the UMLS SPECIALIST Lexicon as a lexical lookup—the linguistic parser extracts noun phrases. Phrases are analyzed and sorted so that each phrase becomes represented as a concept. Phrase and document frequencies are computed and used to weight each phrase. Concept Space is generated in the final step, co-occurrence analysis. Co-occurrence analysis produces a list of weighted, related noun phrases and their individual components (e.g., modifiers, etc.). Noun phrases are semantically tagged by three ontologies: HUGO, GO and the UMLS. The relations between noun phrases represent the relationships in the entire collection of abstracts originally downloaded from MEDLINE. Figure 7-3 illustrates the Genescene process. Future enhancements include an interactive, graphical map display and visual text mining.

By expanding biomedical literature mining beyond simply identifying genes and proteins and by providing a means for researchers and scientists to discover previously undiscovered gene associations, systems like Genescene will play an increasingly critical role in aiding research, knowledge creation and biomedical discovery.

[8] A demo is currently accessible through *http://ai.eller.arizona.edu/*

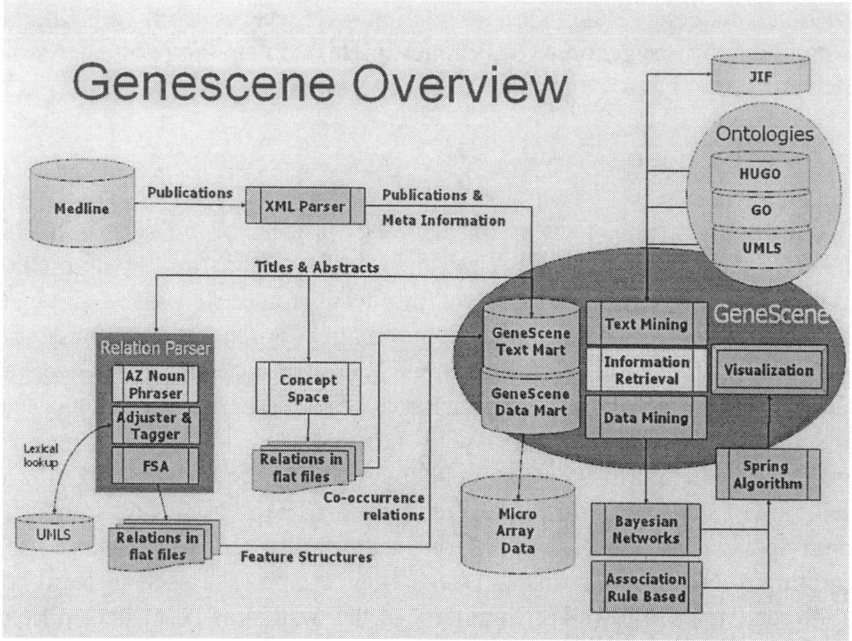

Figure 7-3. Overview of the Genescene architecture Relation Parser. (From: Leroy and Chen, 2005; reproduced with permission of the author.)

3.2 Telemakus

A hybrid system, Telemakus[9] has a broader focus than any system mentioned other than ARROWSMITH. Although Telemakus currently mines the biomedical literature concerned with the biology of aging, the tools and system architecture have been built to handle other biomedical domains as well. Telemakus uses document structure to limit its extraction of research parameters (e.g., age and number of subjects, treatment, etc.) and concept relationships by focusing on the document's methods section and table and figure captions (Fuller et al., 2004; Revere et al., 2004).

In brief, Telemakus processing is initiated by an analyst who runs, reviews and edits as necessary extractions from the document being processed. The process begins similarly to Genescene, although Telemakus utilizes other bibliographic databases in addition to MEDLINE. The first phase is a download of the document's citation details (in XML format) from which are extracted specific bibliographic fields. The electronic document is then processed to extract additional data, including research parameters and the table and figure legends. These document attributes are loaded into a

[9] Telemakus is freely available at *http:www.telemakus.net/*

database for populating the document conceptual schema—a schematic representation or surrogate of the document with extracted representations of research environment, methods and findings (see Figure 7-4). Telemakus employs the UMLS to control the vocabulary used for some fields, for curating domain-specific thesauri and for concept relationship analysis.

Page: 1 of 1

Luchsinger JA, Tang MX, Shea S, Mayeux R(2002). Caloric intake and the risk of Alzheimer disease. [Full Text] Arch Neurol, 59 (8): 1258-63. rpm2@columbia.edu

STUDY DESIGN & CONDUCT

Source of Organisms	Organisms	Age	Sex	Pre-Treat Char	Number	Treat Regimen
Volunteers	Humans	65-80 years	M, F	healthy; residing in geographically defined area of northern Manhattan; no dementia	980	n/a

STUDY OUTCOME

TABLE/FIGURES	RESEARCH FINDINGS
Table 1. Comparison of Characteristics Between Individuals in the 2 Lowest and 2 Highest Quartiles of Daily Total Calorie and Macronutrient Intake, Washington Heights-Inwood Columbia Aging Project, New York, NY, 1991-1996	energy intake - age apolipoproteins - energy intake apolipoproteins - dietary carbohydrates apolipoproteins - dietary protein intake
Table 2. Hazards Ratios of Alzheimer Disease for Individuals in Each Quartile of Total Daily Calorie Intake Using the Lowest Quartile as a Reference	apolipoproteins - dietary fats apolipoproteins - Alzheimer disease dietary carbohydrates - age
Table 3. Hazard Ratios of Alzheimer Disease for Individuals in Each Quartile of Daily Macronutrient Intake Using the Lowest Quartile as a Reference	dietary carbohydrates - energy intake dietary protein intake - age dietary protein intake - energy intake
Table 4. Adjusted Hazard Ratios of Alzheimer Disease per Quartile of Daily Calorie Intake and per Quartile of Fat Intake by Apolipoprotein E (APOE) 4 Status	dietary fats - age dietary fats - energy intake Alzheimer disease - energy intake

* statistically significant finding

Printer-friendly version

Figure 7-4. Telemakus document schematic representation

The relationship analysis procedure is currently in the process of being automated by incorporating MetaMap (Aronson, 2001), NLM's NLP tool, in combination with term co-occurrence analysis. MetaMap maps arbitrary text to concepts in the UMLS Metathesaurus; or, equivalently, it discovers Metathesaurus concepts in text by parsing text into noun phrases, collecting all UMLS terms containing one or more noun phrases or their variants, and ranking the candidate UMLS terms according to their similarities to all the noun phrases in the text.

Concept identification and assignment of concept relationships to individual documents is derived from processing the document's data tables and figures. Concentrating on data tables and figures focuses the concept identification and relationship process and reduces the background noise of the full-text document, making the process tractable.

The Telemakus concept relationship approach is of primary significance. On the document level, the basic motivation behind this analysis is to identify what was actually studied and reported. On the larger, domain level, the basic motivation is to capture concept relationships between the documents that form the domain. The interlinking relationships among concepts are represented graphically as concept maps. Concept mapping is a means of spatially representing knowledge in a visual format and, in the Telemakus system, displays the interrelationships between documents and reported research findings.

Visualization of concept relationships offers significant advantages over a textual listing or graph of relationships in that a spatial representation provides a way for users to interact directly with complex information. Visualization and visual exploration can assist in understanding conceptual relationships across a domain and even assist in identifying previously overlooked potential research connections. A strength of concept mapping is that—even though it does not measure strength of relationship between concepts—by aggregating links to concepts as in a many-to-one relationship, a measure of strength is added. In addition, a visualization of concept relationships may be significant for hypothesis generation, as the lack of linkages (the "non-interactive literatures") is more visually apparent.

The following section demonstrates how an information system that incorporates concept relationships can support knowledge creation and discovery.

3.3 How Can a Concept Relationship System Help with the Researcher's Problem and Questions?

Returning to the scenario at the beginning of this chapter, the AD researcher has a number of questions regarding the potential for treatment of the cognitive disabilities of AD or other neurodegenerative diseases with caffeine and adenosine A2A receptor antagonists (A2A blockers):

- Both caffeine and A2A blockers are reported to have neurotoxicity-blocking effects. Is there any relationship between caffeine consumption and adenosine A2A receptor antagonists?
- Adenosine A2A receptor antagonists have been successfully used to treat Parkinson's Disease. Have they been used to treat AD?
- Has anyone studied the relationship between caffeine and memory loss in AD patients? If yes, how much and over what period of time must caffeine be consumed to slow memory loss? Will a combination of

caffeine and adenosine A2A receptor antagonists shorten that period of time?

- Do the neurotoxicity-blocking effects of caffeine and adenosine A2A receptor antagonists also come into play with other neurodegenerative diseases? What about possible negative associations with other conditions?

Another way to look at these questions is in terms of the concept relationships they represent as listed in Table 7-1. It is notable that not one concept relationship can encapsulate the researcher's information need and that some individual concepts can be related to multiple concepts.

Table 7-1. List of Concepts and Possible Concept Relationships

Concept 1	Concept 2	Possible Semantic Relationship(s)
caffeine	memory loss	associated_with / co-occurs_with
memory loss	Alzheimer's disease	manifestation_of
caffeine	neurodegenerative diseases	treats
caffeine	high blood pressure	associated_with / complicates
caffeine	neurotoxicity-blocking effects	associated_with
caffeine	A2A blockers	associated_with
A2A blockers	neurotoxicity-blocking effects	associated_with
A2A blockers	Parkinson's Disease	treats

As mentioned previously, some of the information the researcher needs will be found in the methods or results sections or in the figures and tables rather than the document's title or abstract. For an information need such as this, a system like Telemakus is an appropriate resource with its extracted representations of the research environment, methods and outcomes of the retrieved documents.

Providing the schematic representations (schemas) of retrieved documents allows the researcher to "browse" the document retrieval space without needing to read the articles in their entirety. In addition, characterizing the concept relationships from each document in a visual format maintains the inter-relationships between documents and reported research findings, as well as assists in understanding conceptual relationships across a domain.

Returning to our scenario, when the AD researcher uses Telemakus for her query, she enters terms similarly to the approach she used with PubMed. However, the list of citations returned provides a pathway to both the content of each document—its methods and research findings—and to interactive concept maps of linked relationships across the group of research reports. Figure 7-5 illustrates the retrieval set interface for a search on

Alzheimer's disease and caffeine. From this list, the researcher can browse the schematic representations or schemas of the content of each document (Figure 7-6).

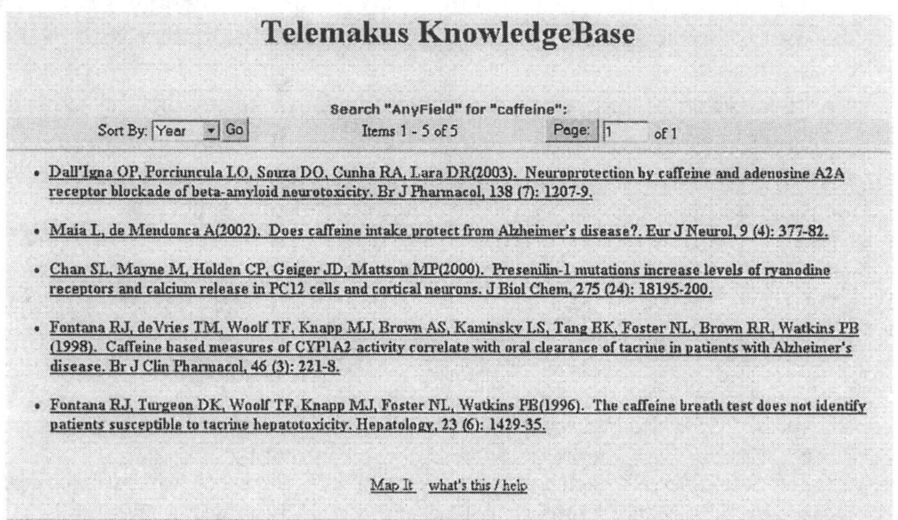

Telemakus KnowledgeBase
Search "AnyField" for "caffeine":
Sort By: Year Go
Items 1 - 5 of 5
Page: 1 of 1
• Dall'Igna OP, Porciuncula LO, Souza DO, Cunha RA, Lara DR(2003). Neuroprotection by caffeine and adenosine A2A receptor blockade of beta-amyloid neurotoxicity. Br J Pharmacol, 138 (7): 1207-9.
• Maia L, de Mendonca A(2002). Does caffeine intake protect from Alzheimer's disease?. Eur J Neurol, 9 (4): 377-82.
• Chan SL, Mayne M, Holden CP, Geiger JD, Mattson MP(2000). Presenilin-1 mutations increase levels of ryanodine receptors and calcium release in PC12 cells and cortical neurons. J Biol Chem, 275 (24): 18195-200.
• Fontana RJ, deVries TM, Woolf TF, Knapp MJ, Brown AS, Kaminsky LS, Tang BK, Foster NL, Brown RR, Watkins PB (1998). Caffeine based measures of CYP1A2 activity correlate with oral clearance of tacrine in patients with Alzheimer's disease. Br J Clin Pharmacol, 46 (3): 221-8.
• Fontana RJ, Turgeon DK, Woolf TF, Knapp MJ, Foster NL, Watkins PB(1996). The caffeine breath test does not identify patients susceptible to tacrine hepatotoxicity. Hepatology, 23 (6): 1429-35.
Map It what's this / help

Figure 7-5. Caffeine and AD search retrieval set

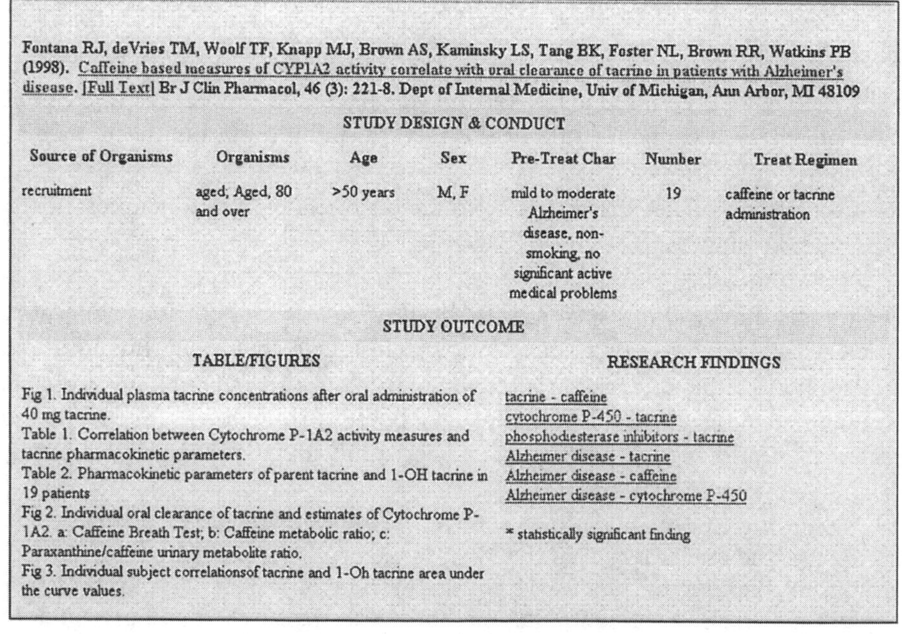

Fontana RJ, deVries TM, Woolf TF, Knapp MJ, Brown AS, Kaminsky LS, Tang BK, Foster NL, Brown RR, Watkins PB (1998). Caffeine based measures of CYP1A2 activity correlate with oral clearance of tacrine in patients with Alzheimer's disease. [Full Text] Br J Clin Pharmacol, 46 (3): 221-8. Dept of Internal Medicine, Univ of Michigan, Ann Arbor, MI 48109
STUDY DESIGN & CONDUCT
Source of Organisms
Organisms
Age
Sex
Pre-Treat Char
Number
Treat Regimen
recruitment
aged; Aged, 80 and over
>50 years
M, F
mild to moderate Alzheimer's disease, non-smoking, no significant active medical problems
19
caffeine or tacrine administration
STUDY OUTCOME
TABLE/FIGURES
Fig 1. Individual plasma tacrine concentrations after oral administration of 40 mg tacrine.
Table 1. Correlation between Cytochrome P-1A2 activity measures and tacrine pharmacokinetic parameters.
Table 2. Pharmacokinetic parameters of parent tacrine and 1-OH tacrine in 19 patients
Fig 2. Individual oral clearance of tacrine and estimates of Cytochrome P-1A2. a: Caffeine Breath Test; b: Caffeine metabolic ratio; c: Paraxanthine/caffeine urinary metabolite ratio.
Fig 3. Individual subject correlationsof tacrine and 1-Oh tacrine area under the curve values.
RESEARCH FINDINGS
tacrine - caffeine
cytochrome P-450 - tacrine
phosphodiesterase inhibitors - tacrine
Alzheimer disease - tacrine
Alzheimer disease - caffeine
Alzheimer disease - cytochrome P-450
* statistically significant finding

Figure 7-6. Schematic representation of a document from searching "caffeine" and "AD" schema

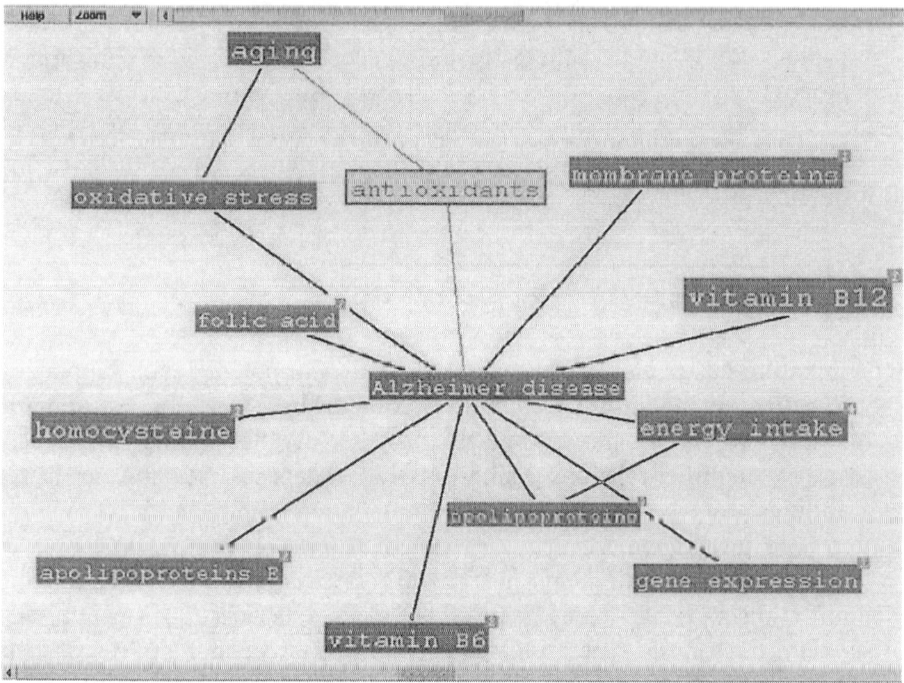

Figure 7-7. Telemakus concept map generated from a search on Alzheimer's disease. Note that edges (lines) of the map signify relationships between concepts, but length does not reflect any weighting scheme.

From the list of retrieved citations, the researcher can also activate the concept mapping function for access to a visualized map of concept relationships for the current retrieval set (Figure 7-7) by selecting "Map It" (at the bottom of the retrieval list in Figure 7-6).

Within each schema, she can access the abstract, document full-text and, by clicking on any blue highlighted item under "Table/Figure," she can link to each table or figure in the full-text article. It is from the schema that the researcher will obtain answers to several of her questions by either examining a table or figure of interest or by browsing the information found under "Study Design and Conduct."

The researcher can also explore individual concept relationships by selecting any pair listed under "Research Findings" in the schema, activating a search of the relationship across all documents in the Telemakus knowledgebase. For example, selecting the pair "caffeine – cell death" will result in a retrieval set listing all documents to which this concept relationship has been assigned. The concept maps generated by Telemakus will help the researcher answer other questions.

Further exploration can be done within the concept map by selecting linkages between concept relationships and by selecting individual concepts.

The iterative nature of the search process and ability to explore research connections from both the schematic and concept map interfaces can support the process of knowledge discovery in a way that mimics the way many scientists work—by providing a means of exploring a variety of types of connections and potentially discovering a new frame of reference for the information problem.

3.4 Summary

Literature-based discovery systems provide the potential for supporting "systematic serendipity." Originally coined by Garfield, systematic serendipity refers to the organized process of discovering previously unknown scientific relations using citation databases, leading to better possibilities for a collaboration of human serendipity with computer-supported knowledge discovery.

Hypothesis generation and testing are critical steps towards making scientific discoveries. Along with the "aha" experience of insight and discovery, hypothesis generation requires prior knowledge. Yet researchers are sometimes unaware of relevant work by others that could be integrated into theirs or are unable to put together enough "pieces" of their domain's jigsaw puzzle to recognize that one is missing or being overlooked. As illustrated in the scenario above, literature-based knowledge discovery systems that include concept relationships and schematic representations are a means of providing additional pathways to these puzzle pieces.

4. CONCLUSIONS AND DISCUSSION

This chapter has presented a variety of approaches that have been used to characterize biomedical concept relationships and document concept interrelationships and some of the ways in which concept relationships have been used in information search and retrieval. We have reviewed a very small number of the innovative approaches utilizing biomedical concept identification and relationships for improved document and information retrieval and analysis. This arena in the knowledge discovery field—utilizing biomedical concept relationships—is fairly young and promises to be a rich and interdisciplinary endeavor.

In our survey of approaches to indexing and retrieval of documents, we have noted that most systems rely on title and/or abstract text to assign or mine concept relationships. In addition, most systems do not exploit the structure of the document itself as a way of more precisely characterizing biomedical relationships. As more and more biomedical literature becomes available electronically, we will likely see an increase in extraction approaches that incorporate the full-text document.

We have assumed, given the rapid expansion of scientific research, that there is great need for creating systems that aid in finding or integrating new domain knowledge. While we have focused on the concept relationship component of knowledge management systems that support biomedical research, we have not thoroughly discussed the role and realistic utility of such systems for creating knowledge. A significant research area that requires attention is evaluation of these systems. Usability research is needed to validate the utility of these approaches for scientists and researchers. There are numerous questions for literature-based concept relationship systems, including the following:

How exactly does including concept relationships in such systems support the creative process of hypothesis development?

How can systems avoid hindering the "eureka" experience of scientific research?

What methods must be employed to answer these questions?

While Swanson and Smalheiser established a discovery framework that has been used by many researchers as a measurement standard, extensive and comprehensive evaluation efforts are still needed for validating the contribution these systems can make for knowledge creation and discovery.

5. ACKNOWLEDGEMENTS

The authors wish to acknowledge the helpful suggestions made by anonymous reviewers.

REFERENCES

Aone, C., Halverson, L., Hampton, T. and Ramos-Santacruz, M. SRA (1998). "Description of the IE2 System Used for MUC-7," in *Proc 7th Message Understanding Conference*.

Aronson, A. R. (2001). "Effective Mapping of Biomedical Text to the UMLS Metathesaurus: the MetaMap Program," in *Proceedings of the AMIA Annual Fall Symposium*, pp. 17-21.

Bush, V. (1945). "As We May Think," *The Atlantic Monthly*, 176*(1)*:101-108.

Chowdury C. G. (1999). Introduction to Modern Information Retrieval. Library Association Publishing.

Craven, T. C. (1978). Linked Phrase Indexing. *Information Processing and Management*, 14(6):469-76.

Farradane, J. (1980). "Relational Indexing, Parts I and II," *Journal of Information Science*, 1(5-6):267-76; 313-24.

Foskett A. C. (1982). *The Subject Approach to Information.* 4th ed. Hamden, CT: Linnet Books.

Fukuda, K., Tamura, A., Tsunoda, T. and Takagi, T. (1988). "Toward Information Extraction: Identifying Protein Names from Biological Papers," in *Proceedings of the Pacific Symposium on Biocomputing*, pp. 707-18.

Fuller, S., Revere, D., Bugni, P. and Martin, G.M. (2004). "A Knowledgebase Information System to Enhance Scientific Discovery: Telemakus," *BMC Digital Libraries*;1*(1)*:2 (21 September 2004).

Gaizauskas, R., Demetriou, G. and Humphreys, K. (2000). "Term Recognition and Classification in Biological Science Journal Articles," in *Proceedings of the Computational Terminology Workshop for Medical and Biological Applications*, pp. 37-44.

Garfield, E. (1966). "The Who and Why of ISI," *Essays of an Information Scientist*, 13:33-37.

Green, R., Bean, C. A. and Myaeng, S. H. (2002). "Introduction," in R. Green, C.A. Bean and S.H. Myaeng (eds.), *The Semantics of Relationships: An Interdisciplinary Perspective*, pp. vii-xviii. Kluwer Academic Publishers.

Green R. (2001). "Relationships in the Organization of Knowledge: An Overview," In C.A. Bean and R. Green (eds.), *Relationships in the Organization of Knowledge*, pp. 3-18. Kluwer Academic Publishers.

Hetzler, B. (1997). "Beyond Word Relations: SIGIR '97 Workshop," *ACM SIGIR Forum*, 31(2):28-33.

Hirschman, L., Park, J. C., Tsujii, J., Wong, L. and Wu, C. H. (2002). "Accomplishments and Challenges in Literature Data Mining for Biology," *Bioinformatics*, 18(12):1553-61.

Hristovski, D., Peterlin, B., Mitchell, J. A. and Humphrey, S. M. (2004; in press). "Using Literature-based Discovery to Identify Disease Candidate Genes," *International Journal of Medical Informatics*.

Leroy, G. and Chen, H. (2005; in press). "Genescene: An Ontology-enhanced Integration of Linguistic and Co-occurrence Based Relations in Biomedical Texts," *Journal of the American Society for Information Science and Technology*.

Liu, F., Jenssen, T.-K, Nygaard, V., Sack, J. and Hovig, E. (2004; in press). "FigSearch: A Figure Legend Indexing and Classification System," *Bioinformatics*.

Ng, S. K. and Wong, M. (1999). "Toward Routine Automatic Pathway Discovery from On-line Scientific Text Abstracts," in *Proceedings of the Genome Informatics Series: Workshop on Genome Informatics*, pp. 104-112.

Pustejovsky, J., Castano, J., Zhang, J., Kotecki, M and Cochran, B. (2002). "Robust Relational Parsing over Biomedical Literature: Extracting Inhibit Relations," in *Proceedings of the Pacific Symposium on Biocomputing*, pp. 362-73.

Revere, D., Fuller, S.S., Bugni, P. and Martin, G.M. (2004). "An Information Extraction and Representation System for Rapid Review of the Biomedical Literature," Accepted for Presentation: *MedInfo*, Sept 2004, San Francisco, CA.

Rindflesch, T. C., Tanabe, L., Weinstein, J. N. and Hunter, L. (2002). "EDGAR: Extraction of Drugs, Genes and Relations from the Biomedical Literature," in *Proceedings of the Pacific Symposium on Biocomputing, p*p. 517-28.

Stapley, B. J. and Benoit, G. (2002). "Biobibliometrics: Information Retrieval and Visualization from Co-occurrences of Gene Names in Medline Abstracts," in *Proceedings of the Pacific Symposium on Biocomputing*, pp. 529-40.

Swanson, D. R. and Smalheiser, N. R. (1997). "An Interactive System for Finding Complementary Literatures: A Stimulus to Scientific Discovery," *Artificial Intelligence*, 91*(2)*:183-203.

Swanson, D. R., Smalheiser, N.R. and Bookstein, A. (2001). "Information Discovery from Complementary Literatures: Categorizing Viruses as Potential Weapons," *Journal of the American Society for Information Science and Technology,* 52(10):797-812.

Taube M. (1953-1959). *Studies in Coordinate Indexing*, vols. 1-5. Washington, DC: Documentation, Inc.

Tolle, K., and Chen, H. (2000). "Comparing Noun Phrasing Techniques for Use With Medical Digital Library Tools," *Journal of the American Society for Information Science,* Special Issue on Digital Libraries, 51(4):352-370.

Wren, J. D., Bekeredjian, R., Stewart, J. A., Shohet, R. V., Garner, H. R. (2004). "Knowledge Discovery by Automated Identification and Ranking of Implicit Relationships," *Bioinformatics*, 20*(3)*:389-98.

Yeh, A. S., Hirschman, L. and Morgan, A. A. (2003). "Evaluation of Text Data Mining for Database Curation: Lessons Learned from the KDD Challenge Cup," *Bioinformatics*, 19*(Suppl 1)*:i331-39.

SUGGESTED READINGS

Adamic, L. A., Wilkinson, D., Huberman, B.A. and Adar, E. "A Literature Based Method for Identifying Gene-disease Connections," in *Proceedings of the IEEE Bioinformatics Conference* 2002, pp. 109-17.

This paper presents a statistical method that can swiftly identify, from the literature, sets of genes known to be associated with given diseases.

Friedman C., Kra, P., Yu H, Krauthammer, M. and Rzhetsky, A. "GENIES: A Natural-language Processing System for the Extraction of Molecular Pathways from Journal Articles." *Bioinformatics* 2001;17*(Suppl 1)*:S74-S82.

A paper describing a system that Extracts and structures information from the biological literature and builds a knowledge base about signal-transduction pathways, the diseases associated with the pathways and drugs that affect them.

Shatkay, H. and Feldman, R. "Mining the Biomedical Literature in the Genomic Era: An Overview." *Journal of Computational Biology* 2003;10*(6)*;821-55.

A review paper describing various biomedical literature-mining methods and the disciplines involved in unstructured-text analysis. Also provides Examples of text analysis methods applied towards meeting some of the current challenges in bioinformatics.

Srinivasan, P. and Sehgal, A. K. "Mining MEDLINE for Similar Genes and Similar Drugs. Technical Report, TR#03-02. School of Library and Information Science, University of Iowa.

This paper presents an application that mines MEDLINE for novel concept connections towards the goal of supporting scientists in hypothesis discovery. The application uses concept profiles as a mechanism for generating concept representations from text collections.

Weeber, M., Vos, R, Klein, H., de Jong-Van den Berg LTW, Aronson, A. and Molema, G. "Generating Hypotheses by Discovering Implicit Associations in the Literature: A Case Report for New Potential Therapeutic Uses for Thalidomide." *JAMIA* 2003;10*(3)*:252-59.

A paper describing a biomedical literature discovery support tool that follows Swanson's model of "disjointed but complementary structures" to systematically analyze the scientific literature in order to generate novel and plausible hypotheses.

ONLINE RESOURCES

ARROWSMITH
http://kiwi.uchicago.edu
http://arrowsmith.psych.uic.edu

Genescene
http://ai.eller.arizona.edu/go/GeneScene/index.html

Telemakus System
http://www.telemakus.net/

Unified Medical Language System
http://www.nlm.nih.gov/research/umls/

QUESTIONS FOR DISCUSSION

1. Compare and contrast traditional indexing and more recent biomedical concept indexing approaches. What are the advantages and disadvantages of each approach?

2. Give five examples of biomedical concept relationships.

3. Why is building systems to promote literature-based knowledge discovery important?

Chapter 8
BIOMEDICAL ONTOLOGIES

Olivier Bodenreider[1] and Anita Burgun[2]

[1]U.S. National Library of Medicine, Bethesda, Maryland 20894 ; USA; [2]Laboratoire d'Informatique Médicale, Université de Rennes I, 35043 Rennes Cedex, France

Chapter Overview

Ontology design is an important aspect of medical informatics, and reusability is a key issue that is determined by the level of compatibility among ontology concepts and among the theories of the biomedical domain they convey. In this article, we examine OpenGALEN, the UMLS Semantic Network, SNOMED CT, the Foundational Model of Anatomy, and the MENELAS ontology as well as descriptions of the biomedical domain in two general ontologies, OpenCyc and WordNet. Using the representation of *Blood* in each system, we examine issues in compatibility among these ontologies. The presence of additional knowledge is also illustrated and some issues in creating and aligning biomedical ontologies are discussed.

Keywords:
biomedical ontology; biomedical knowledge representation; GALEN; UMLS; SNOMED CT; Foundational Model of Anatomy

1. INTRODUCTION

The purpose of biomedical ontology is to study classes of entities (i.e., substances, qualities and processes) in reality which are of biomedical significance. Examples of such classes include substances such as the mitral valve and glucose, qualities such as the diameter of the left ventricle and the catalytic function of enzymes, and processes such as blood circulation and secreting hormones. Unlike biomedical *terminology*, which collects the names of entities employed in the biomedical domain, biomedical *ontology* is concerned with the principled definition of biological classes and the relations among them. In practice, as they are more than lists of terms but do not necessarily meet the requirements of formal organization, the many products developed by biomedical terminologists and ontologists often fall between terminologies and ontologies and constitute an "ontology gradient".

Ontologies may be categorized according to the domain they represent or the level of detail they provide (Figure 8 1). *General ontologies* represent knowledge at an intermediate level of detail independently of a specific task. In such ontologies, upper levels reflect theories of time and space, for example, and provide notions to which all concepts in existing ontologies are necessarily related. *Domain ontologies* represent knowledge about a particular part of the world, such as medicine, and should reflect the underlying reality through a theory of the domain represented. Finally, ontologies designed for specific tasks are called *application ontologies*. Conversely, *reference ontologies* are developed independently of any particular purpose and serve as modules sharable across domains.

Core categories should be sharable across ontologies. Lower levels of upper level ontologies as well as general categories should be compatible with the equivalent semantic areas in the corresponding domain ontologies. For example, *Disease* in a general ontology should be compatible with that concept in a biomedical ontology. In addition, generic theories and meta-level categories should be shared by every type in every ontology. For example, a representation of anatomy should re-use a generic theory of spatial objects. In turn, as anatomy is central to biomedicine and essentially stable, an ontology of anatomy can serve as a reference for ontologies relying on a representation of the human body, e.g., for an ontology of Diseases. In practice, however, these ideals are not always achieved. More generally, constructing biomedical ontologies that accommodate knowledge sharing by both humans and computer systems is challenging.

Ontologies play a fundamental role in medical informatics research (Musen 2002), contributing, for example, to natural language processing (e.g., Hahn et al. 1999), interoperability among systems (e.g., Degoulet et al. 1998), and access to heterogeneous sources of information, including the

Semantic Web (e.g., Pisanelli et al. 2004). Increasingly, ontologies act as enabling resources in a variety of biomedical applications.

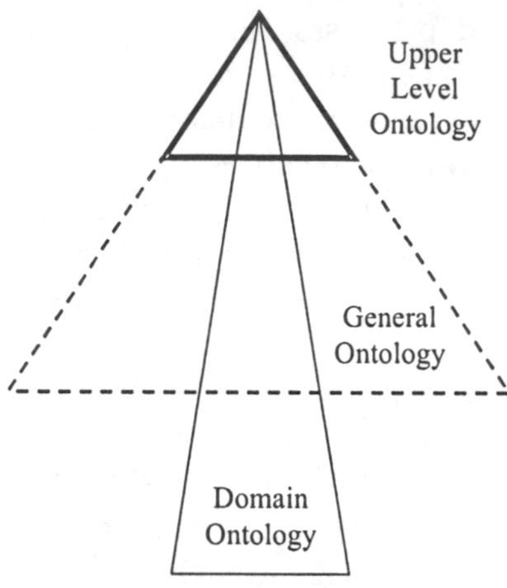

Figure 8-1. Kinds of ontologies.

The objective of this chapter is not to examine how applications benefit from using ontologies, but rather to present the characteristics of some major biomedical ontologies. In particular, we investigate how existing ontologies give differing views of the biomedical domain. First, we examine the representation of biomedicine in general systems such as OpenCyc and WordNet. We then describe three systems in the biomedical domain, GALEN, the UMLS, and SNOMED CT. A reference ontology, the Foundational Model of Anatomy, is also explored. Finally, as an example of an application ontology, we examine the MENELAS project. After a brief presentation of the characteristics of these ontologies, we look at the concept *Blood* in each system to illustrate common features and differences. Issues in building a single, sharable framework for representing biomedical knowledge are discussed.

This study was conducted at the U.S. National Library of Medicine as part of the Medical Ontology Research project (Bodenreider 2001), which focuses on developing methods for acquiring biomedical ontologies from existing resources and for validating them against other knowledge sources. References for the ontologies presented in this chapter are listed in the appendix (Table 8-3) along with a summary of their main characteristics (Table 8-4). It is beyond the scope of this chapter to present the techniques (e.g., description logics and frames) and tools (e.g., Protégé) used for representing ontologies. The interested reader is referred to references such as (Sowa 2000; Brachman and Levesque 2003).

2. REPRESENTATION OF THE BIOMEDICAL DOMAIN IN GENERAL ONTOLOGIES

2.1 OpenCyc

Cyc,® a general ontology developed by Cycorp, Inc., is built around a core of more than 1,000,000 hand-coded assertions (expressed in the formal language CycL) that capture "common sense" knowledge and enable a variety of knowledge-intensive applications. "Microtheories" are groups of assertions sharing a common set of assumptions focused according to a particular parameter, such as domain, level of detail, or time interval. OpenCyc™, the upper level, publicly available part of the ontology contains 6,000 concepts and 60,000 assertions about those concepts.

In OpenCyc as illustrated in Figure 8-2, *Thing*, the universal set, is the collection of everything. *Thing* is partitioned into *Set or collection* vs. *Individual* on the one hand and *Intangible* vs. *Partially tangible* on the other. Entities in OpenCyc are both represented as instances of sets, e.g., *Cancer* is an instance of the type *Disease Type* `(#$isa #$Cancer #$DiseaseType)` and organized in class/subclass hierarchies `(#$genls #$Cancer #$AilmentCondition)`. Further specification may be provided by functions. *CancerFn*, for example, expresses that body parts can be the location of cancers. This function has domain animal body parts and range specific cancers: e.g., `(#$CancerFn #$Throat)`.

Microtheories such as Biology or Ailment are relevant in the biomedical domain and have two primary benefits: (1) some assertions have microtheories as arguments: Everything true in Vertebrate Physiology is also true in Ailment and (2) some entities have distinct representations under distinct microtheories: in Animal Physiology, subordinates of *Sensor* include *Nose*, *Skin*, and *Ear*, while in Naïve Physics they include *Tactile sensor* and *Electromagnetic radiation sensor*.

2.2 WordNet

WordNet® is an electronic lexical database developed at Princeton University (Fellbaum 1999) that serves as a resource for applications in natural language processing and information retrieval.

The core structure in WordNet is a set of synonyms (synset) that represents one underlying concept. Synset formation is based on synonymy (one meaning expressed by several words) and polysemy (one word having several distinct meanings). There are separate structures for each linguistic category covered: English nouns, verbs, adjectives, and adverbs. For example, the adjective "renal" and the noun "kidney," although similar in meaning, belong to two distinct structures, and a specific relationship, "pertainymy," relates the two forms.

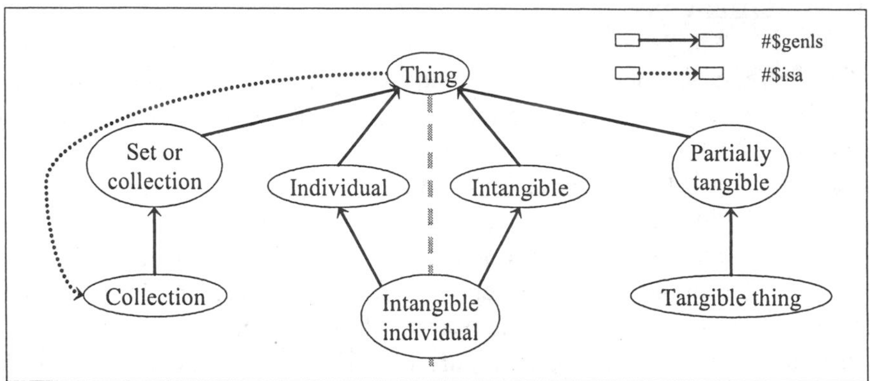

Figure 8-2. Top level in OpenCyc (partial representation).

The current version of WordNet (2.0) contains over 114,000 noun synsets categorized into nine hierarchies, each starting with a "unique beginner" (see Figure 8-3). Each synset in the noun hierarchy belongs to at least one *is-a* tree (hyponymy) and may additionally belong to several *part-of*-like trees (meronymy). Hyponymy relations are established between synsets according to the following definition: A concept represented by the synset {x,x',...} is said to be a hyponym of the concept represented by the synset {y,y',...} if native speakers of English accept sentences constructed from frames such as "An x is a kind of y" (Fellbaum 1999). WordNet has been influenced by cognitive psychology as well as linguistics, and its hierarchies are not based on formal ontology theory. (Gangemi et al. 2001) provide an ontological analysis of WordNet's top level and propose a revised, principled taxonomy.

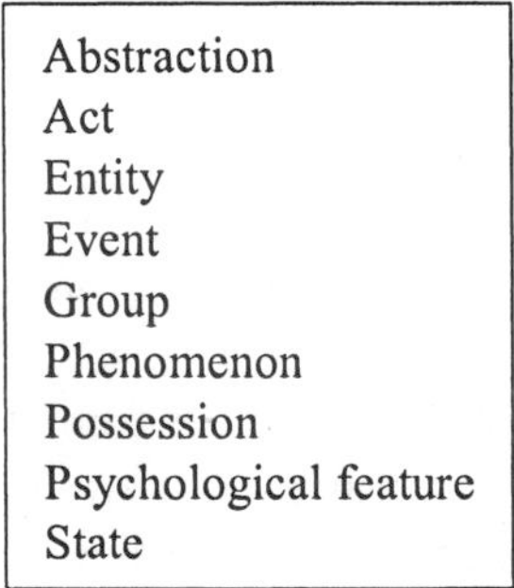

Abstraction
Act
Entity
Event
Group
Phenomenon
Possession
Psychological feature
State

Figure 8-3. Top level in WordNet ("unique beginners").

Many concepts that represent health disorders in medical terminologies, when present in WordNet, are categorized appropriately; for example, *Leukemia* is a hyponym of *Cancer* (Burgun and Bodenreider 2001a; Burgun and Bodenreider 2001b). However, in some instances a medical sign or symptom appears only as a hyponym of a non-medical concept: the hypernym of *Vasoconstriction* (decrease in the diameter of blood vessels) is *Constriction*. This view emphasizes physical mechanism rather than pathology, and as a consequence, there is no formal relationship between *Vasoconstriction* and the biomedical domain in WordNet.

3. EXAMPLES OF MEDICAL ONTOLOGIES

3.1 GALEN

GALEN (Generalised Architecture for Languages, Encyclopaedias, and Nomenclatures in medicine) is a European Union project (1992-1999) that seeks to provide re-usable terminology resources for clinical systems. An ontology, the Common Reference Model, is formulated in a specialized description logic, the GALEN Representation and Integration Language (GRAIL), and is a core feature of GALEN (Rector et al. 1997). This ontology aims to represent "all and only sensible medical concepts," independently of any application. OpenGALEN provides a point of access to the GALEN Common Reference Model and to descriptions and specifications of the GALEN technology.

A key feature of GALEN is that it was constructed by defining the representation formalism and top level knowledge before populating the ontology. In addition, unlike traditional terminological resources whose terms are pre-coordinated, GALEN essentially provides the building blocks required for describing terminologies, as well as a mechanism for combining simple concepts. For example, the concepts *Adenocyte* and *Thyroid gland* are present in GALEN. However, instead of providing an explicit representation for *Adenocyte of thyroid gland*, GALEN indicates that it can be described by a combination of concepts: (*Adenocyte* which < *is structural component of Thyroid gland* >). The current version of OpenGALEN (December 2002) contains about 25,000 concepts. The GALEN ontology has been used for representing complex structures such as descriptions of medical procedures (Trombert-Paviot et al. 2000).

The major division in top level categories (Figure 8-4) is between *Phenomenon*, which subsumes structures, processes and substances, and *Modifier Concept*. The latter notion is used to distinguish concepts that represent things with independent existence (physical objects, for example) from dependent concepts such as modifiers (*Mild severity*), states (*Pathological state*) or roles (*Infective role*).

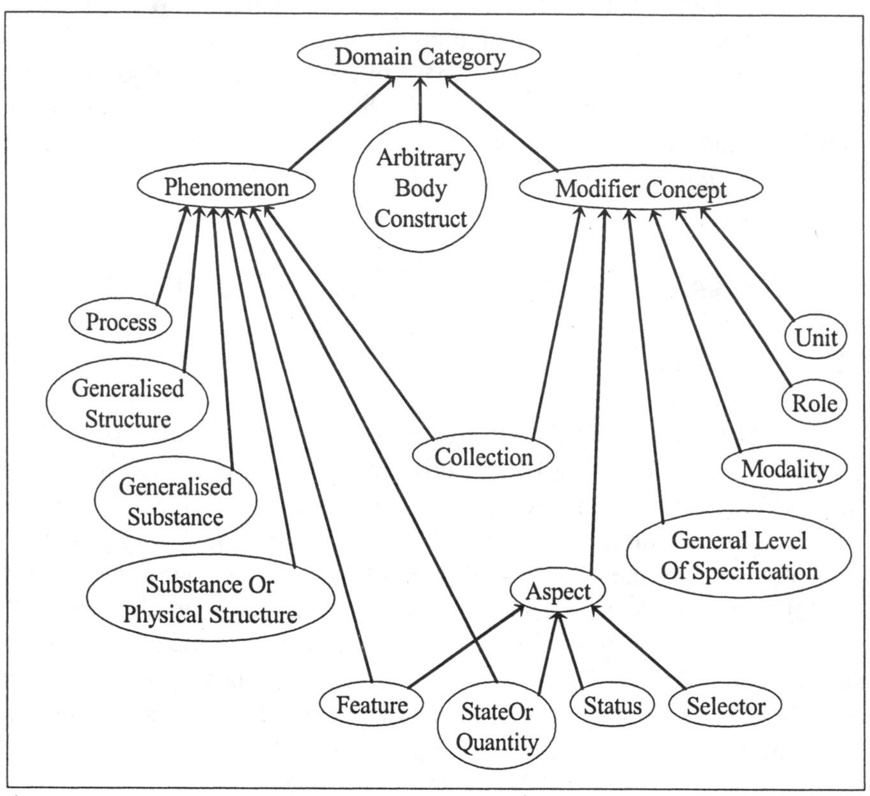

Figure 8-4. Top level in OpenGALEN.

In addition to a hierarchy of categories, GALEN provides a rich hierarchy of associative relationships used to define complex structures. Its representation of partitive relations is particularly developed (Rogers and Rector 2000), including *has surface division* (`Hand has-surface-division Palm`), *has solid division* (`Heart has-solid-division Cardiac Septum`), *has layer* (`Heart has-layer Myocardium`), *has blind pouch division* (`Caecum has-blind-pouch-division Appendix Vermiformis`), *has linear division* (`Intestine has-linear-division Jejunum`), *has specific structural component* (`Knee Joint has-specific-structural-component Patella`), and *is specifically made of* (`Blood Clot is-specifically-made-of Coagulated Blood`).

3.2 Unified Medical Language System

The Unified Medical Language System® (UMLS)® was developed by the National Library of Medicine to help health care professionals and researchers access biomedical information from a variety of sources (Lindberg et al. 1993). The Metathesaurus,® a large repository of concepts, and the Semantic Network, a limited network of 135 semantic types, integrate over one million concepts from more than a hundred vocabularies and terminologies (2004AB version). While the structure of each source is preserved in building the Metathesaurus, equivalent terms are clustered into a semantically unique concept. Interconcept relationships are either inherited from underlying vocabularies or specifically generated. Since the Metathesaurus imposes no restrictions on sources, it cannot provide the kind of organization expected from an ontology. In contrast, the Semantic Network is developed independently of the vocabularies integrated in the Metathesaurus and serves as a basic, high-level ontology for the biomedical domain (McCray 2003). As illustrated in Figure 8-5, semantic types from the Semantic Network are used to categorize all UMLS concepts (McCray and Nelson 1995).

At the highest level, the Semantic Network is organized around the opposition of entities and events, and two single-inheritance hierarchies reflect this distinction. The immediate children of *Entity* are *Physical Object* and *Conceptual Entity*, while *Event* has *Activity* and *Phenomenon or Process* as direct descendants (Figure 8-6). Each semantic type in the network has a textual definition and appears in one of these hierarchies. In addition to the taxonomy, associative relationships in five subcategories are defined between semantic types: physical (e.g., *part_of*, *branch_of*, *ingredient_of*), spatial (e.g., *location_of*, *adjacent_to*), functional (e.g., *treats*, *complicates*, *causes*), temporal (e.g., *co-occurs_with*, *precedes*), and conceptual (e.g., *evaluation_of*, *diagnoses*). Since each Metathesaurus concept is assigned at least one semantic type, relationships between semantic types also define the allowable semantics for relationships between concepts (McCray and Bodenreider 2002).

The categorization of concepts by semantic type is subject to the economy principle (similar to the notion of parsimony developed in (Gruber 1995; Swartout et al. 1996)) and has three key features: (1) Since the most specific semantic type in the taxonomy is assigned to a concept, level of granularity varies across the UMLS (McCray and Hole 1990). (2) Due to single-inheritance tree structure rather than a lattice allowing multiple inheritance, a Metathesaurus concept cross-categorized by two semantic types is assigned to both types. (3) Rather than proliferating semantic types,

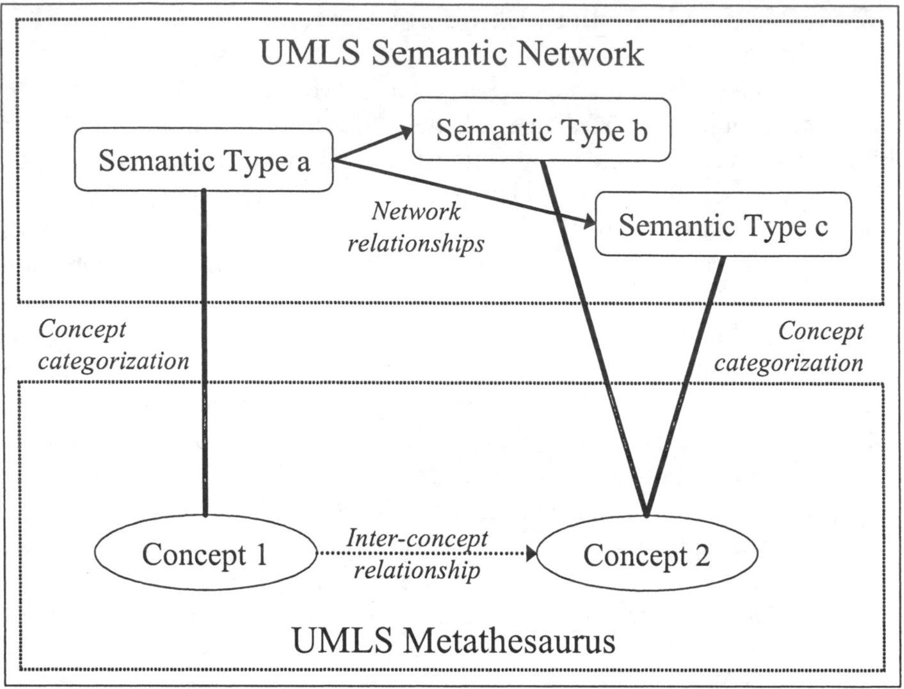

Figure 8-5. The two-level structure in the UMLS.

concepts that cannot be categorized by existing sibling types are assigned their common supertype (McCray and Nelson 1995). The consequences of the economy principle for representing knowledge in the UMLS are discussed elsewhere (Burgun and Bodenreider 2001c).

3.3 The Systematized Nomenclature of Medicine

The Systematized Nomenclature of Medicine (SNOMED®) Clinical Terms® (SNOMED CT), developed by the College of American Pathologists, was formed by the convergence of SNOMED RT and Clinical Terms Version 3 (formerly known as the Read Codes). SNOMED CT is the most comprehensive biomedical terminology recently developed in native description logic formalism. The version described here (January 31, 2004) contains 269,864 classes[1], named by 407,510 names[2]. SNOMED CT is now available as part of the UMLS[3] at no charge for UMLS licensees in the

[1] SNOMED CT has a total of 357,135 classes of which 269,864 are "current"

[2] Among the 957,349 names in SNOMED CT, 407,510 correspond to the 269,864 "current" classes, excluding fully specified names and keeping only names whose status is "current"

[3] http://umlsinfo.nlm.nih.gov/

United States. It is therefore likely to become widely used in medical information systems.

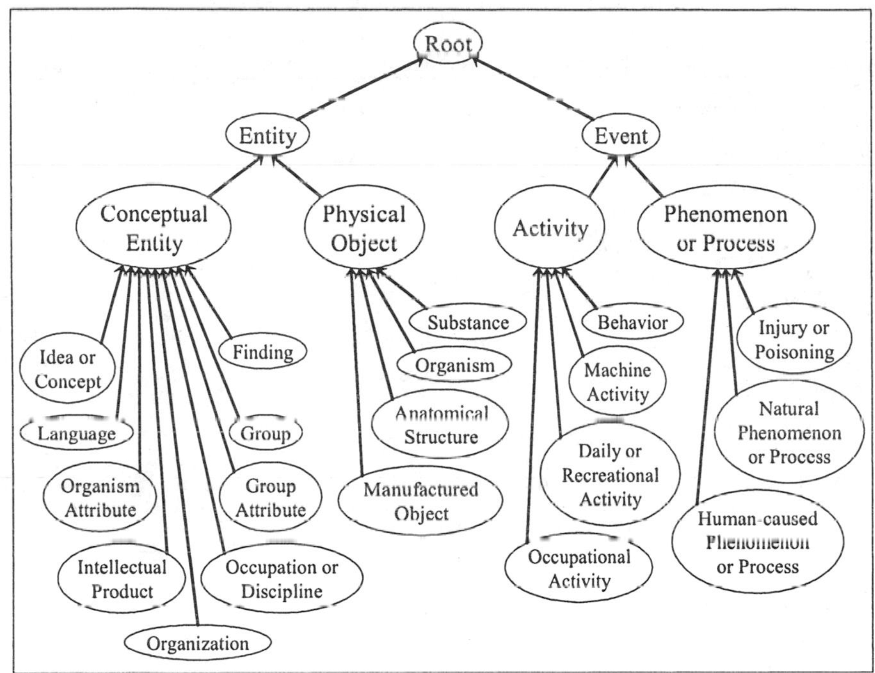

Figure 8-6. Top level in the UMLS Semantic Network.

Each SNOMED CT concept is described by a variable number of elements. For example, the class *Viral meningitis* has a unique identifier (58170007), two parents (*Infective meningitis* and *Viral infections of the central nervous system*), several names (*Viral meningitis*, *Abacterial meningitis*, and *Aseptic meningitis, viral*). The roles (or semantic relations) present in the definition of this concept are listed in Table 8-1.

Table 8-1. Roles present in the definition of *Viral meningitis*

Role	Value
Causative agent	Virus
Associated morphology	Inflammation
Finding site	Meninges structure
Onset	Sudden onset;Gradual onset
Severitiy	Severities
Episodicity	Episodicities
Course	Courses

SNOMED CT consists of eighteen independent hierarchies reflecting, in part, the organization of previous versions of SNOMED into "axes" such as *Diseases*, *Drugs*, *Living organisms*, *Procedures* and *Topography*. The first level concepts are listed in Table 8-2 with their frequency distribution.

Table 8-2. The eighteen top-level concepts in SNOMED CT and their frequency distribution

Top-level concepts	Frequency
Attribute	991
Body structure	30,652
Clinical finding	95,605
Context-dependent categories	3,649
Environments and geographical locations	1,620
Events	87
Observable entity	7,274
Organism	25,026
Pharmaceutical / biologic product	16,867
Physical force	199
Physical object	4,201
Procedure	46,066
Qualifier value	8,134
Social context	4,896
Special concept	178
Specimen	1,053
Staging and scales	1,098
Substance	22,267

3.4 Foundational Model of Anatomy

Development of the Foundational Model of Anatomy (FMA) at the University of Washington grew out of earlier work to enhance the anatomical content of the UMLS. By focusing exclusively on the representation of structure, the FMA expects to serve as a reference ontology, i.e., to allow other ontologies of which anatomy is a component to be aligned with it (Rosse and Mejino 2003). Specifically, the goal of the FMA is to provide a conceptualization of the material objects and spaces that constitute the human body. It integrates an Anatomical Ontology with two much smaller structures: the Physical State Ontology and the Spatial Ontology. The latter represents geometric objects and three-dimensional shape classes, and also distinguishes between bona fide (real) and fiat (virtual) boundaries of volumes, surfaces, and lines. The Anatomical Ontology contains nearly 70,000 concepts originally limited to gross anatomy and is now being extended to cellular and sub-cellular phenomena.

FMA is implemented in Protégé[4], which is a frame-based ontology editing environment developed at Stanford University.

Definitions of physical anatomical entities in the FMA are formulated by specifying constraints (Michael et al. 2001) based on spatial dimension, mass, and inherent three-dimensional shape, as well as the structural units that make up the body. Relationships, however, are constrained to the structural organization of physical anatomical entities. The top level of the taxonomy is *Anatomical entity*, which is divided into *Physical anatomical entity* and *Non-physical anatomical entity* (Figure 8-7). Physical entities have spatial dimension, while non-physical entities, such as *Developmental stage*, do not. Further distinction is made between physical entities that have mass, such as anatomical structures and body substances (*Material physical anatomical entity*), and those that do not, including anatomical spaces, surfaces, lines, and points (*Non-material physical anatomical entity*). The attribute of inherent three-dimensional shape contrasts anatomical structures, which are objects, with body substances.

In addition to the anatomical taxonomy, hierarchies have been formulated using the transitive *part-of* relation as well as two anatomical relations, *branch-of* and *tributary-of*, which represent relationships among tree-like structures such as nerves, arteries, veins, and lymphatic vessels. Moreover, the FMA extends these relationships to boundary, orientation, connectivity, and location; the latter is specified using containment, adjacency, and anatomical coordinates (Mejino et al. 2001).

3.5 MENELAS Ontology

MENELAS, a European Union project for accessing medical records in several European languages (Zweigenbaum 1994), takes a knowledge-based approach to natural language understanding. A pilot application covering coronary artery disease has been developed, and resources (represented as conceptual graphs) include domain-specific syntactic and semantic lexicons as well as an ontology of coronary artery diseases enhanced with structured encyclopedic knowledge for each concept.

The MENELAS ontology (see Figure 8-8 for the top level) has 1,800 concepts and 300 relationship types acquired from several sources, including interviews with physicians, reuse of existing terminological resources, and corpus analysis. It was initially developed as a lattice (Bouaud et al. 1994); however, to avoid ambiguities due to multiple inheritance, the principles of opposition of siblings and unique semantic axis were later adopted, leading to a tree structure (Zweigenbaum et al. 1995). Concept labels in the ontology

[4] http://protege.stanford.edu/

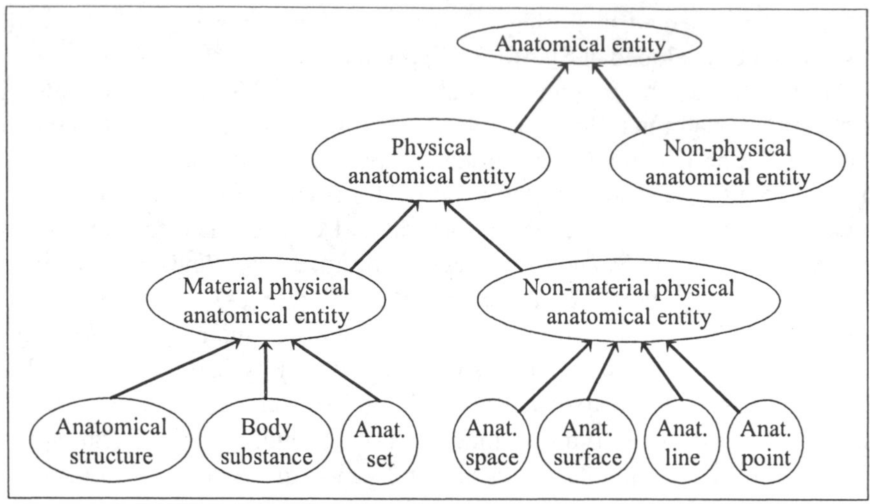

Figure 8-7. Top level in the Foundational Model of Anatomy (Anatomical taxonomy).

are simply mnemonic; the actual meaning of a concept comes from its position in the hierarchy. For example, *Physical object* is a child of *Abstract object*, which in turn is a child of *Substratum*. The latter concept is defined as having instances in the world and is opposed to *Ideal object*: *Apple* is an *Abstract object,* whereas *Two* is an *Ideal object*.

Relations are categorized according to the kinds of concepts they link. Relations between physical objects, for example, link mass objects and countable objects (*contains*, *has for dosage*, and *constituted of*) or real objects and pseudo-objects (*component of*). The *part of* relation links any kind of physical object and has children *part fragment* and *part segment*. There is also a relation, *functional part*, to represent functional viewpoints. Models and schemas provide additional knowledge, which may be limited to the domain-specific and task-oriented context of the MENELAS application. For example, the model for organ component includes the notion of duct in order to accommodate the coronary arteries.

4. REPRESENTATIONS OF THE CONCEPT *BLOOD*

Having discussed the general characteristics and top level organization of several ontologies, we now examine the representation of blood in these systems and analyze the differences among representations. We also show how most ontologies provide a rich representation compared to mere taxonomies by including additional knowledge.

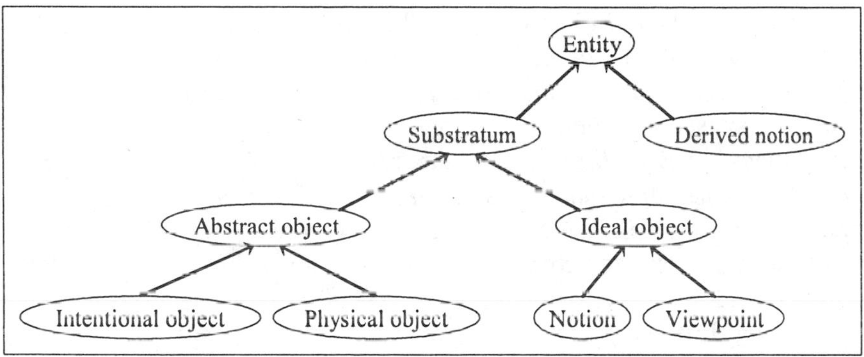

Figure 8-8. Top level in MENELAS.

4.1 *Blood* in Biomedical Ontologies

What makes the representation of *Blood* interesting is the dual nature of blood as both tissue and fluid, a dichotomy reflected in medical dictionary definitions: (1) "the fluid that circulates through the heart, arteries, capillaries, and veins, carrying nutriment and oxygen to the body cells" (*Dorland's Illustrated Medical Dictionary*); and (2) "the 'circulating tissue' of the body; the fluid and its suspended formed elements that are circulating through the heart, arteries, capillaries, and veins" (*Stedman's Medical Dictionary*). In the following discussion, comparison of the ontologies is based on textual and formal definitions of *Blood* as well as ontological properties of that concept.

Although not represented as a type in **OpenCyc**, *Blood* is a specialization of *Mixture*, along with *Mud*, *Air*, and *Carbonated beverage*. (*Blood* referring to lineage is represented separately.) *Mixture* is a subclass of *Partially tangible* and represents a homogeneous, partially tangible thing composed of two or more different constituents which have been mixed. Because its constituents do not form chemical bonds, a mixture may be resolved by a separation event. As a mixture, *Blood* is an element of the collection *Existing stuff type* (`#$isa #$Mixture #$ExistingStuffType`), which implies that division in time or space does not destroy its stuff-like quality. In OpenCyc, *Blood* is represented differently from *Sweat* and *Semen*, which are subordinates of *Bodily secretion*. In addition, *Sweat*, considered as a waste, is also a descendant of *Excretion substance*.

Blood is defined in **WordNet** as "the fluid (red in vertebrates) that is pumped by the heart. Blood carries oxygen and nutrients to the tissues and carries waste products away; the ancients believed that blood was the seat of the emotions." There are five other meanings of "blood," including one referring to temperament or disposition. The direct hypernym of *Blood* is

Liquid body substance. (The complete hierarchy for *Blood* in WordNet is given in Figure 8-9a.) *Blood*, *Sweat*, and *Semen*, are categorized as *Liquid body substance*. Unlike *Blood*, *Sweat* is linked to *Liquid body substance* through the synset *Secretion*.

In **OpenGALEN**, *Blood* is a subordinate of *Soft tissue* as well as *Lymphoid tissue*, *Integument*, and *Erectile tissue*, among others. The hierarchy for *Blood* in GALEN appears in Figure 8-9b. This structure is actually a lattice, since *Substance* is the common subtype of *Generalised substance* and *Substance or physical structure*, both being subtypes of *Phenomenon*. In GALEN, *Blood* is represented differently from *Sweat* and *Semen*, which are subordinates of *Body fluid*.

Blood has the semantic type *Tissue* in the **UMLS Metathesaurus**, which is defined as "An aggregation of similarly specialized cells and the associated intercellular substance. Tissues are relatively non-localized in comparison to body parts, organs or organ components." *Tissue* is a subordinate of *Fully-formed anatomical structure* in the Semantic Network (Figure 8-9c has the entire *is-a* hierarchy for *Blood*). In the UMLS, *Blood* is not assigned the same semantic type as *Sweat* and *Semen*, which are categorized as *Body substance*. Moreover, in the Metathesaurus, ancestors of *Blood* include *Body fluid*, *Body substance*, *Soft tissue* and *Connective tissue*.

In **SNOMED CT**, *Blood* is found in the concept category *Substance* as a subordinate of *Blood material*, as well as *Blood component*. (The hierarchical environment for *Blood* in SNOMED CT is given in Figure 8-9d.) Multiple inheritance allows *Body fluid*, an ancestor of *Blood*, to inherit from both *Body substance* and *Liquid substance*. These two concepts are descendants of the top level category *Substance*. Subordinates of *Body fluid* also include *Sweat* and *Semen*, as well as *Lymph* and *Pus*.

The **Foundational Model of Anatomy** (FMA) represents *Blood* as a subordinate of *Body substance*, which is defined as "a material physical anatomical entity in a gaseous, liquid, semisolid or solid state, with or without the admixture of cells and biological macromolecules; produced by anatomical structures or derived from inhaled and ingested substances that have been modified by anatomical structures as they pass through the body." In addition to *Blood*, this definition covers other cellular fluids, such as *Semen*, as well as secretions (e.g., *Saliva* and *Sweat*), transudates (e.g., *Lymph*, and *Cerebrospinal fluid*), excretions (e.g., *Feces* and *Urine*), along with *Respiratory air* and *Aqueous humor of eyeball*. *Blood* is not considered to be a tissue in the FMA. The complete *is-a* hierarchy for *Blood* is represented in Figure 8-9e, and this lineage is distinct from that of *Tissue*, largely because substances, as defined in the FMA, do not have inherent three-dimensional shape. *Tissue* inherits properties from its ancestor

Anatomical structure, which is a sister of *Body substance* and is differentiated from it by the feature inherent 3D shape.

In **MENELAS**, *Blood* (along with *Lymph*) is a subordinate of *Body fluid*. The ancestors of *Blood* can be found in Figure 8-9f. One of these, *Mass object*, has three subtypes: *Agglomerate* (divided into *Inorganic agglomerate* and *Organic agglomerate*), *Substance* (*Biochemical substance* and *Chemical substance*), and *Tissue* (*Body fluid* and *Connective tissue*). *Blood* as a child of *Body fluid* belongs to a different branch from the one dominated by *Substance*. Furthermore, *Tissue*, defined as a set of cells, is differentiated from *Substance*, defined as a set of molecules. A "model" (which provides additional knowledge) is associated with the concept *Body fluid* and emphasizes one property of fluids, namely viscosity, a feature pertinent to natural language understanding in the MENELAS application. The representation of *Body fluid* as tissue in MENELAS is noncanonical, given that other ontologies separate fluids and substances from tissue. *Semen* is outside the scope of this application ontology for interpreting coronary angiography reports, while *Sweat* is categorized as *Cutaneous sign* (sweating), rather than *Substance*.

4.2 Differing Representations

The differing representations of *Blood* in several systems raise issues about compatibility among ontologies. Obviously, the representation of most concepts is simpler than that of *Blood*, and the ontologies studied often provide compatible views on the biomedical domain. What makes the representation of *Blood* more complex is that two different superordinates are found: *Tissue* and *Body substance*. GALEN and the UMLS Semantic Network categorize *Blood* as *Tissue* while the Foundational Model of Anatomy categorizes it as *Body substance*. In between, WordNet, SNOMED CT and MENELAS categorize *Blood* as *Body fluid*, itself categorized as *Body substance* in WordNet and SNOMED CT, but as *Tissue* in MENELAS. Finally, in GALEN, *Tissue* is a subtype of *Body substance*. A composite representation of *Blood* is shown in Figure 8-10.

Superficially, this dual representation of *Blood*, as both *Tissue* and *Body substance*, does not reveal any major incompatibility, such as circular hierarchical relationships. However, a unified representation in which *Blood* is a common subtype of *Tissue* and *Body substance* would violate the constraint of opposition of siblings. Analyzed more carefully, the definitions of *Tissue* in the Foundational Model of Anatomy (FMA) and the Semantic Network are closely related but not equivalent (the complete definitions are shown in Figure 8-10). In both systems, *Tissue* is a kind of anatomical structure consisting of "similarly specialized cells and intercellular substance/matrix."

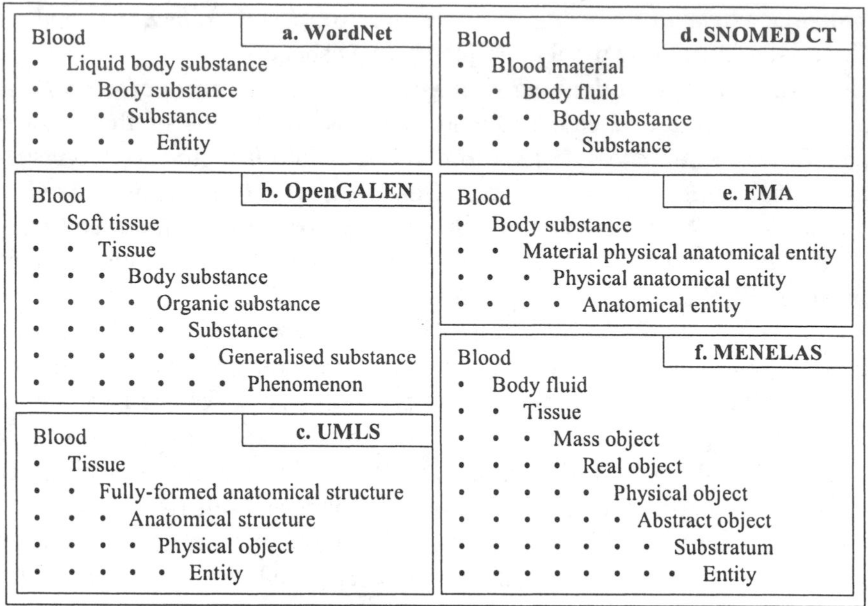

Blood
a. WordNet
• Liquid body substance
• • Body substance
• • • Substance
• • • • Entity
Blood
b. OpenGALEN
• Soft tissue
• • Tissue
• • • Body substance
• • • • Organic substance
• • • • • Substance
• • • • • • Generalised substance
• • • • • • • Phenomenon
Blood
c. UMLS
• Tissue
• • Fully-formed anatomical structure
• • • Anatomical structure
• • • • Physical object
• • • • • Entity
Blood
d. SNOMED CT
• Blood material
• • Body fluid
• • • Body substance
• • • • Substance
Blood
e. FMA
• Body substance
• • Material physical anatomical entity
• • • Physical anatomical entity
• • • • Anatomical entity
Blood
f. MENELAS
• Body fluid
• • Tissue
• • • Mass object
• • • • Real object
• • • • • Physical object
• • • • • • Abstract object
• • • • • • • Substratum
• • • • • • • • Entity

Figure 8-9. Representation of *Blood* in several biomedical ontologies.

The difference between the two systems lies in the precision – found only in the FMA – that this aggregation must follow "genetically determined spatial relationships". Blood cells in suspension in plasma or aggregated after sedimentation are indeed similarly specialized and correspond to the definition of *Tissue* in the UMLS Semantic Network. However, their spatial organization differs from that of an epithelium, muscle tissue and neural tissue in that it is not genetically determined but rather depend on the characteristics of blood circulation. This additional criterion is particularly relevant for disambiguating the classification of *Blood* in the FMA. Moreover, the categorization of *Blood* as *Body substance* rather than *Tissue* in the FMA is consistent with the distinction introduced between *Anatomical structure* (of which *Tissue* is a subtype) and *Body substance* through the property has inherent 3D shape, which is present in *Tissue* and absent in *Body substance.*

The representation of *Blood* illustrates other differences across ontologies. While most ontologies represent the prototypical form of blood (i.e., the fluid circulating in the cardiovascular system), GALEN distinguishes between liquid and coagulated blood. The issue here is that the properties inherent to fluids are inherited by *Blood* in WordNet, SNOMED CT, FMA and MENELAS. As a consequence, if GALEN were integrated with these representations as shown in Figure 8-10, *Coagulated blood*, a descendant of *Blood*, would wrongly inherit such properties. Analogously,

Body substance is likely to represent different entities in FMA and in GALEN. As mentioned earlier, *Body substance* in FMA is a *Material physical anatomical entity* with no inherent three-dimensional shape. In GALEN, *Body substance* is more general, encompassing both *Tissue* and *Body fluid* and defined as an *Organic substance* playing a role in physiology.

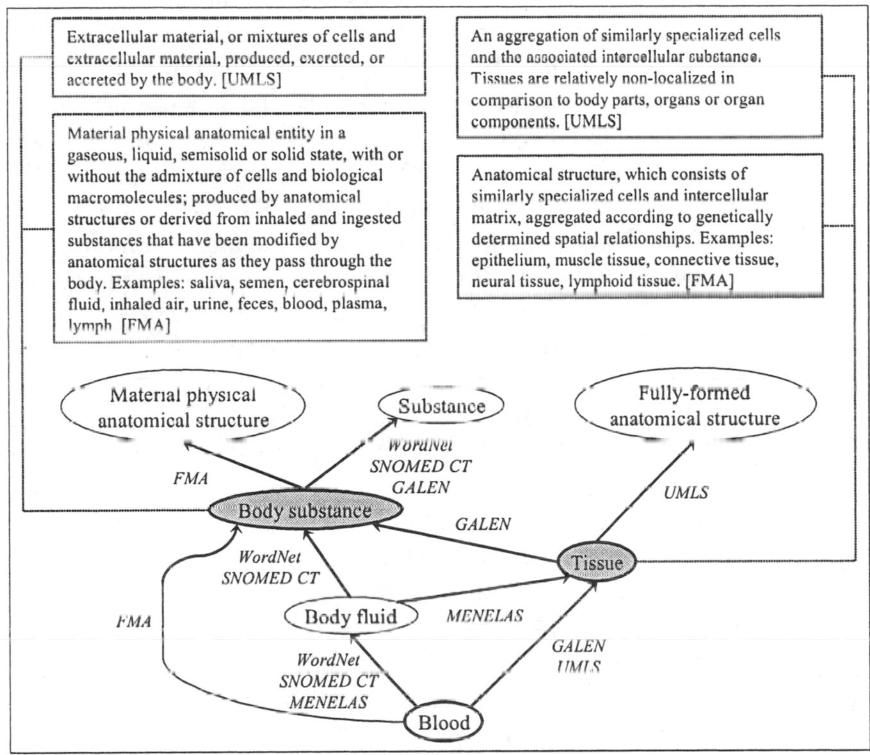

Figure 8-10. Composite representation of *Blood.*

4.3 Additional Knowledge

Taxonomy, i.e., the arrangement of concepts in *is-a* hierarchies, plays a central role in ontologies, of which such hierarchies constitute the backbone. In addition to the relative position of *Blood* in their hierarchies, most ontologies provide additional knowledge about *Blood* through properties attached to this concept and through the associative relations of *Blood* to other concepts. OpenCyc categorizes *Blood* as a *Mixture*, indicating that it can be subject to events such as *Separation mixture*. Erythrocyte sedimentation, resulting from the reversible separation of blood components, is an example of such events. In SNOMED CT, *Blood* is involved in the definition of other concepts through specific roles which provide additional knowledge about it. *Blood* can be analyzed (e.g., `Blood specimen` *has specimen substance* `Blood`), can be the object of medical procedures (e.g., `Transfusion of whole blood` *has direct substance* `Blood` and `Finger-prick sampling` *has direct substance* `Blood`) and can enter in the composition of clinical drugs (e.g., `Antithrombin III preparation` *has active ingredient* `Blood`). As a *Body fluid* in MENELAS, *Blood* acquires the viscosity property. *Blood* is also a subtype of *Mass object* and inherits the general knowledge represented for this type through relations (e.g., *Mass object* may be a component of *Countable object*) and properties (e.g., quantity, expressed with quantitative values and units). GALEN identifies two distinct physical states for *Blood*: *Liquid blood* and *Coagulated blood*, (both represented as descendants of *Blood*). In addition, *Blood* inherits the properties of *Body substance* (e.g., `Body substance` *plays physiological role* `Organic role`). Additionally, GALEN extends the representation of *Blood* through roles such as `Blood` *has countability* `Infinitely divisible`. This role, inherited from *Substance*, expresses that *Blood* is not a discrete object. By categorizing *Blood* as *Tissue* in the UMLS, potential relationships with other kinds of entities can be inferred from the Semantic Network. Relationships of *Tissue* to other Semantic Types, result in predicates including `Tissue` *produces* `Biologically active substance`, `Tissue` *is a location of* `Pathologic function`, `Embryonic structure` *is a developmental form of* `Tissue`, and `Tissue` *surrounds* `Tissue`. In the Foundational Model of Anatomy, *Blood* inherits from *Body substance* the value *False* for the property has inherent 3D shape. The anatomical structures containing *Blood* including *Cavity of cardiac chamber* and *Lumen of cardiovascular system* are represented through relations such as `Blood` *contained in* `Cavity of cardiac chamber`.

5. ISSUES IN ALIGNING AND CREATING BIOMEDICAL ONTOLOGIES

As more biomedical ontologies are created, users might be tempted to integrate these sets of concepts and relations into a single system. However, the analysis of the differences in representation of *Blood* illustrated the limitations of a naïve approach to merging ontologies, even when representations occur within a single theory of the domain (i.e., Western medicine). While difference in granularity is usually not a problem, different naming conventions, the lack of reliable textual definitions and the lack of explicit and consistently applied classificatory principles may result in merging difficulties. Additional difficulty is encountered when attempting to merge ontologies that convey different theories of the domain (e.g., Western and Oriental medicine or modern medical knowledge and pre-scientific representations of the human body). In this case, the target system must be able to clearly identify the underlying theories and to represent them separately. Tools have been developed to assist the ontology developer in merging existing ontologies (Noy and Musen 1999).

Ontology design can benefit from two complementary approaches. First, some methodologies such as the Protégé software engineering methodology, aim at providing a clear division between domain ontologies and domain-independent problem-solvers that, when mapped to domain ontologies, can solve application tasks (Musen 1998). Second, ontologies can be improved by drawing on the results of recent research in philosophy called formal ontology. For example, Guarino et al. (2000) have developed methods built around the fundamental philosophical theories of identity, unity, rigidity and dependence, that can be used to reduce inconsistencies in *is-a* hierarchies. Mereotopology, the theory of parts and boundaries, addresses issues in *part-of* hierarchies. Exploiting these theories helps design principled ontologies. Applied to the biomedical domain, formal ontology addresses, for example, distinctions between a person and its body, or between being a person and being a patient. More generally, formal ontology helps create consistent upper-level ontologies to which domain ontologies can be hooked. For example, the principles of mereotopology have been applied to the representation of anatomical structures and subdivisions of the human body.

6. CONCLUSION

Although general ontologies and limited application ontologies may be useful, biomedical applications (e.g., clinical decision support systems, medical language processing and information retrieval) would benefit from large, principled domain ontologies. We examined some of the biomedical ontologies currently available and found that none of them fully meets the requirements of formal organization. Not surprisingly, we observed a certain lack of compatibility among their representations. Several factors contribute to this situation. First, there is no agreement on an upper level ontology to which a biomedical ontology could hook its concepts. Second, there is no unique theory of the domain, and some characteristics of biomedicine make it particularly difficult to represent (e.g., large number of concepts and vagueness of some concepts). Finally, pragmatic aspects rather than formal principles often prevail in the design of biomedical ontologies. The contribution of formal ontology has been acknowledged and will undoubtedly benefit medical ontology. Meanwhile, we believe that identifying and clarifying the core concepts and relationships of the domain will contribute to improve the sharability of existing ontologies as well as the interoperability of the applications that rely on them.

7. ACKNOWLEDGMENTS

The authors would like to thank Jeremy Rogers, Cornelius Rosse, Jacques Bouaud and Pierre Zweigenbaum for providing valuable insights about the ontologies to which they contributed. Special thanks to Tom Rindflesch for his encouragement, useful comments and invaluable editorial assistance on this manuscript.

This work was done in part while Anita Burgun was a visiting faculty at the Lister Hill National Center for Biomedical Communications, National Library of Medicine, National Institutes of Health, Department of Health and Human Services.

REFERENCES

Bodenreider, O. (2001). *Medical Ontology Research (Report to the Board of Scientific Counselors)*, Lister Hill National Center for Biomedical Communications, Bethesda, Maryland.

Bouaud, J., Bachimont, B., Charlet, J., and Zweigenbaum, P. (1994). "Acquisition and Structuring of an Ontology within Conceptual Graphs," in: *ICCS'94 Workshop on Knowledge Acquisition using Conceptual Graph Theory*, University of Maryland, College Park, MD, pp. 1-25.

Brachman, R.J. and Levesque, H.J. (2003). *Knowledge Representation and Reasoning*, Morgan Kaufmann, Amsterdam; Boston.

Burgun, A. and Bodenreider, O. (2001a). "Comparing Terms, Concepts and Semantic Classes in WordNet and the Unified Medical Language System," in *Proc NAACL Workshop, "WordNet and Other Lexical Resources: Applications, Extensions and Customizations"*:77-82.

Burgun, A. and Bodenreider, O. (2001b). "Mapping the UMLS Semantic Network into General Ontologies," in *Proc AMIA Symp*:81-85.

Burgun, A. and Bodenreider, O. (2001c). "Aspects of the Taxonomic Relation in the Biomedical Domain," in *Collected papers from the Second International Conference "Formal Ontology in Information Systems"* (ed. C. Welty and B. Smith), ACM Press, pp. (222-233.

Degoulet, P., Sauquet, D., Jaulent, M.C., Zapletal, E., and Lavril, M. (1998). "Rationale and Design Considerations for a Semantic Mediator in Health Information Systems," *Methods Inf Med* 37(4-5):518-526.

Dorland's Illustrated Medical Dictionary. Philadelphia, W.B. Saunders [see most current edition].

Fellbaum, C., ed. (1999). *WordNet: An Electronic Lexical Database*, MIT Press, Cambridge, Massachusets.

Gangemi, A. and Oltramari, A. (2001). "A Formal Ontology Approach to Refine Lexical Taxonomies: The Case of WordNet Top Level," in *Collected papers from the Second International Conference "Formal Ontology in Information Systems"* (ed. C. Welty and B. Smith), ACM Press, pp. (285-296.

Gruber, T.R. (1995). "Toward Principles for the Design of Ontologies Used for Knowledge Sharing," *International Journal of Human-Computer Studies* 43(5-6):907-928.

Guarino, N. and Welty, C. (2000). "A Formal Ontology of Properties," in *EKAW-2000: The 12th International Conference on Knowledge Engineering and Knowledge Management*, (ed. R. Dieng and O. Corby), Springer-Verlag, pp. 97-112.

Hahn, U., Romacker, M., and Schulz, S. (1999). "How Knowledge Drives Understanding--Matching Medical Ontologies with the Needs of Medical Language Processing," *Artif Intell Med* 15(1):25-51.

Lindberg, D.A., Humphreys, B.L., and McCray, A.T. (1993). "The Unified Medical Language System," *Methods Inf Med* 32(4):281-291.

McCray, A.T. (2003). "An Upper-level Ontology for the Biomedical Domain," *Comparative And Functional Genomics* 4(1):80-84.

McCray, A.T. and Bodenreider, O. (2002). "A Conceptual Framework for the Biomedical Domain," in *The semantics of relationships: an interdisciplinary perspective* (ed. R. Green, C.A. Bean, and S.H. Myaeng), Kluwer Academic Publishers, Boston, pp. 181-198.

McCray, A.T. and Hole, W.T. (1990). "The Scope and Structure of the First Version of the UMLS Semantic Network," in *Proc Annu Symp Comput Appl Med Care*:126-130.

McCray, A.T. and Nelson, S.J. (1995). "The Representation of Meaning in the UMLS," *Methods Inf Med* 34(1-2):193-201.

Mejino, J.L., Jr., Noy, N.F., Musen, M.A., Brinkley, J.F., and Rosse, C. (2001). "Representation of Structural Relationships in the Foundational Model of Anatomy," *Proc AMIA Symp*:973.

Michael, J., Mejino, J.L., Jr., and Rosse, C. (2001). "The Role of Definitions in Biomedical Concept Representation," *Proc AMIA Symp*:463-468.

Musen, M.A. (1998. "Domain Ontologies in Software Engineering: Use of Protege with the EON Architecture," *Methods Inf Med* 37(4-5):540-550.

Musen, M.A. (2002. "Medical Informatics: Searching for Underlying Components," *Methods Inf Med* 41(1):12-19.

Noy, N.F. and Musen, M.A. (1999). "An Algorithm for Merging and Aligning Ontologies: Automation and Tool Support," in *Proceedings of the 16th National Conference on Artificial Intelligence (AAAI-99) Workshop on Ontology Management*, AAAI Press, Orlando, Florida.

Pisanelli, D.M., Gangemi, A., Battaglia, M., and Catenacci, C. (2004). "Coping with Medical Polysemy in the Semantic Web: The Role of Ontologies," *Medinfo* 2004:416-419.

Rector, A.L., Bechhofer, S., Goble, C.A., Horrocks, I., Nowlan, W.A., and Solomon, W.D. (1997). "The GRAIL Concept Modelling Language for Medical Terminology," *Artif Intell Med* 9(2):139-171.

Rogers, J. and Rector, A. (2000). "GALEN's Model of Parts and Wholes: Experience and Comparisons," *Proc AMIA Symp*:714-718.

Rosse, C. and Mejino, J.L., Jr. (2003). "A Reference Ontology for Biomedical Informatics: the Foundational Model of Anatomy", *J Biomed Inform* 36(6):478-500.

Sowa, J.F. (2000). *Knowledge Representation: Logical, Philosophical, and Computational foundations*, Brooks/Cole, Pacific Grove, Ca.

Stedman's Medical Dictionary. Philadelphia, Lippincott Williams & Wilkins [see most current edition].

Swartout, B., Patil, R., Knight, K., and Russ, T. (1996). "Toward Distributed Use of Large-Scale Ontologies," in *Proceedings of the 10th Workshop on Knowledge Acquisition, Modeling and Management*, (ed. B. Gaines and M.A. Musen), Banff, Canada.

Trombert-Paviot, B., Rodrigues, J.M., Rogers, J.E., Baud, R., van der Haring, E., Rassinoux, A.M., Abrial, V., Clavel, L., and Idir, H. (2000). "GALEN: A Third Generation Terminology Tool to Support a Multipurpose National Coding System for Surgical Procedures," *Int J Med Inf* 58-59:71-85.

Zweigenbaum, P. (1994). "Menelas - an Access System for Medical Records Using Natural-Language," *Computer Methods and Programs in Biomedicine* 45(1-2):117-120.

Zweigenbaum, P., Bachimont, B., Bouaud, J., Charlet, J., and Boisvieux, J.F. (1995). "Issues in the Structuring and Acquisition of an Ontology for Medical Language Understanding," *Methods of Information in Medicine* 34(1-2):15-24.

SUGGESTED READINGS

Pisanelli, D.M (Ed.). (2004. Ontologies in medicine, Studies in Health Technology and Informatics, Vol. 102, IOS Press, Amsterdam; Burke, VA.
The aim of this book is both to review fundamental theoretical issues in ontology and to demonstrate the practical effectiveness of the ontological approach by means of a series of case studies in specific problem areas. This book presents a survey of the most important contributions to the topic of formal ontology in medicine.

Smith, B. (2004. Ontology, in: The Blackwell guide to the philosophy of computing and information (ed. L. Floridi), Blackwell Pub., Malden, MA, pp. 155-166.
This article defines philosophical ontology and discusses relevance to information systems. It provides numerous references to recent studies on formal ontology.

ONLINE RESOURCES

Ontology development resources:

- Protégé (ontology editor, available at http://protege.stanford.edu/)
- N. Noy and D. L. McGuinness. Ontology Development 101: A Guide to Creating Your First Ontology. Stanford University, 2001, Technical report SMI-2001-0880. (Available at http://www-smi.stanford.edu/pubs/SMI_Reports/SMI-2001-0880.pdf)

More than one hundred biomedical vocabularies are integrated in the Unified Medical Language System (UMLS) Metathesaurus, along with the UMLS Semantic Network and the SPECIALIST lexicon and lexical programs (available at http://umlsks.nlm.nih.gov/). The UMLS is available free of charge, but users are required to sign a license agreement.

Biomedical terminology and ontology resources not discussed in the chapter:

- Standards and Ontologies for Functional Genomics (http://sofg.org/)
- Gene Ontology (http://geneontology.org/)
- Open Biological Ontologies (http://obo.sourceforge.net/)

National Cancer Institute's Thesaurus

- http://ncicb.nci.nih.gov/download/index.jsp

QUESTIONS FOR DISCUSSION

1. What are the principal differences between ontologies and controlled vocabularies?

2. What are the different kinds of ontologies?

3. What are the major kinds of relationships represented in biomedical ontologies?

4. What are the major formalisms used to represent biomedical ontologies?

5. What is the main difference between the representation of anatomy in the FMA and in other biomedical ontologies?

6. Is it possible / desirable to merge several biomedical ontologies into a single structure?

7. Why are upper-level ontologies important to biomedical ontologies?

8. How can biomedical ontology benefit from the philosophical principles of formal ontology?

9. What tasks would benefit from biomedical ontologies?

APPENDIX

Table 8-3. References for the ontologies mentioned in this chapter

Ontology	URL
Foundational Model of Anatomy	http://fma.biostr.washington.edu/
MENELAS	http://www.biomath.jussieu.fr/~pz/Menelas/
OpenCyc™	http://www.opencyc.com/
OpenGALEN	http://www.opengalen.org/
SNOMED CT®	http://www.snomed.org/
Unified Medical Language System®	http://umlsks.nlm.nih.gov/ (free UMLS registration required)
WordNet®	http://www.cogsci.princeton.edu/~wn/

Table 8-4. Some characteristics of the ontologies mentioned in this chapter.

Name	**Version**	**Date**	**Scope**	**Objective**	**Formalism**	**Number of concepts**	**Number of relation-ship types**	**Number of assertions (explicitly represented)**
OpenCyc™	0.7	Dec. 2002	General	To support common-sense reasoning	Cycl.	6,000	n/a	60,000
WordNet®	2.0	Aug. 2003	General	Lexical reference	Graph of synsets	152,000	7	344,000
OpenGALEN	6	Dec. 2002	Clinical medicine	To support terminology services	Description logic (GRAIL)	25,000	594	216,000
UMLS® Semantic Network	2004 (AC)	Nov. 2004	Bio-medicine	To provide a consistent categorization of all concepts represented in the UMLS Metathesaurus	Semantic network	135	54	6,864
SNOMED CT®		Jan. 31, 2004	Clinical medicine	Capturing, sharing and aggregating health data	Description logic	270,000	50	1.5 M
Foundational Model of Anatomy		Dec. 2003	Anatomy	Reference ontology	Frame-based	70,000	170	1.5 M
MENELAS	Final	March 1995	Coronary artery diseases	To support natural language processing	Conceptual graphs	1,800	300	n/a

Chapter 9
INFORMATION RETRIEVAL AND DIGITAL LIBRARIES

William R. Hersh

Department of Medical Informatics & Clinical Epidemiology, School of Medicine, Oregon Health and Science University, Portland, OR 97239

Chapter Overview

The field of information retrieval (IR) is generally concerned with the indexing and retrieval of knowledge-based information. Although the name implies the retrieval of any type of information, the field has traditionally focused on retrieval of text-based documents, reflecting the type of information that was initially available by this early application of computer use. However, with the growth of multimedia content, including images, video, and other types of information, IR has broadened considerably. The proliferation of IR systems and on-line content has also changed the notion of libraries, which have traditionally been viewed as buildings or organizations. However, the developments of the Internet and new models for publishing have challenged this notion as well, and new digital libraries have emerged.

Keywords

Information retrieval; digital library; indexing; controlled vocabulary; searching; knowledge-based information

1. OVERVIEW OF FIELDS

IR systems and digital libraries store and disseminate knowledge-based information (Hersh, 2003). What exactly do we mean by "knowledge-based"? Although there are many ways to classify biomedical information, for the purposes of this chapter we broadly divide it into two categories. *Patient-specific information* applies to individual patients. Its purpose is to inform health care providers, administrators, and researchers about the health and disease of a patient. This information typically comprises the patient's medical record. The other category of biomedical information is *knowledge-based information*. This information forms the scientific foundation of biomedicine and is derived and organized from observational and experimental research. In the clinical setting, this information provides clinicians, administrators, and researchers with knowledge that can be applied to individual patients. In the basic science (or really any scientific) setting, knowledge-based information provides the archive of research reports upon which further research builds.

Knowledge-based information is most commonly provided in scientific journals and proceedings but can be published in a wide variety of other forms, including books, clinical practice guidelines, consumer health literature, Web sites, and so forth. Figure 9-1 depicts the "life cycle" of primary literature, which is derived from original research and whose publication is dependent upon the peer review process that insures the methods, results, and interpretation of results meets muster with one's scientific peers. In some fields, such as genomics, there is an increasing push for original data to enter public repositories. In most fields, primary information is summarized in secondary publications, such as review articles and textbooks. Also in most fields, the authors relinquish the copyright of their papers to publishers, although there is increasing resistance to this, as described later in this chapter.

IR systems have usually, although not always, been applied to knowledge-based information, which can be subdivided in other ways. *Primary knowledge-based information* (also called primary literature) is original research that appears in journals, books, reports, and other sources. This type of information reports the initial discovery of health knowledge, usually with either original data or re-analysis of data (e.g., meta-analyses).

Secondary knowledge-based information consists of the writing that reviews, condenses, and/or synthesizes the primary literature. As seen in Figure 9-1, secondary literature emanates from original publications. The most common examples of this type of literature are books, monographs, and review articles in journals and other publications. Secondary literature also includes opinion-based writing such as editorials and position or policy papers.

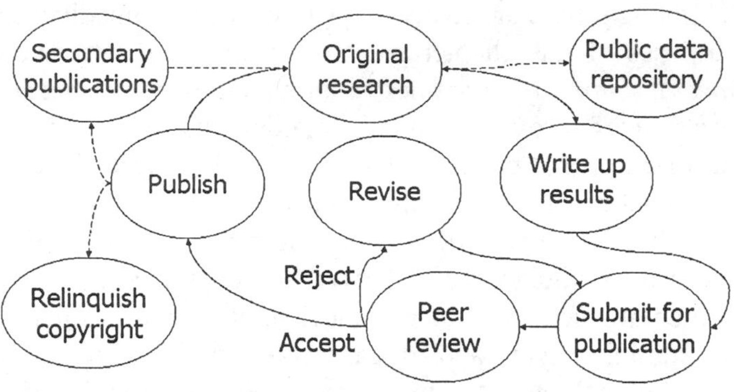

Figure 9-1. The "life cycle" of scientific information.

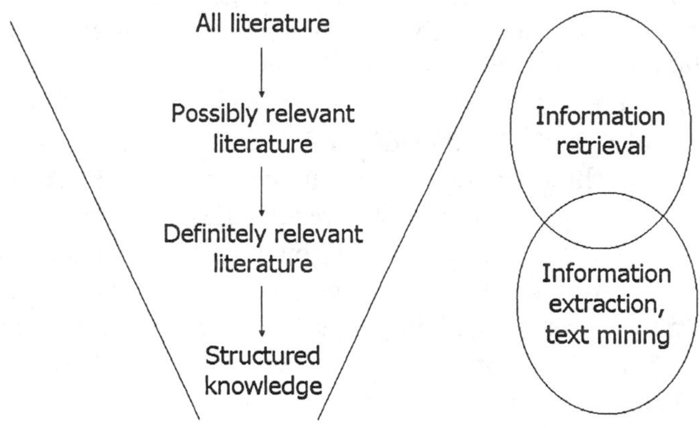

Figure 9-2. Information retrieval and extraction in context.

IR is a distinct process from information extraction (IE), which is covered in many subsequent chapters of this book dealing with vocabularies and ontologies, natural language processing, and text mining. A perspective of the role of IR is provided in Figure 9-2, which shows the flow of extracting knowledge from the scientific literature. IR typically focuses on the initial narrowing of the broad literature, ideally passing off a more focused set of articles for the more intensive processing required for IE and text mining. A goal for the latter processes is often to create structured knowledge resources that can be accessed by other informatics applications.

Libraries have been the historical place where knowledge-based information has been stored.Libraries actually perform a variety of functions including the following:

- Acquisition and maintenance of collections
- Cataloging and classification of items in collections to make them more accessible to users
- Serving as a place where individuals can go to seek information with assistance, including information on computers
- Providing work or studying space (particularly in universities)

Digital libraries provide some of these same services, but they tend to be more focused on content, particularly in digital form, as opposed to a location, although most physical libraries offer increasing amounts of digital library services (Humphreys, 2000).

2. INFORMATION RETRIEVAL

Now that we have had a general overview of knowledge-based biomedical information, we can look in further detail at IR systems. A model for the IR system and the user interacting with it is shown in Figure 9-3 (Hersh, 2003). The ultimate goal of a user of an IR system is to access content, which may be in the form of a digital library. In order for that content to be accessible, it must be described with metadata. The major intellectual processes of IR are *indexing* and *retrieval*. In the remainder of this section, we will discuss content, indexing, and retrieval, followed by an overview of how IR systems are evaluated.

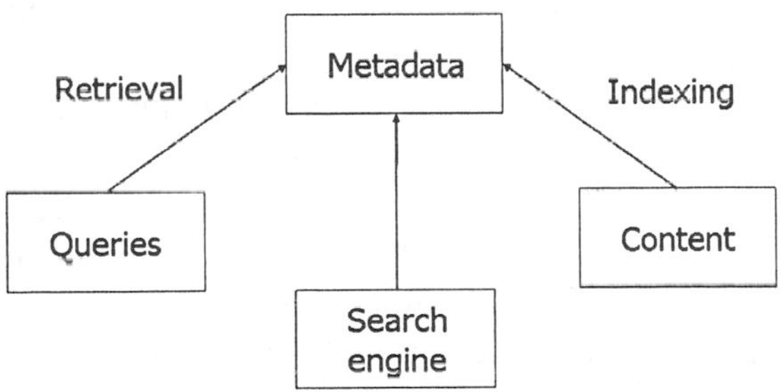

Figure 9-3. A graphic representation of the information retrieval process (Hersh, 2003).

2.1 Content

The ultimate goal of IR systems and digital libraries is to deliver information to users for specific tasks. It is useful to classify the different types of knowledge-based information to better understand the issues in its indexing and retrieval. In this section, we classify content into bibliographic, full-text, database/collection, and aggregated categories and provide an overview of each.

2.1.1 Bibliographic

The first category consists of *bibliographic content.* It includes what was for decades the mainstay of IR systems: literature reference databases. Also called *bibliographic databases*, this content consists of citations or pointers to the medical literature (i.e., journal articles). The best-known and most widely used biomedical bibliographic database is MEDLINE, which is produced by the National Library of Medicine (NLM) and contains bibliographic references to the biomedical articles, editorials, and letters to the editors in approximately 4,500 scientific journals. At present, about 500,000 references are added to MEDLINE yearly. It now contains over 12 million references.

The current MEDLINE record contains up to 49 fields. Probably the most commonly used fields are the title, abstract, and indexing terms. But other fields contain specific information that may be of great importance to smaller audiences. For example, a genomics researcher might be highly interested in the Supplementary Information (SI) field to link to genomic databases. Likewise, the Publication Type (PT) field can be of help to clinicians, designating whether an article is a practice guideline or randomized controlled trial. The NLM also partitions MEDLINE into subsets for users wishing to search on a focused portion of the database, such as *AIDS* or *Complementary and Alternative Medicine.*

MEDLINE is only one of many databases produced by the NLM (Anonymous, 2000c). Other more specialized databases are also available, covering topics from AIDS to space medicine and toxicology. There are a variety of non-NLM bibliographic databases that tend to be more focused on subjects or resource types. The major non-NLM database for the nursing field is CINAHL (Cumulative Index to Nursing and Allied Health Literature, CINAHL Information Systems, www.cinahl.com), which covers nursing and allied health literature, including physical therapy, occupational therapy, laboratory technology, health education, physician assistants, and medical records. Another database is *Excerpta Medica* (Elsevier Science Publishers, www.excerptamedica.com). EMBASE, the electronic version of *Excerpta Medica*, contains over 8 million records dating back to 1974.

A second, more modern type of bibliographic content is the Web catalog. There are increasing numbers of such catalogs, which consist of Web pages containing mainly links to other Web pages and sites. It should be noted that there is a blurry distinction between Web catalogs and aggregations (the fourth category in this classification). In general, the former contain only links to other pages and sites, while the latter include actual content that is highly integrated with other resources. Some well-known Web catalogs include:

- HealthWeb (healthweb.org)—topics maintained by a consortium of 12 midwestern universities (Redman, Kelly et al., 1997)
- HealthFinder (healthfinder.gov)—consumer-oriented health information maintained by the Office of Disease Prevention and Health Promotion of the U.S. Department of Health and Human Services

There are a number of large general Web catalogs that are not limited to health topics. Two examples are Yahoo (www.yahoo.com) and Open Directory (dmoz.org), both of which have significant health components.

The final type of bibliographic content is the specialized registry. This resource is very close to a literature reference database except that it indexes more diverse content than scientific literature. One specialized registry of great importance for clinicians is the *National Guidelines Clearinghouse* (NGC, www.guideline.gov). Produced by the Agency for Healthcare Research and Quality (AHRQ), it is a bibliographic database with exhaustive information about clinical practice guidelines.

2.1.2 Full-text

The second type of content is *full-text content*. A large component of this content consists of the online versions of books and periodicals. A wide variety of the traditional paper-based biomedical literature, from textbooks to journals, is now available electronically. The electronic versions may be enhanced by measures ranging from the provision of supplemental data in a journal article to linkages and multimedia content in a textbook. The final component of this category is the Web site. Admittedly the diversity of information on Web sites is enormous, and sites may include every other type of content described in this chapter. However, in the context of this category, "Web site" refers to a localized collection (that may be large) of static and dynamic pages at a discrete Web location.

Most biomedical journals are now published in electronic form. Electronic publication not only allows easier access, but additional features not possible in print versions. For example, journal Web sites can provide additional data with additional figures and tables, results, images, and even raw data. A journal Web site also allows more dialogue about articles than could be published in a Letters to the Editor section of a print journal. Electronic publication also allows true bibliographic linkages, both to other full-text articles and to the MEDLINE record. The Web also allows linkage directly from bibliographic databases to full text. In fact, the MEDLINE database now has a field for the Web address of the full-text paper.

Several hundred biomedical journals use Highwire Press (www.highwire.org) to provide on-line access to their content. The Highwire system provides a retrieval interface that searches over the complete online contents for a given journal. Users can search for authors, words limited to the title and abstract, words in the entire article, and within a date range. The interface also allows searching by citation by entering volume number and page, as well as searching over the entire collection of journals that use Highwire. Users can also browse through specific issues as well as collected resources.

The most common full-text secondary literature source is the traditional textbook, an increasing number of which are available in electronic form. A common approach with textbooks is to bundle multiple books, sometimes with linkages across them. An early bundler of textbooks was *Stat!-Ref* (Teton Data Systems, www.statref.com), which like many began as a CD-ROM product and then moved to the Web. An early product that implemented linking across books was Harrison's Online (McGraw-Hill, www.harrisonsonline.com), which contains the full text of *Harrison's Principles of Internal Medicine* and the drug reference *Gold Standard Pharmacology*. Another textbook collection of growing stature is the *NCBI Bookshelf*, which contains many volumes on biomedical research topics (http://www.ncbi.nlm.nih.gov/entrez/query.fcgi?db=Books). Some books, such as *On-Line Mendelian Inheritance in Man* (OMIM, http://www.ncbi.nlm.nih.gov/entrez/query.fcgi?db=OMIM) have ceased publishing paper copies.

Electronic textbooks offer additional features beyond text from the print version. While many print textbooks do feature high-quality images, electronic versions offer the ability to have more pictures and illustrations. They also have the ability to use sound and video, although few do at this time. As with full-text journals, electronic textbooks can link to other resources, including journal references and the full articles. Many Web-based textbook sites also provide access to continuing education self-assessment questions and medical news. In addition, electronic textbooks let authors and publishers provide more frequent updates of the information than is allowed by the usual cycle of print editions, where new versions come out only every 2 to 5 years.

As noted above, Web sites are another form of full-text information. Probably the most effective user of the Web to provide health information is the U.S. government. The bibliographic databases of the NLM and AHRQ have already been described. These and other agencies, such as the National Cancer Institute (NCI) and Centers for Disease Control and Prevention (CDC) have also been innovative in providing comprehensive full-text information for healthcare providers and consumers as well. Some of these will be described later as aggregations, since they provide many different types of resources. In addition, a large number of private consumer health Web sites have emerged in recent years. Of course they include more than just collections of text, but also interaction with experts, online stores, and catalogs of links to other sites. There are also Web sites that provide information geared toward healthcare providers as well as scientists.

2.1.3 Databases/Collections

The third category consists of *databases and other specific collections* of content. These resources are usually not stored as freestanding Web pages but instead are often housed in database management systems. This content can be further subcategorized into discrete information types:

- Image databases—collections of images from radiology, pathology, and other areas
- Genomics databases—information from gene sequencing, protein characterization, and other genomics research
- Citation databases—bibliographic linkages of scientific literature
- Evidence-based medicine (EBM) databases—highly structured collections of clinical evidence
- Other databases—miscellaneous other collections

A great number of image databases are available on the Web, particularly those from the "visual" medical specialties, such as radiology, pathology, and dermatology. One collection of note is the Visible Human Project of the NLM, which consists of three-dimensional representations of normal male and female bodies (Spitzer, Ackerman et al., 1996). This resource is built from cross-sectional slices of cadavers, with sections of 1 mm in the male and 0.3 mm in the female. Also available from each cadaver are transverse computerized tomography (CT) and magnetic resonance (MR) images. In addition to the images themselves, a variety of searching and browsing interfaces have been created which can be accessed via the project Web site (http://www.nlm.nih.gov/research/visible/visible_human.html).

Many genomics databases are available across the Web. Some of these are text-based, but even those that are not (such as sequence or structure databases) often contain textual annotations of their data. A key attribute of these databases is their linkage across the Web, such that a record in one database about a gene may have a link to a sequence database with its nucleotide or amino acid sequence, a structure database with the structure of its protein product, or a literature database with papers describing experiments describing the gene. The first issue each year of the journal *Nucleic Acids Research* catalogs and describes these databases (Baxevanis, 2003). At the center of this network of databases are those produced by the National Center for Biotechnology Information (NCBI). All of NCBI's databases are linked among themselves, along with PubMed and OMIM, and are searchable via the Entrez system (http://www.ncbi.nlm.nih.gov/Entrez).

Citation databases provide linkages to articles that cite others across the scientific literature. The best-known citation databases are the *Science Citation Index* (SCI, ISI Thompson) and *Social Science Citation Index* (SSCI, ISI Thompson). A recent development is the *Web of Science*, a Web-based interface to these databases. Another system for citation indexing is the *Research Index* (formerly called *CiteSeer*, citeseer.nj.nec.com) (Lawrence, Giles et al., 1999). This index uses a process called *autonomous citation indexing* that adds citations into its database by automatically processing papers from the Web. It also attempts to identify the context of citations, showing words similar across citations such that the commonality of citing papers can be observed.

EBM databases are devoted to providing synopses of evidence-based information in forms easily accessible by clinicians. Some examples of these databases include the *Cochrane Database of Systematic Reviews*, one of the original collections of systematic reviews (www.cochrane.org), and *Clinical Evidence*, an "evidence formulary" (www.clinicalevidence.com).

There are a variety of other databases/collections that do not fit into the above categories, such as the *ClinicalTrials.gov* database that contains details of ongoing clinical trials sponsored by the National Institutes of Health.

2.1.4 Aggregations

The final category consists of *aggregations* of content from the first three categories. The distinction between this category and some of the highly linked types of content described above is admittedly blurry, but aggregations typically have a wide variety of different types of information serving diverse needs of their users. Aggregated content has been developed for all types of users from consumers to clinicians to scientists.

Probably the largest aggregated consumer information resource is MEDLINEplus (medlineplus.gov) from the NLM (Miller, Lacroix et al., 2000). MEDLINEplus includes all of the types of content previously described, aggregated for easy access to a given topic. At the top level, MEDLINEplus contains health topics, drug information, medical dictionaries, directories, and other resources. MEDLINEplus currently contains over 400 health topics. The selection of topics is based on analysis of those used by consumers to search for health information on the NLM Web site (Miller, Lacroix et al., 2000). Each topic contains links to health information from the NIH and other sources deemed credible by its selectors. There are also links to current health news (updated daily), a medical encyclopedia, drug references, and directories, along with a preformed PubMed search, related to the topic.

Aggregations of content have also been developed for clinicians. *Merck Medicus* (www.merckmedicus.com) was developed by the well-known publisher and pharmaceutical house, is available to all licensed US physicians, and includes such well-known resources as *Harrison's Online*, *MDConsult* (www.mdconsult.com), and *Dxplain* (http://www.lcs.mgh.harvard.edu/dxplain.htm).

There are many aggregations of content for biomedical researchers as well. Probably the best known among these are the *model organism databases* (Perkel, 2003). These databases bring together bibliographic databases, full text, and databases of sequences, structure, and function for organisms whose genomic data has been highly characterized, such as the mouse (Bult, Blake et al., 2004) and *Saccharomyces* yeast (Bahls, Weitzman et al., 2003). Another well-known aggregation of genomics information is the SOURCE (source.stanford.edu) database, which aggregates information from many other sources about individuals genes in species (Diehn, Sherlock et al., 2003).

2.2 Indexing

Most modern commercial IR systems index their content in two ways. In *manual indexing*, human indexers, usually using standardized terminology, assign indexing terms and attributes to documents, often following a specific protocol. Manual indexing is typically done using *controlled vocabularies*, which consist of the set of allowable terms and relationships between them. In *automated indexing*, on the other hand, computers make the indexing assignments, usually limited to breaking out each word in the document (or part of the document) as an indexing term.

Manual indexing is used most commonly with bibliographic databases. In this age of proliferating electronic content, such as online textbooks, practice guidelines, and multimedia collections, manual indexing has become either too expensive or outright unfeasible for the quantity and diversity of material now available. Thus there are increasing numbers of databases that are indexed only by automated means.

2.2.1 Controlled Vocabularies

Before discussing specific vocabularies it is useful to define some terms, since different writers attach different definitions to the various components of thesauri. A *concept* is an idea or object that occurs in the world, such as the condition under which human blood pressure is elevated. A *term* is the actual string of one or more words that represent a concept, such as *Hypertension* or *High Blood Pressure*. One of these string forms is the *preferred* or *canonical* form, such as *Hypertension* in the present example. When one or more terms can represent a concept, the different terms are called *synonyms*.

A controlled vocabulary usually contains a list of terms that are the canonical representations of the concepts. They are also called *thesauri* and contain relationships between terms, which typically fall into three categories:

- Hierarchical—terms that are broader or narrower. The hierarchical organization not only provides an overview of the structure of a thesaurus but also can be used to enhance searching.
- Synonymous—terms that are synonyms, allowing the indexer or searcher to express a concept in different words.
- Related—terms that are not synonymous or hierarchical but are somehow otherwise related. These usually remind the searcher of different but related terms that may enhance a search.

The Medical Subject Headings (MeSH) vocabulary is used to manually index most of the databases produced by the NLM (Coletti and Bleich, 2001). The latest version contains over 21,000 *subject headings* (the word MeSH uses to denote the canonical representation of its concepts). It also contains over 100,000 supplementary concept records in a separate chemical thesaurus. In addition, MeSH contains the three types of relationships described in the previous paragraph:

- Hierarchical—MeSH is organized hierarchically into 15 trees, such as *Diseases*, *Organisms*, and *Chemicals and Drugs*.

- Synonymous—MeSH contains a vast number of entry terms, which are synonyms of the headings.
- Related—terms that may be useful for searchers to add to their searches when appropriate are suggested for many headings.

The MeSH vocabulary files, their associated data, and their supporting documentation are available on the NLM's MeSH Web site (www.nlm.nih.gov/mesh/). There is also a browser that facilitates exploration of the vocabulary (www.nlm.nih.gov/mesh/MBrowser.html).

There are features of MeSH designed to assist indexers in making documents more retrievable (Anonymous, 2000b). One of these is subheadings, which are qualifiers of subject headings that narrow the focus of a term. Under *Hypertension*, for example, the focus of an article may be on the diagnosis, epidemiology, or treatment of the condition. Another feature of MeSH that helps retrieval is check tags. These are MeSH terms that represent certain facets of medical studies, such as age, gender, human or nonhuman, and type of grant support. Related to check tags are the geographical locations in the Z tree. Indexers must also include these, like check tags, since the location of a study (e.g., *Oregon*) must be indicated. Another feature gaining increasing importance for EBM and other purposes is the publication type, which describes the type of publication or the type of study. A searcher who wants a review of a topic may choose the publication type *Review* or *Review Literature*. Or, to find studies that provide the best evidence for a therapy, the publication type *Meta-Analysis*, *Randomized Controlled Trial*, or *Controlled Clinical Trial* would be used.

MeSH is not the only thesaurus used for indexing biomedical documents. A number of other thesauri are used to index non-NLM databases. CINAHL, for example, uses the *CINAHL Subject Headings*, which are based on MeSH but have additional domain-specific terms added (Brenner and McKinin, 1989). EMBASE has a vocabulary called EMTREE, which has many features similar to those of MeSH (www.elsevier.nl/homepage/sah/spd/site/locate_embase.html).

2.2.2 Manual Indexing

Manual indexing of bibliographic content is the most common and developed use of such indexing. Bibliographic manual indexing is usually done by means of a controlled vocabulary of terms and attributes. Most databases utilizing human indexing usually have a detailed protocol for assignment of indexing terms from the thesaurus. The MEDLINE database is no exception. The principles of MEDLINE indexing were laid out in the two-volume MEDLARS Indexing Manual (Charen, 1976; Charen, 1983). Subsequent modifications have occurred with changes to MEDLINE, other databases, and MeSH over the years (Anonymous, 2000a). The major concepts of the article, usually from two to five headings, are designed as central concept headings, and designated in the MEDLINE record by an asterisk. The indexer is also required to assign appropriate subheadings. Finally, the indexer must also assign check tags, geographical locations, and publication types.

Few full-text resources are manually indexed. One type of indexing that commonly takes place with full-text resources, especially in the print world, is that performed for the index at the back of the book. However, this information is rarely used in IR systems; instead, most online textbooks rely on automated indexing (see below). One exception to this is MDConsult, which uses back-of-book indexes to point to specific sections in its online books.

Manual indexing of Web content is challenging. With several billion pages of content, manual indexing of more than a fraction of it is not feasible. On the other hand, the lack of a coherent index makes searching much more difficult, especially when specific resource types are being sought. A simple form of manual indexing of the Web takes place in the development of the Web catalogs and aggregations as described above. These catalogs make not only explicit indexing about subjects and other attributes, but also implicit indexing about the quality of a given resource by the decision of whether to include it in the catalog.

Two major approaches to manual indexing have emerged on the Web, which are not mutually incompatible. The first approach, that of applying metadata to Web pages and sites, is exemplified by the Dublin Core Metadata Initiative (DCMI, www.dublincore.org). The second approach, to build directories of content, is further described below.

Table 9-1. Elements of Dublin Core Metadata.

Element	Definition
DC.title	The name given to the resource
DC.creator	The person or organization primarily responsible for creating the intellectual content of the resource
DC.subject	The topic of the resource
DC.description	A textual description of the content of the resource
DC.publisher	The entity responsible for making the resource available in its present form
DC.date	A date associated with the creation or availability of the resource
DC.contributor	A person or organization not specified in a creator element who has made a significant intellectual contribution to the resource but whose contribution is secondary to any person or organization specified in a creator element

continued

Element	Definition
DC.type	The category of the resource
DC.format	The data format of the resource, used to identify the software and possibly hardware that might be needed to display or operate the resource
DC.identifier	A string or number used to uniquely identify the resource
DC.source	Information about a second resource from which the present resource is derived
DC.language	The language of the intellectual content of the resource
DC.relation	An identifier of a second resource and its relationship to the present resource
DC.coverage	The spatial or temporal characteristics of the intellectual content of the resource
DC.rights	A rights management statement, an identifier that links to a rights management statement, or an identifier that links to a service providing information about rights management for the resource

The goal of the DCMI has been to develop a set of standard data elements that creators of Web resources can use to apply metadata to their content (Weibel, 1996). The specification has defined 15 elements, as shown in Table 9-1 (Anonymous, 1999). The DCMI has been anointed a standard by the National Information Standards Organization (NISO) with the designation Z39.85 (Anonymous, 2001a).

While Dublin Core Metadata was originally envisioned to be included in HTML Web pages, it became apparent that many non-HTML resources exist on the Web and that there are reasons to store metadata external to Web pages. For example, authors of Web pages might not be the best people to index pages or other entities might wish to add value by their own indexing of content. An emerging standard for cataloging metadata is the *Resource Description Framework* (RDF) (Miller, 1998). A framework for describing and interchanging metadata, RDF is usually expressed in XML. Increasingly XML is being used to interchange data between databases and has been designated the preferred interchange format in the *Clinical Document Architecture* of the Health Level-7 (HL7, www.hl7.org) standard (Dolin, Alschuler et al., 2001). RDF also forms the basis of what some call the future of the Web as a repository not only of content but also knowledge, which is also referred to as the *Semantic Web* (Lassila, Hendler et al., 2001). Dublin Core Metadata (or any type of metadata) can be represented in RDF (Beckett, Miller et al., 2000).

Another approach to manually indexing content on the Web has been to create directories of content. The first major effort to create these was the Yahoo! search engine, which created a subject hierarchy and assigned Web sites to elements within it (www.yahoo.com). When concern began to emerge that the Yahoo! directory was proprietary and not necessarily representative of the Web community at large (Caruso, 2000), an alternative movement sprung up, the Open Directory Project.

Manual indexing has a number of limitations, the most significant of which is inconsistency. Funk and Reid (Funk and Reid, 1983) evaluated indexing inconsistency in MEDLINE by identifying 760 articles that had been indexed twice by the NLM. The most consistent indexing occurred with check tags and central concept headings, which were only indexed with a consistency of 61 to 75%. The least consistent indexing occurred with subheadings, especially those assigned to non-central concept headings, which had a consistency of less than 35%. Manual indexing also takes time. While it may be feasible with the large resources the NLM has to index MEDLINE, it is probably impossible with the growing amount of content on Web sites and in other full-text resources. Indeed, the NLM has recognized the challenge of continuing to have to index the growing body of biomedical literature and is investigating automated and semi-automated means of doing so (Aronson, Bodenreider et al., 2000).

2.2.3 Automated Indexing

In automated indexing, the work is done by a computer. Although the mechanical running of the automated indexing process lacks cognitive input, considerable intellectual effort may have gone into building the automated indexing system. In this section, we will focus on the automated indexing used in operational IR systems, namely the indexing of documents by the words they contain.

Some may not think of extracting all the words in a document as "indexing," but from the standpoint of an IR system, words are descriptors of documents, just like human-assigned indexing terms. Most retrieval systems actually use a hybrid of human and word indexing, in that the human-assigned indexing terms become part of the document, which can then be searched by using the whole controlled vocabulary term or individual words within it. Indeed, most MEDLINE implementations have always allowed the combination of searching on human indexing terms and on words in the title and abstract of the reference. With the development of full-text resources in the 1980s and 1990s, systems that only used word indexing began to emerge. This trend increased with the advent of the Web.

Word indexing is typically done by taking all consecutive alphanumeric characters between white space, which consists of spaces, punctuation, carriage returns, and other nonalphanumeric characters. Systems must take particular care to apply the same process to documents and the user's queries, especially with characters such as hyphens and apostrophes. Some systems go beyond simple identification of words and attempt to assign weights to words that represent their importance in the document (Salton, 1991).

Many systems using word indexing employ processes to remove common words or conflate words to common forms. The former consists of filtering to remove stop words, which are common words that always occur with high frequency and are usually of little value in searching. The *stop list*, also called a *negative dictionary*, varies in size from the seven words of the original MEDLARS stop list (*and, an, by, from, of, the, with*) to the 250 to 500 words more typically used. Examples of the latter are the 250-word list of van Rijsbergen (vanRijsbergen, 1979), the 471-word list of Fox (Fox, 1992), and the PubMed stop list (Anonymous, 2001c). Conflation of words to common forms is done via *stemming*, the purpose of which is to ensure words with plurals and common suffixes (e.g., *-ed, -ing, -er, -al*) are always indexed by their stem form (Frakes, 1992). For example, the words *cough, coughs*, and *coughing* are all indexed via their stem *cough*. Stop word removal and stemming also reduce the size of indexing files and lead to more efficient query processing.

A commonly used approach for term weighting is TF*IDF weighting, which combines the inverse document frequency (IDF) and term frequency (TF). The IDF is the logarithm of the ratio of the total number of documents to the number of documents in which the term occurs. It is assigned once for each term in the database, and it correlates inversely with the frequency of the term in the entire database. The usual formula used is:

$$IDF(term) = \log \frac{\text{number of documents in database}}{\text{number of documents with term}} + 1 \qquad (1)$$

The TF is a measure of the frequency with which a term occurs in a given document and is assigned to each term in each document, with the usual formula:

$$TF(term, document) = \text{frequency of term in document} \qquad (2)$$

In TF*IDF weighting, the two terms are combined to form the indexing weight, WEIGHT:

$$WEIGHT(term, document) = TF(term, document) * IDF(term) \qquad (3)$$

Another automated indexing approach generating increased interest is the use of link-based methods, fueled no doubt by the success of the Google (www.google.com) search engine. This approach gives weight to pages based on how often they are cited by other pages. The *PageRank* algorithm is mathematically complex, but can be viewed as giving more weight to a Web page based on the number of other pages that link to it (Brin and Page, 1998). Thus, the home page of the NLM or a major medical journal is likely to have a very high PageRank (and presumed to be more "authoritative"), whereas a more obscure page will have a lower PageRank. A whole industry has evolved around improving the PageRank scores of one's Web sites (Anonymous, 2003).

Similar to manual indexing, word-based automated indexing has a number of limitations, including:

- Synonymy—different words may have the same meaning, such as *high* and *elevated.* This problem may extend to the level of phrases with no words in common, such as the synonyms *hypertension* and *high blood pressure.*
- Polysemy—the same word may have different meanings or senses. For example, the word *lead* can refer to an element or to a part of an electrocardiogram machine.
- Content—words in a document may not reflect its focus. For example, an article describing *hypertension* may make mention in passing to other concepts, such as *congestive heart failure,* that are not the focus of the article.
- Context—words take on meaning based on other words around them. For example, the relatively common words *high*, *blood*, and *pressure*, take on added meaning when occurring together in the phrase *high blood pressure.*
- Morphology—words can have suffixes that do not change the underlying meaning, such as indicators of plurals, various participles, adjectival forms of nouns, and nominalized forms of adjectives.
- Granularity—queries and documents may describe concepts at different levels of a hierarchy. For example, a user might query for *antibiotics* in the treatment of a specific infection, but the documents might describe specific antibiotics themselves, such as *penicillin.*

2.3 Retrieval

There are two broad approaches to retrieval. *Exact-match searching* allows the user precise control over the items retrieved. *Partial-match searching*, on the other hand, recognizes the inexact nature of both indexing and retrieval, and instead attempts to return the user content ranked by how close it comes to the user's query. After general explanations of these approaches, we will describe actual systems that access the different types of biomedical content.

2.3.1 Exact-match

In exact-match searching, the IR system gives the user all documents that exactly match the criteria specified in the search statement(s). Since the Boolean operators AND, OR, and NOT are usually required to create a manageable set of documents, this type of searching is often called *Boolean searching*. Furthermore, since the user typically builds sets of documents that are manipulated with the Boolean operators, this approach is also called *set-based searching*. Most of the early operational IR systems in the 1950s through 1970s used the exact-match approach, even though Salton was developing the partial-match approach in research systems during that time (Salton and Lesk, 1965). In modern times, exact-match searching tends to be associated with retrieval from bibliographic databases, while the partial-match approach tends to be used with full-text searching. A more detailed example of an exact-match searching system, PubMed, is provided below.

Typically the first step in exact-match retrieval is to select terms to build sets. Other attributes, such as the author name, publication type, or gene identifier (in the secondary source identifier field of MEDLINE), may be selected to build sets as well. Once the search term(s) and attribute(s) have been selected, they are combined with the Boolean operators. The Boolean AND operator is typically used to narrow a retrieval set to contain only documents about two or more concepts. The Boolean OR operator is usually used when there is more than one way to express a concept. The Boolean NOT operator is often employed as a subtraction operator that must be applied to another set. Some systems more accurately call this the ANDNOT operator.

Some systems allow terms in searches to be expanded by using the *wild-card character*, which adds all words to the search that begin with the letters up until the wild-card character. This approach is also called *truncation*. Unfortunately there is no standard approach to using wild-card characters, so syntax for them varies from system to system. PubMed, for example, allows a single asterisk at the end of a word to signify a wild-card character. Thus the query word `can*` will lead to the words *cancer* and *Candida*, among others, being added to the search. The AltaVista search engine (www.altavista.com) takes a different approach. The asterisk can be used as a wild-card character within or at the end of a word but only after its first three letters. For example, *col*r* will retrieve documents containing *color*, *colour*, and *colder*.

2.3.2 Partial-match

Although partial-match searching was conceptualized in the 1960s, it did not see widespread use in IR systems until the advent of Web search engines in the 1990s. This is most likely because exact-match searching tends to be preferred by "power users" whereas partial-match searching is preferred by novice searchers, the ranks of whom have increased substantially with the growth and popularity of the Web. Whereas exact-match searching requires an understanding of Boolean operators and (often) the underlying structure of databases (e.g., the many fields in MEDLINE), partial-match searching allows a user to simply enter a few terms and start retrieving documents.

The development of partial-match searching is usually attributed to Salton (Salton, 1991). Although partial-match searching does not exclude the use of nonterm attributes of documents, and for that matter does not even exclude the use of Boolean operators (e.g., see (Salton, Fox et al., 1983)), the most common use of this type of searching is with a query of a small number of words, also known as a *natural language query*. Because Salton's approach was based on vector mathematics, it is also referred to as the *vector-space model* of IR. In the partial-match approach, documents are typically ranked by their closeness of fit to the query. That is, documents containing more query terms will likely be ranked higher, since those with more query terms will in general be more likely to be relevant to the user. As a result this process is called relevance ranking. The entire approach has also been called *lexical-statistical retrieval*.

The most common approach to document ranking in partial-match searching is to give each a score based on the sum of the weights of terms common to the document and query. Terms in documents typically derive their weight from the TF*IDF calculation described above. Terms in queries are typically given a weight of one if the term is present and zero if it is absent. The following formula can then be used to calculate the document weight across all query terms:

$$Document weight = \sum_{all\ query terms} Weight of\ term\ in\ query * Weight of\ term\ in\ document \quad (4)$$

This may be thought of as a giant OR of all query terms, with sorting of the matching documents by weight. The usual approach is for the system to then perform the same stop word removal and stemming of the query that was done in the indexing process. (The equivalent stemming operations must be performed on documents and queries so that complementary word stems will match.)

2.4 Evaluation

There has been a great deal of research over the years devoted to evaluation of IR systems. As with many areas of research, there is controversy as to which approaches to evaluation best provide results that can assess their searching and the systems they are using. Many frameworks have been developed to put the results in context. One of these frameworks organizes evaluation around six questions that someone advocating the use of IR systems might ask (Hersh and Hickam, 1998):

- Was the system used?
- For what was the system used?
- Were the users satisfied?
- How well did they use the system?
- What factors were associated with successful or unsuccessful use of the system?
- Did the system have an impact on the user's task?

A simpler means for organizing the results of evaluation, however, groups approaches and studies into those which are system-oriented, i.e., the focus of the evaluation is on the IR system, and those which are user-oriented, i.e., the focus is on the user.

2.4.1 System-oriented

There are many ways to evaluate the performance of IR systems, the most widely used of which are the relevance-based measures of recall and precision. These measures quantify the number of relevant documents retrieved by the user from the database and in his or her search. They make use of the number of relevant documents (Rel), retrieved documents (Ret), and retrieved documents that are also relevant (Retrel). *Recall* is the proportion of relevant documents retrieved from the database:

$$Recall = \frac{Retrel}{Rel} \qquad (5)$$

In other words, recall answers the question, For a given search, what fraction of all the relevant documents have been obtained from the database?

One problem with Eq. (5) is that the denominator implies that the total number of relevant documents for a query is known. For all but the smallest of databases, however, it is unlikely, perhaps even impossible, for one to succeed in identifying all relevant documents in a database. Thus most studies use the measure of *relative recall*, where the denominator is redefined to represent the number of relevant documents identified by multiple searches on the query topic.

Precision is the proportion of relevant documents retrieved in the search:

$$Precision = \frac{Retrel}{Ret} \quad (6)$$

This measure answers the question, For a search, what fraction of the retrieved documents are relevant?

One problem that arises when one is comparing systems that use ranking versus those that do not is that nonranking systems, typically using Boolean searching, tend to retrieve a fixed set of documents and as a result have fixed points of recall and precision. Systems with relevance ranking, on the other hand, have different values of recall and precision depending on the size of the retrieval set the system (or the user) has chosen to show. For this reason, many evaluators of systems featuring relevance ranking will create a recall-precision table (or graph) that identifies precision at various levels of recall. The "standard" approach to this was defined by Salton (Salton, 1983), who pioneered both relevance ranking and this method of evaluating such systems.

To generate a recall-precision table for a single query, one first must determine the intervals of recall that will be used. A typical approach is to use intervals of 0.1 (or 10%), with a total of 11 intervals from a recall of 0.0 to 1.0. The table is built by determining the highest level of overall precision at any point in the output for a given interval of recall. Thus, for the recall interval 0.0, one would use the highest level of precision at which the recall is anywhere greater than or equal to zero and less than 0.1. An approach that has been used more frequently in recent times has been the *mean average precision* (MAP), which is similar to precision at points of recall but does not use fixed recall intervals or interpolation (Voorhees, 1998). Instead, precision is measured at every point at which a relevant document is obtained, and the MAP measure is found by averaging these points for the whole query.

No discussion of IR evaluation can ignore the *Text REtrieval Conference* (TREC, trec.nist.gov) organized by the U.S. National Institute for Standards and Technology (NIST, www.nist.gov) (Voorhees and Harman, 2000). Started in 1992, TREC has provided a testbed for evaluation and a forum for presentation of results. TREC is organized as an annual event at which the tasks are specified and queries and documents are provided to participants. Participating groups submit "runs" of their systems to NIST, which calculates the appropriate performance measure, usually recall and precision. TREC is organized into tracks geared to specific interests. Voorhees recently grouped the tracks into general IR tasks (Voorhees and Harman, 2001):

- Static text—Ad Hoc
- Streamed text—Routing, Filtering
- Human in the loop—Interactive
- Beyond English (cross-lingual)—Spanish, Chinese, and others
- Beyond text—OCR, Speech, Video
- Web searching—Very Large Corpus, Web
- Answers, not documents—Question-Answering
- Retrieval in a domain—Genomics

TREC has been an initiative for the general IR community and, as such, has mostly newswire, government, and Web (i.e., non-biomedical) content. However, a recent track has been formed using biomedical data, the TREC Genomics Track (http://medir.ohsu.edu/~genomics). The first year of the track featured tasks in both IR and IE (Hersh and Bhupatiraju, 2003). Further iterations of the track will feature more advanced approaches to evaluation of retrieval as well as user studies. Another advantage of this track has been to bring the IR and bioinformatics research communities into more contact.

Relevance-based measures have their limitations. While no one denies that users want systems to retrieve relevant articles, it is not clear that the quantity of relevant documents retrieved is the complete measure of how well a system performs (Swanson, 1988; Harter, 1992). Hersh (Hersh, 1994) has noted that clinical users are unlikely to be concerned about these measures when they simply seek an answer to a clinical question and are able to do so no matter how many other relevant documents they miss (lowering recall) or how many nonrelevant ones they retrieve (lowering precision).

What alternatives to relevance-based measures can be used for determining performance of individual searches? Many advocate that the focus of evaluation put more emphasis on user-oriented studies, particularly those that focus on how well users perform real-world tasks with IR systems. Some of these studies are described in the next section, while a series of biomedically focused user studies by Hersh and colleagues are presented later.

2.4.2 User-oriented

A number of user-oriented evaluations have been performed over the years looking at users of biomedical information. Most of these studies have focused on clinicians. One of the original studies measuring searching performance in clinical settings was performed by Haynes et al. (Haynes, McKibbon et al., 1990). This study also compared the capabilities of librarian and clinician searchers. In this study, 78 searches were randomly chosen for replication by both a clinician experienced in searching and a medical librarian. During this study, each original ("novice") user had been required to enter a brief statement of information need before entering the search program. This statement was given to the experienced clinician and librarian for searching on MEDLINE. All the retrievals for each search were given to a subject domain expert, blinded with respect to which searcher retrieved which reference. Recall and precision were calculated for each query and averaged. The results (Table 9-2) showed that the experienced clinicians and librarians achieved comparable recall, although the librarians had statistically significantly better precision. The novice clinician searchers had lower recall and precision than either of the other groups. This study also assessed user satisfaction of the novice searchers, who despite their recall and precision results said that they were satisfied with their search outcomes. The investigators did not assess whether the novices obtained enough relevant articles to answer their questions, or whether they would have found additional value with the ones that were missed.

Table 9-2. Recall and precision of MEDLINE searchers.

Users	Recall (%)	Precision (%)
Novice clinicians	27	38
Experienced clinicians	48	49
Medical librarians	49	58

A follow-up study yielded some additional insights about the searchers (McKibbon, Haynes et al., 1990). As was noted, different searchers tended to use different strategies on a given topic. The different approaches replicated a finding known from other searching studies in the past, namely, the lack of overlap across searchers of overall retrieved citations as well as relevant ones. Thus, even though the novice searchers had lower recall, they did obtain a great many relevant citations not retrieved by the two expert searchers. Furthermore, fewer than 4% of all the relevant citations were retrieved by all three searchers. Despite the widely divergent search strategies and retrieval sets, overall recall and precision were quite similar among the three classes of users.

Other user-oriented evaluation has looked at how well users complete tasks with IR systems. Egan et al. (Egan, Remde et al., 1989) evaluated the effectiveness of the Superbook application by assessing how well users could find and apply specific information. Mynatt et al. (Mynatt, Leventhal et al., 1992) used a similar approach in comparing paper and electronic versions of an online encyclopedia, while Wildemuth et al. (Wildemuth, deBliek et al., 1995) assessed the ability of students to answer testlike questions using a medical curricular database. The TREC Interactive Track has also used this approach. This work showed that some algorithms found effective using system-oriented, relevance-based evaluation measures did not maintain that effectiveness in experiments with real users (Hersh, 2001).

2.5 Research Directions

A steady stream of research continues to look at new approaches to IR, a detailed discussion of which is beyond the scope of this chapter. The NLM sponsors biomedical IR research both internally and externally. Its biggest internal project is the *Indexing Initiative*, which is investigating new approaches to automated and semi-automated indexing, mostly based on tools using the UMLS and natural language processing tools (Aronson, Bodenreider et al., 2000).

Other approaches to research have focused on improving aspects of automated indexing and retrieval. A number of these have been found to improve retrieval performance in the TREC environment, including:

- Improved approaches to term weighting, such as Okapi (Robertson and Walker, 1994), pivoted normalization (Singhal, Buckley et al., 1996), and language modeling (Ponte and Croft, 1998)
- Passage retrieval, where documents are given more weight in the ranking process based on local concentrations of query terms within them (Callan, 1994)
- Query expansion, where new terms from highly ranking documents are added to the query in an automated fashion (Srinivasan, 1996; Xu and Croft, 1996)

Additional work has focused on improving the user interface for the retrieval process by organizing the output better. An example of this is *Dynacat*, a system for consumers which uses UMLS knowledge and MeSH terms to organize search results (Pratt, Hearst et al., 1999). The goal is to present search results with documents clustered into topical groups, such as the treatments for a disease or the tests used to diagnose it. Another approach is to make the search system vocabulary more understandable in context. The *Cat-a-Cone* system provides a means to explore term hierarchies by using *cone trees*, which rotate the primary term of interest to the center of the screen and show conelike expansion of other hierarchically related terms nearby (Hearst and Karadi, 1997).

3. DIGITAL LIBRARIES

Discussion of IR "systems" thus far has focused on the provision of retrieval mechanisms to access online content. Even with the expansive coverage of some IR systems, such as Web search engines, they are often part of a larger collection of services or activities. An alternative perspective, especially when communities and/or proprietary collections are involved, is the digital library. Digital libraries share many characteristics with "brick and mortar" libraries, but also take on some additional challenges. Borgman (Borgman, 1999) notes that libraries of both types elicit different definitions of what they actually are, with researchers tending to view libraries as content collected for specific communities and practitioners alternatively viewing them as institutions or services.

3.1 Access

Probably every Web user is familiar with clicking on a Web link and receiving the error message: *HTTP 404 - File not found.* Digital libraries and commercial publishing ventures need mechanisms to ensure that documents have persistent identifiers so that when the document itself physically moves, it is still obtainable. The original architecture for the Web envisioned by the Internet Engineering Task Force was to have every uniform resource locator (URL), the address entered into a Web browser or used in a Web hyperlink, linked to a uniform resource name (URN) that would be persistent (Sollins and Masinter, 1994). The combination of a URN and URL, a uniform resource identifier (URI), would provide persistent access to digital objects. The resource for resolving URNs and URIs was never implemented on a large scale.

One approach that has begun to see widespread adoption by publishers, especially scientific journal publishers, is the digital object identifier (DOI, www.doi.org) (Paskin, 1999). The DOI has recently been given the status of a standard by the National Information Standards Organization (NISO) with the designation Z39.84. The DOI itself is relatively simple, consisting of a prefix that is assigned by the IDF to the publishing entity and a suffix that is assigned and maintained by the entity. For example, the DOI for articles from the *Journal of the American Medical Informatics Association* have the prefix *10.1197* and the suffix *jamia.M####*, where #### is a number assigned by the journal editors. Likewise, all publications in the Digital Library of the Association for Computing Machinery (http://www.acm.org/dl) have the prefix *10.1145* and a unique identifier for the suffix (e.g., *345508.345539*) for the paper. Publishers are encouraged to facilitate resolution by encoding the DOI into their URLs in a standard way, e.g., http://doi.acm.org/10.1145/345508.345539.

3.2 Interoperability

As noted throughout this chapter, metadata is a key component for accessing content in IR systems. It takes on additional value in the digital library, where there is desire to allow access to diverse but not necessarily exhaustive resources. One key concern of digital libraries is interoperability (Besser, 2002). That is, how can resources with heterogeneous metadata be accessed? Arms et al. (Arms, Hillmann et al., 2002) note that three levels of agreement must be achieved:

- Technical agreements over formats, protocols, and security procedures
- Content agreement over the data and the semantic interpretation of its metadata
- Organizational agreements over ground rules for access, preservation, payment, authentication, and so forth

One approach to interoperability gaining increasing use is the Open Archives Initiative (OAI, www.openarchive.org) (Lagoze and VandeSompel, 2001). While the OAI effort is rooted in access to scholarly communications, its methods are applicable to a much broader range of content. Its fundamental activity is to promote the "exposure" of archives' metadata such that digital library systems can learn what content is available and how it can be obtained. Each record in the OAI system has an XML-encoded record. The OAI Protocol for Metadata Harvesting (PMH) then allows selective harvesting of the metadata by systems. Such harvesting can be date-based, such as items added or changed after a certain date, or set-based, such as those belonging to a certain topic, journal, or institution. A growing number of biomedical resources have adopted OAI (McKiernan, 2003).

3.3 Preservation

Another concern for digital libraries is the preservation of content, especially with the growing trend towards electronic subscriptions to journals that result in fewer physical copies (electronic or printed) being produced. Also a concern is the longevity of digital materials (Lesk, 1997). Of all media, the longevity is the least for magnetic materials, with the expected lifetime of magnetic tape being 5 to 10 years. Optical storage has somewhat better longevity, with an expected lifetime of 30 to 100 years depending on the specific type. Ironically, paper has a life expectancy well beyond all these digital media. A growing concern is that with the increasing move towards electronic publishing, there are fewer copies of journal material produced using media that have lesser longevity.

As such, there is an imperative to preserve documents of many types, whatever their medium (Tibbo, 2001). For society in general, there is certainly impetus to preserve historical documents in an unaltered form. And in all of science, certainly biomedicine, there is need to preserve the archive of scientific discoveries, particularly those presenting original experiments and their data. A number of initiatives have been undertaken to insure preservation of digital information. These include the National Digital Information Infrastructure Preservation Program (NDIIPP, www.digitalpreservation.gov) of the US Library of Congress (Friedlander, 2002) and the Digital Preservation Coalition in the United Kingdom (Beagrie, 2002).

4. CASE STUDIES

In this section, we will explore three case studies or examples of IR in further detail. These include a retrieval system, user-oriented evaluation, and issues surrounding electronic publishing.

4.1 PubMed

Probably the best known and most widely used biomedical IR system is PubMed (pubmed.gov) from the NLM. (Unless one considers Google to be a biomedical IR system, for which a tenable case can be made!) PubMed searches MEDLINE and other bibliographic databases from the NLM. Although presenting the user with a simple text box, PubMed does a great deal of processing of the user's input to identify MeSH terms, author names, common phrases, and journal names (Anonymous, 2001c). In this automatic term mapping, the system attempts to map user input, in succession, to MeSH terms, journal names, common phrases, and authors. Remaining text that PubMed cannot map is searched as text words (i.e., words that occur in any of the MEDLINE fields).

PubMed allows the use of wild-card characters. It also allows phrase searching in that two or more words can be enclosed in quotation marks to indicate they must occur adjacent to each other. If the specified phrase is in PubMed's phrase index, then it will be searched as a phrase. Otherwise the individual words will be searched. PubMed allows specification of other indexing attributes via the PubMed "Limits" screen. These include publication types, subsets, age ranges, and publication date ranges.

As in most bibliographic systems, users search PubMed by building search sets and then combining them with Boolean operators to tailor the search. Consider a user searching for studies assessing the reduction of mortality in patients with *congestive heart failure (CHF)* through the use of medications from the *angiotensin-converting (ACE) inhibitors* class of drugs. A simple approach to such a search might be to combine the terms *ACE Inhibitors* and *CHF* with an AND. The easiest way to do this is to enter the search string *ace inhibitors AND CHF.* (The operator *AND* must be capitalized because PubMed treats the lowercase and as a text word, since some MeSH terms, such as *Bites and Strings*, have the word and in them.) Figure 9-4 shows the PubMed History screen such a searcher might develop.

PubMed also has a *Clinical Queries* interface, where the subject terms are limited by search statements designed to retrieve the best evidence based on principles of EBM. There are two different approaches. The first uses strategies for retrieving the best evidence for the four major types of clinical question. These strategies arise from research assessing the ability of MEDLINE search statements to identify the best studies for therapy, diagnosis, harm, and prognosis (Haynes, Wilczynski et al., 1994). The second approach to retrieving the best evidence aims to retrieve evidence-based resources including meta-analyses, systematic reviews, and practice guidelines. When the Clinical Queries interface is used, the search statement is processed by the usual automatic term mapping and the resulting output is limited (via AND) with the appropriate statement.

PubMed is actually part of the larger Entrez system at NLM that provides access to the entire range of on-line content (http://www.ncbi.nlm.nih.gov/entrez/query.fcgi). Another interface that searches over the range of NLM content is the NLM Gateway (http://gateway.nlm.nih.gov/gw/Cmd).

4.2 User-oriented Evaluation

Recognizing the limitations of recall and precision for evaluating clinical users of IR systems, Hersh and colleagues have carried out a number of studies assessing the ability of systems to help students and clinicians answer clinical questions. The rationale for these studies is that the usual goal of using an IR system is to find an answer to a question. While the user must obviously find relevant documents to answer that question, the quantity of such documents is less important than whether the question is successfully answered. In fact, recall and precision can be placed among the many factors that may be associated with ability to complete the task successfully.

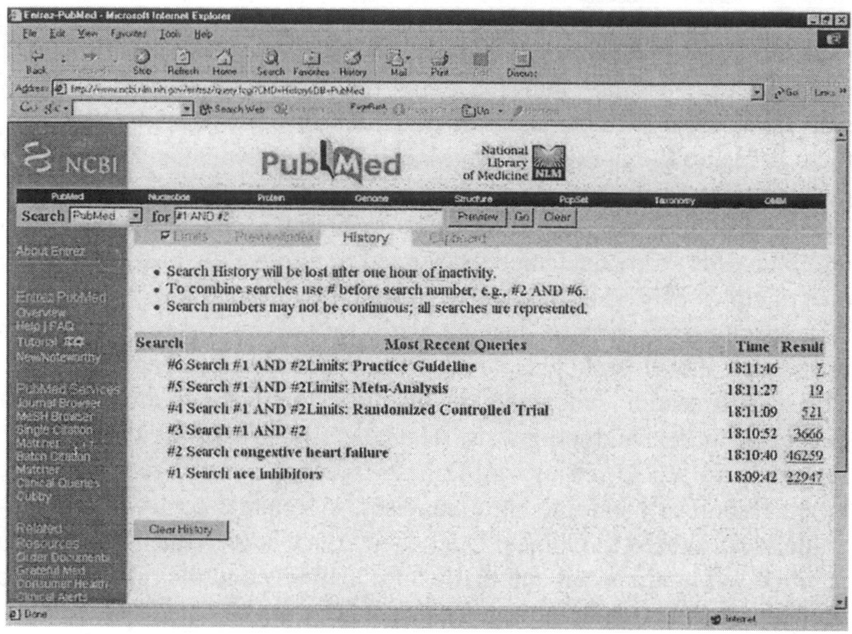

Figure 9-4. PubMed History screen. (Courtesy of NLM.)

The first study by this group using the task-oriented approach compared Boolean versus natural language searching in the textbook *Scientific American Medicine* (Hersh, Elliot et al., 1994). Thirteen medical students were asked to answer 10 short-answer questions and rate their confidence in their answers. The students were then randomized to one or the other interface and asked to search on the five questions for which they had rated confidence the lowest. The study showed that both groups had low correct rates before searching (average 1.7 correct out of 10) but were mostly able to answer the questions with searching (average 4.0 out of 5). There was no difference in ability to answer questions with one interface or the other. Most answers were found on the first search of the textbook. For the questions that were incorrectly answered, the document with the correct answer was actually retrieved by the user two-thirds of the time and viewed more than half the time.

Another study compared Boolean and natural language searching of MEDLINE with two commercial products, CD Plus (now Ovid, www.ovid.com) and Knowledge Finder (Aries Systems, www.ariessystems.com) (Hersh, Pentecost et al., 1996). These systems represented the ends of the spectrum in terms of using Boolean searching on human-indexed thesaurus terms (CD Plus) versus natural language searching on words in the title, abstract, and indexing terms (Knowledge Finder). Sixteen medical students were recruited and randomized to one of the two systems and given three yes/no clinical questions to answer. The students were able to use each system successfully, answering 37.5% correct before searching and 85.4% correct after searching. There were no significant differences between the systems in time taken, relevant articles retrieved, or user satisfaction. This study demonstrated that both types of system can be used equally well with minimal training.

The most comprehensive study looked at MEDLINE searching by medical and nurse practitioner (NP) students to answer clinical questions. A total of 66 medical and NP students searched five questions each (Hersh, Crabtree et al., 2002). This study used a multiple-choice format for answering questions that also included a judgment about the evidence for the answer. Subjects were asked to choose from one of three answers:

- Yes, with adequate evidence
- Insufficient evidence to answer question
- No, with adequate evidence

Both groups achieved a presearching correctness on questions about equal to chance (32.3% for medical students and 31.7% for NP students). However, medical students improved their correctness with searching (to 51.6%), whereas NP students hardly did at all (to 34.7%).

This study also assessed what factors were associated with successful searching. A number of factors, such as age, gender, computer experience, and time taken to search, were not associated with successful answering of questions. However, successful answering was associated with answering the question correctly before searching, spatial visualization ability (measured by a validated instrument), searching experience, and EBM question type (prognosis questions easiest, harm questions most difficult). An analysis of recall and precision for each question searched demonstrated their complete lack of association with ability to answer these questions.

4.3 Changes in Publishing

Any discussion of IR systems and digital libraries cannot ignore the larger context of the political and economic aspects of publishing. While a complete discussion is beyond the scope of a chapter like this, some of the high points can and should be elucidated, if for no other reason than that they impact access to content for the kinds of innovations and research described in this book.

The Internet and WWW have had profound impact in the publishing of knowledge-based information. The technical impediments to electronic publishing of journals have largely been solved. Most scientific journals are published electronically in some form already. Journals that do not publish electronically likely could do so easily, since most of the publishing process has already been converted to the electronic mode. A modern Internet connection is sufficient to deliver most of the content of journals. Indeed, a near turnkey solution is already offered through Highwire Press, which has an infrastructure that supports journal publishing from content preparation to searching and archiving.

There is great enthusiasm for electronic availability of journals, as evidenced by the growing number of titles to which libraries provide access. Likewise, since most scientists have the desire for widespread dissemination of their work, they have incentive for their papers to be available on the Web. Indeed, it has been shown, at least in the computer science domain, that papers freely available on the Web have a higher likelihood of being cited by other papers than those which are not (Lawrence, 2001). As citations are important to authors for academic promotion and grant funding, authors have incentive to maximize the accessibility of their published work.

The technical challenges to electronic scholarly publication have been replaced by economic and political ones (Hersh and Rindfleisch, 2000; Anonymous, 2001b). Printing and mailing, tasks no longer needed in electronic publishing, comprised a significant part of the "added value" from publishers of journals. There is, however, still value added by publishers, such as hiring and managing editorial staff to produce the journals and managing the peer review process. Even if publishing companies as they are known were to vanish, there would still be some cost to the production of journals. Thus, while the cost of producing journals electronically is likely to be less, it is not zero, and even if journal content is distributed "free," someone has to pay the production costs.

The economic issue in electronic publishing, then, is who is going to pay for the production of journals. This introduces some political issues as well. One of them centers on the concern that much research is publicly funded through grants from federal agencies such as the National Institutes of Health (NIH) and the National Science Foundation (NSF). In the current system, especially in the biomedical sciences (and to a lesser extent in nonbiomedical sciences), researchers turn over the copyright of their publications to journal publishers. The political concern is that the public funds the research and the universities carry it out, but individuals and libraries then must buy it back from the publishers to whom they willingly cede the copyright (McCook, 2004). This problem is exacerbated by the general decline in funding for libraries that has occurred over the last couple decades (Boyd and Herkovic, 1999; Meek, 2001).

Some have proposed models of scholarly publishing that keep the archive of science freely available. One of these is *open access* publishing, where authors and their institutions pay the cost of production of manuscripts up front after they are accepted through a peer review process. It has been suggested that this cost could even be included in the budgets of grant proposals submitted for funding agencies. After the paper is published, the manuscript becomes freely available on the Web. The first publisher to take this approach has been Biomed Central (BMC, www.biomedcentral.com). Another highly visible open access approach is the Public Library of Science (PLOS, www.plos.org). Although legislation has been proposed requiring research funded by government agencies to be open access (McLellan, 2003), at least one journal editor has expressed caution that the untested model may not work as well as advertised, especially for the major biomedical journals that devote substantial resources to insuring the quality of high-profile biomedical research (DeAngelis and Musacchio, 2004).

Another model is that of PubMed Central (PMC, pubmedcentral.gov), which provides free access to published literature but allows publishers to maintain copyright as well as optionally keep the papers on their own servers. A lag time of up to 6 months is allowed so that journals can reap the revenue that comes with initial publication. The number of journals submitting their content to PMC has been modest, and there are currently about 100 that contribute to its repository.

5. ACKNOWLEDGEMENTS

The author's research has been generously funded by the National Library of Medicine, Agency for Healthcare Quality and Research, and National Science Foundation over the years. He is particularly grateful to the NLM for its strong leadership in promoting research and education in the field of medical informatics.

REFERENCES

Anonymous. (1999). Dublin Core Metadata Element Set, Version 1.1: Reference Description. Dublin Core Metadata Initiative, http://www.dublincore.org/documents/dces/.

Anonymous. (2000a). Cataloging Practices. National Library of Medicine, http://www.nlm.nih.gov/mesh/catpractices.html.

Anonymous. (2000b). Features of the MeSH Vocabulary. National Library of Medicine, http://www.nlm.nih.gov/mesh/features.html.

Anonymous. (2000c). Organization of National Library of Medicine Bibliographic Databases. National Library of Medicine, http://www.nlm.nih.gov/pubs/techbull/mj00/mj00_buckets.html.

Anonymous. (2001a). The Dublin Core Metadata Element Set. Dublin Core Metadata Initiative, http://www.niso.org/standards/resources/Z39-85.pdf.

Anonymous. (2001b). "The Future of the Electronic Scientific Literature," *Nature,* 413: 1-3.

Anonymous. (2001c). PubMed Help. National Library of Medicine, http://www.ncbi.nlm.nih.gov/entrez/query/static/help/pmhelp.html. Accessed: July 1, 2002.

Anonymous. (2003). The Google Ranking Report. Sedona, AZ, Cyberdifference Corp., http://www.mseo.com/google_ranking_report.html.

Arms, W., Hillmann, D., et al. (2002). "A Spectrum of Interoperability: The Site for Science Prototype for the NSDL," *D-Lib Magazine,* 8, http://www.dlib.org/dlib/january02/arms/01arms.html.

Aronson, A., Bodenreider, O., et al. (2000). "The NLM Indexing Initiative," in *Proceedings of the AMIA 2000 Annual Symposium*, Los Angeles, CA. Hanley & Belfus, 17-21.

Bahls, C., Weitzman, J., et al. (2003). "Biology's Models," *The Scientist.* June 2, 2003. 5, http://www.the-scientist.com/yr2003/jun/feature_030602.html.

Baxevanis, A. (2003). "The Molecular Biology Database Collection: 2003 update," *Nucleic Acids Research,* 31: 1-12.

Beagrie, N. (2002). "An Update on the Digital Preservation Coalition," *D-Lib Magazine,* 8, http://www.dlib.org/dlib/april02/beagrie/04beagrie.html.

Beckett, D., Miller, E., et al. (2000). Using Dublin Core in XML. Dublin Core Metadata Initiative, http://dublincore.org/documents/dcmes-xml/.

Besser, H. (2002). "The Next Stage: Moving from Isolated Digital Collections to Interoperable Digital Libraries," *First Monday,* 7(6), http://www.firstmonday.dk/issues/issue7_6/besser/

Borgman, C. (1999). "What are Digital Libraries? Competing Visions," *Information Processing and Management,* 35: 227-244.

Boyd, S. and Herkovic, A. (1999). Crisis in Scholarly Publishing: Executive Summary. Stanford Academic Council Committee on Libraries, http://www.stanford.edu/~boyd/schol_pub_crisis.html.

Brenner, S. and McKinin, E. (1989). "CINAHL and MEDLINE: A Comparison of Indexing Practices," *Bulletin of the Medical Library Association,* 77: 366-371.

Brin, S. and Page, L. (1998). "The Anatomy of a Large-scale Hypertextual Web Search Engine," *Computer Networks,* 30: 107-117.

Bult, C., Blake, J., et al. (2004). "The Mouse Genome Database (MGD): Integrating Biology with the Genome," *Nucleic Acids Research,* 32: D476-481.

Callan, J. (1994). "Passage Level Evidence in Document Retrieval," in *Proceedings of the 17th Annual International ACM SIGIR Conference on Research and Development in Information Retrieval*, Dublin, Ireland. Springer-Verlag. 302-310.

Caruso, D. (2000). "Digital Commerce; If the AOL-Time Warner Deal is about Proprietary Content, Where Does that Leave a Noncommercial Directory It Will Own?" *New York Times*. January 17, 2000.

Charen, T. (1976). *MEDLARS Indexing Manual, Part I: Bibliographic Principles and Descriptive Indexing, 1977*. Springfield, VA: National Technical Information Service.

Charen, T. (1983). *MEDLARS Indexing Manual, Part II*. Springfield, VA: National Technical Information Service.

Coletti, M. and Bleich, H. (2001). "Medical Subject Headings Used to Search the Biomedical Literature," *Journal of the American Medical Informatics Association,* 8: 317-323.

DeAngelis, C. and Musacchio, R. (2004). "Access to JAMA," *Journal of the American Medical Association,* 291: 370-371.

Diehn, M., Sherlock, G., et al. (2003). "SOURCE: A Unified Genomic Resource of Functional Annotations, Ontologies, and Gene Expression Data," *Nucleic Acids Research,* 31: 219-223.

Dolin, R., Alschuler, L., et al. (2001). "The HL7 Clinical Document Architecture," *Journal of the American Medical Informatics Association,* 8: 552-569.

Egan, D., Remde, J., et al. (1989). "Formative Design-evaluation of Superbook," *ACM Transactions on Information Systems,* 7: 30-57.

Fox, C. (1992). "Lexical Analysis and Stop Lists," in Frakes, W. and Baeza-Yates, R., eds. *Information Retrieval: Data Structures and Algorithms*, Englewood Cliffs, NJ: Prentice-Hall, pp.102-130,

Frakes, W. (1992). "Stemming Algorithms," in Frankes, W. and Baeza-Yates, R., eds. *Information Retrieval: Data Structures and Algorithms*, Englewood Cliffs, NJ: Prentice-Hall, pp. 131-160.

Friedlander, A. (2002). "The National Digital Information Infrastructure Preservation Program: Expectations, Realities, Choices, and Progress to Date," *D-Lib Magazine,* 8, http://www.dlib.org/dlib/april02/friedlander/04friedlander.html.

Funk, M. and Reid, C. (1983). "Indexing Consistency in MEDLINE," *Bulletin of the Medical Library Association,* 71: 176-183.

Harter, S. (1992). "Psychological Relevance and Information Science," *Journal of the American Society for Information Science,* 43: 602-615.

Haynes, R., McKibbon, K., et al. (1990). "Online Access to MEDLINE in Clinical Settings," *Annals of Internal Medicine,* 112: 78-84.

Haynes, R., Wilczynski, N., et al. (1994). "Developing Optimal Search Strategies for Detecting Clinically Sound Studies in MEDLINE," *Journal of the American Medical Informatics Association,* 1: 447-458.

Hearst, M. and Karadi, C. (1997). "Cat-a-Cone: An Interactive Interface for Specifying Searches and Viewing Retrieval Results Using a Large Category Hierarchy," in *Proceedings of the 20th Annual International ACM SIGIR Conference on Research and Development in Information Retrieval*, Philadelphia, PA. ACM Press. 246-255.

Hersh, W. (1994). "Relevance and Retrieval Evaluation: Perspectives from Medicine," *Journal of the American Society for Information Science,* 45: 201-206.

Hersh, W. (2001). "Interactivity at the Text Retrieval Conference (TREC)," *Information Processing and Management,* 37: 365-366.

Hersh, W. (2003). *Information Retrieval: A Health and Biomedical Perspective.* Second Edition. New York: Springer-Verlag, http://www.irbook.org.

Hersh, W. and Bhupatiraju, R. (2003). "TREC Genomics track overview," in *The Twelfth Text Retrieval Conference: TREC 2003*, Gaithersburg, MD. National Institute of Standards & Technology, http://trec.nist.gov/pubs/trec12/papers/GENOMICS.OVERVIEW3.pdf.

Hersh, W., Crabtree, M., et al. (2002). "Factors Associated with Success for Searching MEDLINE and Applying Evidence to Answer Clinical Questions," *Journal of the American Medical Informatics Association,* 9: 283-293.

Hersh, W., Elliot, D., et al. (1994). "Towards New Measures of Information Retrieval Evaluation," in *Proceedings of the 18th Annual Symposium on Computer Applications in Medical Care*, Washington, DC. Hanley & Belfus. 895-899.

Hersh, W. and Hickam, D. (1998). "How Well Do Physicians Use Electronic Information Retrieval Systems? A Framework for Investigation and Review of the Literature," *Journal of the American Medical Association,* 280: 1347-1352, http://jama.ama-assn.org/cgi/content/full/280/15/1347.

Hersh, W., Pentecost, J., et al. (1996). "A Task-oriented Approach to Information Retrieval Evaluation," *Journal of the American Society for Information Science,* 47: 50-56.

Hersh, W. and Rindfleisch, T. (2000). "Electronic Publishing of Scholarly Communication in the Biomedical Sciences," *Journal of the American Medical Informatics Association,* 7: 324-325.

Humphreys, B. (2000). "Electronic Health Record Meets Digital Library: A New Environment for Achieving an Old Goal," *Journal of the American Medical Informatics Association,* 7: 444-452.

Lagoze, C. and VandeSompel, H. (2001). "The Open Archives Initiative: Building a Low-barrier Interoperability Framework," in *Proceedings of the First ACM/IEEE-CS Joint Conference on Digital Libraries*, Roanoke, VA. ACM Press. 54-62.

Lassila, O., Hendler, J., et al. (2001). "The Semantic Web," *Scientific American,* 284(5): 34-43, http://www.scientificamerican.com/article.cfm?articleID=00048144-10D2-1C70-84A9809EC588EF21&catID=2.

Lawrence, S. (2001). "Online or Invisible?" *Nature,* 411: 521.

Lawrence, S., Giles, C., et al. (1999). "Digital Libraries and Autonomous Citation Indexing," *Computer,* 32: 67-71.

Lesk, M. (1997). *Practical Digital Libraries - Books, Bytes, and Bucks.* San Francisco: Morgan Kaufmann.

McCook, A. (2004). "Open Access to US Govt Work Urged," The Scientist, http://www.biomedcentral.com/news/20040721/01.

McKibbon, K., Haynes, R., et al. (1990). "How Good Are Clinical MEDLINE Searches? A Comparative Study of Clinical End-user and Librarian Searches," *Computers and Biomedical Research,* 23(6): 583-593.

McKiernan, G. (2003. "Open Archives Initiative Service Providers. Part I: Science and Technology," *Library Hi Tech News,* 20(9): 30-38, http://www.public.iastate.edu/~gerrymck/OAI-SP-I.pdf.

McLellan, F. (2003). "US Bill Says Government Funded Work Must Be Open Access," *Lancet,* 362: 52.

Meek, J. (2001). "Science World in Revolt at Power of the Journal Owners," *The Guardian*, http://www.guardian.co.uk/Archive/Article/0,4273,4193292,00.html.

Miller, E. (1998). "An Introduction to the Resource Description Framework," *D-Lib Magazine,* 4, http://www.dlib.org/dlib/may98/miller/05miller.html.

Miller, N., Lacroix, E., et al. (2000). "MEDLINEplus: Building and Maintaining the National Library of Medicine's Consumer Health Web Service," B*ulletin of the Medical Library Association,* 88: 11-17.

Mynatt, B., Leventhal, L., et al. (1992). "Hypertext or Book: Which Is Better for Answering Questions?" in *Proceedings of Computer-Human Interface 92.* 19-25.

Paskin, N. (1999). "DOI: Current Status and Outlook," *D-Lib Magazine,* 5, http://www.dlib.org/dlib/may99/05paskin.html.

Perkel, J. (2003). "Feeding the Info Junkies," *The Scientist.* June 2, 2003. 39, http://www.the-scientist.com/yr2003/jun/feature14_030602.html.

Ponte, J. and Croft, W. (1998). "A Language Modeling Approach to Information Retrieval," in *Proceedings of the 21st Annual International ACM SIGIR Conference on Research and Development in Information Retrieval*, Melbourne, Australia. ACM Press. 275-281.

Pratt, W., Hearst, M., et al. (1999). "A Knowledge-based Approach to Organizing Retrieved Documents," in *Proceedings of the 16th National Conference on Artificial Intelligence*, Orlando, FL. AAAI. 80-85.

Redman, P., Kelly, J., et al. (1997). "Common Ground: The HealthWeb Project as a Model for Internet Collaboration," B*ulletin of the Medical Library Association,* 85: 325-330.

Robertson, S. and Walker, S. (1994). "Some Simple Effective Aproximations to the 2-Poisson Model for Probabilistic Weighted Retrieval," in *Proceedings of the 17th Annual International ACM SIGIR Conference on Research and Development in Information Retrieval*, Dublin, Ireland. Springer-Verlag. 232-241.

Salton, G. (1983). *Introduction to Modern Information Retrieval.* New York: McGraw-Hill.

Salton, G. (1991). "Developments in Automatic Text Retrieval," *Science,* 253: 974-980.
Salton, G., Fox, E., et al. (1983). "Extended Boolean Information Retrieval," *Communications of the ACM,* 26: 1022-1036.
Salton, G. and Lesk, M. (1965). "The SMART Automatic Document Retrieval System: An Illustration," *Communications of the ACM,* 8: 391-398.
Singhal, A., Buckley, C., et al. (1996). "Pivoted Document Length Normalization," in *Proceedings of the 19th Annual International ACM SIGIR Conference on Research and Development in Information Retrieval*, Zurich, Switzerland. ACM Press. 21-29.
Sollins, K. and Masinter, L. (1994). Functional Requirements for Uniform Resource Names. Internet Engineering Task Force, http://www.w3.org/Addressing/rfc1737.txt.
Spitzer, V., Ackerman, M., et al. (1996). "The Visible Human Male: A Technical Report.," *Journal of the American Medical Informatics Association,* 3: 118-130.
Srinivasan, P. (1996). "Query Expansion and MEDLINE," *Information Processing and Management,* 32: 431-444.
Swanson, D. (1988). "Historical Note: Information Retrieval and the Future of an Illusion," *Journal of the American Society for Information Science,* 39: 92-98.
Tibbo, H. (2001). "Archival Perspectives on the Emerging Digital Library," *Communications of the ACM,* 44(5): 69-70.
vanRijsbergen, C. (1979). *Information Retrieval.* London. Butterworth.
Voorhees, E. (1998). "Variations in Relevance Judgments and the Measurement of Retrieval Effectiveness," in *Proceedings of the 21st Annual International ACM SIGIR Conference on Research and Development in Information Retrieval*, Melbourne, Australia. ACM Press. 315-323.
Voorhees, E. and Harman, D. (2000). "Overview of the Sixth Text REtrieval Conference (TREC)," *Information Processing and Management,* 36: 3-36.
Voorhees, E. and Harman, D. (2001). "Overview of TREC 2001," in *Proceedings of the Text Retrieval Conference 2001*, Gaithersburg, MD. 1-15.
Weibel, S. (1996). "The Dublin Core: A Simple Content Description Model for Electronic Resources," *ASIS Bulletin,* 24(1): 9-11, http://www.asis.org/Bulletin/Oct-97/weibel.htm.
Wildemuth, B., DeBliek, R., et al. (1995). "Medical Students' Personal Knowledge, Searching Proficiency, and Database Use in Problem Solving," *Journal of the American Society for Information Science,* 46: 590-607.
Xu, J. and Croft, W. (1996). "Query Expansion Using Local and Global Document Analysis," in *Proceedings of the 19th Annual International ACM SIGIR Conference on Research and Development in Information Retrieval*, Zurich, Switzerland. ACM Press. 4-11.

SUGGESTED READINGS

Baeza-Yates, R. and Ribeiro-Neto, B., eds. 1999. *Modern Information Retrieval.* New York. McGraw-Hill. A book surveying most of the automated approaches to information retrieval.

Frakes, W.B., Baeza-Yates, R. *Information Retrieval: Data Structures and Algorithms*, Englewood Cliffs, NJ: Prentice-Hall, 1992. A textbook on implementation of information retrieval systems. Covers all of the major data structures and algorithms, including inverted files, ranking algorithms, stop word lists, and stemming. There are plentiful examples of code in the C programming language.

Hersh, W.R. *Information Retrieval, A Health and Biomedical Perspective* (Second Edition), New York: Springer-Verlag, 2003. A textbook on information retrieval systems in the health and biomedical domain that covers the state of the art as well as research systems.

Humphreys, B., Lindberg, D., et al. 1998. *The Unified Medical Language System: an informatics research collaboration.* Journal of the American Medical Informatics Association, 5: 1-11. A paper describing the motivation and implementation of the National Library of Medicine's Unified Medical Language System.

Miles, W.D. *A History of the National Library of Medicine*, Bethesda, MD: U.S. Dept. of Health and Human Services, 1982. A comprehensive history of the National Library of Medicine and its forerunners, covering the story of Dr. John Shaw Billings and his founding of Index Medicus to the modern implementation of MEDLINE.

Salton, G. Developments in automatic text retrieval, *Science*, 253: 974-980, 1991. The last succinct exposition of word-statistical retrieval systems from the person who originated the approach.

ONLINE RESOURCES

Biomed Central
http://www.biomedcentral.com

Highwire Press
http://www.highwire.org

National Center for Biotechnology Information
http://www.ncbi.nlm.nih.gov

National Library of Medicine
http://www.nlm.nih.gov

ACM Digital Library
http://www.acm.org/dl

CiteSeer
http://citeseer.ist.psu.edu

D-Lib Magazine
http://www.dlib.org

MEDLINEplus consumer health information resource
http://medlineplus.gov

PubMed access to MEDLINE
http://pubmed.gov

TREC
http://trec.nist.gov

QUESTIONS FOR DISCUSSION

1. With the advent of full-text searching, should the National Library of Medicine abandon human indexing of citations in MEDLINE? Why or why not?

2. Explain why open access publishing is or is not a good idea.

3. Devise a curriculum for teaching clinicians, researchers, or patients the most important points about searching for health-related information.

4. What are the limitations of recall and precision as evaluation measures and what alternatives would improve upon them?

5. Describe how one might devise a system that achieved a happy medium between protection of intellectual property and barrier-free access to the archive of science.

6. How might IR systems be developed to lower the effort it takes for clinicians to get to the information they need rapidly in the busy clinical setting?

7. Can standards be developed for digital libraries that facilitate interoperability but maintain ease of use, protection of intellectual property, and long-term preservation of the archive of science?

Chapter 10

MODELING TEXT RETRIEVAL IN BIOMEDICINE

W. John Wilbur

National Center for Biotechnology Information, National Library of Medicine, 8600 Rockville Pike, Bethesda, MD 20894

Chapter Overview

Given the amount of literature relevant to many of the areas of biomedicine, researchers are forced to use methods other than simply reading all the literature on a topic. Necessarily one must fall back on some kind of search engine. While the Google PageRank algorithm works well for finding popular web sites, it seems clear one must take a different approach in searching for information needed at the cutting edge of research. Information which is key to solving a particular problem may never have been looked at by many people in the past, yet it may be crucial to present progress. What has worked well to meet this need is to rank documents by their probable relevance to a piece of text describing the information need (a query). Here we will describe a general model for how this is done and how this model has been realized in both the vector and language modeling approaches to document retrieval. This approach is quite broad and applicable to much more than biomedicine. We will also present three example document retrieval systems that are designed to take advantage of specific information resources in biomedicine in an attempt to improve on the general model. Current challenges and future prospects are also discussed.

Keywords

relevance; probability; ranking; term weighting; vector model; unigram language model; smoothing

1. INTRODUCTION

Most papers written on the subject of text retrieval begin with the observation that the digital age has brought a deluge of natural language text and we are more or less overwhelmed by the amount of text available on the web and even in the specialized databases of interest to researchers. Certainly this is true in the field of biomedicine. The serious question is whether a researcher must personally read everything remotely related to her field of interest, or can technology rule some texts as not useful for her purposes and allow her to concentrate her search on a few texts where her efforts will have high yield. The answer is that, with a certain risk, technology can reduce the work load for information access in textual databases.

To better understand the risk of using technology in lieu of reading all the documents for oneself, it is helpful to think in terms of a simple model. Let the user be denoted by X, the information need state of X be denoted by S, and finally let the query by which X has expressed their information need be denoted by q. Typically for users of a search engine, q consists of one to three words or a short phrase. Such a short query is naturally quite inadequate to represent the need state S. In fact, an important study by Furnas et al. (1987) found that common objects are generally referred to by, on the average, five different names over a sample of references by different people. This same fact is underlined by the famous Blair and Maron (1985) study of retrieval in the area of legal documents. They discovered that legal experts, after a careful search using keywords and Boolean queries, felt they had found most of the relevant material pertaining to a case, but in fact more extensive and careful search showed they had only found less than 20% of what was relevant to the case. It proved impossible to predict the words people would use to describe relevant material. Further evidence on this point is provided by a study of MEDLINE® indexing. Funk et al. (1983) found inter-indexer consistencies ranging from 0.3 to 0.6 for different types of MeSH® term assignments. But to the extent we are unable to predict the indexing we are also unable to use it effectively for retrieval.

While we might try to improve indexing by expending more human effort on the process, there is an even more fundamental barrier to perfect retrieval. This stems from the variation observed in what people judge to be relevant to a query. Different judges agree on what is relevant to a query from 40% to 75% of the time (Saracevic, 1991; Swanson, 1988). This is true even for queries as long as the title and abstract of a MEDLINE document (Wilbur, 1998). If a human processing the query q can only find material relevant to user X with a precision of 75%, then that says something very important. An algorithm that is as "smart" as a typical human is also going to find relevant material with a precision no better than 75%. Of course we do not have algorithms that can perform at a human level and probably will not for the foreseeable future (Shieber, 1994). Now one can lengthen the query q and thereby decrease the ambiguity. One way to accomplish this is to allow the user to choose a document that represents what they would like to see. Our research (Wilbur and Coffee, 1994) shows that one can make a substantial improvement in the retrieval process by this method. The chosen document becomes a query in its own right which is generally much longer and more detailed than one a user is willing to write. One can carry this idea even further by applying relevance feedback. Here a user makes judgments for the top few documents retrieved and these judgments are used to improve the ranking of the documents the user has not yet examined. Results of one study (Wilbur, 1998) show that if a user makes judgments on the top 50 documents, then a machine learning algorithm can convert that additional information into retrieval on new documents at a level at least as good as a human agent could accomplish based on the original query. Of course, even this is not as good as the user can do for himself and probably not as good as a human agent could do given the additional information consisting of the user's judgments. Furthermore there are practical limits in getting users to make multiple relevance judgments and the method has seen little use.

Based on the above described limitations it seems unlikely that any algorithm can ever remove the risk of missing important information. On the other hand it is clear that users must rely on algorithms because there are few topics in biomedicine where one could hope to read all the literature available. The best algorithms are those that minimize the risk of information loss.

2. LITERATURE REVIEW

One could say that the field of modern information retrieval began with the work of Maron and Kuhns (Maron and Kuhns, 1960) describing how to calculate probability of relevance of documents to a query. Their approach required that the individual documents have index terms assigned, each with a probability that if that document were retrieved this index term would be the term used to retrieve it. While such probabilistic indexing is doable in principle it is not very practical. However the approach clearly showed the way to a probabilistic treatment of information retrieval as reflected in later work by Sparck Jones and Robertson (Robertson and Sparck Jones, 1976; Sparck Jones, 1972). One of the problems with the probabilistic approach to information retrieval is a lack of the specific information needed to give the best possible estimates of the probabilities involved. Work by Croft and

Harper (Croft and Harper, 1979) pointed the way to giving reasonable estimates without detailed relevance information. We believe the traditional probabilistic approach to information retrieval has achieved its most mature statement in (Sparck Jones et al., 2000a, 2000b).

At about the same time as the efforts to understand information retrieval in terms of probability theory just described, there was a significant initiative to perform computerized retrieval experiments at Cornell University under the direction of Gerard Salton (Salton (Ed.), 1971). A retrieval system called the SMART system was under development and a number of different algorithms were tested on a variety of test databases to assess the value of different approaches to retrieval. Out of this effort came $tf \times idf$ term weighting (Salton, 1975) and the vector retrieval approach (Salton et al., 1975). Here tf stands for a factor related to the frequency of a term in a specific document (local factor) and idf stands for a factor related to the frequency of the term throughout the database (global factor). The vector approach assumes that each document can be represented by a vector in a space with as many dimensions as there are unique keywords throughout the database. A vector representing a particular document will have a co-ordinate value corresponding to a particular keyword that is the $tf \times idf$ weight for that keyword in that particular document. As a general rule tf (and hence $tf \times idf$) is zero when the keyword does not occur in the document. Thus documents are represented by sparse vectors in the vector space model.

Though the vector space model may seem to be fundamentally different than the probabilistic model, in the final analysis the probabilistic model may be seen to be a special case of the vector space model by simply choosing the formulas for tf and idf to be the values they receive in the probabilistic model. In fact, the vector space model is quite general and allows for the possibility of many different forms depending on how the tf and idf formulas are chosen. We shall subsequently give forms for these quantities which we have found very useful for retrieval in the PubMed database. However, Witten, Moffat, and Bell (Witten et al., 1999) make the following significant observations. Whatever formula is used for tf, within a single document a term with a higher frequency within that document should have a tf at least as great as any term with lower frequency. Likewise, globally a term with a lower frequency throughout the database should have an idf value at least as great as any term of higher frequency. They point out that hundreds of formulas that obey these constraints have been tested on the TREC (Text REtrieval Conference, http://trec.nist.gov/) data (Zobel and Moffat, 1998) and no one formula is best. Rather the choice of formula is a matter of taste and perhaps of the idiosyncrasies of the particular type of data at hand.

In 1998 Ponte and Croft (Ponte and Croft, 1998) introduced a new approach to information retrieval with what they termed language modeling. The idea is that given a query and a document in the database, one may use the frequencies of words in the document to estimate the probabilities of the words in the query and hence the likelihood that the query came from the same source as the document. It is assumed that the document which assigns the highest probability to the words in the query is the document most likely to be relevant to the query. One of the difficulties faced by the method is that not all the words in the query necessarily appear in the document. This is solved by a process of smoothing, which relies on the frequencies of words throughout the database to estimate the probability of seeing words in the query that do not occur in the document. This smoothing process is the real tie to language modeling. The language modeling approach is competitive with other methods and is an active area of research (Kurland and Lee, 2004; Zaragoza et al., 2003; Zhai and Lafferty, 2004). It remains to be seen whether it offers an advantage over other methods and whether there is one best way to do it, or many ways that each offer some small advantage for a particular type of text or a particular database.

In what follows we present an ideal model of information retrieval and then show how the different methods we have described can be seen as special cases of this ideal model. Finally we give some examples of systems that attempt to use particular resources and aspects of biology to advantage to provide a more convenient or more effective approach to information retrieval in limited subdomains.

3. AN IDEAL MODEL

Because of the inherent limitations of information retrieval from natural language texts the problem is most conveniently formulated in terms of probability theory. It is helpful to approach the problem by first describing an ideal retrieval model. We assume that people can be in any one of a set of mutually exclusive states of information need. Such a set of states can be denoted by $\{S_i\}$. Then given a document d there are three different probabilities that are important to consider:

$p(S_i)$ - The prior probability that the randomly chosen human X is in the information need state S_i. This is global information and has nothing to do with a particular person or a particular document.

$p(d|S_i)$ - The probability that a person in the information need state S_i would consider d relevant to that information need. This is local information about S_i and the probability that d gives useful information about the concern expressed by S_i.

The third probability can be expressed in terms of the first two through application of Bayes' theorem

$$p(S_i \mid d) = \frac{p(d \mid S_i)p(S_i)}{\sum_{j=1}^{N} p(d \mid S_j)p(S_j)}. \tag{1}$$

This is the probability that if a person has judged the document d relevant, that person is in information need state S_i.

Now we make an assumption about information need states. Namely, the need state of a user contains all the information about that user's need and once the need state is known the relevance of different documents to the need state become independent events. The probability that a person who sees document d as relevant will also see document e as relevant is an important quantity in the theory. By use of the probabilities just considered we may write this probability.

$$p(e \mid d) = \sum_{j=1}^{N} p(e \mid S_j)p(S_j \mid d) = \frac{\sum_{j=1}^{N} p(e \mid S_j)p(d \mid S_j)p(S_j)}{\sum_{j=1}^{N} p(d \mid S_j)p(S_j)} \tag{2}$$

Here the equality on the left follows from the assumed independence of the relevance of e and d given the information need state S_j. The right side equality follows from substitution of Eq. (1) into the middle term of Eq (2). It is illegal to substitute d in place of e in Eq. (2) because the formula is only derivable if e and d are independent as assumed and d cannot be independent of itself. The value $p(d \mid d)$ is of course 1, while if one incorrectly substitutes d for e in Eq. (2) one generally obtains a number less than 1.

The information that a person has observed the document d could change the state of information need, but that is not dealt with in this model. It would require some modification of Bayes' formula as it appears in Eq. (1). To deal with this one can make the distinction of transient and stable states of information need. This introduces the concept of dynamics into the problem. In our formulation here we deal with only stable states. A person can deal with a change in state in a search for neighbors by simply dropping the search and perhaps taking up another thread of interest where there is still a need which is described by a different state. In this approach the human deals with the issue and it is not necessary to introduce this complexity into the computer model.

If we are given a document d and the knowledge that a user has found it relevant to their information need, then we may wish to find other documents most likely also relevant. For this purpose we may apply Eq. (2) to rank all the other documents. In this process d is constant and all we are concerned about is the relative ratings. Thus we may simplify the formula to

$$\begin{aligned} sim(e,d) &= \sum_{j=1}^{N} p(e \mid S_j) p(d \mid S_j) p(S_j) \\ &= \left(p(e \mid S_1)\sqrt{p(S_1)}, \ldots, p(e \mid S_N)\sqrt{p(S_N)} \right) \cdot \\ &\quad \left(p(d \mid S_1)\sqrt{p(S_1)}, \ldots, p(d \mid S_N)\sqrt{p(S_N)} \right) \end{aligned} \quad (3)$$

This formula has the advantage that it is symmetric in its arguments and can be written as a vector dot product of vectors that represent the two documents involved. These vectors are not normalized in general because they come from probabilities which need not obey such rules. On the other hand the possibility that they are normalized is not excluded.

The formula Eq. (3) may be applied to find the documents related to a given document d or it may be applied more generally when d is understood to represent some query text q. The key to its application is to identify some meaningful set of information need states that can represent the set $\{S_i\}$. How this may be done is the subject of the next section.

4. GENERAL TEXT RETRIEVAL

In general text retrieval, two kinds of information have proven useful. First, the frequency of a term throughout the database carries information about the general usefulness of the term. The less frequent the term is overall, the more informative that term tends to be. Second, the frequency of a term in a document and the overall size of the document combine to give an indication of the importance of the term within the document. The higher the frequency of a term relative to the frequency of other terms in the document the more important the term is likely to be in representing the document's subject matter. These two kinds of information together provide the raw material from which need states may be constructed. There are two important ways that this has been done.

4.1 Vector Models

The vector model assumes that each keyterm is weighted by a global weight gw_t for the term t and by a local weight that relates the term to the document and may be denoted by lw_{td}. For any document d we can then construct a vector

$$v_d = (lw_{td} \times gw_t)_{t \in T} \tag{4}$$

where T represents the set of all keyterms used in the database. Typically the local weight lw_{td} is zero if the term t does not appear in the document. With this representation the similarity between two documents is given by

$$sim(d,e) = v_d \cdot v_e \tag{5}$$

Equations (3) and (5) will correspond if we identify the set of states of information need with the set of keyterms T and define the probabilities by

$$\begin{aligned} p(t) &= (gw_t)^2 \\ p(d \mid t) &= lw_{td} \end{aligned} \tag{6}$$

With these identifications we have an exact correspondence between the two equations. There is one minor problem with the correspondence. That is that $(gw_t)^2$ may not be a number between zero and one and further the sum of all such numbers may not be one. Both these problems can be corrected easily by making the definition

$$p(t) = (gw_t)^2 / \sum_{t' \in T} (gw_{t'})^2 . \tag{7}$$

This has no effect on the ranking because the normalization factor is a constant, but it endows the numbers with the correct formal properties to be probabilities. Thus the typical vector retrieval formula can be derived from the state space paradigm by making the correct identification of the probabilities involved.

One must ask how realistic it is to identify the set of states of information need with the set of keyterms. There are several pieces of evidence that favor this interpretation. First, it finds some justification in the fact that in search engines people typically express their information need with one or a very few terms (Silverstein and Henzinger, 1999). Thus in many cases a single word will express an information need effectively. Second, the formulation provides a natural probabilistic interpretation to vector retrieval, which has been viewed as ad hoc and empirical (Salton, 1991). Third, some of the local weight formulas that prove to be very effective in practice produce a number between 0 and 1 which is readily interpretable as a probability. This grows out of work by Harter who hypothesized (Harter, 1975) that important and unimportant terms follow two different Poisson distributions in their occurrence within documents. While this hypothesis did not initially lead directly to an advantage in information retrieval, Robertson and Walker

(Robertson and Walker, 1994) used the basic idea to design formulas for the local weighting of terms in documents. One of their more effective formulas appeared in (Ponte and Croft, 1998)

$$lw_{td} = tf_d / (tf_d + 0.5 + 0.5 * dlen / avedlen) \tag{8}$$

Here tf_d is the number of occurrences of t in d and $dlen$ is the length of d and equals the number of tokens in d while $avedlen$ is the average length of documents over the whole collection. Our own formulation is based on a more direct application of Harter's idea. Assuming two different rate constants, λ_i for important words in a document and λ_u for unimportant words, the probability that a word is important is given by

$$lw_{td} = \left[1 + Ce^{(\lambda_i - \lambda_u)dlen} \left(\lambda_u / \lambda_i\right)^{tf_d - 1}\right]^{-1} = \left[1 + e^{0.0044 dlen} \left(0.7\right)^{tf_d - 1}\right]^{-1}. \tag{9}$$

Here the constants are determined by the data to obtain good performance. We find a slight (not statistically significant) advantage with Eq. (9) on our test data and also prefer it because of its sound theoretical basis in probability theory. It is used in computing the related documents in PubMed. For the global weight we use the traditional *IDF* weighting formula $\log(N / n_t)$ and set

$$gw_t = \sqrt{\log(N / n_t)} \tag{10}$$

where N is the total number of documents in the database and n_t the number of documents that contain the term t.

4.2 Language Models

Beginning with the seminal paper by Ponte and Croft (1998), unigram language models have become an important approach to textual information retrieval. Typically a language model is estimated from some corpus of text and used to estimate the probability of some new piece of text that is not a part of the corpus used to produce the language model. Bigram or trigram models involve the frequencies of word pairs or triples, respectively. They are useful in speech recognition or spelling correction tasks where one uses the most recent word or pair of words in an attempt to predict the next word. In a unigram model one simply uses the frequencies of words in an attempt to estimate the probability of seeing each word in a piece of text and thereby the probability of that piece of text. This approach to computing the probability of a piece of text naturally fits the paradigm of Eq. (3) provided we identify the states of information need with the possible language models that would be used to describe text in the area of need. This approach to information retrieval has been articulated by Zaragoza et al. (2003). Assuming a Dirichlet prior distribution $p(S)$ and assuming a unigram language model (multinomial) the distribution $p(S\,|\,d)$ has a natural interpretation as the conjugate Dirichlet distribution. They are able to use this approach to compute $p(q\,|\,d)$.

$$p(q\,|\,d)=\frac{1}{p(d)}\int_S p(q\,|\,S)p(d\,|\,S)p(S)dS \tag{11}$$

This equation is just a form of Eq. (2) when one recognizes that the integral is a generalized sum. For ranking purposes this is equivalent to $p(d\,|\,q)$ (assuming a flat prior distribution $p(d)$). For further details we refer the reader to the original paper.

The more typical approach in language modeling for retrieval is to assume the distribution $p(S\,|\,d)$ is all concentrated in the single language model that maximizes $p(d\,|\,S)p(S)$. In this calculation the prior distribution $p(S)$ is assumed to be Dirichlet and is based on the collection frequencies of all terms. The resulting maximum likelihood language model blends the term counts in d with the collection frequencies and produces probabilities for individual terms given by

$$p(t)=\frac{tf_d+\mu p(t\,|\,C)}{dlen+\mu} \tag{12}$$

Here $p(t\,|\,C)$ is the fraction of tokens in the collection C that are t. This formula blends the estimate that would be based on the term counts in the document with the estimate that comes from the whole database. Terms that occur in the document would otherwise have their probabilities over estimated while terms that did not occur would have their probabilities under estimated. The result of the formula is a correction for this and is known as smoothing. Typically the parameter μ is several hundred to a few thousand (Zaragoza et al., 2003; Zhai and Lafferty, 2004). A method to automatically choose μ for good performance has been proposed in (Zhai and Lafferty, 2004).

Given the probabilities of individual words as in Eq. (12), the probability of a query text, $q=q_1q_2\ldots q_n$, is computed as

$$p(q\,|\,d)\propto\prod_{i=1}^{n}p(q_i). \tag{13}$$

Such numbers are equivalent to $p(d\,|\,q)$ (because $p(d)$ is assumed to be a constant over documents) and are used to rank the documents for retrieval.

One may naturally ask which approach to text retrieval, vector or language modeling, is best? We are not aware of any definitive comparison of the two techniques. Researchers reporting on the language modeling approach have found it to perform well and it seems to be competitive with the more traditional vector approach of single term weighting. There are some differences in the two theories, in particular relating to how a user's information need state S is conceived (Robertson and Hiemstra, 2001; Sparck Jones, 2001). On the other hand it can be shown that in practice the way the two models are implemented produces results that are closely related (Zhai and Lafferty, 2004) and smoothing in the language model produces the equivalent of *IDF* weighting in the vector model.

We believe progress is possible in the general retrieval model, provided one can find a more realistic model for the information need states of a user. One can imagine that a more realistic way to represent an information need is in terms of concepts. However, it has not yet proved practical to represent the full scope of needs for a user of a large database with concepts. Concepts tend to be difficult to define and require a good deal of human curation. Even the concepts defined in the Unified Medical Language System (Humphreys et al., 1998) are not sufficient to represent all the different ideas that come into play in medical literature. An automatic way of finding concepts could lead to progress in this area.

5. EXAMPLE TEXT RETRIEVAL SYSTEMS SPECIALIZED TO A BIOLOGICAL DOMAIN

Given a large database in a medical or biological field as opposed to a general text collection such as the Brown Corpus or a collection of news articles from the Wall Street Journal, one might expect that there would be methods of retrieval in the area of biology in general that would work better for biology than for other areas. However, there is no approach that we are aware of that really makes information retrieval in the biological area better than general retrieval. This is true because the area of biology is simply too broad to allow any simplifications specific to biology. Just about any kind of text construction or topic that can appear in a large database of text in the field of biology can appear in any other collection, though the frequency of some types of text is less in documents on biology. As a consequence the PubMed (http://web.ncbi.nlm.nih.gov/PubMed/) search-engine-related documents function is based on a version of vector retrieval as outlined in the previous section. However, there are attempts to create databases in specialty areas of biology and medicine where retrieval can improve on the general model. We will describe several of those systems here.

5.1 Telemakus

The Telemakus system developed by S. Fuller and colleagues (Fuller et al., 2004) at the University of Washington represents research reports schematically with twenty-two fields or slots that contain information describing the research in different ways. Twelve slots are bibliographic and filled from PubMed, one is the Telemakus ID, and the remainder are extracted from the full text of the document. Among the fields filled from the document are items from the Methods section of a report that describe how the research was performed. Perhaps of most significance is the field that holds research findings. These are extracted especially from the captions of figures and tables and the extraction process makes use of the fact that the language in such captions is somewhat restricted and easier to process. Telemakus uses automated extraction to initially produce the schematic surrogate for a document. Then this automatically produced surrogate is displayed along with a marked up version of the original report so that a human expert can correct errors and finalize the schematic representation of the document.

Once data representing research reports has been entered into the system a user can access this information by keyword searching or in some cases browsing an index. When a particular study has been displayed, figures and tables can be accessed directly as can the full text document if available. Research findings are displayed for the study and can be queried for other studies reporting the same finding. In addition concepts that are represented in the database can be displayed in a window as a concept map. Such a map displays the concept along with other related concepts (measured by co-occurrence in research reports). One can then navigate by clicking on different concepts to search for concepts related to the original but perhaps more specific to the information need. A concept map for "neoplasms" is illustrated in Figure 10-1.

The Telemakus system is available at http://www.telemakus.net/ and currently comprises a database of research reports on Caloric Restriction and the Nutritional Aspects of Aging. A strength of the system is the ease with which one can examine the important findings in a report without having to read the whole report. A potential weakness is the need for a subject expert to examine each surrogate for a report and correct mistakes. Ways are being sought to make the system more nearly automatic. Currently the system is tied to the area of biology as a number of Unified Medical Language resources are used in its processing. However, there is in principle nothing to preclude its application in a wider context.

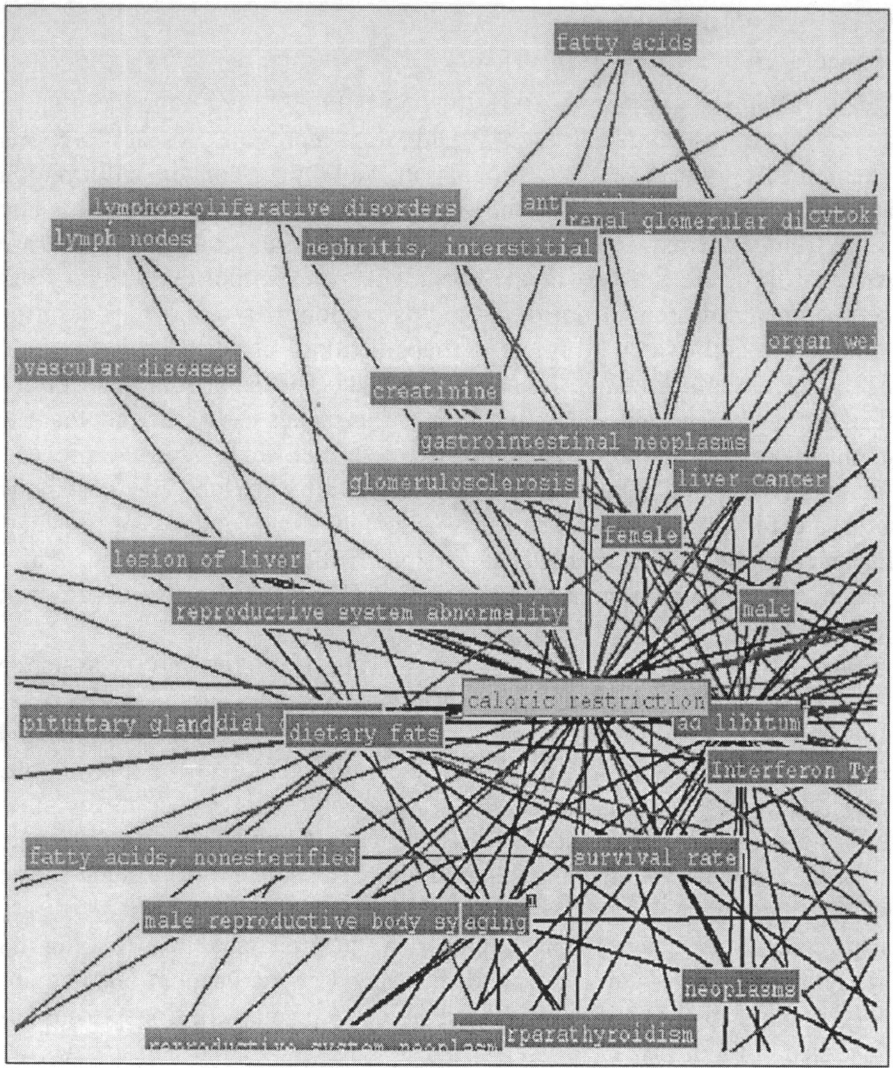

Figure 10-1. A concept map of research findings linked to "caloric restriction" (links reported by authors as statistically significant are in blue in original).

5.2 XplorMed

XplorMed is a system developed by Perez-Iratxeta and colleagues (Perez-Iratxeta et al., 2002; Perez-Iratxeta et al., 2001, 2003) for browsing the MEDLINE literature database. Given a set of MEDLINE abstracts (up to 500) the system computes the words with the strongest relations with other words (stop words excluded) as the keywords for the set. One keyword is said to include another at a level greater than α if the presence of the first keyword in an abstract implies the presence of the second with a probability of at least α based on the data in the set of abstracts under analysis. The inclusion relation is used to define chains of words that all occur together in at least some of the documents. Keywords, included words, and word chains are displayed to the user during a session.

To begin a session the user of XplorMed may define a subset of MEDLINE abstracts by running a PubMed query or importing an already defined subset from another source or application. In the first step the set of abstracts is broken up into mutually exclusive subsets based on a set of broad MeSH categories. The user may then select a combination of the documents in any number of these categories to include in his analysis. Once the set of abstracts is finalized the system extracts keywords and computes their relationships to each other. Keywords are displayed as a list to the user who then has a number of options including asking to see a particular keyword in context in the abstracts in which it occurs, asking for the words implied by or included with a given keyword, or asking for all the word chains involving the keywords. Given a word chain one can then ask for a ranked list of those documents that contain the word chain. At the same time one may request to display links from the resulting set of documents to other databases such as OMIM, SwissProt, etc. It is also possible to display a listing of the MeSH terms that are contained in the resulting set of documents. Finally one may take the current set of documents and start the analysis cycle over. One also has the option at this stage to enlarge the set by pulling in related documents using the PubMed related documents function and to restart the analysis with this larger set.

A strength of the XplorMed system is its application of simple statistics on word use to find useful relationships between text words without reliance on a controlled vocabulary. A weakness is that such processing cannot guarantee that the relationships found are useful at the level that defined relationships between elements of a thesaurus are useful.

5.3 ABView:HivResist

In order to focus on a small set of the MEDLINE literature, Belew and Chang (2004) have developed a system called ABView:HivResist. This system is designed to provide an enhanced environment for the study of HIV drug resistance and the mutations that produce it. Currently the system contains 9,190 MEDLINE abstracts in the area of HIV protease inhibitors. Focusing on a limited set such as this reduces the ambiguity of terms, especially abbreviations, and makes practical the construction of a thesaurus which captures much of the synonymy in the domain. In particular the thesaurus includes alternative ways of referring to mutations in the HIV protease molecule and the different names applied to the drugs studied for their inhibitory effects on this molecule. Relatedness of documents within the set may be assessed based on the citation of one by the other or by their relatedness as computed in the related documents function from PubMed.

The user of ABView:HivResist interacts with the system through a GUI and the results of a Boolean query appear in the main window (see Figure 10-2).

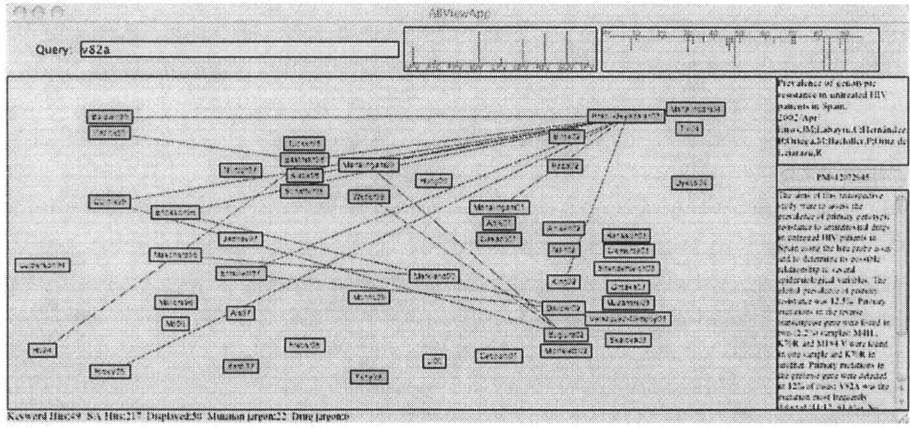

Figure 10-2. ABView:HivResist GUI allowing user to see the results of a search for documents that mention a mutation replacing valine by alanine at position 82 in the HIV-1 protease molecule.

Each document is represented as an icon and the most relevant documents appear highest in the window, while horizontal position in the window denotes time of publication over the past ten years. The citation of one document by another is denoted by an arrow from one icon to the other in the display. Documents directly retrieved by the query are displayed in dark green and those present only by virtue of a related document link to a direct hit are displayed in light green. There are also two histograms displayed, one showing which residues of HIV protease are mentioned and the other which inhibitory drugs are mentioned in the documents displayed. Clearly one of the strengths of this system is the narrow focus which allows one to develop a thesaurus tailored to a particularly important research problem. The drawback is the human effort required to develop this enhanced facility.

5.4 The Future

The three systems presented here each have the objective of using specialized information resources to enhance retrieval. We believe that such approaches are the future of information retrieval. However, there are several problems that stand in the way of progress in improving information retrieval for specialized domains. The first problem is related to the cognitive effort required to create the structures that allow improved retrieval. Such cognitive effort is required in both Telemakus and ABView:HivResist. In Telemakus a human operator must examine the automatically produced surrogate document and make any necessary corrections. Ideally one would produce a completely reliable surrogate automatically, but computers do not have the necessary capabilities to understand language. Thus it is not possible to tell reliably what is a research finding without human intervention. Likewise ABView:HivResist required a human to describe the different ways a mutation is indicated in text. This is a specialized task and could not be done automatically based on the UMLS Thesaurus, for example. On the other hand the UMLS Thesaurus may be helpful in indicating alternative names for drugs used to treat drug resistant HIV. However, again one could not rely on any pre-constructed source to reveal just which drugs are important in treating drug resistance in HIV infection. The point is simply that in order to construct such specialized access tools a significant human effort is required and this will limit the more global application of such methods until the time we have more reliable automatic language processing tools.

A second problem standing in the way of progress is the lack of understanding of what the human mind is actually doing when one is searching for information. If we understood this we might be able to leverage the computer's strengths to help in the processing. Computers are very good at certain tasks, such as rapidly processing huge amounts of data looking for matching strings or strings satisfying simple criteria. A computer also has the ability to perfectly remember large amounts of information. But what are people actually doing when they look for information? We do not really know the answer. This is a problem in human cognition and its solution promises at some level to give guidance in how to perform better retrieval. For example, XplorMed uses simple statistics to provide terms that may be useful in refining a search. Is this really something that fits well what a user is trying to accomplish when he is searching? We do not know the answer to this, but it clearly would be helpful to know. We may hope that future research on human cognition will provide some answers.

Finally one of the important unsolved problems in this area of research is how to measure success. How can one accurately measure the utility of such a complicated system? Clearly successful usage will depend to a large extent on the knowledge and skill of the operator. Also one strategy for the use of a system may not be as good as another. Such heterogeneity makes comparison of different systems difficult. As far as we are aware there are no published formal evaluations of the systems presented here.

REFERENCES

Belew, R. K. and Chang, M. (2004). "Purposeful Retrieval: Applying Domain Insight for Topically-focused Groups of Biologists," Paper presented at the *Search and Discovery in Bioinformatics: SIGIR 2004 Workshop.*

Blair, D. C. and Maron, M. E. (1985). "An Evaluation of Retrieval Effectiveness for a Full-text Document-retrieval System," *Communications of the ACM*, 28(3), 289-299.

Croft, W. B. and Harper, D. J. (1979). "Using Probabilistic Models of Document Retrieval Without Relevance Information," *Journal of Documentation*, 35(4), 285-295.

Fuller, S., Revere, D., Bugni, P., and Martin, G. M. (2004). "Telemakus: A Schema-based Information System to Promote Scientific Discovery," *Journal of the American Society for Information Science and Technology*, In press.

Funk, M. E., Reid, C. A., and McGoogan, L. S. (1983). "Indexing Consistency in MEDLINE," *Bulletin of the Medical Librarians Association*, 71(2), 176-183.

Furnas, G. W., Landauer, T. K., Gomez, L. M., and Dumais, S. T. (1987). "The Vocabulary Problem in Human-System Communication," *Communications of the ACM*, 30(11), 964-971.

Harter, S. P. (1975). "A Probabilistic Approach to Automatic Keyword Indexing: Part I. On the Distribution of Specialty Words in a Technical Literature," *Journal of the American Society for Information Science*, 26, 197-206.

Humphreys, B. L., Lindberg, D. A., Schoolman, H. M., and Barnett, G. O. (1998). "The Unified Medical Language System: An Informatics Research Collaboration," *Journal of the American Medical Informatics Association*, 5(1), 1-11.

Kurland, O. and Lee, L. (2004). "Corpus Structure, Language Models, and Ad Hoc Information Retrieval," Paper presented at the *ACM SIGIR 2004.*

Maron, M. E. and Kuhns, J. L. (1960). "On Relevance, Probabilistic Indexing and Information Retrieval," *Journal of the ACM*, 7(3), 216-243.

Perez-Iratxeta, C., Keer, H. S., Bork, P., and Andrade, M. A. (2002). "Computing Fuzzy Associations for the Analysis of Biological Literature," *BioTechniques*, 32, 1380-1385.

Perez-Iratxeta, C., Perez, A. J., Bork, P., and Andrade, M. A. (2001). "XplorMed: A Tool for Exploring MEDLINE Abstracts," *TRENDS in Biochemical Sciences*, 26(9), 573-575.

Perez-Iratxeta, C., Perez, A. J., Bork, P., and Andrade, M. A. (2003). "Update on XplorMed: A Web Server for Exploring Scientific Literature," *Nucleic Acids Research*, 31(13), 3866-3868.

Ponte, J. M. and Croft, W. B. (1998). "A Language Modeling Approach to Information Retrieval," Paper presented at the *SIGIR98*, Melbourne, Australia.

Robertson, S. and Hiemstra, D. (2001). "Language Models and Probability of Relevance," Paper presented at the *First Workshop on Language Modeling and Information Retrieval*, Pittsburgh, PA.

Robertson, S. E. and Sparck Jones, K. (1976). "Relevance Weighting of Search Terms," *Journal of the American Society for Information Science*, May-June, 129-146.

Robertson, S. E. and Walker, S. (1994). "Some Simple Effective Approximations to the 2-Poisson Model for Probabilistic Weighted Retrieval," Paper presented at the *17th Annual International ACM SIGIR Conference on Research and Development in Information Retrieval.*

Salton, G. (1975). *A Theory of Indexing* (Vol. 18). Bristol, England: J. W. Arrowsmith, Ltd.

Salton, G. (1991). "Developments in Automatic Text Retrieval," *Science*, 253, 974-980.

Salton, G., Wong, A., and Yang, C. S. (1975). "A Vector Space Model for Automatic Indexing," *Communications of the ACM*, 18, 613-620.

Salton, G. (Ed.). (1971). *The SMART Retrieval System: Experiments in Automatic Document Processing*, Englewood Cliffs, NJ: Prentice-Hall, Inc.

Saracevic, T. (1991). "Individual Differences in Organizing, Searching, and Retrieving Information," Paper presented at the *Proceedings of the 54th Annual ASIS Meeting*, Washington, D.C.

Shieber, S. M. (1994). "Lessons from a Restricted Turing Test," *Communications of the ACM*, 37(6), 70-78.

Silverstein, C. and Henzinger, M. (1999). "Analysis of a Very Large Web Search Engine Query Log," *SIGIR Forum*, 33(1), 6-12.

Sparck Jones, K. (1972). "A Statistical Interpretation of Term Specificity and its Application in Retrieval," *The Journal of Documentation*, 28(1), 11-21.

Sparck Jones, K., Walker, S., and Robertson, S. E. (2000a). "A Probabilistic Model of Information Retrieval: Development and Comparative Experiments (Part 1)," *Information Processing and Management*, 36, 779-808.

Sparck Jones, K., Walker, S., and Robertson, S. E. (2000b). "A Probabilistic Model of Information Retrieval: Development and Comparative Experiments (Part 2)," *Information Processing and Management*, 36, 809-840.

Sparck-Jones, K. (2001). "LM vs PM: Where's the Relevance?" Paper presented at the *First Workshop on Language Modeling and Information Retrieval*, Pittsburgh, PA.

Swanson, D. R. (1988). "Historical Note: Information Retrieval and the Future of an Illusion," *Journal of the American Society for Information Science*, 39(2), 92-98.

Wilbur, W. J. (1998). "The Knowledge in Multiple Human Relevance Judgments," *ACM Transactions on Information Systems*, 16(2), 101-126.

Wilbur, W. J. and Coffee, L. (1994). "The Effectiveness of Document Neighboring in Search Enhancement," *Information Processing and Management*, 30(2), 253-266.

Witten, I. H., Moffat, A., and Bell, T. C. (1999). *Managing Gigabytes* (Second ed.), San Francisco: Morgan-Kaufmann Publishers, Inc.

Zaragoza, H., Hiemstra, D., and Tipping, M. (2003). "Bayesian Extension to the Language Model for Ad Hoc Information Retrieval," Paper presented at the *SIGIR'03*, Toronto, Canada.

Zhai, C. and Lafferty, J. (2004). "A Study of Smoothing Methods for Language Models Applied to Information Retrieval," *ACM Transactions on Information Systems*, 22(2), 179-214.

Zobel, J. and Moffat, A. (1998). "Exploring the Similarity Space," *ACM SIGIR Forum*, 32(1), 18-34.

SUGGESTED READINGS

van Rijsbergen, C. J. (1979). *Information Retrieval*, Second Edition, London: Butterworths.

A classic in the field. Gives a highly readable account of fundamental topics such as indexing, file structures, clustering, term dependencies, probabilistic methods, and performance evaluation.

Salton, G. (1989). *Automatic Text Processing*, New York: Addison-Wesley.

The book has four parts and the third part of the book consists of three chapters on various aspects of information retrieval. The emphasis is on the vector model and tfxidf weighting, methods largely developed by the author and his students at Cornell University.

Sparck Jones, K. and Willet, P. (Eds.). (1997). *Readings in Information Retrieval*, San Francisco: Morgan Kaufman.
An important resource reprinting many of the most important papers detailing significant advances in the science of information retrieval over the years.

Witten, I. H., Moffat, A., and Bell, T. C. (1999). *Managing Gigabytes*, Second Edition, San Francisco: Morgan Kaufmann.
A very good treatment of basic vector information retrieval for text. Also an important resource for those who must manage large text files as it emphasizes compression methods and their practical implementation to construct digital libraries.

Baeza-Yates, R. and Ribeiro-Neto, B. (1999). *Modern Information Retrieval*, New York: Addison-Wesley.
A wide ranging coverage of all the basic approaches to information retrieval except the language modeling approach. Includes extensive treatment of user interfaces, the internet, and digital libraries.

Belew, Richard K. (2000). *Finding Out About*, Cambridge: Cambridge University Press.
A good introductory text with an emphasis on the World Wide Web and artificial intelligence.

ONLINE RESOURCES

Information Retrieval Links: Lists many resources related to the field of information retrieval. Included are links to access the software for the SMART retrieval system developed by Gerard Salton and his students:
http://www-a2k.is.tokushima-u.ac.jp/member/kita/NLP/IR.html

Information Retrieval Software: This site provides links to information retrieval software (some as freeware), to internet search engines and web directories, and to search engine optimization sites: http://www.ir-ware.biz

The Apache Jakarta Project: Jakarta Lucene is a high-performance, full-featured text search engine library written entirely in Java. It is a technology suitable for nearly any application that requires full-text search, especially cross-platform. Jakarta Lucene is an open source project available for free download from Apache Jakarta:
http://jakarta.apache.org/lucene/docs/index.html

QUESTIONS FOR DISCUSSION

1. Describe how one might use the World Wide Web to construct a representation for the information need states $\{S_i\}$. How might one estimate $p(S_i)$ and $p(d \mid S_i)$ used in equation (2) for such a model?

2. In choosing the features to represent documents in the vector method of retrieval it is generally found that single words work as well or better than single words plus phrases. Provide what you believe could be an explanation for this phenomenon.

3. In the MEDLINE database each document has on the average about a dozen MeSH headings assigned to help characterize the subjects discussed in the document. These MeSH terms make useful features for retrieval but they involve a significant expense and human effort to assign. Describe what you think the barriers are to making these MeSH assignments more useful.

4. The MEDLINE record of a document does not contain the list of citations or references which generally appear at the end of a document. However, some databases do have such information. Describe how these citations could be used as features in a vector retrieval system along with the words in the text. How would you weight them?

5. Describe one method that you feel would be appropriate to evaluate the Telemakus system and how this method could be used to make decisions about use of the system.

6. Describe one method that you feel would be appropriate to evaluate the XplorMed system and how this method could be used to make decisions about use of the system.

7. Some retrieval systems, such as ABView:HivResist, attempt to use graphical displays of documents in space to convey information. What do you see as problems with this approach? How do you think the relationships between documents could best be represented graphically or otherwise?

Chapter 11
PUBLIC ACCESS TO ANATOMIC IMAGES

George R. Thoma

Communications Engineering Branch, Lister Hill National Center for Biomedical Communications, National Library of Medicine, Bethesda, Maryland 20894

Chapter Overview

Described here is an R&D project at the National Library of Medicine with the goal of creating systems to (a) provide the lay public images of the human anatomy, specifically high resolution color cryosections from NLM's Visible Human Project and 3D images of anatomic structures created from these cryosections; (b) enhance text-based information services with relevant anatomic images. To accomplish these objectives, investigations into advanced techniques and technologies were conducted, including multi-tier system architectures, database design, design of suitable image viewers, image compression, and use of the Unified Medical Language System (UMLS), among others. This research has contributed to the design and development of AnatQuest, a system released for use by the lay public. It has also helped define the system architecture and essential functions required to link biomedical terms in documents to relevant anatomic structures in our database through UMLS concepts and relationships and to display these to the reader. In this chapter, we describe how our research has informed the overall goal to explore and implement new and visually compelling ways to bring anatomic images from the Visible Human dataset to the general public.

Keywords

anatomic images; AnatQuest system; Visible Human project; multimedia; object-oriented database; client-server architecture; Unified Medical Language System; text-to-image linking; public access; Java applet; servlet

1. INTRODUCTION

Ever since the Internet and the World Wide Web became ubiquitous, the lay public, as much as the scientific community, has taken for granted easy and reliable access to information of all kinds. This expectation continues to be met by commercial database providers and, increasingly, by national institutions such as the Library of Congress and the U.S. National Library of Medicine (NLM). Expressly stated as a goal in NLM's long range plan formulated in 2000 is to "Encourage use of high quality information by health professionals and *the public*." In addition, among the high priority new initiatives identified by NLM's Board of Regents in 2001 is Health Information for the Public (National Library of Medicine, 2001). In implementing this vision, NLM has created such services as MedlinePlus®, ClinicalTrials.gov, and NIHSeniorHealth, all primarily for the public rather than for its more traditional constituencies, the biomedical clinical and research communities (National Library of Medicine website).

It is in this same spirit that the AnatQuest project discussed here has been organized. Our focus is to provide the lay public images of the human anatomy, specifically high resolution color cryosections from NLM's Visible Human Project (National Library of Medicine, 2003; Ackerman, 1998) and 3D images of anatomic structures created from these cryosections. By enabling public access to these images, we contribute to the increasingly important mission of the NLM to "universalize" access to biomedical information.

This effort, however, requires investigation into advanced techniques and technologies, e.g., multi-tier system architectures, database design, design of suitable image viewers, image compression, and others. Research in these areas inform our overall goal which is to explore and implement new and visually compelling ways to bring anatomic images from the Visible Human (VH) dataset to the general public. Specific objectives are to:

1. Develop a system, also called AnatQuest, to let users query the VH image database via visual and textual navigation, and retrieve and display high resolution images, assuming minimal bandwidth requirements.

2. Investigate techniques to extend text-based information services for the lay public (e.g., MedlinePlus) to include access to anatomic images.

3. Explore options to segment and label the *high resolution* VH cross-section images beyond the thorax, the only region currently segmented and labeled, to enable the creation of images of 3D structures from all anatomic regions of the Visible Male and Female datasets.

Since its availability, the VH image set has inspired many projects and applications worldwide. Of these many applications there are a few that meet three conditions we consider important for public access: Internet accessibility via browsers; the provision of at least some labels for anatomic structures in each cryosection slice; and acceptable user interaction in a low bandwidth environment. One is the Workshop Anatomy for the Internet (WAI) from the Johannes Gutenberg University in Mainz, Germany. WAI contains both labeled and unlabeled cryosections, correlated CT and MRI images (also part of the VH image set), animations, and a vocabulary of gross anatomy. Another is the Visible Human Web Server from Ecole Polytechnique Federale de Lausanne in Switzerland. This application offers services for extracting labeled slices, surfaces, and animations; real-time navigation through the body; constructing 3D anatomic structures; and creating teaching modules. A third application, Net Anatomy, is a multimodal teaching tool for anatomy from Scholar Educational Systems, Inc.

While these are effective means for the public to view and use VH images, none of them offer the *high resolution* version of the image set. The high resolution set was created by digitizing 70mm film frames captured during cryosectioning to a resolution of 4K x 6K, and cropped to 4K x 2.7K. The "standard" resolution set was directly captured by a charge coupled device (CCD) camera at 2048 x 1216 pixels. The latter is universally used by application developers, but we use the high resolution images with the expectation that fine, subtle structures possibly missed in the CCD-captured images would be evident in these (with four times the number of pixels).

This chapter is organized as follows. In Section 2 we briefly describe earlier in-house projects as background. In Section 3 we present the design tradeoffs that underlie the development of the online AnatQuest system, and the design considerations in creating a kiosk version of AnatQuest for onsite exhibits. In Section 4 we describe ongoing work that should provide the lay public greater access to anatomic images. All work described here has been done at the Communications Engineering Branch of the Lister Hill National Center for Biomedical Communications, an R&D division of the NLM.

2. BACKGROUND

2.1 Previous Work

Previous in-house work undergirding the AnatQuest project includes the VHSystems and 3DSystems projects. First, the VHSystems project established a means for the bulk transfer of Visible Human data, mainly the color cross-sections, over the Internet. Both high resolution images as well as the original CCD-captured files were disseminated by an FTP server that continues to deliver an increasing amount of data, as seen in Figure 11-1. Demand for these images is widespread, as evident from the geographic and domain distribution of the recipients (Table 11-1).

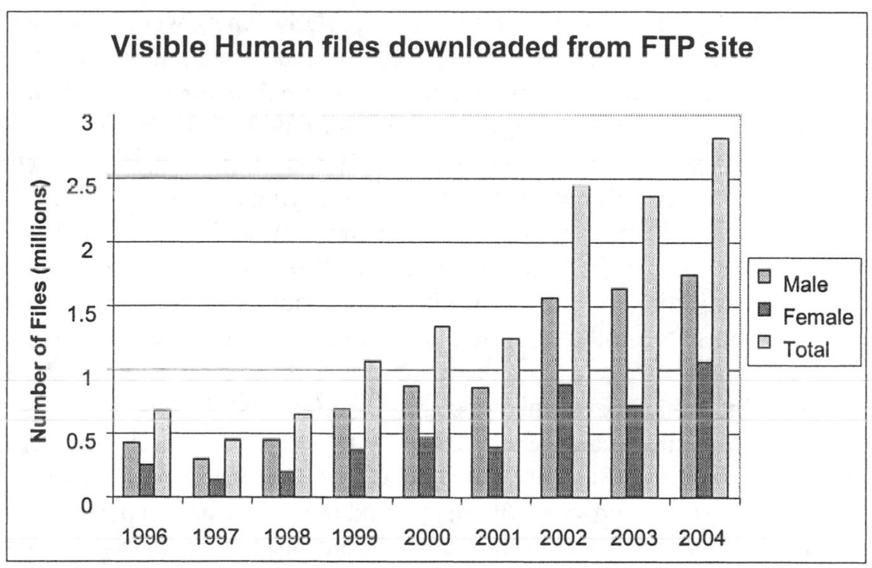

Figure 11-1. Visible Human files downloaded from FTP site.

Table Chapter 11-1. Domain distribution of Visible Human files downloaded.

Educational (.edu)	1,942,254	Brazil	51,709	South Africa	11,660
Network (.net)	770,223	Belgium	47,288	Switzerland	10,311
Commercial (.com)	648,003	Italy	45,716	Slovenia	9,021
Japan	623,679	India	45,150	Romania	6,529
Germany	365,791	Hong Kong	41,262	United States (.us)	5,831
Canada	254,616	Mexico	35,542	Colombia	5,514

continued

Domain distribution of Visible Human files downloaded, continued

United Kingdom	228,301	Portugal	32,243	Greece	5,340
Korea (South)	103,289	Sweden	26,366	Malaysia	4,069
Czech Republic	100,405	Spain	25,948	Norway	3,872
Taiwan	91,247	Israel	23,447	Austria	3,739
Netherlands	77,967	Finland	22,607	Denmark	2,710
France	75,735	Ireland	16,384	Military (.mil)	646
US Government (.gov)	75,352	Venezuela	14,787	Thailand	140
Singapore	61,360	Chile	14,188	Egypt	93
Poland	60,750	Hungary	14,011	New Zealand	13
China	53,759	Organization (.org)	13,793		
Australia	52,984	Iceland	13,589		

The main goals of the second project, 3DSystems, were to create a database out of the VH dataset and to create suitable data and file structures so that access would be provided to the original images as well as to derivative products. A non-proprietary file format, VHI (Henderson et al., 1998), was created to accommodate a wide range of image types: color cross-sections, MRI, CT, segment masks (identifying contours and labels on anatomic structures in the cross-sectional images), and volume of interest (VOI) image stacks. The targeted uses for these images were: rendered images for education, e.g., for curriculum development; the segment masks and VOI stacks for product development, e.g., to create surface and volume rendered images of organs; and the color cross-sections for research into the design of algorithms for segmentation, registration, and rendering.

To achieve the goals of the 3Dsystems project, a prototype image management system, AnatLine (Strupp-Adams and Henderson, 2000), was developed to import and store the images and to retrieve and export them. As a means for validating the design and implementation of the database and image management system, example 3D rendered images and VOI stacks were required. To create these, the thorax region of the male (consisting of 411 slices out of the total 1,878) was segmented and labeled, and a tool VHVis (Zhou et al., 1998), was developed to use these labeled segments to surface-render selected anatomic objects.

Also developed were tools needed to use AnatLine: VHParser and VHDisplay. The first is for unpacking the VHI data files into individual components (cross-section images, byte masks, coordinate and label tables, etc.). VHDisplay is for displaying both cross-sectional and rendered images. Also, VHDisplay is augmented to audibly voice the names of anatomic structures as the images are displayed. These tools may be downloaded from the AnatQuest Website: anatquest.nlm.nih.gov/anatline/.

2.2 Prologue: Database Design

Since the AnatQuest system inherits the image and data repository developed for the AnatLine system as part of the earlier 3Dsystems project, here we present the principal design considerations leading to AnatLine. We focus on: the conceptual data model, the choice of the object-oriented framework for the design of the database, and the selection of a specific object-oriented DBMS. Each of these is discussed next.

a. Data

In addition to the raw color cryosection images and the CT and MRI images from the data collection process, the data to be stored also includes annotated rendered images, as well as segmented and labeled images from the Visible Male's thorax region. Concepts and relations in the Metathesaurus of NLM's Unified Medical Language System (UMLS) were used to assign labels to the segments, and anatomical relationships among the segmented structures were identified. The x-y-z coordinates of the segmented structures were used to compute their spatial relationships. The anatomical and spatial relationships transform the database of structures to a form essential for navigating through the body. These entities and their relationships are described next in the data model.

b. Data Model

The data model, as shown in the object relationship diagram in Figure 11-2, is organized around a number of objects representing: the body (i.e., the male or female cadaver), body regions (e.g., thorax, abdomen, etc.), the anatomical structures (e.g., organs and their parts), and the images. Note that the data model is general enough to accommodate future data from other cadavers or their parts.

The relationships among these objects may be either hierarchical or not. Hierarchical relationships allow for inheritance of shared properties. For example, the male and female body objects inherit generic descriptions of a heart and cardiovascular system, but have different reproductive systems. Non-hierarchical relationships (that do not exhibit inheritance properties) represent the anatomical relationships among anatomical objects. These are modeled in terms of <u>part-of</u> and <u>contains</u> (or <u>has-part</u>) relationships as shown in the conceptual model in Figure 11-2.

Each of the objects is described in terms of its attributes. Simple attributes are brief, such as a name. More complex are the relational ones that contain instances of other objects which define a <u>part-of /contains</u> relationship between two objects.

As seen in the figure, the anatomical structure object consists of attributes for its name, superstructure, the region of which it is a part, and the physiologic systems to which it belongs. The "superstructure" attribute of an anatomical structure object refers to a second anatomical structure object which contains the first object. For example the *right-ventricle*'s superstructure is *heart,* which means that the *right ventricle* structure is part-of the *heart* structure. The "body region" attribute points to a body-region object which contains the anatomical structure object. For example, *heart* belongs to the *thoracic* region. A physiologic system is a function of the body in which a structure is a component. For example, *heart* is a component of the *cardiovascular* system.

In addition, each of these database objects points to image metadata objects in the database representing data descriptive of an available image (whether MRI, CT, a color slice, a rendered image, a segment mask, etc.), including its size and the name of the image file in the Visible Human file server where the images actually reside.

The male and female body objects are represented in terms of their basic characteristics such as age, gender, and race.

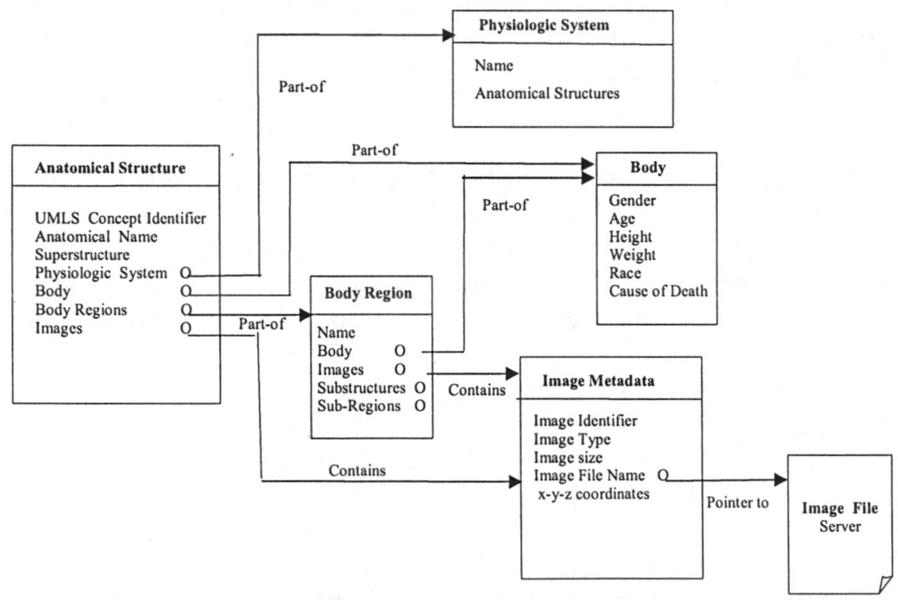

Figure 11-2. Database conceptual model.

c. Object-oriented Framework

The choice of the object-oriented framework for programming and data modeling was influenced by a number of factors including the structure of the data and the requirement for efficient data representation and retrieval, as discussed below.

- Data structure: Anatomical data consists of both spatial and structural information, which in turn are represented by other complex objects. For example, a *heart* object consists of four other objects, each describing one of its chambers in terms of other substructures, e.g., *right atrium* containing the *right auricular appendage*. Such nested structures form graphs of objects that are represented more readily in object-oriented environments.
- Efficient data representation: An object-oriented programming framework provides efficient data representation through instantiation and inheritance. Male and female hearts, for example, are instances of the same structure which can be described once but *instantiated* twice. To give another example, a degenerative heart *is a* normal heart with extra conditions associated with it. This is-a relationship forms a class hierarchy which allows specialized objects to *inherit* characteristics of their exemplar objects, thereby providing a more efficient representation for reuse and maintenance. Since the "healthy" heart object is described by the common characteristics of heart, the "degenerative" heart only needs to add the extra conditions, because it inherits the generic information from its exemplar object (the healthy heart). These examples demonstrate the benefits of the instantiation and inheritance of objects in the object-oriented framework allowing efficient data representation, modeling, and maintenance.
- Efficient data and image retrieval: A goal of the 3Dsystems project was to enable the development of human atlases in which users may explore the body by navigating through structures, substructures, and sub-sub-structures by using the spatial and structural relationships among anatomical parts. This requirement is best supported with the graph navigation property of the object-oriented framework.

d. Object-oriented Database

The advantages of the object-oriented framework listed above determined the selection of an object-oriented database over a relational alternative. Object-oriented databases offer the following advantages over their relational counterparts in the object-oriented programming environment used in developing AnatLine:

- Transparent persistence: The objects within the programming environment are automatically saved in, and retrieved from, the database. This makes the database an extension of the computer memory and hence transparent to the programmer. For example, when navigating through a *heart* object, its *right ventricle* is retrieved from the database transparently when needed, with no additional programming. For relational databases additional modules would be required to provide the necessary mapping between the database and the object in memory.

- Unified model: A single model for representation of data in the object-oriented programming language as well as the object-oriented database is preferable from the point of view of reduced effort in development and maintenance and eliminating mapping between two disparate models.
- Ease of navigation: The navigation through objects corresponds to a graph or tree, thereby fitting the requirement for a navigable human image atlas.

In light of these factors, ObjectStore® was selected as the database management system (DBMS). It was found to provide efficient data management, good performance, concurrency, and multithreading. Unlike relational or object-relational DBMS which retrieve related rows of data by executing joins at runtime, ObjectStore stores and manages data components and objects with their relationships intact. Also, as in any DBMS, ObjectStore offers concurrency (allowing multiple users and applications to simultaneously access and update the database) and a multithreading feature (allowing the use of kernel threads, asynchronous I/O and shared memory).

3. THE ANATQUEST SYSTEM

3.1 Need for Public Access

In this section we give reasons for developing the AnatQuest system followed by a discussion of its design. Basically, AnatQuest became necessary because AnatLine was not suitable for the lay public. To begin with, the principal users of AnatLine were expected to be scientists interested in testing algorithms for image processing (e.g., registration, segmentation, feature extraction) or constructing 3D anatomical objects. For such expert users it was considered reasonable to have them download software for disassembling the retrieved files and displaying them. However, this approach posed significant barriers for the lay public. First, the requirement to download and install VHParser and VHDisplay was neither desired nor easily done by novice users. The second barrier was the lack of immediate visual feedback, a consequence of the large size of the VHI files and the difficulty of unbundling them quickly. Some files are on the order of 1 GB when segmented slice images and bitmap overlays for some of the larger organs are packaged. But lossy compression, which would have yielded significant file size reduction, was not considered in order to preserve the complete content of these files. Furthermore, to transfer such large files to a user site in a reasonable time requires connections that not only are high speed links, but need to be reliable to avoid resending because of intermittent connections. Such robust high speed links are often not available to the lay user.

As a consequence of these factors, AnatLine found relatively low use overall and almost none at all by the general public. This led to our development of the AnatQuest system.

3.2 AnatQuest: Design Considerations

The design of AnatQuest retained AnatLine's database system, but its goal was to provide widespread access to the VH images for lay users with special attention to those with low speed connections as well. AnatQuest offers users thumbnails of the cross-sectional, sagittal, and coronal images of the Visible Male, from which detailed (full-resolution) views may be accessed. Low bandwidth connections are accommodated by a combination of user-adjustable viewing areas and image compression done on the fly as images are requested. Users may zoom and navigate through the images. As shown in Figure 11-3, the number of hits for AnatQuest far exceeds that for AnatLine.

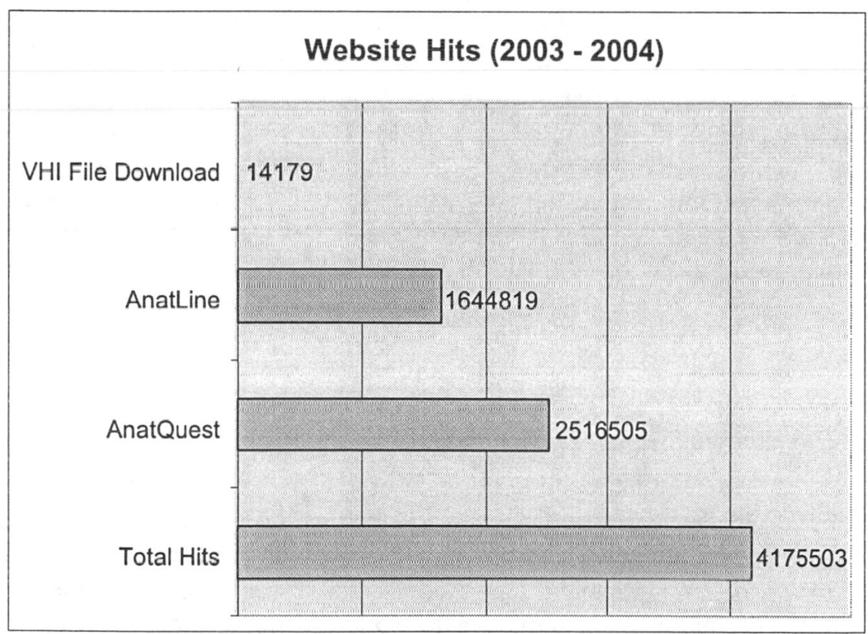

Figure 11-3. Website hits, June 2003 - July 2004.

In addition to its main purpose, AnatQuest serves as an entry point for both the FTP server for bulk downloading of VH files as well as all the functions of AnatLine. Through AnatQuest the user may also retrieve more than 400 surface-rendered objects created at the Lister Hill Center as well as a few samples from outside sources.

Since a key goal is to accommodate low bandwidth users, we had to address the file size problem. The large size of the VH images, both in totality and as individual files (7.5 MB for the CCD files; 33.2 MB for the scanned 70mm photographs), motivated parallel in-house research in the compression and transmission of these images (Thoma and Long, 1997; Meadows et al., 1997; Long, 1996; Mitra et al., 1996; Pemmaraju et al., 1996; and Long et al., 1995). Studies were conducted in both lossy and lossless compression techniques. Among the lossy techniques investigated were JPEG and Digital Wavelet Transform (DWT) followed by scalar and vector quantization. For equivalent compression ratios (CR), DWT was found to yield better quality, artifact-free, decompressed images, but was computationally too expensive for real-time operation. Moreover, DWT would require client-side plug-ins, while JPEG is accommodated by most Web browsers.

Lossless techniques (Unix compress and its variations) yield low CR, on the order of 2 to 3. We combined background removal with a lossless method (arithmetic coding) and achieved a CR of about 9. Though a considerable improvement, this figure was deemed too low for practical use in transmitting VH images to the AnatQuest user. Our final choice was to remove the background (i.e., convert the blue background of all the slice images to uniform black), followed by JPEG compression. Other details on improving transmission rate are given later in this section.

The main effort in developing AnatQuest focused on creating suitable image viewers and server-side image processing modules, each dictating different sets of development tools.

Two image viewers are provided in the AnatQuest GUI: A Rendered Image Viewer (RIV) to display rendered images in 2D projection, and a Cutaway Viewer (CAV) to display thumbnail as well as detailed-view images of two-dimensional slices of the front (coronal), side (sagittal), and top (axial) views of the body. Both image viewers are viewed through a standard Web browser and do not require the installation of any plug-ins.

The image viewers were designed to serve as the clients in a three-tier client-server architecture (Figure 11-4), in which a tier is a logical partitioning of an application across client and server. The image viewers have only the graphical user interface, while the middle tier contains the application logic, and the third tier consists of the image server.

The three-tier client-server architecture, also called "thin-client architecture," was chosen over a two-tier approach for better scalability and maintainability. The three-tier approach has the advantage that any changes made to the application logic (middle tier) do not require changes to the client (first tier). By contrast, in a two-tier architecture, the client typically contains the application logic that sends requests to the server or database and processes the returned results sent on to the user. This architecture is commonly referred to as "fat-client" because most of the application logic resides in the client. Although this is easier to build than the three-tier architecture, the graphical user interface (here the image viewers) would be closely tied to the application, and any changes to the application would require modification of the client as well, thus making two-tier applications less scalable and maintainable.

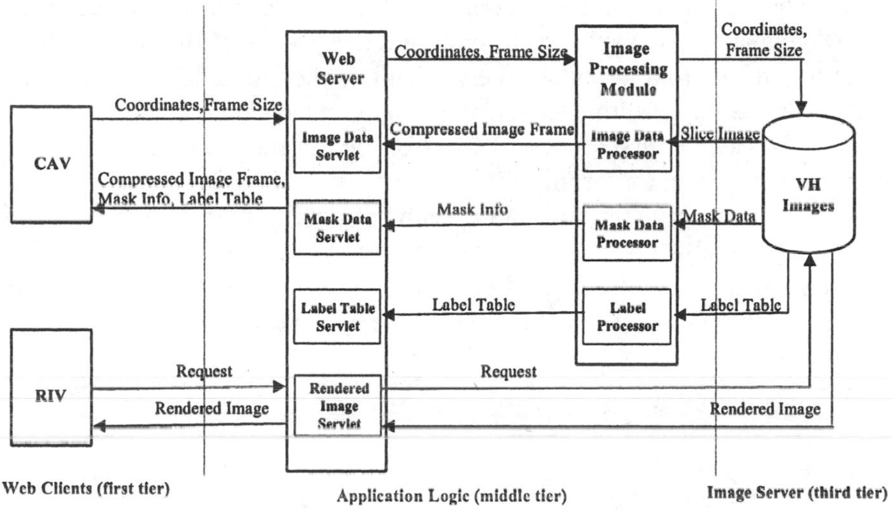

Figure 11-4. AnatQuest's 3-tier client-server architecture.

We chose to write both image viewers as Java applets rather than servlets (programs that reside in a Web server's servlet engine), though either would meet our objective of allowing user access to the system via a browser, such as Microsoft's Internet Explorer or Netscape Navigator, without having to download and install plug-ins in the browser. In other words, since the standard browsers have the built-in Java Virtual Machine (JVM) necessary to run any Java program, the image viewers could be either applets or servlets. Our choice was dictated by empirical testing that proved that applets were superior to servlets in allowing users to scale and manipulate the horizontal and vertical hairlines and field-of-view controls that float over each of the three view ports in an intuitive and pleasing way.

At the middle tier of the system, however, the application logic was written as servlets. These servlets transfer the burden of computing the size of, and retrieving, an image to the server side. They process the request for a new image on the server, retrieve the image, compress it on the fly, and return it to the applet for display. This interaction between applets at the first tier and servlets at the middle tier provides the fast response necessary for a wide user community equipped with low speed connections. (Running the Web server and the servlet engine in this middle tier are Apache and Java Jakarta-Tomcat, respectively.)

The Cut-away Viewer allows the user to dynamically navigate the Visible Human body via a Web browser along three dimensions: along the x, y, and z axes. As shown in Figure 11-5, the user sees three side-by-side view ports that occupy the left part of the Web page, each view port containing a thumbnail representation of a sagittal, coronal, or axial slice. A user-movable hairline controls the location of the three axes (planes) in each view port. To dynamically display a different cut-away view of the body in three directions, the user moves a hairline vertically or horizontally across any one of the view ports, resulting in updating the corresponding spatially-related one. In this way, the first view port (control source) and the second view port (control target) form a coordinated pair.

Two parameters determine the amount of data transferred to the client interface, thereby accommodating users with varying bandwidth connections: the size of the field-of-view which is user-adjustable, and the degree of image compression, also selectable by the user.

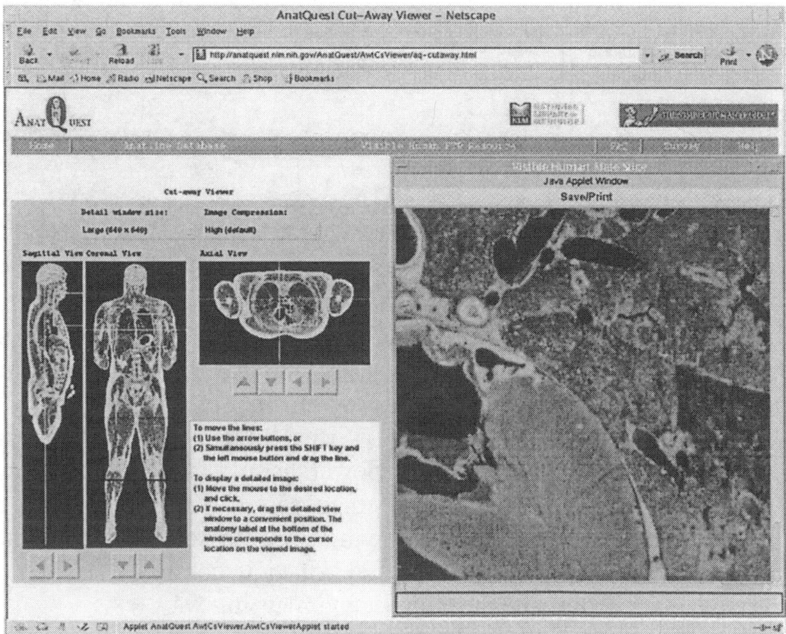

Figure 11-5. AnatQuest Cut-away Viewer.

When the user clicks on the rectangular field-of-view box in any of the view ports, a detailed view of the section of the image encompassed by the box is displayed. The user may select the size of this detailed view (and therefore the number of bytes received from the server) from a pull-down menu in the GUI. The three field-of-view sizes are: 448 x 320 pixels (for low bandwidth connections); 640 x 448 pixels and 640 x 640 pixels, for higher bandwidth connections. At the user's click, the coordinates and size of the selected field-of-view are sent to ImageDataServlet on the server.

This servlet validates the parameters and passes the information to another server-side program, called ImageDataProcessor, which extracts the cryosection image file that contains the requested detailed image frame according to the specified position and dimensions. It then uses the JPEG compression scheme in Java Advanced Imaging to compress the image data at a user-specified compression ratio. The generated image data is then returned to the ImageDataServlet, which sends the data to the applet at the client to be displayed. Similarly, mask information is retrieved through MaskDataServlet and MaskDataProcessor, and label table data (a table of anatomical terms) through the LabelTableServlet and LabelTableProcessor. The label table and mask information enable the detailed view window to display the label (name) of an anatomic structure under the mouse cursor.

In order to have the region of interest efficiently encompass the detailed image, i.e., at their boundaries, we picked view port sizes that are divisible by 64. Also, since the files are stored as tiled TIFF images and the average user's screen resolution is expected to be 1024 x 768, we set the maximum window size to 640 x 640 pixels to enable users to display an image portion that is large enough to include a significant part of the image without taking up the entire screen.

To increase the transmission speed of the detailed image portions from the AnatQuest server to the user's browser, the images are compressed on the fly using the JPEG compression scheme, as mentioned earlier. The user may select one of five compression levels from a pull-down menu in the GUI: high (the default value), medium/high, medium, low/medium, or low. The approximate compression ratios (sample averages) range from 60 at the high level to 11 at the low. To achieve a reasonable response time for users with low speed modems, we set the default window size to 448 x 320 pixels and the default image compression option to high. For example, assuming an average transfer speed of 33.2 kbps commonly available through such modems, these default values allow a JPEG image of approximately 7,168 bytes to be retrieved and displayed in the user's browser in 1.7 seconds.

The GUI of the Rendered Image Viewer applet consists of a user-selectable list of anatomic structures in available rendered images. Selecting an item in this list results in the display of its thumbnail image, together with an option to view the full image. When the user clicks on the thumbnail, the full-size rendered image is displayed in a separate window. Figure 11-6 shows the components of the RIV.

As noted earlier and summarized in the following, several aspects of the AnatQuest design exploited the functionality offered by Java Advanced Imaging. First, we exploited JAI's file handling capabilities to locally store very large image files in the tiled TIFF format that would be too resource intensive for most users to download. JAI was also used to extract only those tiles required by the user (as defined by the coordinates of the selected region of interest) and to compress the image data to a JPEG file, as mentioned earlier.

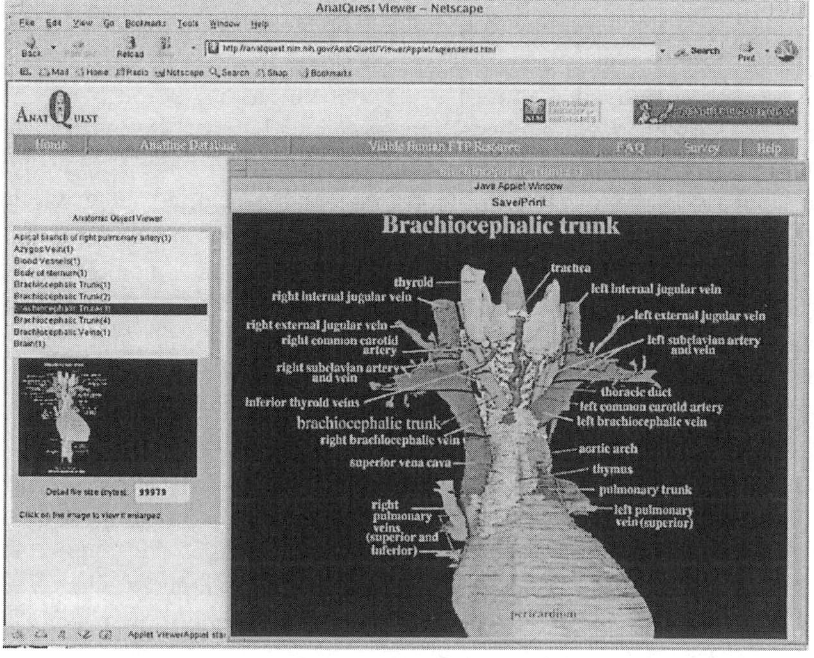

Figure 11-6. AnatQuest Rendered Image Viewer.

As a class library, JAI supports generalized image processing functionality built as an extension to the Java programming language. In a simple programming model, JAI provides a rich set of imaging capabilities that can be readily used in applications without undue programming overhead. JAI encapsulates image data formats and remote method invocations within reusable image data objects, so that an image file would be processed the same way, whether in local storage or across networks. JAI also provides cross-platform imaging APIs, allows distributed imaging, and comes as an object-oriented API that is flexible and extensible. In addition, it is device independent and provides high quality performance on various platforms.

3.3 AnatQuest for Onsite Visitors

While reliable and rapid online access promotes the use of anatomic images by the public, onsite displays in exhibits are further opportunities to reach another public constituency: visitors to the library. One such exhibit was Dream Anatomy (National Library of Medicine, 2002), installed at the NLM for a year spanning 2002-2003. To serve as an onsite display in this exhibit, AnatQuest was modified to take advantage of the particular characteristics of this environment. Different issues come into play in the design of this modified system we call AnatQuestKiosk.

First, as an *onsite* system, AnatQuestKiosk did not need to rely on a Web browser and could therefore be designed as a standalone application. Second, an exhibit visitor in close proximity to the screen is inclined to navigate by touch; hence, a touch screen monitor is provided (Figure 11-7). The design implications of these factors are discussed below.

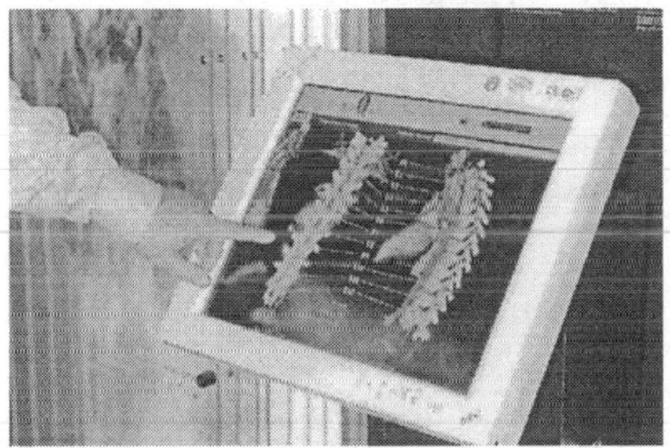

Figure 11-7. AnatQuestKiosk: Touch screen version of AnatQuest.

Eliminating the need for a Web browser allowed both the Cut-away Viewer and the Rendered Image Viewer in AnatQuestKiosk to be designed as Java applications rather than as applets. Compared to applets, Java applications allow the user interface to be built with a greater variety of Java visual control class libraries (as in Java Swing), giving the interface a more polished look than possible with the Abstract Windowing Toolkit (AWT) libraries used in AnatQuest. Using AWT in the online AnatQuest system is necessary since these libraries are built into Web browsers, while Swing libraries are not. Swing libraries have to be downloaded into the Web browser as plug-ins, adding to the download time for an online system. Since this is not a problem with a standalone onsite system, the use of Swing in

AnatQuestKiosk is an advantage.

While touch screen monitors are attractive for onsite applications, in designing applications to run on them one must take into account the size of the GUI controls. For example, buttons and sliders must be large enough to respond to the touch of a finger. To support this, we replaced the thin vertical and horizontal user-movable hairlines that control the thumbnail images in AnatQuest with large sliders. This required the modification of the Java library slider controls to achieve a tailor-made look-and-feel.

To ensure that AnatQuestKiosk runs continuously and covers the entire screen, it is necessary to prevent users from inadvertently stopping or starting the application, or from coming in contact with the operating system. We met these requirements by using Java's Fullscreen Exclusive Mode API, a new feature in JDK 1.4. This API supports high-performance graphics by suspending the operating system's windowing function so that the application takes full control of the contents of video memory and draws directly to the screen.

Finally, as an exhibit display it is desirable for the AnatQuestKiosk application to be unaffected by network outages. To guarantee uninterrupted use, the image files are locally stored in the system to avoid continuous fetching from a remote site. However, the application is designed to fetch images remotely as well by enabling the parameters passed to its startup script to specify the image site as a URL for the image database server. This feature is useful in the event that new images, particularly newly rendered 3D structures, are added to the database.

4. NEXT STEPS

There are a number of interesting directions that the AnatQuest project is taking to further serve the lay public. Among these are: (1) Increasing the contents of the image database; and (2) Extending information systems designed for the public, e.g., MedlinePlus, to provide anatomic images. The full realization of the second goal depends on adequately addressing the first, as discussed below.

4.1 Increasing Content

We believe the lay public would be well served with more image content in the database, and in different forms, e.g., 3D volume and surface renderings and animated sequences. To our knowledge, existing images from third-party sources have been created from the CCD-captured cross-sections and not from the high resolution version, thereby possibly precluding the most detailed structures.

To date, we have produced a limited number of 3D surface-rendered structures, mainly as exemplar images to evaluate the design of the database and the processing tools. These images were created from the segmented and labeled *high resolution* cross-sections of the Visible Male thorax region, with the assumption that this dataset would allow rendering of more detailed structures than possible with the lower resolution CCD-captured data. Currently, AnatQuest provides access to these images, numbering about 200 (showing about 400 anatomic structures), as well as some produced by other organizations.

However, the fact remains that other regions of the high resolution Visible Male, and all of the Visible Female, have neither been segmented nor labeled. If one is to make structures in all regions of the human anatomy available, this would be a necessary task.

User feedback suggests a demand for rendered images of structures in all regions, but particularly in the head and abdomen. That these regions are of highest interest is evident not only from anecdotal evidence, but also from the cumulative demand statistics shown in Figure 11-8.

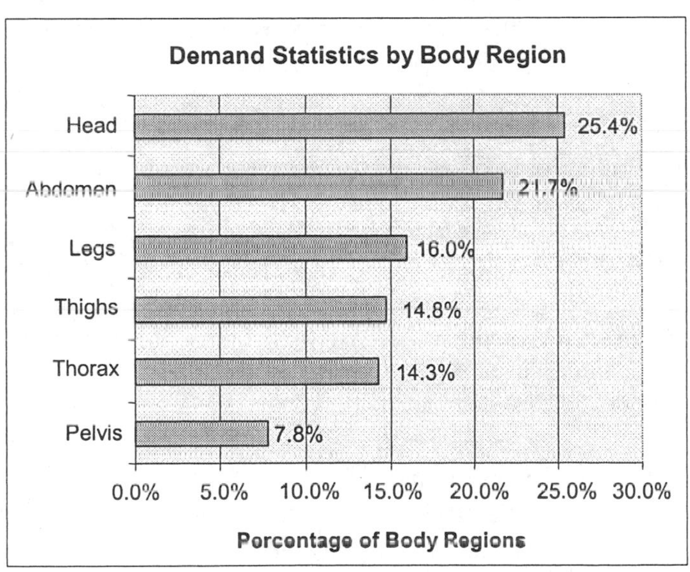

Figure 11-8. Demand statistics by body region.

Meeting this demand would require these regions to be segmented and labeled. The approach taken in segmenting and labeling the high resolution slice images of the thorax region, though computer-assisted, had a manual component and was therefore labor intensive. Since there appears to be work done in segmenting the low resolution (CCD-captured) images (e.g., Gold Standard), it is possible to conceive of more efficient techniques based on pixel statistics and interpolation to use the low resolution data to segment the high resolution images. The availability of a complete set of segmented and labeled high resolution slice images for both the Visible Male and Female would offer developers material for creating and testing tools for surface and volume rendering, and would consequently supply the lay user with rendered images of objects anywhere in the human body.

We additionally note that successful rendering requires the correct registration (alignment) of adjacent slice images. The high resolution images are found to be misaligned to some degree, whereas the low resolution slices have been shown to be correctly aligned. It is possible to conceive of techniques to use the low resolution set as a reference to correcting the registration of the high resolution slices.

4.2 Linking Text Resources to Image Database

Here we come to one of our most important objectives. A long term goal of the Visible Human Project is to transparently link the print library of *functional-physiological* knowledge with the image library of *structural-anatomic* knowledge into a single, unified resource for health information. Indeed this has been echoed several times in the past, including the NLM's Board of Regents Planning Panel whose recommendations as far back as 1989 stated: "The NLM should encourage... research into methods for representing and linking spatial and textual information..."

In this section we present our early research in this area. We explore the steps required to link text from a biomedical document to the relevant images and apply the concepts to the design of a prototype linking a search of MedlinePlus, a text-based document source popular with the lay public, to anatomic images in our database (described in Section 4.3). We define the following four functions to implement such a linkage:

1. Document Analyzer: Identifying biomedical terms in a document.
2. Term Mapper: Identifying the relevant anatomical terms.
3. Image Locator: Identifying the images in the image database.
4. Link Assembler: Linking the identified terms to the images.

Figure 11-9 shows the top level architecture of the system and its components. Actual implementations might utilize the components differently. For example, the document analyzer and the term mapper can be provided as independent services for indexing documents outside of AnatQuest. For simplicity, all the above text and image services can be made available through a single API interface.

The AnatQuest API is a URL-based parametric API in which a desired image can be described in terms of its metadata in the URL statement. In addition, the user can specify: (a) whether mapping the terms through a vocabulary (e.g., UMLS) should be utilized, and (b) the form of the output desired (e.g., XML file, image file). The following is an example of the API for the metadata in XML for a heart image:

```
http://image1.nlm.nih.gov/pm/servlet/ImageLogic?name=
heart&resultFormat=xml
```

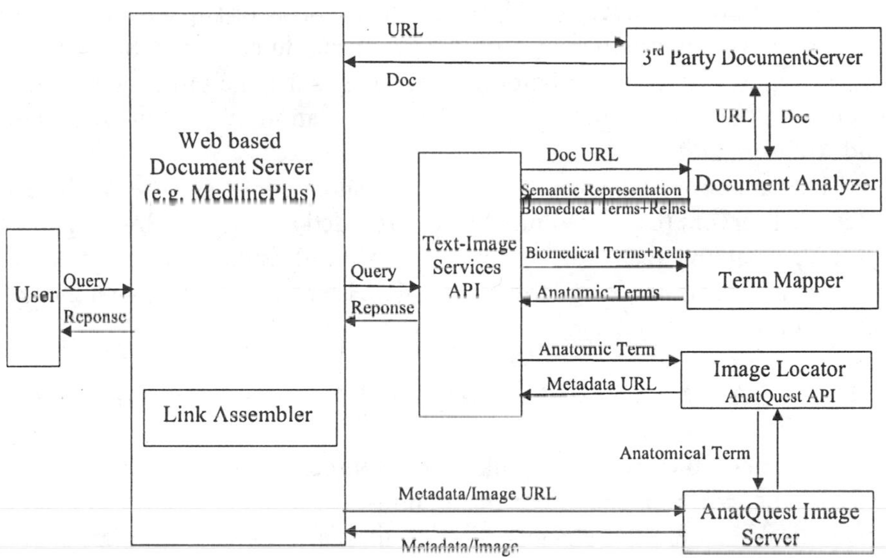

Figure 11-9. Architectural components for linking text and images.

4.2.1 Document Analyzer: Identifying Biomedical Terms in a Document

The document analyzer (parser) creates a representation of document content in the form of a list of keywords or, in a more complex way, as a semantic network representation of the contents.

Alternative approaches to implementing this function are:

- Word frequency vector
- Pre-assigned vocabulary terms
- Vocabulary-based term identification
- Vocabulary-based term and relation identification
- Text understanding

These alternatives range from the relatively simple (a word-based analyzer) to the most sophisticated (a full fledged text understanding system). The latter would be able to infer from the context, for example, whether the term "brain" should point to a male or female brain.

We focus on an approach that is reasonably practical at present: vocabulary-based term identification. This approach processes the text to identify the occurrences of the vocabulary terms in it. As implemented in the in-house MetaMap program (Aronson, 2001), this takes into account such factors as lexical variances, word order variances, and synonymy. The accuracy of this approach would depend on the level of sophistication of the analyzer. An alternative would be to use domain-specific semantic processing to identify the relationships as well as the anatomical terms in the document, thereby generating a richer representation of content (Rindflesch and Aronson, 2000).

The parsing of documents can be initiated either by the Web-based document servers (e.g., MedlinePlus) or provided as a service by AnatQuest. Note that parsing may not be necessary if the documents already possess vocabulary-based indexing terms (e.g., MeSH terms assigned by an indexer) as part of their metadata.

4.2.2 Term Mapper: Identifying the Relevant Anatomical Terms

Since our objective is to link text resources to *anatomic* images, the question is whether the biomedical terms identified in a document are anatomical. In fact, the chances are that they are not explicitly anatomical terms. However, using the concept relationships in NLM's Unified Medical Language System (UMLS) Metathesaurus we can map a biomedical term in the document to a related anatomical one. For example, the term *pneumonia* (a disease) could be mapped to *lung*, the underlying organ with the disease. Other mappings are also possible, e.g., mapping a particular anatomical term, for which there is no image in the AnatQuest image database, to a more general anatomical term for which there is an image in the database. But to demonstrate the concept we focus on the <u>location-of</u> relationship in our prototype system linking a search of MedlinePlus to images. Figure 11-10 shows the functions that are part of the term mapper.

The following steps are taken by the term mapper:

1. Biomedical terms are mapped to UMLS concepts through the Knowledge Source Server, UMLS-KSS, resulting in their concept unique identifiers (CUI) and semantic types (STY). The mapping takes into consideration lexical variants and synonymy.
2. A term identified as anatomic by its semantic type is returned by the mapper as is.

3. For terms that are not anatomical (e.g., a disease name), the server uses the UMLS Metathesaurus concept relations to obtain the related anatomical term. As shown, a disease term would be mapped via the location-of relationship to the corresponding anatomical terms. This mapping appears to suit MedlinePlus since most health topic pages to which it points possess MeSH designators that are disease terms.

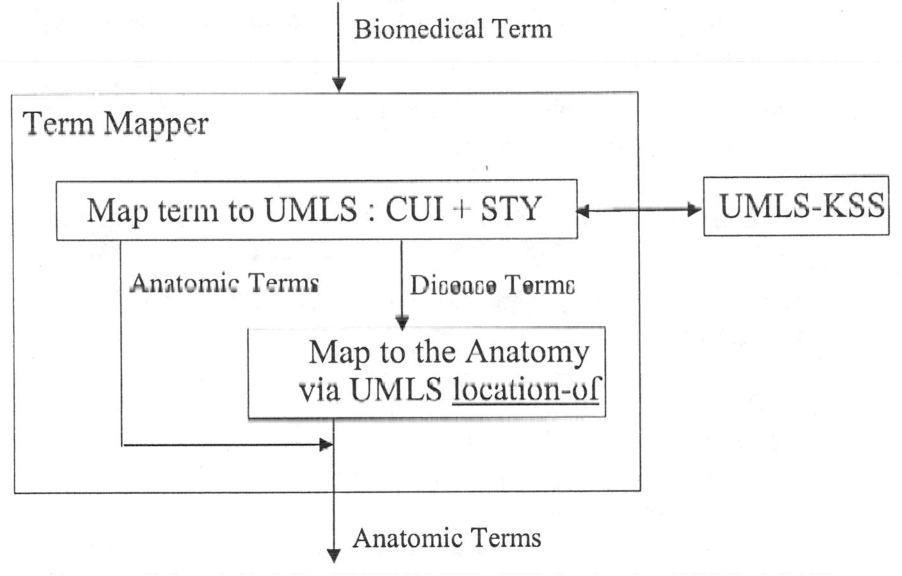

Figure 11-10. Term Mapper components.

While our current prototype implementation of the mapper uses the location-of relationship in the UMLS for mapping mostly disease terms to their underlying anatomical structure, we note that this relationship is only one among 88 available in the Metathesaurus. Of these we have identified 22 that could potentially lead to relevant anatomical structures and hence images. Some of these relationships are: Is-a, part-of, has-part, branch-of, has-branch, has-tributary, tributary-of, manifestation-of, has-manifestation, broader, diagnosed-by.

While a chain of such relationships can link a biomedical term to a possible underlying anatomic structure, some types of relationship chains will not be successful, as shown in examples below. [Note that since in the UMLS Metathesaurus concept relation table the relation names are from the second concept to the first, we are displaying the relation names in the chains in *reverse* order.] The following are some findings and suggested heuristics for the selection of chains and their ranking:

- Combination of mapped-to and mapped-from creates a sibling relationship which might map to unrelated concepts ("nose") in the following path generated for heart attack ("myocardial infarction"):

  Nose location-of --> *Necrosis of nose* mapped-to --> *Necrosis* mapped-from --> *Myocardial Infarction*

- Combining part-of and has-location relations provide substructures which might not be quite relevant, so they should be ranked lower, or deleted. Example:

  Mitral Valve part-of --> *Heart* location-of --> *Heart Diseases* mapped-from --> *Myocardial Infarction*

- The mapped-from relation provides a broader concept than mapped-to, suggesting that mapped-from should be ranked higher. Example:

  Necrosis of ovary mapped-to --> *Necrosis* mapped-from --> *Myocardial Infarction*

Other heuristics suggested in the course of our investigation are:

- Shorter chains should be ranked higher than longer ones.
- The combination of is-a and broader should be avoided. They create a sibling relationship that may yield unrelated concepts.
- The mapped-from relation should be ranked higher than the broader relation, because the former is more specific and the latter too general.

On evaluating a number of sample paths for a given input term we have identified some relations and their combinations which generate the most promising mapping of biomedical terms to their related anatomical structure. The following are relations which result in successful identification of anatomic terms (in descending order of effectiveness): location-of, mapped-from, is-a, broader, has-part, has-branch, has-tributary.

Our implementation of the relation-based term mapper consists of a look up of a mapped-table in which each entry contains a biomedical concept and its related anatomical structure. This is followed by ranking the entries and inserting the highest ranked ones into the document content.

While we have focused on the relation-based mapping strategy as outlined above, we are also exploring two other methods: image-based and model-based. The image-based strategy, like the relation-based method, uses the Metathesaurus to map biomedical terms to anatomical structures, but employs a different ranking approach. To rank the related anatomical concepts and their images, this method relies on the *number* of mapped concepts which are labeled on an image. For example, a heart image labeled with four mapped concepts of *Right-atrium, Left-atrium, Right-ventricle,* and

Left-ventricle would be assigned a ranking of 4, whereas an isolated mapped concept labeled on an image would be given a lower rank of 1.

The third strategy, model-based mapping, is based on clustering the mapped anatomical concepts using the Metathesaurus concept relation table. The mapped concepts in the most concentrated clusters will be assigned higher rankings. These techniques will be examined in research to follow.

4.2.3 Image Locator: Identifying the Images in the Image Database

There are three possible approaches to identifying and accessing the images, each influenced by the design of the image database:

- Database-brokered access via a URL-based parametric API
- XML file-based access with searchable metadata
- Integrated image and metadata file format

The first approach is based on a private database system which brokers access to the images within a controlled environment and custom-made interfaces designed for particular systems. The AnatQuest Web browser falls in this category. Here users can access the images through the GUI menus, the database is local to the system, and the database query language is not accessible to the users. In order to relax this constraint, as mentioned earlier, we have defined a URL-based parametric API in which a desired image can be described in terms of its metadata in the URL expression.

The second approach is to eliminate the database and store the metadata for each image in separate files that are made public for search engines to crawl and index. These metadata files, in addition to the information about the image, contain the URL to the actual image file. The metadata encoding can be as simple as unstructured text included in the ALT text of the IMG tag of an HTML document. Alternatively, the metadata can be represented in a more formal XML structure with data elements that can be searched individually. Search engines can index specific XML data elements, in order to provide results with high precision and recall. Alternatively, search engines may ignore the XML structure of the data and index all the metadata page contents. We have defined a detailed XML schema describing the structure of the image files which may be crawled and indexed by search engines. Figure 11-11 shows a fragment of the XML structure for the image metadata file.

In this example the web crawlers may be instructed by their owners to index the contents of the VHI-image.term.name nested field when it comes across an XML file with the root element <VHI-image>. Here the value of the name field is *Heart,* allowing the crawler to add the URL of this example XML file to its list of index entries for the term *Heart*. Thereafter, when the search engine is queried for *Heart* the URL for this XML file will be the name and the thumbnail for the image.

The third alternative solution for the image locator is to combine an image with its metadata in a single file. This has the advantage of preventing dangling links between the image and its metadata when they are stored in separate files and moved individually. This approach requires a standard file format that encompasses both pieces, as well as a standard set of metadata. A number of groups are attempting this. For example, the PNG2000 (W3C, 2003) and extensions to the Dublin Core-based metadata activities (Dublin Core Metadata Initiative) seem to be moving towards this objective.

```
<?xml version="1.0"?>
<VHI-image xmlns:xlink="http://www.w3.org/1999/xlink"
  xmlns:VHI="http://anatquest.nlm.nih.gov/vhi">
      <term anatomicalType="Structure">
            <cui>C0018787</cui>
            <name>Heart</name>
      </term>
      <specimen>
            <id>1</id>
            <name>Visible Human Male</name>
            <owner>National Library of Medicine (NLM)</owner>
            <sex>Male</sex>
            <race>Caucasian</race>
            <age>38</age>
      </specimen>
      <image size="59008" format="jpg" modality="70mm" capturedBy="NLM">
            <rendered segmentedBy="EAI" renderedBy="NLM">
                  <url>...</url>
                  <title>Anterior View of Heart</title>
                  <modifier>1</modifier>
                  <thumbnail size="5073" format="jpg" dimensions="">
                        <url>...</url>
                  </thumbnail>
            </rendered>
      </image>
</VHI-image>
```

Figure 11-11. XML structure of image data.

This approach also requires modifications to search engine crawlers to open a compressed image for retrieving and indexing its metadata.

A PNG-enabled search engine should be able to decompress a PNG file format, as well as know about the metadata in order to extract and index the data elements. An example of such an implementation is the PNG-enabled version of HotMeta (Distributed Systems Technology Centre, 1999).

4.2.4 Link Assembler: Linking the Identified Terms to the Images

This function is intended to incorporate into the document links that point to the images associated with the biomedical terms found in the document. Ways to present the links depend on the design of the GUI of the Web-based document server. Possible approaches are:

- Hot linking anatomical terms in a document
- Related-Images button or menu option, based on document similarity criteria
- Portal approach

In the hot-linking approach, each term identified in the parsing phase is converted to a hot link that indirectly points to a relevant image in the Visible Human Image database through the AnatQuest API. The degree to which such a link points to a specific image correctly depends on the ability of the document analysis system (parser) to resolve ambiguities. For example, by default the links would point to images of the Visible Male unless the parser can understand from the document's context that the topic is about the female. Needless to say, this is a difficult task with a high probability of error. Figure 11-12 provides the flow diagram for the hot-linking approach, with numbers representing the sequence of actions following a user query.

As shown in the diagram, steps C1 and C2 require the Web-based document server (e.g., MedlinePlus) to compute and save the offsets of the anatomical terms within the documents at the indexing phase so that hot links may be inserted in the third-party documents at the time of retrieval. The Web-based document server would not store these documents but would act as a proxy server for them in order to insert the hot links.

An advantage of this approach is that the anatomical terms are turned into hot links which the user can easily click on while reading the document. A drawback is the possibility of mismatch between the text and the image due to parsing inaccuracies. Further, the server needs to maintain the offsets to all the anatomical terms within the document, requiring additional storage and processing to include the hyperlinks for each retrieved document.

In the Related-Images approach to displaying the links the original text of the document is displayed for viewing by the user. The GUI provides a "related anatomical images" button in proximity to the text. This button, when clicked, will send the set of vocabulary terms, identified earlier in the

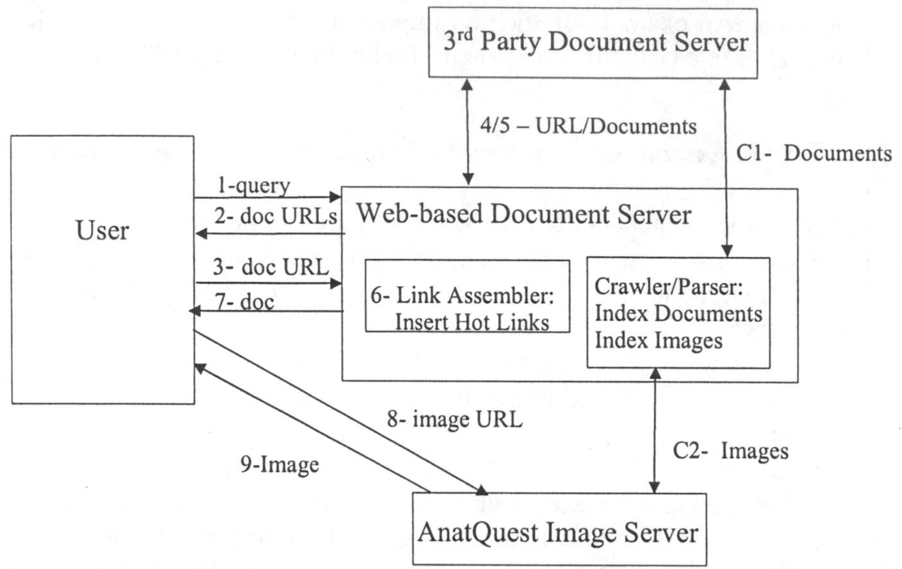

Figure 11-12. Link Assembler – hot linking anatomical terms.

parsing phase and already associated and stored with the document metadata, as query terms to the image database. Thumbnails of the returning (matched) images will then be displayed on a sidebar next to the main document for the user to view. Selection of each thumbnail will open up a detailed view of the image. Alternatively, the thumbnails of related images may be shown on a sidebar when the document is first displayed instead of providing the Related-Images button.

An advantage of this approach is that it reduces possible confusion caused by a mismatch between terms and images, since the links or thumbnails are offered as "related images" rather than "exact images" suggested by hot linked terms in the text (as in the previous approach).

In the portal approach to displaying the links, the documents need not be analyzed at all, nor are changes made to them. We simply make the Visible Human image database Website open to the Web crawlers of the Web-based document servers (e.g., MedlinePlus). These crawlers can then build an index to the images in the database and simply treat the images as individual documents. Through metadata, the image database would provide the anatomical terms and possibly their synonyms to be indexed by the crawlers. Advantages of this approach are that: (a) no changes need be made to existing Web-based document servers; and (b) since the documents do not need to be parsed, parser accuracy would not be an issue. The flow diagram for the portal approach is given in Figure 11-13. The interaction consists of:

1. The user enters a term at the query prompt of the document server.

2. When the document server is MedlinePlus, the browser displays a list of (third party) documents grouped under a number of categories. For each document, a one-line summary is displayed together with the hot linked URL to access the full document. One of the categories is "Images," under which appears a list of anatomical terms the crawler has found to be relevant to the user query.
3. A document URL points directly to a third party document server.
4. A returned document from this server is displayed for the user.
5. An image URL points to the AnatQuest Image server.
6. An image or its metadata file is returned from this server.

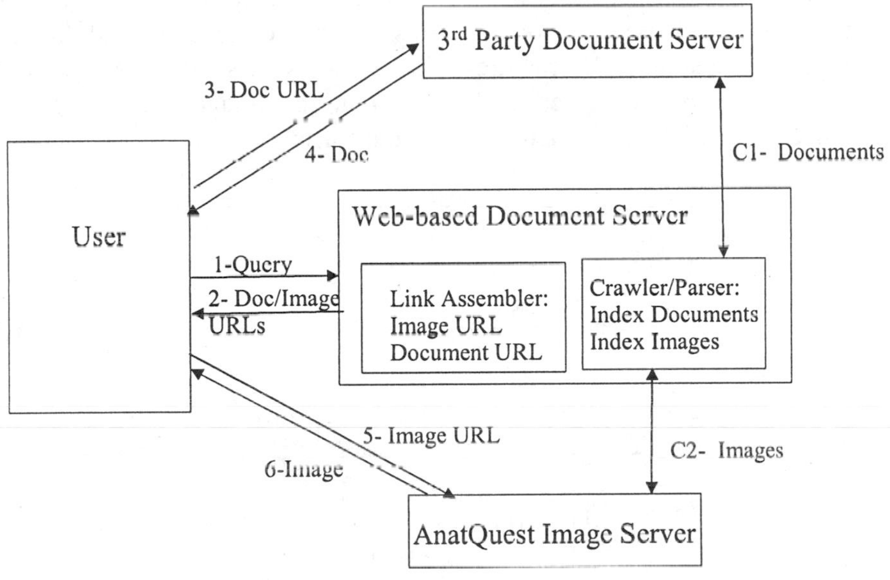

Figure 11-13. Link Assembler - portal approach.

As shown in the diagram, steps C1 and C2 consist of crawling documents in the third party document server as well as the image files and generating indexes for both images and documents. These indexes are used by the link assembler component.

The advantage of this approach is that it requires no changes to Web-based document servers such as MedlinePlus. A drawback is that only images related to the user query are provided, and not those related to the contents (or "meaning") of the documents.

A comparison of the three alternatives for linking terms to images suggests that the best approach might be to combine the portal and the Related-Images methods. This approach not only provides the images matching the user queries (portal approach) but also allows the user to view the images related to the contents of the selected documents. Further, when the metadata in the relevant documents include MeSH headings, this approach avoids the parsing requirement of the hot linking approach.

4.3 Implemented Prototype: MedlinePlus Proxy Server

We have implemented a prototype of a text-to-image linking system based on the MedlinePlus health information Web server. As shown in Figure 11-14, a proxy server has been developed to intercept the user request to MedlinePlus. The proxy server first retrieves the MedlinePlus page that satisfies the user query, and in parallel sends the user query to the AnatQuest image server which uses the UMLS Knowledge Source Server and a term mapper module to map the query terms (mostly disease names) to the corresponding anatomical structures. The links to these images are then inserted by the proxy server as hot links in the image section of the MedlinePlus page, which is then returned to the user.

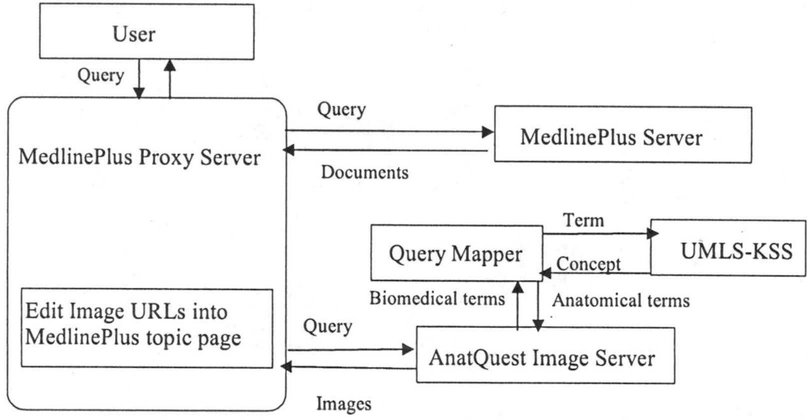

Figure 11-14. Prototype implementation.

In a variation of the prototype, instead of parsing the text we could use the MeSH terms attached to the document, if these exist, since these terms may be assumed to represent the focus or topic of the document. The document is then linked to the images through the MeSH terms.

Our initial prototype serves to demonstrate the feasibility of three of the four functions necessary for text-to-image linkage: viz., the term mapper, image locator, and link assembler. It does not address the document analyzer stage. Instead, the user query, rather than the contents of the returned document, is mapped to the anatomic term which then is linked to the image. Even the functions included are not addressed comprehensively, as for example, the term mapper uses only the location-of relationship. Nevertheless, this prototype provides a platform on which a more extensive testbed may be designed to address research issues related to the larger problem of linking text in documents to images. Some of these issues are:

1. Relevance of images to the topics in a document. This is influenced by the types of mappings used by the term mapper. For example, would the location-of relationship used in this prototype result in the display of appropriate images (of a diseased lung rather than those of a healthy lung) for a document about pneumonia? This measure is subjective, but may be quantifiable by user studies.

2. Precision and recall of images. These well-defined and quantifiable measures reflect the performance of the image retriever searching the metadata in order to locate desired images.

3. Accuracy and precision of the links. This is of particular importance to the hot linking approach for the display of links where document terms are hot linked to the images. This approach might raise user expectations for an "exact image" rather than a "relevant image." An image may be seen as relevant but not necessarily "accurate."

4. Term mapper performance. Considering other promising relationship mappings (e.g., is-a, part-of, etc.) should increase recall. In addition, a ranking of the individual mappings should provide a measure for judging the relevance of the resulting anatomical terms.

5. Parser performance. In addition to extracting the anatomical terms, identifying the explicit anatomical relationships in the document enables formulation of more refined search queries against the image database, which should increase relevancy.

6. System performance. Some approaches require more resources than others, which could translate into higher turnaround time. For example, the hot linking approach requires modifying each document before it is sent to the user. Turnaround time, excluding the network delay, should provide a measure of system performance.

Addressing these research issues should allow us to identify the best combination of the alternative solutions in each of the four functions described here for an optimum solution for linking document text to images.

5. SUMMARY

The goal of providing the public ready access to anatomic images, in particular, "real" human anatomy from the Visible Human Project, is one aspect of universalizing access to biomedical information. This goal is being implemented in the AnatQuest project in which a system has been developed that provides Internet accessibility to *high resolution* Visible Human images via browsers, labels for anatomic structures in each cryosection slice as well as 3D rendered images, and acceptable user interaction in a low bandwidth environment. Besides the AnatQuest system for Web access by resource-limited end users, we have also developed the AnatQuestKiosk system for onsite visitors to the National Library of Medicine, and a prototype system demonstrating the linking of queries by MedlinePlus users to anatomic images. We have also defined an architecture and functional components to further investigate the linking of the text content of documents retrieved by MedlinePlus and other document sources to anatomic images in our database.

6. ACKNOWLEDGEMENTS

The extensive R&D and implementation activities in the AnatQuest project are the work of a team at the Lister Hill Center's Communications Engineering Branch. In particular, Dr. Amir Razi has made significant contributions in research toward the architecture design and component definition of a system linking textual document sources to anatomic images; Mr. James Seamans is responsible for an automated system for the registration of Visible Human cryosection images; Ms. Annette Strupp-Adams and Ms. Jinghong Gao, as well as Mr. Seamans, are responsible for the successful implementation of the Web and kiosk versions of the AnatQuest system and the prototype system for text to image linkage. Each of them has made important contributions to the research reported in this chapter.

REFERENCES

Ackerman, M.J. (1998). "The Visible Human Project," in *Proc. IEEE,* 86(3): 504-11.

Aronson, A.R. (2001). "Effective Mapping of Biomedical Text to the UMLS Metathesaurus: The MetaMap Program," in *Proc AMIA Symp,* 17-21.

Distributed Systems Technology Centre (1999). HotMeta. http://archive.dstc.edu.au/RDU/HotMeta/png/index.html

Dublin Core Metadata Initiative, http://dublincore.org/

Henderson, E., Seamans, J., Strupp-Adams, A. (1998). "VHIF: A Prototype File Format for Anatomical Images," *in Proc. Visible Human Conference* (available on CD-ROM from the National Library of Medicine).

Long, L.R. (1996). "Transmission of Medical Images over Wide Area Networks," in *Proc. Visible Human Conference* (available on CD-ROM from the National Library of Medicine).

Long, L.R., Berman, L.E., Neve, L., Thoma, G.R. (1995). "An Applications-level Technique for Transmission of Large Images on the Internet," in *Proc. SPIE: Multimedia Computing and Networking*, vol. 2417, San Jose, CA, February 1995, 116-29.

Meadows, S., Thoma, G.R., Long, L.R., Mitra, S. (1997). "Entropy Encoding of Difference Images from Adjacent Visible Human Digital Color Photographic Slices for Lossless Compression," in Kim, Yongmin (editor), *Medical Imaging 1997: Image Display*. SPIE, vol. 3031, 749-55.

Mitra, S., Long, L.R., Pemmaraju, S., Muyshondt, R., Thoma, G.R. (1996). "Color Image Coding using Wavelet Pyramid Coders," in *Proc. SSIAI'96*, San Antonio, TX, April 1996; 52-63.

National Library of Medicine website, http:// www.nlm.nih.gov

National Library of Medicine (2001). *Programs and Services, Fiscal Year 2001*. NLM/NIH, Bethesda, MD, p. 1.

National Library of Medicine (2002). Dream Anatomy. Technologies of Anatomical Representation, http://www.nlm.nih.gov/exhibition/dreamanatomy/da_technology.html

National Library of Medicine (2003). The Visible Human Project®. http://www.nlm.nih.gov/research/visible/visible_human.html

Pemmaraju, S., Mitra, S., Long, L.R., Shieh, Y.-Y., Roberson, G. (1996). "An Adaptive Vector Quantization with Fuzzy Distortion Measure for Image Coding," in *Proc. SPIE Medical Imaging '96*, Newport Beach, CA, February 1996: 112-5.

Rindflesch, T.C., Aronson, A.R. (2000). "Semantic Processing for Enhanced Access to Biomedical Knowledge," in Kashyap, V., Shklar, L. (eds.). *Real World Semantic Web Applications*, IOS Press. 157-72.

Strupp-Adams, A., Henderson, E. (2000). "Retrieving High Resolution Images over the Internet from an Anatomical Image Database," in *Proc. SPIE: Internet Imaging:* 3964, 259-65.

Thoma, G.R., Long, L.R. (1997). "Compressing and Transmitting Visible Human Images," *IEEE MultiMedia*, April-June, 4(2): 36-45.

W3C (World Wide Web Consortium) (2003). Portable Network Graphics, http://www.w3.org/Graphics/PNG

Zhou, R., Henderson, E., Seamans, J. (1998). "Visualization of Visible Human Anatomic Images," in *Proc. Visible Human Conference*. (CD-ROM available from National Library of Medicine).

SUGGESTED READINGS

Thoma, G.R., Long, L.R. (1997). "Compressing and Transmitting Visible Human Images," *IEEE MultiMedia*, April-June; 4(2): 36-45.
Compares the performance of different compression techniques applied to the Visible Human images.

Aronson, A.R. (2001). "Effective Mapping of Biomedical Text to the UMLS Metathesaurus: The MetaMap Program," in *Proc AMIA Symp*, 2001:17-21.

Outlines techniques for identifying terms in text as biomedical ones.

Rindflesch, T.C., Aronson, A.R. (2000). "Semantic Processing for Enhanced Access to Biomedical Knowledge," in Kashyap, V., Shklar, L. (eds.). *Real World Semantic Web Applications*, IOS Press, 157-72.
Describes NLP techniques for identifying biomedical terms.

Russ, John C. *The Image Processing Handbook.* Boca Raton: CRC Press. 1995.
Describes techniques for image capture, enhancement, segmentation, and other stages in image processing.

ONLINE RESOURCES

AnatQuest Website: http://anatquest.nlm.nih.gov

The Visible Human Project: http://www.nlm.nih.gov/research/visible/visible_human.html

Dublin Core Metadata Initiative: http://dublincore.org

NLM Website: http://www.nlm.nih.gov

QUESTIONS FOR DISCUSSION

1. What descriptive metadata are most useful for comprehensively describing anatomic images for search and retrieval?

2. What technical metadata are most useful for describing anatomic image file characteristics for future migration and display in a long-term preservation (archiving) system?

3. Are content-based image retrieval (CBIR) techniques useful for extracting metadata based on shape, color, or texture for anatomic images?

4. What would be an effective technique to segment and label high resolution anatomic images, given a fully segmented and labeled low resolution set?

5. What are the most effective strategies to link biomedical terms found in a textual document to the most relevant anatomic images?

Chapter 12

3D MEDICAL INFORMATICS

Information Science in Multiple Dimensions

Terry S. Yoo

Office of High Performance Computing and Communications, Lister Hill National Center for Biomedical Communications, National Library of Medicine, Bethesda, MD 20894

Chapter Overview

This chapter describes the emerging discipline of 3D Medical Informatics. While text-based informatics has a distinguished history and accepted fundamental linguistic principles, the use of 2D and 3D data in informatics has emerged relatively recently as computing capabilities have rapidly advanced, imaging and modeling standards have been established, and as high performance networking has made possible the sharing of the large and complex data that images, volumes, and models represent. This chapter outlines the early developments of this discipline, presents some examples of how image data is managed and presented, and suggests some of the leading research challenges in this young field.

Keywords

content-based retrieval; visualization; volume rendering; digital image archiving

1. INTRODUCTION

What is 3D informatics? Simply, this term applies to the study of informatics or information sciences associated with images, volume data, and other dimensional data in addition to the text-based metadata that surrounds such information. Libraries and other archival institutions are fast becoming repositories for complex information, non-print materials including audio recordings, film and video collections, and sophisticated scientific data. Tools to index these collections are rudimentary today, relying solely on textual descriptions of their contents and routine indexing of the annotated bibliographic information. Today's advanced computing and networking environments enable the distribution and display of high dimensional data beyond simple text. 3D informatics is the study of how to manipulate and manage these complex data.

> 3D Informatics: the science concerned with the gathering, manipulation, classification, storage, retrieval, representation, navigation, and display of complex, high-dimensional data. This data may include more than three independent dimensions, including position, time, and scale. The data may also represent many more than one dependent dimensions, including multichannel data (e.g., the RGB values of the Visible Human Project color cryosection data).

There are strong analogs between text-based informatics and the concepts of visual and 3D informatics. For example, in 1992, the U.S. National Science Foundation held a workshop to "identify major research areas that should be addressed by researchers for visual information management systems that would be useful in scientific, industrial, medical, environmental, educational, entertainment, and other applications." Participants helped to shape research positions in the many disciplines involved in framing the overall field (Smeulders, 2000). These disciplines include shape segmentation, object recognition, feature extraction, indexing, similarity metric development, display and feedback. These elements, while very different in their implementations and founding scientific principles, fill similar roles as their counterparts in text information management systems. Figure 12-1 shows the basic algorithmic components for information retrieval from text-based data collections. Compare and contrast this view of data flow in text-based informatics with a similar structure in Figure 12-2 showing the data flow of query by pictorial example in a visual information management system.

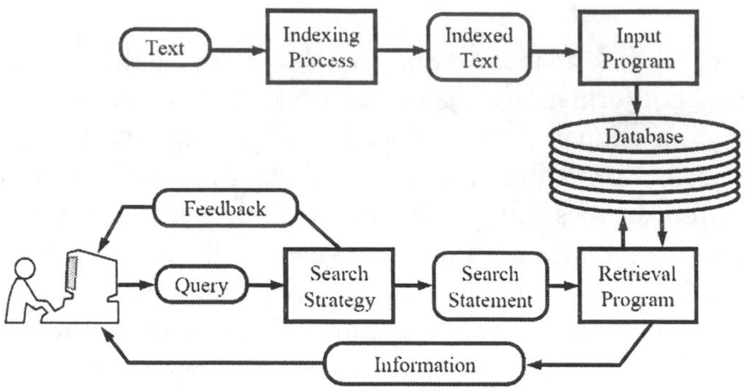

Figure 12-1. A simplified informatics view of the data flow in text information retrieval.

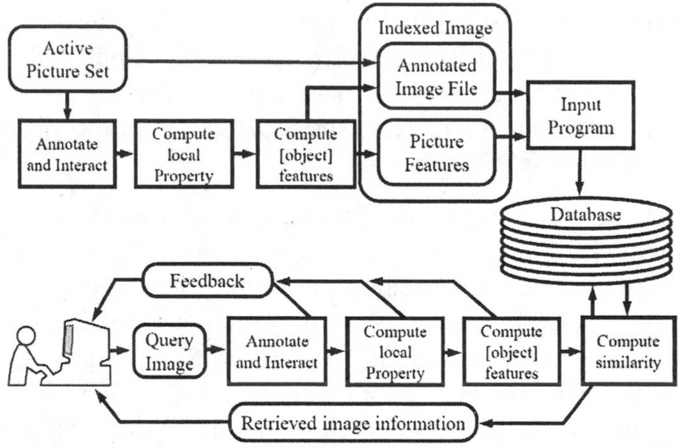

Figure 12-2. Basic algorithmic components of query by pictorial example. This data-flow diagram has many features in common with the data retrieval methodology described in Figure 12-1. (Adapted from (Smeulders, 2000)).

In the area of high-dimensional data management systems, developing and understanding the processes of deriving syntactic and semantic knowledge from the complex data are active areas of research. Where images are involved, deriving local visual properties, detecting shapes and deriving their features, and exploiting those features as indexing tools are problems at the forefront of 3D informatics research (Castelli, 2002). Shape features derived from objects identified in the images or data often require complex analyses arising from the topology or geometry of the object under study. Hilaga, et al., describe the use of topological abstractions known as

Reeb graphs to index collections of 3D graphical models. Given a representative shape, their methods permit the retrieval of comparable shapes from the collection (Hilaga, 2001). Funkhouser, et al., have been using mathematical methods such as spherical harmonics and automatic symmetry detection on 3D graphical models to index large collections aggregated from the World Wide Web (Funkhouser, 2003). Their sketch-based retrieval system permits query-by-sketch, attempting to match primitive 2D sketches of objects to collections of complex 3D models; it has been online since 2001.

The areas of informatics that are based on textual data are far ahead of their visual analogs. Linguistics and computer science in the form of artificial intelligence and natural language processing are driven to understand text, investigating techniques in lexical analysis, parsing, and semantic discovery. Two-dimensional or volume image informatics have not had similar intensive explorations. What has been achieved are a series of early models for extracting features from images using transfer functions and low-level image processing, partitioning datasets into cohesive, contiguous regions using segmentation, studying the shape properties and generating digital models of objects, and presenting these objects to expert users for study and knowledge integration. Developing these early separate methods into knowledge discovery systems is the current challenge.

Despite the emphasis on image processing, shape recognition, and automated indexing, 3D informatics is not a simple outgrowth of fields of computer graphics or image processing. 3D informatics incorporates additional research areas including content-based retrieval, image understanding, indexing, data mining, and data management. When such methods of high dimensional data analysis are applied to problems involving 2D images and 3D volume data in medicine, the result is 3D medical informatics.

The remaining sections of this chapter concentrate on this specific area. The next section begins with an overview of 3D medical informatics, including the roots of the discipline as well as some examples of motivating applications for why it is an important area of exploration. Two subsequent sections are in-depth descriptions of examples of 3D medical informatics. The last two sections are a statement of some of the grand challenges in this research area as well as a summary and conclusions for this chapter.

2. OVERVIEW – 3D MEDICAL INFORMATICS

Consider a possible scenario in 3D medical informatics. Some time in the not-too-distant future, a pathologist looks through his stereo microscope at a tissue sample, taken in a biopsy of a patient. There is something distinctive about the architecture of the tissue that he is viewing, but what he is seeing is beyond his experience. He calls a colleague and she places the sample in a 3D confocal microscope to acquire a digital volume image of the sample. Together, they perform some simple computer operations on the resulting data to derive some fundamental image metrics, and then use the sample 3D image and the derived metrics to search a distributed public data repository for comparable data. The request can be paraphrased as, "I'm sending you a picture. Send back pictures like this one along with clinical information, diagnosis, treatment, and prognosis of similar patients."

Within minutes, five comparable cases have been found. The pathologist reviews the images and decides that only four of the five are relevant, and the pathologist proceeds to download and read the case histories of the comparable patients. In three of the remaining cases, a diagnosis of "Disease A" was made, and the condition was resolved through medical treatment with the patients making full recoveries. The fourth case could not be controlled with medical treatment, so additional CT scans were taken, 3D reconstructions of the affected tissue were made and interventions planned with computer assistance, and subsequently, surgery and other therapies were attempted. Ultimately, the diagnosis was modified to a completely different disease, "Disease B," the treatment switched, and the patient made a full recovery. The pathologist reviews the five cases and their histories complete with accompanying volumetric image data.

Armed with this information, the pathologist recommends to his patient and the original referring physician that the prognosis is likely to be good, that a most likely diagnosis would be "Disease A," but the primary care physician should take special care to order additional tests to rule out a diagnosis of "Disease B."

This hypothetical example is a scenario that utilizes query by pictorial example to solve a medical problem. Both automated and visual comparisons are used to ascertain the relevance of the retrieved cases. An initial prognosis is given based on the number of comparable cases returned and the general consensus among the related cases that the condition is treatable and all five had full recoveries. Secondary findings are also indicated, and the referring physician is warned to rule out other causes. Also assumed in this discussion are facilities to acquire high dimensional data at multiple resolutions, the existence of methods to display and manipulate multidimensional data to develop surgical plans, and multiple means of indexing and retrieving complex image and text data using both images and text as sources of query information. While not routine today, the time is not far off where such capabilities and facilities are commonplace; the algorithms, methods, and technologies are currently being developed.

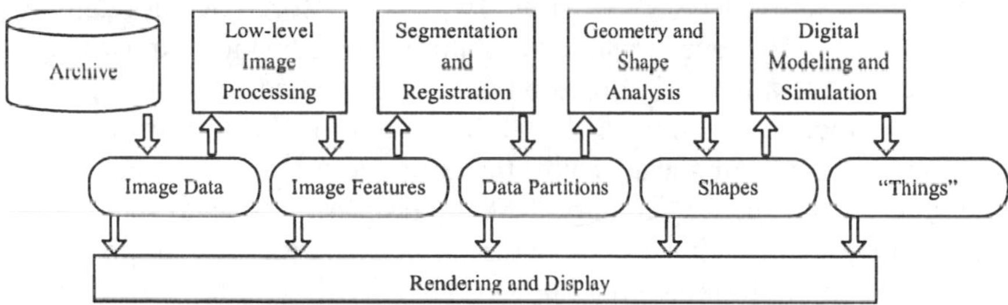

Figure 12-3. An idealized view of the medical visualization pipeline. This part of 3D Informatics encompasses the study the process of medical visualization by research in the four core areas of Volume Image Archives, Low-level Image Processing, Geometry and Shape Analysis, and Digital Modeling and Simulation. Note that the results of many of the subfields can be directly rendered, revealing the information that is hidden within the dataset at different levels of abstraction. One attribute of any informatics program is the progressive abstraction of volume information from image to model.

2.1 From Data to Knowledge

The greater part of this chapter will consider the enterprise of medical visualization from data storage and retrieval to the rendering of the final visual presentation; conceived as a pipeline, the process is seen to be connected and yet divisible into separate subfields that can be studied and improved (See Figure 12-3). Every process in a 3D medical informatics pipeline has the potential for interactive feedback and display. This treatment will not ignore the problems of data retrieval, but rather concentrate more heavily on those elements not shared with text informatics.

Informatics is often a building of abstractions from which knowledge can be synthesized. For instance, a taxonomy of life forms builds from the characteristics of individuals sharing traits, to species, to genus, eventually to the basic abstraction of kingdoms such as plant or animal.

A similar progression is seen in 3D medical informatics. Consider the simplified pipeline in Figure 12-3. As information is passed from left to right, it is refined and higher-level abstractions are created to describe the concepts contained in the volume data files. The progressive refinement of volume data and other multi-dimensional information from raw image, to embedded shapes, to recognized objects, to anatomical models is the focus of this treatment of 3D medical informatics.

2.2 History

3D medical informatics is a very young discipline. As with all information science, technology is the key driving force that accelerates its development. However, while the study of semantic knowledge in linguistics and text-based information had been studied for decades prior to the development of computer technology, 3D medical informatics was essentially created with the advent of the computer.

At the turn of the 20th century, Roentgen discovered x-rays and applied them to imaging the human body, creating a revolution in both physics and medicine, earning him the Nobel prize. This revolution did not extend to a comparable development in imaging informatics. While publishers of textual information had centuries of developed experience in printing, publishing, and distribution, imaging sciences had no comparable experience in these endeavors. Moreover, the media for storing the resulting medical images and data were fragile, flammable, difficult to copy, and impossible to distribute widely. Thus, while there was intense development in using medical images, there was little advance in indexing, cataloging, and understanding the information captured in early radiographs.

CAT scans, or more precisely X-ray computed tomography only arrived much later in the century. The mathematical principles for tomographic reconstruction were first published by Johann Radon in 1917. More than fifty years transpired before the technology and engineering of x-ray detectors and computers matured sufficiently to enable the creation of medical scanners. For this engineering feat, Allan Cormack and Geoffrey Hounsfield were independently awarded the Nobel Prize in 1979. The resulting systems are routine tools in medicine today and constitute a multi-billion dollar industry. Practical Magnetic Resonance Imaging (MRI) awaited affordable superconducting magnets, but now is an essential tool in many hospitals and radiology practices. The development of practical high-resolution 3-dimensional scanning set the cornerstone for modern imaging in medicine.

Simultaneously, emerging computer and network technologies facilitated the storage, reproduction and distribution of complex medical image data. An industry of Picture Archiving and Communications Systems (PACS) technologies rapidly established itself, requiring that more and more sophisticated data indexing and retrieval methods be applied to medical image collections. The creation of standards for DIgital COmmunications in Medicine (DICOM) have helped to make these areas more interoperable. Computers and their associated technologies have enabled all of these developments. Although the revolution in imaging sciences was started by

Roentgen, imaging informatics was finally enabled by the advent of widespread, interconnected, digital computing.

In 1986, the National Science Foundation (NSF) held a workshop on Visualization in Scientific Computing (McCormick, 1987). In their report, the panelists outlined a programmatic need for new developments in image understanding, networking, standards, visualization algorithm design, computer graphics hardware, and other related technologies. In the years following the NSF report on visualization, commercial and university interests have applied themselves to many of the tasks foreshadowed by the panelists. Since that time, the academic and industrial research communities have achieved some dramatic successes.

For example, while the domain of illuminating 3D data in medicine remains a conceptually difficult problem, the technical aspects for rendering multidimensional scalar data have been largely overcome. The human visual system is not accustomed to seeing both the surfaces of things as well as their interiors, so new metaphors for transparent and textured surfaces were needed and the methods and algorithms to generate them from volume data. In 1987, Lorensen and Kline published the Marching Cubes algorithm, setting the standard for extraction of surfaces from volume data (Lorensen, 1987). Shortly afterward in 1988, Drebin, et al., and Levoy independently

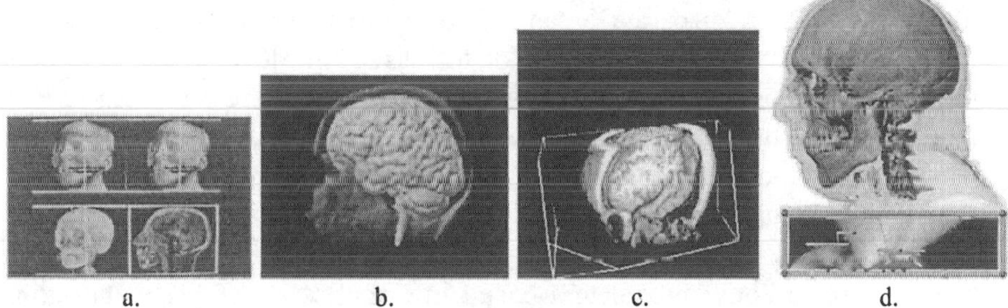

Figure 12-4. Volume rendering: A chronological progression of interactive volume rendering techniques: (a) Splatting [Westover, 1989]; (b) parallel raycasting with interactive segmentation [Yoo, 1992]; (c) texture-based volume rendering (Circa, 1997) [Cabral, 1994]; (d) interactive multidimensional transfer functions on PC hardware [Kniss, 2001]. The progression shows a trend from software to hardware, from special to general purpose, and from interactive speeds of 0.25 to 20 frames per second.

developed raycasting methods for directly rendering shaded views of volume data (Levoy and Drebin, 1988). Alternate approaches to raycasting volume rendering were soon proposed (Westover, 1989) and accelerated parallel methods also introduced (Yoo, 1992). Volume rendering techniques were merged with computer graphics texturing hardware in 1994, enabling truly interactive visualization of volume data (Cabral, 1994). Navigation and exploration of volume data is now possible at rates of 20 frames per second on PC hardware with dynamic user control over viewpoint, transparency, and illumination (Kniss, 2001). This progression shows a successful transition over more than a decade from the development of new methods, through their engineering refinement, to product development, and finally the release of these methods as commodity tools for a broad medical audience.

Algorithmic developments are not the only successes for 3D medical informatics. Pilot studies in data collection, distribution, indexing, and content-based retrieval have advanced significantly in the last decade, and are partnered with the emergence of accepted public sources of information, common conventions, and shared tools. The Visible Human Project™ provided one of the most advanced studies in human gross anatomy to be shared widely among the 3D medical informatics research community. The availability of common data helped to accelerate the growth of visualization systems in medical settings (Ackerman, 1998). However, two datasets are not sufficient to study the nature of collections, and work on large volume medical data collections has grown in recent years (*e.g.*, Tagare, 1997, Leiman, 2003).

In the areas of segmentation, object recognition, and image understanding, new efforts in consolidation and interoperability are helping to unify research in disparate areas. Public tools for analyzing complex imaging data have recently emerged to assist in developing common conventions for processing, indexing, and understanding volumes. For example, the National Library of Medicine has sponsored the Insight Toolkit (ITK) as a common API for the segmentation and registration of 3D and higher dimensional data (Yoo, 2004). Common tools and conventions for image understanding are as essential in 3D medical informatics as shared ontologies are to conventional text-based informatics as well as cataloging systems are to the library sciences.

2.3 Why study 3D Medical Informatics?

3D Medical Informatics has compelling and innately satisfying motivations. From its early introductions, the value of X-ray CT scanning as immediately and intuitively recognized, and hospitals worldwide began investing in CT scanners. Medical visualization and 3D informatics similarly generate an immediate resonance with surgeons, developmental biologists, and other disciplines requiring navigation of the human condition where knowledge of spatial relationships is required.

Anatomy instruction is one area where 3D medical informatics has strong justification and sound foundations. The Visible Human Project™ data fostered a range of publications and educational products targeting all instructional levels from secondary school to graduate medical education. Beyond books and dissection software, there have been significant attempts to integrate 3D interactive file formats with multiple imaging modalities and existing anatomy ontologies. K.H. Höhne and the Voxel-Man software development team at the University of Hamburg have been using advanced rendering techniques to display and navigate complex information and have released digital products through a commercial publisher at a cost comparable to traditional textbooks (Höhne, 2000). Multimedia products such as these may represent the future of anatomy instruction.

Beyond instruction, 3D medical informatics has direct applications in surgical planning and intervention, transforming information into knowledge, providing clarity for critical decisions. Advanced instrumentation and rapid 3D analysis of surgical situations are becoming routine, integrating acquisition technology with interactive analysis and dynamic feedback during surgery (Jolesz, 1997). Operating rooms with integrated scanning devices and computer displays are now commonplace.

This effort is not limited to just three dimensions, and researchers routinely invoke time varying data and multimodal, multiscale radiological information in their investigations. Many diseases involve chronic, degenerative conditions that must be tracked over time. It is the progression of the condition as much as the immediate situation that is of concern to patient and physician alike. Critical areas of 3D medical informatics research include the recall of previous relevant examinations, the fusion of previous studies of the same patient with current cases, comparing the progress or the contraction of the disease. One example of the tracking of time-varying chronic diseases, is the study of multiple sclerosis, a condition involving repeated scarring of the insulating layers of nerve cells. These inflammatory attacks flare up and subside over time, and tracking the condition using spatial references to show where new lesions are appearing as opposed to older scar tissue using modern imaging allows clinicians to study the physical aspects of the disease as well as the neurological effects. These studies require the comparison of multiple images of the same patient, taken over the course of years, and align and compare the results (Kikinis, 1999). Such studies are giving new hope to patients and clinical scientists that some means may be found for controlling the disease.

In addition, the processes of 3D medical informatics can be applied in other domains. A study of topological image metrics based on juxtaposition and arrangement was published as a means of indexing medical image collections (Tagare, 1995). These metrics are invariant with respect to size and color, making them suitable for comparing images of arbitrary resolution, for instance, permitting the comparison of pediatric with adult cases. The study of such size and illumination invariant metrics has been now successfully applied to tracking marine mammal migrations where cetaceans are tracked by photographs taken of their fins. Identifying an animal from a photo of its dorsal fin by comparing the image with a veritable mug-shot catalog of dorsal fins is tedious and error-prone work. Adapting the work of Tagare, et al., this process has been streamlined from hours of visual canvassing to seconds of automated searching. Size and illumination invariant content-based retrieval techniques are changing the way marine biologists conduct their research (Hillman, 1998).

3. EXAMPLE: 3D MODELS AND MEASUREMENT OF NEUROANATOMY ACROSS SUBJECTS

Recall from Figure 12-3 that 3D medical informatics can be considered as a progression of processes applied to datasets. What are these stages? What are the input data and output data for each of these processes?

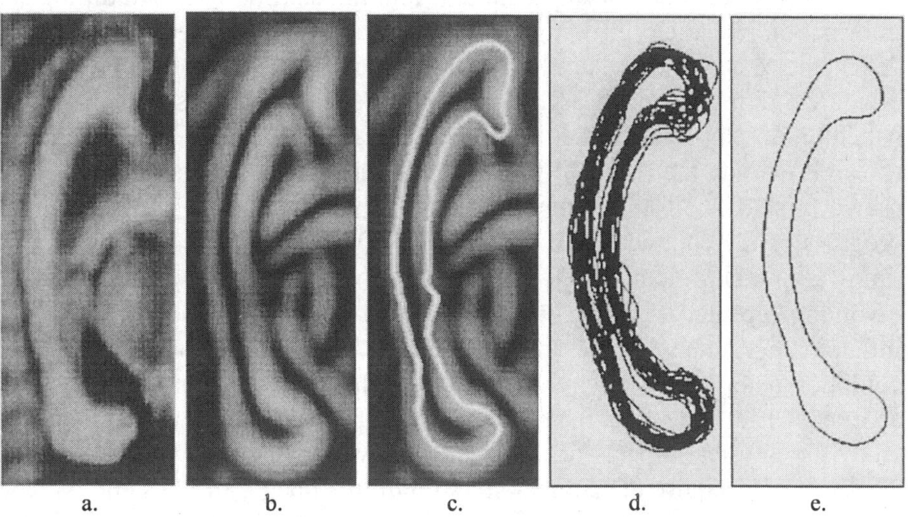

Figure 12-5. A progressive view of modeling the corpus callosum across multiple subjects. 5a: mid-saggital MR image of a human brain cropped to show the corpus callosum; 5b: *Low-level image processing* – cropped image from (5a) after a simple gradient magnitude edge detector has been applied; 5c: *Segmentation*: shape segmented using the edge strength image in (5b); 5d: *Registration*: outlines of the same structure registered across 30 subjects showing the normal (expected) variation in shape within a sample population; 5e: the average or mean shape of a corpus callosum. (Adapted with permission from Szekely, et al. (Szekely, 1996)).

The answers depend on the particular application and the methods being applied. In this section, we will explore these questions through the processes of modeling and measurement of deep brain structures across multiple subjects.

This example is largely described in terms of 2D models and has been drawn with permission from the work of Szekely, et al. (Szekely, 1996). Derivative work has been applied to full 3D models and has been used to study structural trends in schizophrenic patients across multiple subjects (Shenton, 2002). While this work concentrates on relatively familiar frequency-based Fourier shape descriptors, later related work has explored deformations of medial shape descriptions as well as statistical moments of point distribution models.

3.1 Indexing Images with 3D Medical Informatics

Following the progression in Figure 12-5, consider the corpus callosum, the dense white matter of connecting nerve fibers in the center of the brain that bridges the two hemispheres. If only the middle plane of the brain is considered, the vertical saggital plane that denotes the bilateral symmetry of the human body, the resulting image of the corpus callosum is an elongated arcing shape (see Figure 12-5a). Architectural defects in this structure are often linked to mental health disorders, and tumors in this region are difficult to treat unless detected early. Finding this structure in a brain scan and comparing it with "healthy" examples requires first detecting the edges in the image (low-level image processing) (see Figure 12-5b), then partitioning of the image into shapes and the matching or registration of shapes across multiple samples (segmentation and registration) (see Figure 12-5c,d). Once a shape is found, an analysis of the shape extracts features of the object (geometry and shape analysis). Based on this analysis of object features (usually across multiple samples), an object similarity metric can be created and dynamic models or templates of the structure can be formed for object detection and comparison (digital modeling and simulation).

Once segmented and aligned, the 30 cases in the training set are further analyzed for the normal or expected variation between these healthy subjects. A decomposition of the 2D shapes is performed, ordering the primary forms (eigenmodes) of deformation needed to match one shape to another. An average shape can be extracted from the set, and similarity or differences among the shapes can be derived (see Figure 12-6).

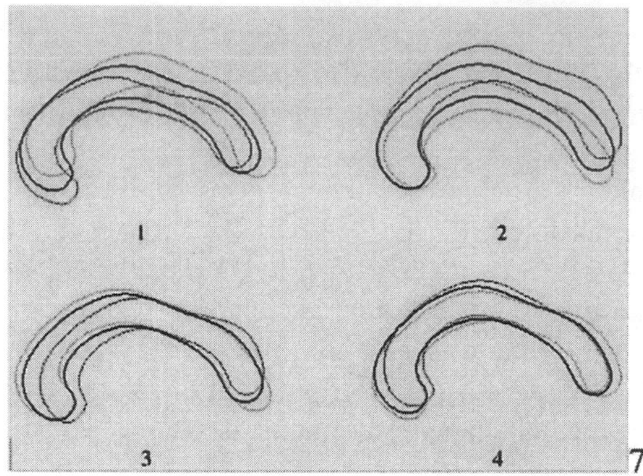

Figure 12-6. The first four eigenmodes of the deformations of the 30 objects in the training set. The calculations are based on contours represented by Fourier descriptors, which are normalized only with respect to the choice of the starting point. The deformation range amounts to eigenvalues. (Adapted with permission from Szekely, et al. (Szekely, 1996)).

The resulting metrics can be used to improve the automated segmentation of new subject data, and they can also be used as indexing tools for collections of MR scans of human brain structures.

3.2 Generalizing Elastic Deformable Models to 3D

This work has been generalized to 3 dimensions, permitting the analysis of solid shapes rather than simple 2D figures. Figure 12-7 shows the same strategy applied using active surfaces (as opposed to active contours) and spherical harmonics (as opposed to circular harmonics) to the caudate nucleus, another deep brain organ. The resulting digital models and the deformation metrics provide shape indices that can be used to provide quantitative comparisons of human anatomy and to catalog collections of volume data of human subjects.

An in-depth description of the work of Szekely, Kelemen, Brechbüler and Gerig is beyond the scope of the lay introduction intended here in this chapter. Their process includes a sophisticated automatic segmentation technique that uses both a frequency-based Fourier decomposition as well as an energy-minimizing elastically-deformable active-contour approach for reliably finding the corpus callosum. For a full description, see their comprehensive paper (Szekely, 1996) or their later work (Kelemen, 1999).

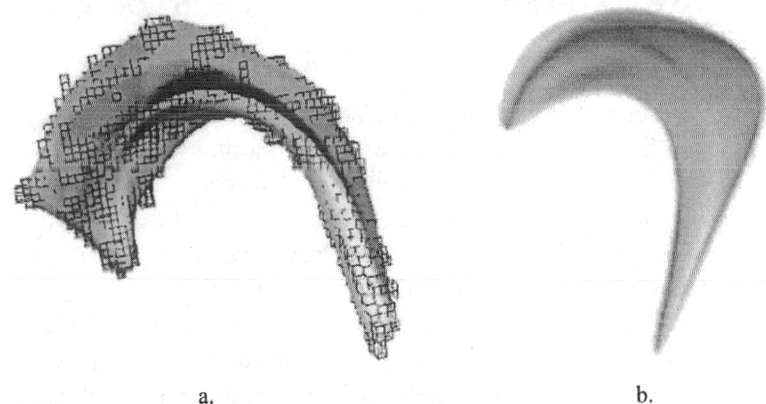

Figure 12-7. An example of this technique in 3D dimensions showing a parameterized description by spherical harmonics of the caudate nucleus, a deep brain organ. 7a: model of up to degree 8 showing the original voxel object overlaid as a wire-frame structure of the voxel edges; 7b: a segmented caudate nucleus using the elastically deformed model from (7a) using spherical harmonics up to degree 5 (108 parameters). (Adapted with permission from Szekely, et al. (Szekely, 1996)).

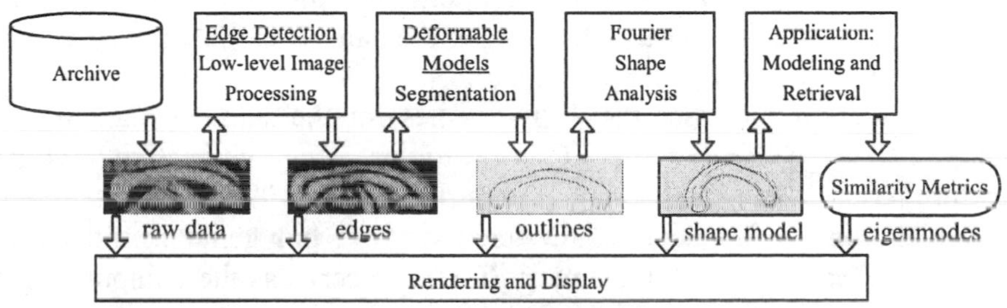

Figure 12-8. A 3D medical informatics pipeline using elastic deformable models of the corpus callosum as a case study. The data transforms left to right from raw image information to image features (edges), to segmented shapes (outlines), to controlled shapes, to similarity measures that can be used to improve segmentation, to capture normal (expected) human variation, or to index image collections.

Using this example as a case study, we can revisit the view in Figure 12-3 and map some of these methods to the stages of the processing pipeline. The resulting view is show in Figure 12-8. These or related processes are required elements of any 3D informatics application.

4. SURGICAL TEMPLATES: A CASE STUDY IN 3D INFORMATICS

3D medical informatics can have a direct influence on intervention and treating humans. Some of the most interesting rendering technologies today are appearing as 3D printing devices capable of directly creating tangible objects as "rapid prototypes" of instruments or devices. In this example, a team at the NLM's Office of High Performance Computing and Communications has combined surgical planning techniques with computer aided manufacturing systems to create custom surgical aids for orthopedic surgery. Our application area is planning and controlling the trajectories of pedicle screws for spine surgery. The goal is to manufacture a physical jig that conforms to the contours of the patient's vertebrae. The jig is constructed with holes that correspond to the trajectories of the pedicle screws. These holes guide the drill placement and depth, increasing the accuracy and precision of screw placement (Figure 12-9). These devices are created for each individual patient, one per segment, and improve accuracy without the introduction of navigation tools or increased fluoroscopic radiation dose. The intent is to use a jig to transfer the surgical plan directly to the operating room without introducing additional technology. The complexities of computer-assisted surgery remain in the laboratory without intruding into the operating room.

This section addresses the design issues for the surgical planning and template design workstation. Our prototype is an interactive modified texture-based volume rendering program (Cabral, 1994) augmented with physical user interface devices, 3D stereo viewing, polygonal primitives, and tools for constructive solid geometry (CSG) to serve as the computer aided design foundation for modeling templates.

4.1 Background and Related work

Figure 12-9a shows the basic problem faced in spine procedures. Appliances such as plates and rods require fixation through narrow channels called pedicles. Trajectories through these narrow isthmuses of bone have optimal placement and limited tolerances. Complications can arise when the screws accidentally enter epidural or spinal spaces and transect the spinal cord, constrict the emerging nerve roots arising from the ganglion, drift through a disc, or emerge through the anterior surface and cut the aorta.

Figure 12-9b shows the goal of our project, the creation of custom drill guides designed to mate closely with individual vertebrae that limit depth and provide for precise control of the screw path. Other groups have also pursued templates for pedicle screw placement. Radermacher and Birnbaum have reported favorable results using numerically controlled (NC) machine tools to create plastic templates (Birnbaum, 2001).

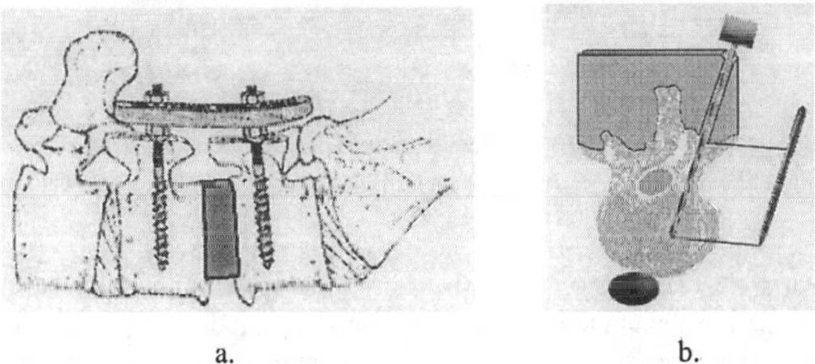

a. b.

Figure 12-9. Patient Specific Surgical Instrumentation (PSSI) for the precise placement of pedicle screws required for many spine procedures. Figure 9a: lumbar plate secured with pedicle screws. Figure 9b: Our goal – a proposed template, guiding the drill path and depth. Structures such as the spinal cord, nerve roots, and the aorta must be avoided.

Our approach differs from theirs in our use of Fused Deposition Modeling (FDM), an alternate technology for creating the templates. Many affordable NC machines are limited to three axes of control. This prevents many NC machines from supporting the complex geometries required to create custom templates. By contrast, FDM is the successor to stereolithography in rapid prototyping technology. It is flexible and accurate, capable of creating a wide variety of geometries, well beyond those needed in pedicle screw placement procedures. Stratasys, a manufacturer of these rapid prototyping devices, has secured U.S. FDA approval for the generation of 3D models for diagnostic purposes. The requirements of high fidelity reproduction necessary for diagnosis are equally important in intervention. They also supply a production material that can be sterilized for use in medical procedures.

4.2 Design and Software Tools for Template Planning Workstation

We built our surgical planning workstation around an interactive volume rendering system. Recent trends in graphics workstations have led to the emergence of 3D transparent textures, enabling interactive volume rendering using conventional graphics primitives (Cabral, 1994). Figure 12-10 shows the process and the console.

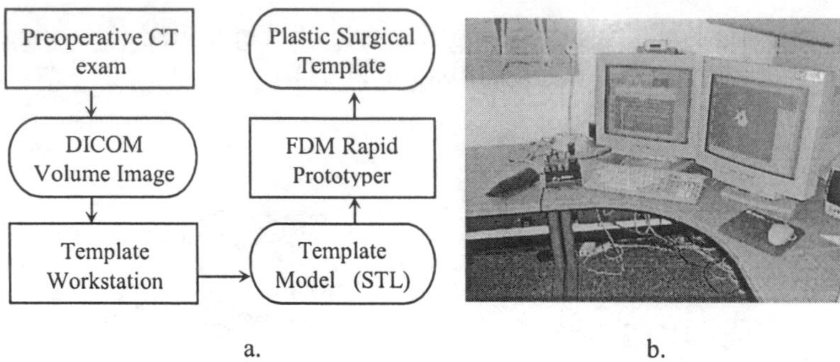

Figure 12-10. Patient Specific Surgical Instrumentation procedure and console. 10a: The pipeline from pre-operative exam to template. 10b: This section describes the template design workstation: Joysticks, stereo viewing glasses, pushbuttons, and physical sliders provide more natural human interfaces than overloaded mouse controls.

To augment the interactive visualization of the vertebrae and the placement of tool paths, the workstation has recently been augmented with physical I/O devices including a 3D joystick, supplementing the overloaded mouse controls. A mouse is by definition a 2D input device, limiting its use for viewpoint control. Tactile coherence can be improved by the use of physical input devices.

Volren 6.1, a texture-based volume rendering system running on a dual 250 MHz CPU Onyx2 with dual Reality Graphics™ raster managers was modified into a surgical planning workstation enabling the modeling of objects through constructive solid geometry (CSG). Texture based volume rendering naturally combines clipping planes and polygonal objects in a simplified volume rendering pipeline. The hardware-accelerated graphics systems necessary to support these methods are available in PC cards today. Clip planes, polygonal models, and volume rendering combine to make a natural graphical interface for planning screw placement.

4.3 Results and Discussion

As a test of the technology and its precision, we selected a dry, dissected lumbar vertebrae and created a surgical plan for pedicle screw placement. Thin section CT scans (1mm apart and 1mm thick) of 5 individual dry lumbar vertebrae were obtained on a GE Genesis High Speed RP Scanner. A block was modeled to fit tightly to the posterior surface of the vertebrae. Cylinders, that would ultimately be the drill guides, were then modeled through the block. The positioning of the cylinders, or trajectory planning, was accomplished with the aid of clipping planes and interactive control of the volume rendering transfer functions. This assured the authors that the planned trajectory was through the isthmus and along the axis of the pedicle, as shown in Figure 12-11. The 3-D drill guide block with trajectories was then divided into 2-D slices and converted to DICOM files. The 2-D slices were imported into Mimics v.6.3 (Materialise) and converted to STL files. The STL files were then used to generate the tool paths for the Fused Deposition Modeler (FDM) 2000 (Stratasys).

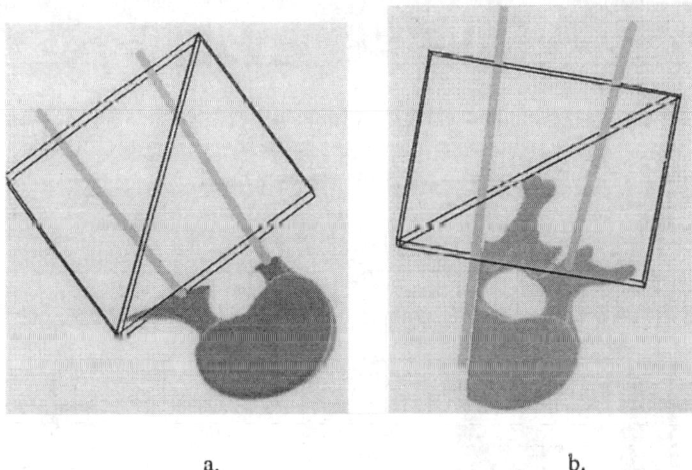

a. b.

Figure 12-11. The template design, taken from the texture-based volume rendering based surgical planning workstation. Clipping planes, polygonal primitives, and volume data are easily combined using advanced graphics.

We have achieved frame rates on the order of 10 frames per second for interactive template design including rendering the medical volume. Drawing from experience with molecular modeling systems, physical I/O tools such as joysticks were added improving the intuitive feel of the workstation. Some overloading of the input devices still occurs, leading to occasional confusion. Stereo viewing does not appear to speed template design. The use of Open GL as a programming base significantly reduced software development costs and permitted the fast integration of CSG.
A drill guide was produced by the FDM 2000 with approximately 125.06 cm^3 of non-medical ABS plastic at a slice interval of 0.2540 cm. The block, as designed, had an intricate area reserved for the posterior elements of the chosen lumbar vertebra. This first attempt was flipped in the x-axis due to an image format discrepancy. A second template was produced, correcting the defect, and pedicle pilot holes were drilled into the dry vertebrae. A CT scan was conducted to verify the placement of the pilot holes (Figure 12-12). Physical templates transfer the power of surgical planning workstations to the operating room without the need for complex technology. Confidence in the path planning will increase accuracy, speed procedures and reduce patient radiation dose by decreasing fluoroscopic verification (Yoo, 2001b).

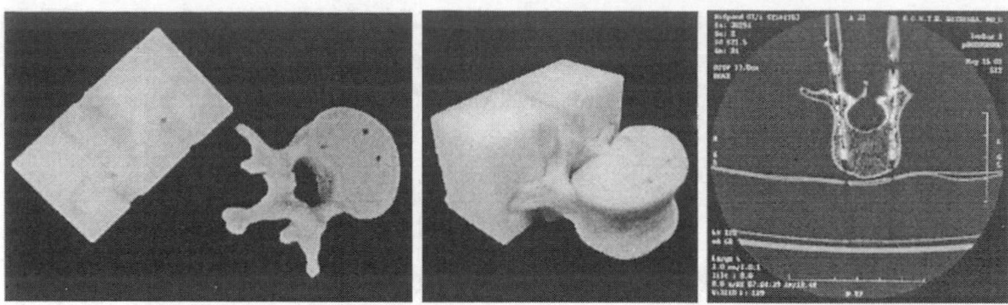

Figure 12-12. The prototype template and the dry spine test with a validating CT exam. (The pinholes in the top of the vertebrae are incidental, and leftover from the string connecting multiple vertebrae for its former use as an anatomy teaching tool.).

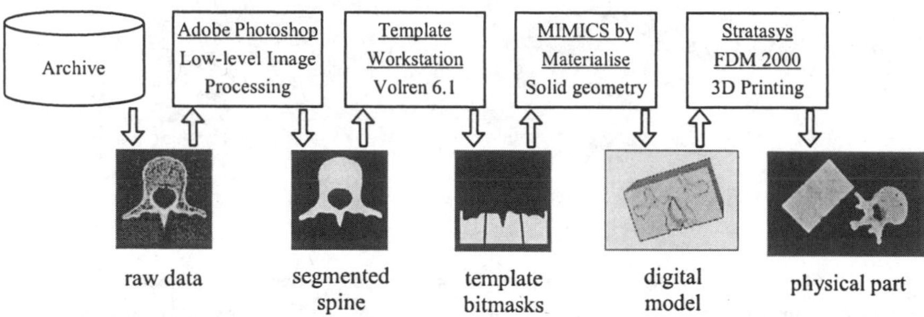

Figure 12-13. A 3D medical informatics pipeline using patient specific surgical instrumentation as a case study. The data transforms from raw image images to segmented objects (bitmasks), to designed shapes (embedded trajectories in mold bitmasks), to digital models (computer graphics models in STL format) to physical plastic parts manufactured with a rapid prototyping system. As the data is transformed, it is modified from raw images to physical instruments, encapsulating information in progressively higher-level abstractions.

The problems associated with designing and executing patient specific surgical instrumentation are easily cast as part of a 3D medical informatics data flow. Essentially, the idea of patient-specific surgical instrumentation in this example is the communication of information from a complex digital setting to precision surgical environment through a physical device. Intuitively, this means acquiring specific information about the patient's anatomy, planning the trajectories of the pedicle screws, developing spatial relationships between the plan and the patient, and embedding that information in an instrument. Like other problems in 3D medical informatics, this can be seen to be a series of processes that refines information from raw data through abstractions such as positive and negative images of the spine to digital models of solid geometry and eventually to a device (See Figure 12-13). Beyond information retrieval, 3D medical informatics can play a vital and direct role in medical intervention.

5. GRAND CHALLENGES IN 3D MEDICAL INFORMATICS

What are the grand challenges in 3D medical informatics? Whenever asking such questions, it is useful to phrase the notion using three interrogatives: What? How? If? Taking these points in order, "*What* areas of medical practice and public health can we affect through 3D medical informatics?" 3D medical informatics can provide new frontiers in medical education in anatomy, physiology, molecular biology, and other disciplines. In addition, research in this area has already had a profound impact on emerging technologies in computer assisted surgical planning and image-guided interventions. The areas where 3D medical informatics may have its greatest influence is in the areas of computer-aided diagnosis in early detection and progressive tracking of chronic, degenerative diseases where the pathology has important expressions in time and spatial domains.

How can we influence these public health concerns? We can open entire search strategies for finding information based on visual data. New databases based on volumetric data collections may permit rapid disease identification and provide indices to other sources of textual and digital information. Many diseases are not characterized by their expression in a single point in time, but rather by the chronology of the changes in their shape or function. Spiral chest CT exams and MRI breast imaging are currently being considered as alternatives to traditional 2D chest X-rays and 2D mammography as screening tools for early cancer detection. 3D imaging will become commonplace in medical diagnosis and intervention. Detection and dynamic modeling of anatomy and physiology represents new medicine.

We can make this leap to 3D informatics only *if* new techniques and mathematics emerge to describe, index, and characterize complex, high-dimensional data. We do not yet have the visual linguistic tools to decompose image information into subject, object, predicate, modifier, or prepositional phrase. Even the visual alphabets of pixels and polygons are not sufficiently rich to support the power of images, much less 3D scans. What is required is the formation of a comprehensive foundation in image linguistics built upon reproducible object detection, computable shape descriptors, and common indexing metrics for organizing human thought represented in visual and volumetric data. Some of the specific challenges include:

- The development of new mathematical foundations for data representation, not based on polygons, voxels, or patches of curved surfaces. Shapes are not local collections of such atomic data elements, but rather the overall complex composition of them. Aggregating pixels or patches into shapes is akin to parsing language in a text-based system.

Mathematical formulations using implicit methods are emerging to describe anatomical structures through equations, but they lack sufficient maturity yet to influence the field (Yoo, 2001a).

- A scientifically grounded foundation in data decomposition, perceptual analysis, and ontology generation of complex multidimensional information built upon computable, reproducible image metrics. This challenge includes broad fundamental investigations into data segmentation and registration, image statistics, and image understanding to answer the need for repeatable, comparable methods for decomposing, parsing, and extracting semantic knowledge from images.
- Establishment of public repositories of medical volume data. Libraries and museum collections have been some of the greatest incubators for the taxonomies and ontologies necessary to index and catalog human knowledge. Photographs are less than 200 years old, and the enabling digital technologies for 3D medical informatics are younger still. However, our society has an opportunity to begin to explore visual communication at a fundamental level, and archival collections of complex data will help to accelerate the creation of new forms of reasoning about visual and volumetric information.

Linguistics and text-based informatics have strong historical roots, and the expression of human thought in language has descended and evolved through the ages. There are museums and libraries dedicated to collecting and understanding text. There is even a museum of the alphabet that celebrates the constellation of alphabets used worldwide. Visual and digital analogs to these institutions do not yet exist. Yet there is incontrovertible potential for creating visual lexicons, parsing high-dimensional data, and developing semantic understanding of complex data. There may come a time when the routine representation of information in human endeavors similarly includes and celebrates images and volume data.

6. CONCLUSION

3D medical informatics is a synthesis of medical image processing, visualization and computer graphics, modeling, and data storage and retrieval. This concept is built up of multiple strengths with extensive study in particular core areas. Like text-based informatics, visual information is refined through a variety of methods creating from raw data abstractions of greater and greater sophistication, permitting the collection, indexing and storage, display, and manipulation of complex data. The tools for this work are still rudimentary, but they have strong analogs in the text domain. New collections including complex spatial and temporal data will continue to challenge researchers to refine and invent methods for handling these issues, likely borrowing heavily from the linguistics and informatics communities. The future of this discipline is uncertain but bright with possibilities.

7. ACKNOWLEDGEMENTS

Special thanks to Gabor Szekely and Guido Gerig for providing information and materials for the section on volumetric image comparisons and data analysis. This work is supported by the National Library of Medicine, Office of High Performance Computing and Communications.

REFERENCES

Ackerman, M.J. (1998). "The Visible Human Project," in *Proceedings of the IEEE*, 86(3),504-511.

Birnbaum, K., Schkommodau, E., Decker, N., Prescher, A., Klapper U., and Radermacher, K. (2001). "Computer-assisted Orthopedic Surgery with Individual Templates and Comparison to Conventional Operation Method," *Spine*, 26(4), (2001 Feb 15), 365-70.

Cabral, B., Cam, N., and Foran, J. (1994). "Accelerated Volume Rendering and Tomographic Reconstruction Using Texture Mapping Hardware," in *Proceedings of the 1994 Symposium on Volume Visualization* (Tysons Corner, Virginia), ACM Press, 91-98.

Castelli, V. and Bergman, L.D. (2002). *Image Databases: Search and Retrieval of Digital Imagery*, Wiley Interscience: New York.

Drebin, R., Carpenter, L., and Hanrahan, P. (1988). "Volume Rendering," *Computer Graphics* (Proceedings of ACM SIGGRAPH 88), 22(4), 65–74.

Funkhouser, T., Min, P., Kazhdan, M., Chen, J., Halderman, A., Dobkin, D., and Jacobs, D. (2003). "A Search Engine For 3D Models," *ACM Transactions on Graphics (TOG)*, 22(1) (January 2003), 83—105.

Hilaga, M., Shinagawa, Y., Kohmura, T., and Kunii. T.L. (2001). "Topology Matching for Fully Automatic Similarity Estimation of 3D Shapes," in *Proceedings of the ACM SIGGRAPH*, Los Angeles, CA, USA, 203-212.

Hillman, G.R., Tagare, H.D., Elder, K., Drobyshevski, A., Weller, D., and Würsig, B. (1998). "Shape Descriptors Computed from Photographs of Dolphin Dorsal Fins for Use as Database Indices," in *Proceedings of the IEEE Eng. Medicine Biology Society*.

Höhne, K.H., E*t al.* (2000). *Voxel-man 3D Navigator*, Springer: Berlin.

Jolesz, F. (1997). "Image-guided Procedures and the Operating Room of the Future," *Radiology*, 204, 601-612.

Kelemen, A., Szekely, G., and Gerig, G. (1999). "Elastic Model-based Segmentation of 3-D Neuroradiological Data Sets," *IEEE Trans. on Medical Imaging*, 18(10), 828-839.

Kikinis, R., Guttmann, CRG., Metcalf, D., Wells, W.M., Ettinger, G.J., Weiner, H.L., and Jolesz, F.A. (1999). "Quantitative Follow-up of Patients with Multiple Sclerosis Using MRI: Technical Aspects," *Journal of Magnetic Resonance Imaging*, 9(4), 519-530.

Kniss, J., Kindlmann, G., and Hansen, C. (2001). "Interactive Volume Rendering Using Multi-dimensional Transfer Functions and Direct Manipulation Widgets," in *Proceedings IEEE Visualization 2001).* (October 2001), 255-262.

Leiman, D.A, Twose, C., Lee, T.Y.H, Fletcher, A., and Yoo, T.S. (2003). "Rendering an Archive in Three Dimensions," *Medical Imaging 2003, Visualization. Image Guided Processing and Display* (February 16-21, 2003), San Diego, CA, Proc. SPIE, 5029, 9-17.

Levoy. M. (1988). "Display of Surfaces from Volume Data," *IEEE Computer Graphics and Applications*, 8(3)(May 1988), 29–37.

Lorensen, W.E., and Cline, H. (1987). "Marching Cubes: A High Resolution 3D Surface Construction Algorithm," in *Computer Graphics* (Proc. SIGGRAPH 87*)*, 21, 163-169.

McCormick, B., DeFanti, T., and Brown, M., (*eds.*), 1987. "Visualization in Scientific Computing," ACM SIGGRAPH, New York.

Shenton, M.E., Gerig, G., McCarley, R.W. Szekely, G., and Kikinis, R. (2002). "Amygdala-hippocampus Shape Differences in Schizophrenia: The application of 3D Shape Models to Volumetric MR Data," *Psychiatry Research Neuroimaging*, 115, 15-35.

Smeulders, A.W.M, Worring, M., Santini, S., Gupta, A., and Jain, R. (2000). "Content-based Image Retrieval: The End of the Early Years," *IEEE Trans. Pattern Anal. Machine Intell.* 22(12), 1349–1380.

Szekely, G., Kelemen, A., Brechbuehler, Ch. and Gerig, G. (1996). "Segmentation of 3D Objects from MRI Volume Data Using Constrained Elastic Deformations of Flexible Fourier Surface Models," *Medical Image Analysis* (*MEDIA)*, 1(1)(March 1996), 19-34.

Tagare, H.D., Vos, F., Jaffe, C.C., and Duncan, J.S. (1995). "Arrangement: A Spatial Relation Between Parts for Evaluating Similarity of Tomographic Section," *IEEE Trans. Pattern Anal. Machine Intell. (1*7(9), 880–893.

Tagare, H.D., Jaffe, C.C., and Duncan, J.S. (1997). "Medical Image Databases: A Content-based Retrieval Approach," *Journal of the American Medical Informatics Association.*, 4(3), 184–198.

Westover, L. (1989) "Interactive Volume Rendering," in *Proceedings of Volume Visualization Workshop* Department of Computer Science, University of North Carolina, Chapel Hill, NC., May 18-19, Pp. 9-16.

Yoo, T.S., Neumann, U., Fuchs, H., Pizer, S., Cullip, T., Rhoades, J., and Whitaker, R. (1992). "Direct Visualization of Volume Data," *IEEE Computer Graphics and Applications,"* 12(4)(July 1992), 63-71.

Yoo, T.S., Morse, B., Subramanian, K.R., Rheingans, P., and Ackerman, M.J. (2001a). "Anatomic Modeling Ffom Unstructured Samples Using Variational Implicit Surfaces," in *Studies in Health Technology and Informatics,* 81 (*Proceedings of Medicine Meets Virtual Reality 2001.* J. D. Westwood, *et al., eds.*), Amsterdam, IOS Press: 594-600.

Yoo, T.S., Morris, J., Chen, D.T., Burgess, J., and Richardson, A.C. (2001b). "Template Guided Intervention: Interactive Visualization and Design for Medical Fused Deposition Models," in *Proceedings of the Workshop Interactive Medical Image Visualization and Analysis* (18 October 2001, Utrecht, the Netherlands), 45-48.

Yoo, T.S. and Ackerman, M.J. (2004). "Engineering a Scientific Rendezvous: The Insight Toolkit for Medical Image Processing and Visualization," *Communications of the ACM.*

SUGGESTED READINGS

Christopher Johnson and Charles Hansen, *eds.* 2004. *The Visualization Handbook.* Elsevier, Academic Press.

Visualization involves constructing graphical interfaces that enable humans to understand complex data sets; it helps humans overcome their natural limitations in terms of extracting knowledge from the massive volumes of data that are now routinely connected. This book is a new resource on advanced visualization edited by two of the best known people in the world on the subject with chapters contributed by authoritative experts.

Vittorio Castelli and Larwence D. Bergman. 2002. *Image Databases: Search and Retrieval of Digital Imagery*, Wiley Interscience: New York.
This book is a broad introduction to the topic of image databases with in-depth analyses provided by some of the leading researchers in the field. It includes an introduction on the basics of image databases and the sources of such data as well as the technologies to store, compress, transmit and search large image-data collections.

Arnold W. M. Smeulders, Marcel Worring, Simone Santini, Amarnath Gupta, and Ramesh Jain. Content-based image retrieval: the end of the early years. *IEEE Trans. PAMI*, 22 - 12:1349 -- 1380, 2000.
This is an essential review of over 200 peer-reviewed publications on the topic of content-based image retrieval. The article describes methodologies and motivations for pursuing this domain and the intersection of this area with other parts of informatics, engineering, and science.

ONLINE RESOURCES

The National Library of Medicine's Visible Human Project™ (NLM VHP) has sponsored the collection and distribution of anatomical data of two human subjects. In addition, the NLM VHP has also created and managed the program to create the Insight Toolkit (ITK), the public open-source software collection for high-dimensional image segmentation and registration. The URL for these projects are respectively:
http://www.nlm.nih.gov/research/visible/visible_human.html
http://www.itk.org

The Princeton Shape Retrieval and Analysis Group has a web site for their 3D Model Search Engine. This system permits the indexing and retrieval of thousands of 3D computer graphics models based on query by sketch. You can try your hand at sketching and retrieving models at:
http://shape.cs.princeton.edu/search.html

The Voxel-man group at the University of Hamburg has published digital collections of Quicktime-VR representations of human anatomy linking of ontologies/taxonomies with image data in multiple dimensions. These CD sets show what is possible today using modern technologies in medical scanning, image generation, and careful presentation of digital anatomy resources. These CD sets are published by Springer and are available from bookstores and online vendors. Look for:
Höhne, K.H., *et al.*, 2000, Voxel-man 3D Navigator

QUESTIONS FOR DISCUSSION

1. It is said that, "A picture is worth a thousand words." Compare the notions of text-based informatics and image-based informatics. What are

the similarities? What are the salient differences? What foundations exist in text-based informatics that are missing among images? What areas of fundamental cognitive and information science research does this suggest? Discuss.

2. How do images play a role in society? How does the growth of volume data affect medicine and other disciplines such as meteorology, seismology, and oceanography? What burdens are these growing computational needs going to place on medical informatics? What opportunities? Discuss.

Chapter 13

INFECTIOUS DISEASE INFORMATICS AND OUTBREAK DETECTION

Daniel Zeng[1], Hsinchun Chen[1], Cecil Lynch[2], Millicent Eidson[3], and Ivan Gotham[3]

[1]Management Information Systems Department, Eller College of Management, University of Arizona, Tucson, Arizona 85721; [2]Division of Medical Informatics, School of Medicine, University of California, Davis, California 95616; also with California Department of Health Services; [3]New York State Department of Health, Albany, New York 12220; also with School of Public Health, University at Albany

Chapter Overview

Infectious disease informatics is an emerging field that studies data collection, sharing, modeling, and management issues in the domain of infectious diseases. This chapter provides an overview of this field with specific emphasis on the following two sets of topics: (a) the design and main system components of an infectious disease information infrastructure, and (b) spatio-temporal data analysis and modeling techniques used to identify possible disease outbreaks. Several case studies involving real-world applications and research prototypes are presented to illustrate the application context and relevant system design and data modeling issues.

Keywords

infectious disease informatics; data and messaging standards; outbreak detection; hotspot analysis; data visualization

1. INTRODUCTION

Infectious or communicable diseases have been a risk for human society since the onset of the human race. The large-scale spread of infectious diseases often has a major impact on the society and individuals alike and sometimes determines the course of history (McNeill, 1976). Infectious diseases are still a fact of modern life. Outbreaks of new diseases such as AIDS are causing major problems across the world and known, treatable diseases in developing countries still pose serious threats to human life and exact heavy tolls from nations' economies (Pinner et al., 2003). For instance, the estimated economic cost of Tuberculosis (TB) is 3 billion US dollars per year in India (http://www.healthinitiative.org).

In addition to diseases that occur naturally, there has been increasing concern that terrorists may choose to attack by deliberate transmission of infectious disease using biological agents. With greatly expanded trade and travel, infectious diseases, either naturally occurring or caused by bioterrorism attacks, can spread at a rapid rate, resulting in potentially significant loss of life, major economic crises, and political instability (Chang et al., 2003).

Information systems play a central role in developing an effective comprehensive approach to prevent, detect, respond to, and manage infectious disease outbreaks of plants, animals, and humans (Damianos et al., 2002; Buehler et al., 2004). Currently, a large amount of infectious disease data is being collected by various laboratories, health care providers, and government agencies at local, state, national, and international levels (Pinner et al., 2003). Furthermore, many agencies have developed information access, analysis, and reporting systems of varying degrees of sophistication. For example, in its role as the key agency responsible for human reportable diseases in the U.S., the Centers for Disease Control and Prevention (CDC) has developed computerized reporting systems for local and state health departments. Similarly, the U.S. Department of Agriculture (USDA) is enhancing data systems for certain animal diseases (e.g., mad cow disease and foot-and-mouth disease), and the U.S. Geological Survey (USGS), through its National Wildlife Health Center (NWHC) and numerous partners, manages databases for wildlife diseases. Databases may also be available at other federal and state or local health, agriculture, and environment/wildlife agencies and laboratories.

In addition to infectious disease-related data sources, the research and public health communities have developed a wide array of analytical and statistical models targeted at analyzing disease data for surveillance and outbreak prediction purposes. For instance, such models have been employed to predict outbreaks of West Nile Virus (WNV) (Eidson, 2001; Eidson et al., 2001; Julian et al., 2002; Guptill et al., 2003; Mostashari et al., 2003; Ruiz et al., 2004; Wonham et al., 2004) and of influenza (Hyman and LaForce, 2003).

This chapter introduces infectious disease informatics (IDI), an emerging subfield of biomedical informatics that systematically studies these information management and analysis issues in the domain of infectious diseases. More specifically, the objective of IDI research can be summarized as the *development of the science and technologies needed for collecting, sharing, reporting, analyzing, and visualizing infectious disease data and for providing data and decision-making support for infectious disease prevention, detection, and management.* IDI research directly benefits public health agencies in their infectious disease surveillance activities at all levels of government and in the international context. It also has important applications in law enforcement and national security concerning potential bioterrorism attacks (Siegrist, 1999).

Aimed at providing an overview of the emerging field of IDI, this chapter emphasizes the technical side of IDI research with detailed discussions on (a) the design and various system components of an infectious disease information infrastructure and (b) an important class of IDI data analysis techniques concerning the identification of possible outbreaks. In order to provide the readers with a concrete sense of IDI application contexts and relevant system design choices, we discuss both infectious disease information systems and standards that have been deployed in real-world applications and research prototypes for illustrative purposes. The majority of these case studies involve system development and deployment, and research projects in which we have been actively participating.

The rest of the chapter is structured as follows. Section 2 presents an overview of IDI. It discusses practical challenges arising when managing infectious disease data and presents the main technical components of an infectious disease information infrastructure and related research and system development issues. It also provides the readers with an introduction to analytical models and tools useful for infectious disease data analysis, in particular, outbreak prediction. Section 3 contains three case studies based on real-world applications and research prototypes to illustrate how to apply and synthesize in practice IDI system development and data analysis techniques introduced in Section 2. Section 4 concludes the chapter by summarizing the main learning objectives of this chapter and discussing future directions in IDI research and practice.

2. INFECTIOUS DISEASE INFORMATICS: BACKGROUND AND OVERVIEW

2.1 Practical Challenges and Research Issues

In practice, infectious disease data collection and analysis are complex and in most cases involve a multi-stage process with multiple stakeholders across organizational boundaries. Due to the nature of epidemics, there is also a critical need for timely data collection and processing.

The traditional approach to infectious data collection, dissemination, and reporting is a paper-based system that depends on telephone conversations for transmission of data and multiple personnel entering or updating paper-based case report forms. This approach leads to many problems related to information processing. Figure 13-1 illustrates a typical paper-based state and local data management approach for Botulism, a rare but significant infectious disease caused by a disease agent that can be used in a bioterrorism attack. This figure summarizes the information flow between various roles and organizations for Botulism case reporting and demonstrates the multiple areas of potential breakdown in communication (indicated by the ⊗ sign.)

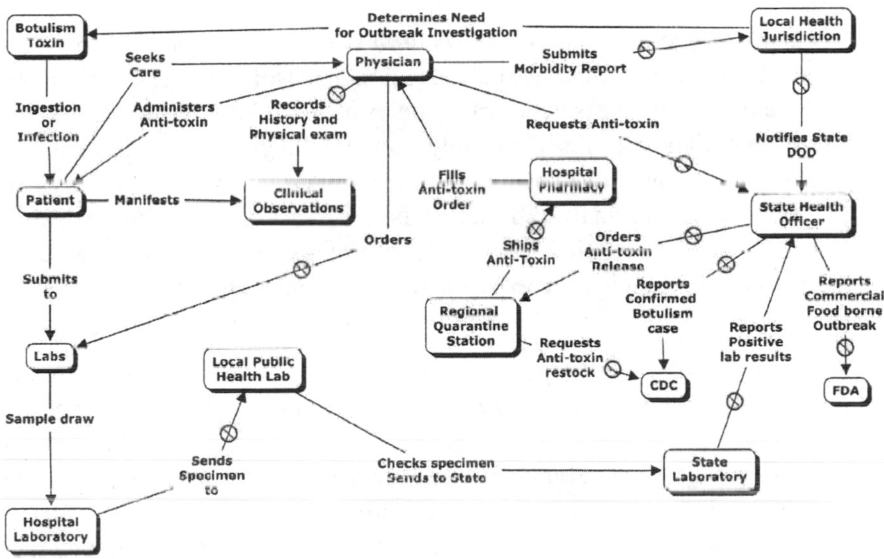

Figure 13-1. Paper-based Botulism Case Reporting

Many of these areas of potential breakdown in communication could be overcome with an automated process of data flow provided by a computerized infectious disease information system. Having historical data collected on outbreaks that are readily available for predictive modeling would also lead to improved surveillance activities, fewer data entry errors, and better public health data. Data in an information system would facilitate the required reporting of infectious diseases from local public health jurisdictions to state organizations and from state organizations to national surveillance entities such as the CDC. Additionally, information systems would streamline the reporting requirements for bioterrorism agents to the Department of Justice. Such systems also simplify dramatically the required chain of custody analysis for samples being tested for biological or chemical hazards. Infectious disease information systems also provide the ability for application of the ever expanding collection of statistical algorithms to data in real time that would not be possible without such systems.

Increasingly, infectious disease data are being collected in an electronic form by various laboratories, health care providers, and government agencies at local, state, national, and international levels (Pinner et al., 2003). There has been a steady trend for public health agencies to develop in-house infectious disease information systems tailored for their access, analysis, and reporting needs. However, the development and deployment of such systems do not automatically guarantee the effective collection and use of infectious data in broader contexts (Kay et al., 1998). Additional technical and policy challenges need to be met. Some of these challenges are summarized below.

- *Existing infectious disease information systems do not fully interoperate.* Most existing systems have been developed in isolation (Kay et al., 1998). As such, when disease-control agencies need to share information across systems, they may resort to using nonautomated approaches such as e-mail attachments and manual data (re)entry. In addition, much of the search and data analysis function is only accessible to internal users. Real-time data sharing, especially of databases across species, could enhance expert scientific review and rapid response using input and action triggers provided by multiple government and university partners.

- *The information management environment used to analyze infectious disease data and develop predictive models needs major improvements in areas such as system scalability, user friendliness, and model sophistication.* The amount of information necessary to collect and analyze for any public health outbreak has exceeded the capacity of epidemiologists to work without the assistance of a computerized system. Even with the assistance of statistical packages and geographical information systems, the lack of integration of current data collection, analysis, and visualization activities still leaves a significant gap in the ability to timely process the information acquired by field investigators. Current infectious disease information systems provide very limited support to professionals analyzing data and developing predictive models. An integrated environment that offers functionalities such as geocoding, advanced spatio-temporal data analysis and predictive modeling, and visualization is critically needed. Having an information system where data are integrated from the point of collection to the level of modeling and visualization-facilitated analysis would lead to higher quality of data input, more timely analysis, better predictive analysis of outbreaks, and potentially improved disaster incident management.

- *An efficient reporting and alerting mechanism across organizational boundaries is lacking.* Certain infectious disease information needs to be quickly propagated through the chain of public health agencies and shared with law enforcement and national security agencies in a timely manner. Certain models exist within the human public health community (e.g., CDC's ArboNet and Epi-X) and within certain states (e.g., New York State's Health Information Network (HIN)) (Gotham et al., 2002). However, in general the current reporting and alerting mechanism is far from complete and efficient, and often involves extensive and error-prone human interventions.

- *Data ownership, confidentiality, security, and other legal and policy-related issues need to be closely examined.* When infectious disease datasets are shared across jurisdictions, important access control and data security issues need to be resolved between the involved data providers and users. Subsets of such data are also governed by relevant health care and patient-related laws and regulations.

2.2 Infectious Disease Informatics Research Framework

IDI is an emerging field of study that systematically examines information management and analysis issues related to infectious disease prevention, detection, and management. IDI research is inherently interdisciplinary, drawing expertise from a number of fields including but not limited to various branches of information technologies such as data sharing, security, Geographic Information Systems (GIS), data mining and visualization, and other fields such as biostatistics and bioinformatics. It also has a critical policy component dealing with issues such as data ownership and access control, privacy and data confidentiality, and legal requirements. Because of its broad coverage, IDI research can be most successful through broad participation and partnership between various academic disciplines, public health and other disease surveillance, management, diagnostic, or research agencies at all levels, law enforcement and national security agencies, and related international organizations and government branches.

Figure 13-2 illustrates the major components of IDI research and summarizes related research and system development challenges.

As discussed in Section 2.1, the basic and necessary functions required of any infectious disease information system are data entry, storage, and query, typically implemented as computerized record management systems deployed by local public health agencies or at local hospitals and laboratory facilities.

To enable information sharing and reporting across jurisdictions and among record management systems maintained at different sites, system designers and operators need to agree on a common technical approach. This technical approach should include data sharing protocols based on interoperable standards such as Extensible Markup Language (XML) and a Web-enabled distributed

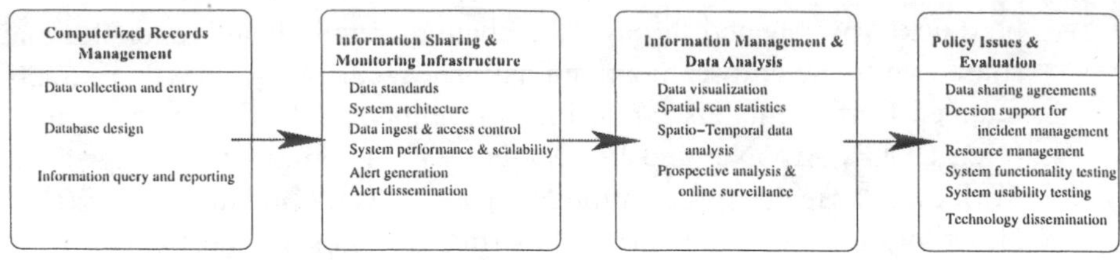

Figure 13-2. Infectious Disease Informatics Research Framework

data store infrastructure to allow easy access. It also needs to provide a scalable and effective reporting and alerting mechanism across organizational boundaries and provide geocoding and GIS-based visualizations to facilitate infectious disease data analysis.

To maximize the potential payoff of an infectious disease information system, advanced information management and data analysis capabilities need to be made available to the users. Such information management capabilities include visualization support that has been proven to be an effective tool to facilitate understanding and summarization of large amounts of data. An important aspect of data analysis is concerned with outbreak detection and prediction in the form of spatio-temporal data analysis and real-time surveillance. New "privacy-conscious" data mining techniques also need to be developed to better protect privacy and patient confidentiality (Wylie and Mineau, 2003; Kargupta et al., 2003; Ohno-Machado et al., 2004).

From a policy perspective, there are mainly four sets of issues that need to be studied and related guidelines developed. The first set is concerned with legal issues. There exist many laws, regulations, and agreements governing data collection, data confidentiality and reporting, which directly impact the design and operations of IDI systems. The second set is mainly related to data ownership and access control issues. The key questions are: Who are the owner(s) of a particular dataset and derivative data? Who is allowed to input, access, aggregate, or distribute data? The third set concerns data dissemination and alerting: What alerts should be sent to whom under what circumstances? The policy governing data dissemination and alerting needs to be made jointly by organizations across jurisdictions and has to carefully balance the needs for information and possibility of information overflow. The fourth set is concerned with data sharing and possible incentive mechanisms. To facilitate fruitful sharing of infectious disease data on an ongoing basis, all contributing parties need to have proper incentives and benefit from the collaboration.

To summarize, from an application standpoint, the ideal infectious disease information system would include a field deployable electronic collection instrument that could be synchronized with server-based information systems in public health departments. Biological specimen processing would be handled in laboratory information systems that were integrated with epidemiological and demographic information collected from the field and with the electronically submitted data from non-public health clinical laboratory information systems. The integrated laboratory, demographic, and epidemiological information would be available to analyze statistically and by GIS in real time as the data are collected. The data collected would be available to authorized users of a system that would protect identifying information of any individuals using role-based user access and permissions. Data could be shared across public health jurisdictions and between public health and nonpublic health agencies where such sharing was appropriate and where only appropriate data were provided. The ideal data system would use standards for metadata, terminologies, messaging formats, and security to maintain true semantic interoperability. Decision support and data analysis would be integrated into the data stream for data validation, message routing, and data de-duplication. GIS visualization and alerting capacity based on tested and dynamic algorithms would round out the necessary components of this idealized architecture.

The rest of Section 2 focuses on two sets of IDI research issues that are critical to the design and development of effective infectious disease information systems and have been drawing significant ongoing research attention. The first set is concerned with key system challenges and design decisions related to the development of the infectious disease information sharing infrastructure. The second set is centered around outbreak detection and related predictive modeling.

2.3 Infectious Disease Information Sharing Infrastructure

The following technical decisions are central to an infectious disease information sharing infrastructure: data standards, system architecture and messaging standards, and data inject and access control. In this section, we discuss them in turn.

2.3.1 Data Standards

Data standards represent the cornerstone of interoperability between information systems involved in disease reporting and surveillance. Likewise, data standards are critical to provide unambiguous meaning to data and form the foundation that enables data aggregation as well as data discrimination in data mining applications. The often repeated quote "the nice thing about standards is that there are so many to choose from" certainly rings true when data standards in health care and public health informatics are discussed. Fortunately, the swarm of data standards applicable to infectious disease informatics is beginning to narrow to a manageable group by the combined efforts of the National Center for Vital Health Statistics (NCVHS) and the Consolidated Health Informatics (CHI) E-Gov initiative (see Table 13-1) (Goldsmith et al., 2003). While these standards are currently required only for federal government information systems, in all likelihood, data standards adopted by the federal government will be assimilated and adopted by private industry over a relatively short period of time due to the combination of payor (Medicare, Medicaid, the Civilian Health and Medical Program of the Uniformed Services (CHAMPUS)) pressures, the sheer size of the federal government health care sector, and the need for private industry to communicate with these government systems. The Health Resources and Services Administration (HRSA) has also provided funds for encoding hospital laboratory information systems with the intention of helping migrate the systems from local code sets or CPT4 and ICD-9 code systems to LOINC and SNOMED codes that will allow interoperability with local, state, and federal health information systems adhering to CHI standards.

Table 13-1. CHI Standards Applicable to IDI

CHI Adopted Standard	Domain
Health Level 7 (HL7) messaging	messaging
Laboratory Logical Observation Identifiers Names and Codes (LOINC)	laboratory test orders
SNOMED CT	laboratory result contents; non-laboratory interventions and procedures, anatomy, diagnosis and problems
RxNORM	describing clinical drugs
HL-7 clinical vaccine formulation (CVX) and manufacturer codes (MVX)	immunization registry, terminology

In the public health sector, the CDC has led the way in the push for data standardization through the National Electronic Disease Surveillance System (NEDSS) and the Public Health Information Network (PHIN) initiatives. These initiatives define a set of vocabularies, messaging standards, message and data formats as well as the architectural components required for public health jurisdictions utilizing the federal bioterrorism grants for funding information system development. The National Library of Medicine brokered contract with the American College of Pathologists for the United States licensure of the SNOMED vocabulary, the naming of the first National Health Information Technology Coordinator, and the ongoing work on the National Health Information Infrastructure (NHII) provides the means for accelerating the pace for data standardization.

There are several barriers to the implementation of the CHI adopted coding systems.

1. There is a lack of trained personnel to perform the tedious transition from local code sets to LOINC and SNOMED. There are no automated

tools that can completely perform the transition; therefore, accurate coding still requires significant time and resources for direct one-on-one communication between the end-user, coder, and the information systems vendor.

2. Mapping from CPT4 codes to LOINC codes is oftentimes a one-to-many relationship that requires intervention by the personnel performing the laboratory tests to provide additional detail about methodology so that the choices of LOINC codes are sufficiently narrowed.

3. Many of the more detailed genomic tests utilizing polymerase chain reaction technology performed by the CDC and several advanced state labs for confirmation of bioterrorism agents do not have corresponding LOINC codes or CPT codes and require new local codes to be instituted.

4. Certain terms that are common in public health investigations are not represented in SNOMED because of the restriction of the SNOMED model to the more classical medical vocabulary domain. For example, there is no SNOMED code for the term "soil" that would be used to express the reservoir for an organism such as *Bacillus anthracis* or "bath house" used to describe a risk factor (location) for acquisition of HIV infection.

5. The domain of infectious disease surveillance is broad, spanning a range of terms not contained in any one terminology system.

2.3.2 System Architecture and Messaging

Messaging standards have been fewer in number with significantly deeper penetrance in the health care marketplace. From the perspective of IDI, Health Level 7 (HL7) is the dominant messaging standard for transfer of clinical information. Almost all hospital information systems exchange HL7 messages and the majority of large private clinical labs have adopted the HL7 standard as well. The current ANSI-approved version of HL7 is 2.5; however, several new Version 3 messages for public health reporting have been developed and are being reviewed for implementation as a normative standard. The CDC has set a goal of using Version 3 messages for morbidity reporting from states to the CDC. Additionally, the HL7 Clinical Document Architecture (CDA) standard is being considered in a variety of reporting and data collection scenarios including the CDC Outbreak Management System. The HL7 Version 3 specification represents a paradigm shift from the flat file structure of Version 2.x to an object-oriented Reference Information Model (RIM) foundation that provides the necessary structure to disambiguate the detailed information in the message and maintain contextual relationships between data elements. Support for such structures is critical in infectious disease and bioterrorism system-to-system communication. For example, structured complex post-coordination is handled by the Concept Descriptor (CD) data type in the Version 3 specification, which is not possible in the Version 2.x specifications.

To communicate the semantic content of infectious disease messages, Version 3 of the HL7 standard is mandatory. The implementation barriers are numerous, however.

1. The standard is still in evolution. While portions have been accepted as an ANSI standard, there is much work to be done to completely adopt the many sections of the balloted standard.

2. There is a large base of installed Version 2.x users whose communication needs are already met. They have little incentive to migrate to Version 3.

3. The expertise in Version 3 is limited and there are far fewer tools for implementation and support of Version 3 than for Version 2.x.

4. Implementation requires migration to new standard vocabularies and value sets, some still being developed and others being expanded.

5. The messages for infectious disease communication are very limited at present and complex in nature, representing a challenge for vendor implementation. Figure 13-3 shows the expression of a single clinical observation using the XML implementation of the CD data type.

Given the complexity of the HL7 Version 3 standard, it will be several years before we can expect the standard to be fully approved by ANSI and implemented widely.

2.3.3 Data Ingest and Access Control

An important function of an infectious disease information sharing infrastructure is data ingest and access control. Data ingest control is responsible for checking the integrity and authenticity of data feeds from the underlying information sources. Access control is responsible for granting and restricting user access to potentially sensitive data.

Data ingest and access control is particularly important in IDI applications because of obvious data confidentiality concerns and data sharing requirements imposed by data contributors. Although ingest and access control issues are common in many application domains, IDI poses some unique considerations and requirements. In most other applications, a user is either granted or denied access to a particular information item. In IDI applications, however, user access privilege is often not binary. For instance, a local public health official has full access to data collected from his or her jurisdiction but typically does

```
<Observation>
  <id root="10.23.4573.90009" />
    <code code="39154008" codeSystem="2.16.840.1.113883.6.96"
    codeSystemName="SNOMED CT" codeSystemVersion="0307core"
    displayName="clinical diagnosis" />
  <value xsi:type="CD" code="26544005" codeSystem=
"2.16.840.1.113883.6.96"
codeSystemName="SNOMED" displayName="muscle weakness">
  <originalText>bilateral symmetrical muscle weakness: Present
  </originalText>
 <qualifier>
    <name code="78615007" codeSystem=
"2.16.840.1.113883.6.96"
codeSystemName="SNOMED CT" codeSystemVersion="0307core" displayName=
"with laterality" />
    <value code="51440002" codeSystem="2.16.840.1.113883.6.96"
codeSystemName="SNOMED CT" codeSystemVersion="0307core"
displayName="bilateral" />
 </qualifier>
 <qualifier>
 <name code="103372001" codeSystem="2.16.840.1.113883.6.96"
codeSystemName="SNOMED CT" codeSystemVersion="0307core" displayName=
"with pattern" />
 <value code="255473004" codeSystem="2.16.840.1.113883.6.96"
codeSystemName="SNOMED CT" codeSystemVersion="0307core"
displayName="Symmetrical" />
 </qualifier>
</Observation>
```

Figure 13-3. An XML Sample for A Single Clinical Observation

not have the same access level to data from neighboring jurisdictions. However, it does not necessarily mean that this official has no access at all to such data from neighboring jurisdictions. Often he or she can be granted access to such data in some aggregated form (e.g., monthly or county-level statistics). Such granularity-based data access requirements warrant special treatment when designing an infectious disease information system.

2.4 Infectious Disease Data Analysis and Outbreak Detection

2.4.1 Background

In IDI applications, measurements of interest such as disease cases are often made at various locations both in space and in time. The cases of disease reported to the CDC through its National Notifiable Diseases Surveillance System are collected with timestamps at various places across the entire nation. Similar disease case reporting practices exist at state and local jurisdictions, typically with cases identified with specific geolocations.

Recent years have seen increasing interest in answering the following central questions of great practical importance arising in spatio-temporal data analysis and related predictive modeling: (a) How to identify areas having exceptionally high or low measures? (b) How to determine whether the unusual measures can be attributed to known random variations or are statistically significant? In the latter case, how to assess the explanatory factors? and (c) How to identify any statistically significant changes (e.g., in rates of health syndromes) in a timely manner in geographic areas?

Two types of approaches have been developed in the academic literature to address some of these questions. The first type of approach falls under the general umbrella of *retrospective* models (Kulldorff, 1997). It is aimed at testing statistically whether a disease is randomly distributed over space and time for a predefined geographical region during a predetermined time period. The second type of approach is *prospective* in nature, with repeated time periodic analyses targeted at identification of statistically significant changes in an online context (Rogerson, 1997; Sonesson and Bock, 2003).

This section discusses both retrospective and prospective approaches in IDI. We are primarily concerned with disease outbreak detection methods that consider both spatial and temporal data elements. From a modeling and computational perspective, two distinct types of spatio-temporal data and hotspot analysis techniques have been developed in the literature. The first type is based on various kinds of scan statistics and is used with increasing frequency in public health and infectious disease studies (Kulldorff, 2001). The second type is based on data clustering and its variations and has found successful application in crime analysis (Levine, 2002) with potential to be effectively applied in IDI. At the end of this section, we also point the readers to a rich set of epidemiological models of much longer history that studies epidemic phenomena and control strategies.

2.4.2 Scan Statistics-Based Techniques

Although a wide range of methods have been proposed for spatio-temporal data analysis in both retrospective and prospective cases, the scan statistic, in particular, the space scan statistic, has become one of the most popular methods for detection of disease clusters, and is being widely used by many public health departments and researchers (Kulldorff, 2001). For instance, using the spatial scan approach in New York, researchers found that persons residing in or near a dead crow cluster in the current or prior 1–2 weeks were 2–3 times more likely to become a WNV case than those not residing in or near such clusters (Johnson et al., 2004). In this section we provide a brief survey of the spatial scan statistic (as a retrospective method) and its spatio-temporal variation (as a prospective method).

The spatial scan statistic has been developed to test for geographical clusters and to identify their approximate location (Kulldorff, 1997). The number of events, e.g., incident cases, may be assumed to be either Poisson or Bernoulli distributed. Depending on the availability of data, the spatial scan statistic can be used for either aggregated data or for the special case of precise geographical coordinates, where each area contains only one person at risk. Algorithmically, the spatial scan statistic method imposes a circular window on the map under study and lets the center of the circle move over the area so that at different positions the window includes different sets of neighboring cases. Different sizes of circular windows will be examined. Over the course of data analysis, the method creates a large number of distinct circular windows (other shapes such as rectangular and ellipse have also been used), each with a different set of neighboring areas within it and each a possible candidate for containing a cluster of events. Conditioning on the observed total number of cases, N, the spatial scan statistic S is defined as the maximum likelihood ratio over all possible circles Z

$$S = \frac{\max_{Z} L(Z)}{L_0} \qquad [1]$$

where $L(Z)$ is the likelihood for circle Z, indicating the likelihood of the observed data given a differential rate of events within and outside the zone and where L_0 is the likelihood function under the null hypothesis. As this likelihood ratio is maximized over all the circles, it identifies the circle that constitutes the most likely cluster. In the hypothesis testing stage, because the exact distribution of the test statistic cannot be determined, Monte Carlo simulation is used as follows: Random datasets are generated under the known null hypothesis and the value of the scan statistic is calculated for both the real dataset and the simulated random datasets. If the scan statistic for the real dataset is among the highest 5%, then the detected cluster is significant at the 0.05 level.

Prospective spatio-temporal models needed by online surveillance are based on statistical surveillance concepts and techniques (Rogerson, 1997; Sonesson and Bock, 2003). Statistical surveillance refers to the on-line monitoring of a stochastic process with the aim of detecting a "significant" change in the process at an unknown time point as quickly and as accurately as possible. Many different disease surveillance situations have been studied in public health and a wide range of methods have been suggested for the surveillance, including different types of cumulative sums (Jacquez et al., 1996) and various scan statistic methods (Kulldorff, 2001). Note that the spatial context leads to a multivariate surveillance situation where methods of multivariate surveillance must be used. To evaluate these methods, different types of measures are used to characterize the behavior both when the process is in control and out of control. A critical trade-off that has to be considered is between false alarms and delay times for legitimate alarms. As in hypothesis testing, this trade-off is typically dealt with by evaluation of statistical power for detection of true associations under varying assumptions of type I error—the probability of falsely concluding there is an association.

The spatial scan statistic has been extended to the spatio-temporal domain to perform prospective data analysis. Instead of a circular window in two dimensions, the space-time scan statistic uses a cylindrical window in three dimensions (Kulldorff, 2001). The base of the cylinder represents geographical data while height represents time. The cylinder is flexible in its circular geographical base as well as in its starting date, independently of each other. The likelihood ratio test statistic is constructed in the same way as for the purely spatial scan statistic. The algorithm for calculating the likelihood of each window can also be extended to work in three rather than two dimensions.

2.4.3 Clustering-Based Analysis Techniques

The spatial scan statistic and its variations have been widely used in public health research. Despite their success, there are a number of limitations associated with such approaches (Patil and Tailie, 2004). First, their efficient operation depends on the use of simple, fixed symmetrical shapes of regions. As a result, when the real underlying clusters do not conform to such shapes, the identified regions are often not well localized. They either are larger than the real ones or leave out significant data points because incorporating them will bring in other non-significant data points. Second, it is difficult to customize and fine-tune the clustering results using scan statistic approaches. For different types of analysis, the users often have different needs as to the level of granularity and number of the resulting clusters, and they have different degrees of tolerance regarding outliners.

We now briefly introduce two alternative modeling approaches that have the potential to alleviate these problems with scan statistics-based approaches: *Risk-adjusted Nearest Neighbor Hierarchical Clustering* (RNNH) and *Support Vector Machines* (SVMs).

RNNH. Developed for crime hotspot analysis (Levine, 2002), RNNH is based on the well-known nearest neighbor hierarchical clustering (NNH) method (Jain et al., 1999), combining the hierarchical clustering capabilities with kernel density interpolation techniques. As with the scan statistic approach, in New York the kernel density approach also identified areas of 2–3 times greater risk for human WNV cases based on clusters of dead crow sightings in the current or prior 1–2 weeks (Johnson et al., 2004). The standard NNH approach identifies clusters of data points that are close together (based on a threshold distance). Many such clusters, however, are due to some background or baseline factors (e.g., a population that is not evenly distributed over the entire area of interest). RNNH is primarily designed to identify clusters of data points *relative* to the baseline factor. Algorithmically, it dynamically adjusts the threshold distance inversely proportional to some density measure of the baseline factor (e.g., the threshold should be lower in regions where the population is high). Such density measures are computed using kernel density based on the distances between the location under study and some or all other data points. We summarize below the key steps of the RNNH approach.

- Define a grid over the area of interest; Calculate the kernel density of baseline points for each grid cell; Rescale such density measures using the total number of cases
- Calculate the threshold distances between data points for hierarchical clustering purposes using the following formula:

$$0.5\sqrt{\frac{A_i}{Ni}} \pm t\frac{0.26136}{\sqrt{\frac{N_i^2}{A_i}}} \qquad [2]$$

  where A_i indicates the area of grid cell i, N_i the estimated number of data points in grid cell i based on kernel density estimations, and t the probability level for the distance between two points under the null assumption that they are random distributed.

- Perform the standard NNH clustering based on the above distance threshold (Jain et al., 1999).

RNNH has been shown to be a successful crime analysis tool in the context of retrospective data analysis. We argue that its built-in flexibility of incorporating any given baseline information and computational efficiency also make it a strong candidate for analyzing spatial-temporal data in other applications such as IDI and bioterrorism data analysis. In addition, compared to spatial scan techniques, RNNH supports clusters with flexible shapes and allows the user to specify preferences regarding the granularity of clusters.

SVMs. SVMs are the most well-known of a class of algorithms that use the idea of kernel substitution (Bennett and Campbell, 2000; Vapnik, 1998). Motivated by statistical learning theory, SVMs are a systematic approach with well-defined optimization formulations which have no local minima to complicate the learning process and can be solved using well-established computational methods. It also has a clean geometric interpretation. As a linear discriminant to separate data points with binary labels in a d-dimensional input space, an SVM-based approach finds the solution by either maximizing the margin between parallel supporting planes separating the data points of different labels or, equivalently, by bisecting closest points in the convex hulls encompassing data points of the same label. Both objective functions lead to a quadratic program that can be solved rather efficiently. Using Hilbert-Schmidt kernels, the above linear classification algorithm can be extended to handle nonlinear cases. Conceptually, a nonlinear classification problem can be converted into a linear one by adding additional attributes to the data that are nonlinear functions of the original data. This expanded attribute space is called the feature space. Computationally, however, through the use of kernels, this nonlinear mapping method can be implemented without even knowing how the mapping from the original input space to the expanded feature space is done. As a result, the same efficient and robust optimization-based training method for the linear case can be readily applied to produce a general nonlinear algorithm.

Although the above nontechnical description uses classification to illustrate the basic ideas of SVMs, they can be applied in a wide range of other types of machine learning and data mining problems (Bennett and Campbell, 2000). Among them, SVM-based data description and novelty detection (DDND) is particularly relevant to our research. SVM-based DDND methods are aimed at identifying the *support* of a data distribution (Ben-Hur et al., 2001; Tax and Duin, 2004). In a simple application, for instance, these methods can estimate a binary-valued function that is 1 in those regions of input space where the data predominantly lie and cluster together and 0 elsewhere. Informally, these methods proceed as follows: First, they map implicitly the input data to a high-dimensional feature space defined by a kernel function (typically the Gaussian kernel $K(\mathbf{x}_i, \mathbf{x}_j) = e^{-q||\mathbf{x}_i - \mathbf{x}_j||^2}$ with width parameter q). Second, these methods find a hypersphere in the feature space with a minimal radius which contains most of the data. Slack variables can be introduced to allow for datasets outside the sphere. The problem of finding this hypersphere can be formulated as a quadratic or linear program depending on the distance function used. Third, the function estimating the support of the underlying data distribution is then constructed using the kernel function and the parameters learned in the second step.

SVM-based methods have several attractive modeling and computational properties, making them suitable for IDI applications. SVM-based DDND methods can single out data clusters in complex shapes and have been well-tested in complex, noisy domains (e.g., handwritten symbol recognition). Through control parameters such as the width parameter in the Gaussian kernel and the slack variables used to include the outliners, the user can easily control the behavior of the algorithm to satisfy different modeling needs. Also, the introduction of slack variables allows overlapping clusters to be generated. SVM-based density estimation methods are powerful approaches producing actual density estimations which contain more information than what other methods can offer. As in the case for SVM-based DDND, using a limited number of parameters, the user can easily experiment with density estimations of varying properties to meet their modeling needs.

The standard version of SVM-based DDND does not take into consideration baseline data points and therefore cannot be directly used in spatio-temporal data analysis. In our recent research, we have developed a *risk-adjusted* variation based on ideas similar to those in RNNH (Zeng et al., 2004). In this new approach we first compute the kernel density estimations using the baseline data points and then adjust width parameter q in the Gaussian kernel function based on such density estimations. The basic adjustment idea is as follows: When the baseline density is high, we use larger q, which makes it harder for points to be clustered together.

2.4.4 Mathematical Epidemiology

The methods discussed in the previous two sections are "weak" models in the sense that they do not reply on any structured understanding of the underlying disease transmission mechanism. Rather, they provide a method to detect any unusual clusters of disease cases or relevant sightings/observations that could indicate infectious disease outbreaks. Such a method is generically applicable to any disease type and serves as an important component of disease surveillance and early warning systems.

The research and public health communities have also developed another rich set of epidemiological models of a different nature (Brauer and Castillo-Chavez, 2001). Such models provide mathematical constructs to model known disease transmission mechanisms and population characteristics. The underlying mathematical modeling framework is typically based on biologically relevant ordinary differential equations. Developing such epidemiological models is not easy. The specific understanding of the transmission mechanism is required, making these models typically disease-specific. Many parameters need to be estimated to operationalize these models. The mathematical tools used in model building are also relatively sophisticated, making it hard for practitioners to understand and apply. However, these models offer many advantages as to their predictive power and their capability of analyzing and evaluating disease control strategies.

We briefly summarize a recent epidemiological study to illustrate how such models have been used in practice (Wonham et al., 2004). Infectious diseases such as WNV present many ecological and public health challenges. In particular, WNV exhibits a complex seasonal ecology because of its vector (primarily mosquitoes) and host (primarily birds) relationship. Differential equation models can be developed to capture the dynamic relations between vectors and hosts. Such models can then be analyzed and used to evaluate various disease control strategies. For instance, in one study (Wonham et al., 2004), the authors have developed a WNV model based on the classical susceptible-infectious-removed (SIR) model originally created for malaria transmission and showed that this model produced disease level predictions consistent with independent observational data. In addition, the authors are able to establish and evaluate indicators for WNV outbreaks based on their models and demonstrate that mosquito control decreases, but bird control increases, the chance of a WNV outbreak. This result is certainly not entirely intuitive. If verified, such an analysis can have important public health implications.

To learn more about epidemiological models and their applicability, the readers are encouraged to browse through standard mathematical epidemiology references such as (Brauer and Castillo-Chàvez, 2001; Mazumdar, 1999) and related new studies on homeland security applications (Hyman and LaForce, 2003).

3. INFECTIOUS DISEASE INFORMATION INFRASTRUCTURE AND OUTBREAK DETECTION: CASE STUDIES

3.1 New York State's Health Information Network System

The emergence of WNV in the Western Hemisphere was reported first in New York State in late summer 1999. This unprecedented event required rapid mobilization and coordination of hundreds of public health workers, expenditure of millions of dollars on an emergency basis, and immediate implementation of massive disease surveillance and vector control measures. The New York State Department of Health's (NYSDOH) Health Information Network (HIN) was used by New York to enable rapid and effective response to the WNV crisis (Gotham et al., 2001). The HIN is an enterprise-wide information infrastructure for secure Web-based information interchange between NYSDOH and its public health information trading partners, including local health departments and the New York State Department of Agriculture and Markets, New York State Department of Environmental Conservation, and the United States Department of Agriculture's Wildlife Services New York office. This system currently supports 20,000 accounts and 100 mission critical applications, cross-cutting all key public health response partners in the state of New York (Gotham et al., 2002). It implements sophisticated data access and security rules, allowing for real-time use of the data within the state while protecting confidentiality and scientific integrity of the data. The infrastructure is well suited to public health response, as illustrated by New York's ability to rapidly incorporate it into its plan to respond to the WNV outbreak in NY in 1999–2000. The system has evolved into an integrated surveillance system containing large quantities of real-time data related to WNV including (a) detailed human cases, (b) dead bird surveillance data, (c) asymptomatic bird surveillance data, (d) detailed reports on veterinarian, owner/residence, necropsy, clinical outcome for mammals, and (e) mosquito surveillance data. The functionalities have since been leveraged into a reporting system for any animal illnesses possibly related to bioterrorism or emerging diseases. For instance, this existing system was heavily utilized for rapid sharing of suspect rabid animal reports when raccoon rabies first became established in Long Island in 2004.

3.2 The BioPortal System

The second case study is based on our ongoing IDI research project, which is aimed at developing an integrated, cross-jurisdiction infectious disease information infrastructure called the BioPortal system.

One of the technical objectives of the BioPortal system has been to integrate infectious disease datasets on WNV and Botulism (BOT) from New York, California, and several federal data sources. The system also provides a set of data analysis, predictive modeling, and information visualization tools tailored for these two diseases. Figure 13-4 summarizes these datasets and intended users of BioPortal. CADHS stands for the California State Department of Health Services.

As illustrated in Figure 13-5, from a systems perspective, BioPortal is loosely-coupled with the state public health information systems in that the state systems will transmit WNV/BOT information through secure links to the portal system using mutually-agreed protocols. Such information, in turn, will be stored in the internal data store maintained by BioPortal. The system also

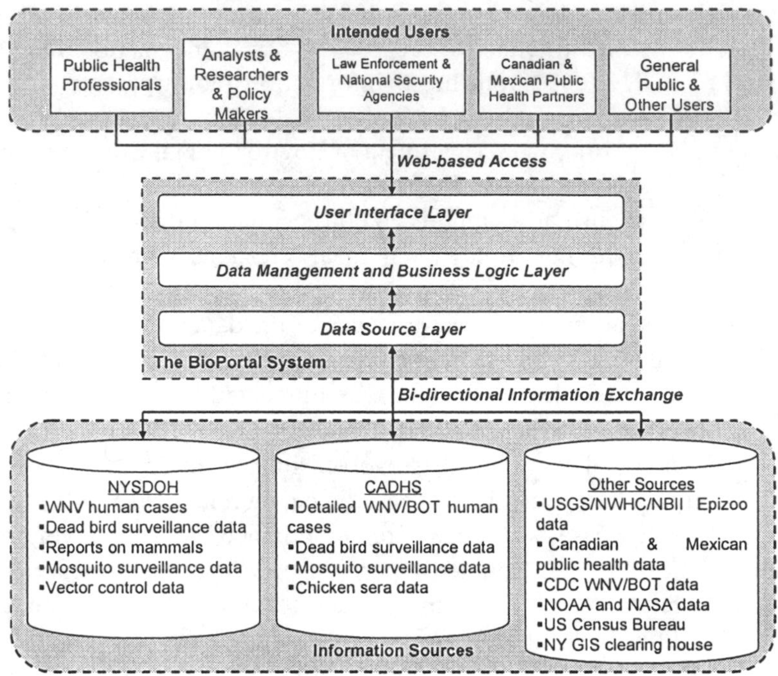

Figure 13-4. Data Sources and Intended Users of BioPortal

automatically retrieves data items from sources such as those from USGS and stores them in the internal data store.

Architecturally, BioPortal consists of three major components: a Web portal, a data store, and a communication backbone. Figure 13-5 illustrates these components and shows the main data flows between them and the underlying WNV/BOT data sources. The Web portal component implements the user interface and provides the following main functionalities: (a) searching and querying available WNV/BOT datasets, (b) visualizing WNV/BOT datasets using spatial-temporal visualization, (c) accessing analysis and prediction functions, and (d) accessing the alerting mechanism. The remainder of this section discusses the design philosophy and technical details of the data store layer, the communication backbone, data access control, and the visualization module.

Portal Data Store. A main objective of BioPortal is to enable users from partnering states and organizations to share data. Typically data from different organizations have different designs and are stored in different formats. To enable data interoperability, we use HL7 standards as the main storage format. In our approach, contributing data providers transmit data to BioPortal as HL7-compliant XML messages (through a secure network connection if necessary).

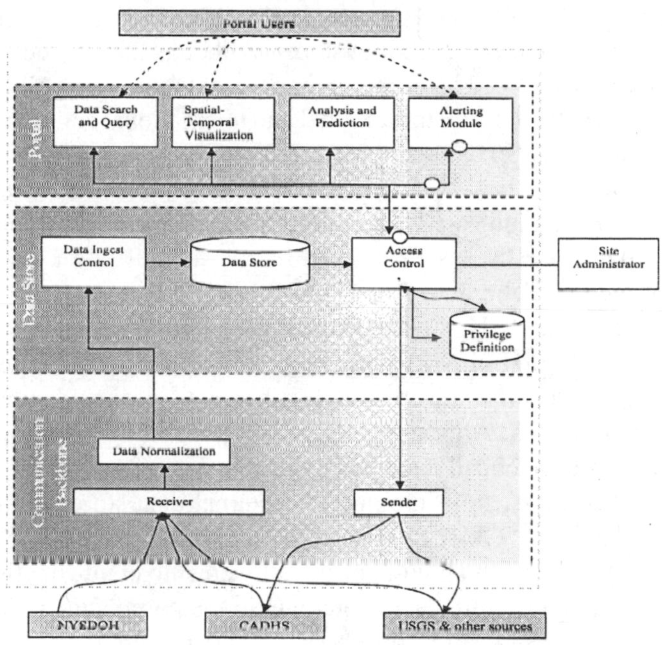

Figure 13-5. Overall Architecture of BioPortal

After receiving these XML messages, BioPortal will store them directly in its data store. This HL7 XML-based design provides a key advantage over an alternative design based on a consolidated database. In a consolidated database design, the portal data store has to consolidate and maintain all the data fields for all datasets. Whenever an underlying dataset changes its data structure, the portal data store needs to be redesigned and reloaded to reflect the changes. This severely limits system scalability and extensibility. Our HL7 XML-based approach does not have these limitations. To alleviate potential computational performance problems associated with this approach, we have identified a core set of data fields based on which queries will be performed frequently. These fields are extracted from all XML messages and stored in a separate database table to enable fast retrieval.

Communication Backbone. The communication backbone component enables data exchanges between BioPortal and the underlying WNV/BOT sources. Several federal programs have been recently created to promote data sharing and system interoperability in the health care domain. The CDC's NEDSS initiative is particularly relevant to our research. It builds on a set of recognized national standards such as HL7 for data format and messaging protocols and provides basic modeling and ontological support for data models and vocabularies. NEDSS and HL7 standards are having a major impact on the development of disease information systems. Although these standards have not yet been tested in cross-state sharing scenarios, they provide a solid foundation for data exchange standards in the national and international contexts. BioPortal heavily utilizes NEDSS/HL7 standards.

The communication backbone component uses a collection of source-specific "connectors" to communicate with underlying sources. We use the connector linking NYSDOH's HIN system and BioPortal to illustrate a typical design of such connectors. The data from HIN to the portal system is transmitted in a "push" manner. HIN sends secure Public Health Information Network Messaging System (PHIN MS) messages to the portal at pre-specified time intervals. The connector at the portal side runs a data receiver daemon listening for incoming messages. After a message is received, the connector will check for data integrity syntactically and normalize the data. Then the connector will store the verified message in the portal's internal data store through its data ingest control module. Other data sources (e.g., those from USGS) may have "pull"-type connectors which will periodically download information from the source Websites and examine and store data in the portal's internal data store. In general, the communication backbone component provides data receiving and sending functionalities, source-specific data normalization, as well as data encryption capabilities.

Data Confidentiality and Access Control. Data confidentiality, security, and access control are among the key research and development issues for the BioPortal project. In regard to system development, programming and rules already developed for New York's HIN system discussed above constitute the main sources of design and implementation ideas. Because there was no precedence for extending access to a data system across state lines, new access rules need to be developed for BioPortal. New security and user agreement forms were developed for both organizations/agencies with proposed access and individuals within those organizations. In addition, a Memorandum of Understanding (MOU) is pending between the agencies developing the BioPortal system, prior to sharing of real data. However, system development in such applications can proceed using test data pending final arrangements and signed user agreements. Responsibilities for participating organizations and individuals with access include:

- Establishment and definition of roles within the agency for access, and determination of individuals who would fill those roles, including systems for termination of access when individuals no longer fill those roles.

- Security of data physically located on, or transported over, the organization's network. This includes: validation of users who are granted access to BioPortal, physical security of computers on its network, security of removable data, and immediate notification of BioPortal when the status of the authorized individual user changes because of reassignment of duties or change in employment.

- Prohibitions against making unauthorized entry to other communication devices or resources; interfering with or disrupting other users; using the data for any illegal purpose; propagating computer worms or viruses; or infringing upon any copyright protections applicable to programs and/or data.
- Protections for confidentiality of all data accessed, with prohibitions against disclosure of personal or health data to any other agency, person, or public media outlet.
- Recognition of ownership rights of parties that have provided data: the originating party retains ownership and control of its data, and the data may not be provided to any other agency, individual, or public media outlet, or utilized for data analyses or summaries for presentations, abstracts, reports, or publications, without the originating party's consent.

The types of data that need to be addressed separately in regards to access are data from humans or owned animals, that require the highest levels of confidentiality, data from free-ranging wildlife, and data from other systems such as vectors (e.g., mosquitoes for WNV), land use, etc. The need for maximum access for disease tracking must be balanced against the confidentiality concerns and risks of jeopardizing reporting of data to the system.

For BioPortal, access to individual data records are separated for key neighbors and neighbors (see Figure 13-6). Key neighbors include the neighboring State Epidemiologist, Public Health Veterinarian, and Arthropod-Borne Disease Director, all key roles in surveillance and control of vector-borne diseases. Neighbors include the Key Neighbors with the addition of the neighboring State Laboratory Director and Environmental Health Director. The data elements to be included are date, species, age, signs, age/gender, case status/lab results, and location. The most sensitive information in regards to confidentiality is the point location of the data. On the other hand, data location is also critical for detecting geographic patterns of disease. For BioPortal, access to the zipcode and county of individual records is proposed for Neighbors. Only Key Neighbors would have access to the specific address or coordinates for Wildlife and Mosquito data. Although the animals/vectors themselves are not owned, the properties on which they are located are owned, and property owners and disease control officials have an interest in whether specific disease reports are tied to those individual properties. For humans and owned animals, only the origination organization would have access to the specific address or coordinates, but the system would allow others with access to request that information for specific needs from the originating organization. Such protection is critical to avoid any potential breaches of confidentiality that are mandated often in state laws, regulations, or agreements with local health agencies, health care providers, and other agencies.

Element	*Human/Owned Animal*	*Wildlife*	*Mosquito*
Date	Neighbors	Neighbors	Neighbors
Species	General types	Neighbors	Neighbors
Signs	Neighbors	Neighbors	Not applicable
Age/Gender	Neighbors	Not available	Not applicable
Case Status/ Lab Results	Neighbors	Neighbors	Neighbors
ZIP/County	Neighbors	Neighbors	Neighbors
Address/Lat/Long	CONFIDENTIAL	Key Neighbors	Key Neighbors

CONFIDENTIAL = Within Jurisdiction need to know access.

Key Neighbors = Neighboring State Epidemiologist, Public Health Vet, Arbo Director

Neighbors = Key neighbors + Neighboring State Laboratory Director, Environmental Health Director

Figure 13-6. Types of Data Elements and Access to Individual Data Records

For summary tables of data records, access is granted for all BioPortal participants to key summary data elements (see Figure 13-7). These include the number confirmed, probable, negative, and reported (no laboratory testing for confirmation done). These numbers are available in summary tables by species (general categories), for time periods of the users choice (week, month, year), and general location (county, zipcode). Specific unusual species are not named both to insure confidentiality (if there is an unusual animal, people may be likely to know who owns it) and to streamline reports. For captive wildlife species in private ownership or at zoos, only "captive" is listed in the summary tables. Zoologic institutions are a key partner in disease surveillance and have an existing reporting system established during the WNV outbreak (see http://www.aza.org/Newsroom/PRWestNileVirus/). Provision of information that would lead to identification of specific cases at individual zoos can lead to unwarranted public concern about their safety in visiting.

BioPortal participants have the choice of logging onto the system and generating real-time summary tables per their individual needs by clicking on specific icons ("pull"), or receiving alerts generated by specific triggers programmed into the system ("push"). In addition to the Key Neighbors and Neighbors previously specified for individual record access, the BioPortal participants can include federal disease control agencies (including those responsible for human, animal, or ecosystem health), neighboring directors of public health preparedness, directors of local health units in the state and neighboring states, State Epidemiologists in non-neighbor states, key state and federal law enforcement agencies, and academic centers participating in disease studies.

Element	*Human/Owned Animal*	*Wildlife*	*Mosquito*
Number confirmed	Portal Participants	Portal Participants	Portal Participants
Number probable	Portal Participants	Portal Participants	Portal Participants
Number negative	Portal Participants	Portal Participants	Portal Participants
Number reported (no lab testing)	Undetermined	Portal Participants	Portal Participants

Key neighbors = Neighboring State Epidemiologist, Public Health Vet, Arbo Director

Neighbors = Key Neighbors + Neighboring State Lab Director, Environmental Health Director

Undetermined = Availability of data still to be determined

Portal Participants = Neighbors + Federal Disease Control Agencies, Neighboring Directors of Public Health Preparedness and Local Health Units, State Epidemiologists in non-neighbor states, State and Federal Law Enforcement, Academic Centers

Note: Data elements will be available for summary by species (general), time period (week, month, year), and location (county, ZIP code)

Figure 13-7. Types of Data Elements and Access to Summary Data Records

Data Visualization. The role of visualization techniques in the context of large and complex dataset exploration is to organize and characterize the data visually to assist users in overcoming the information overload problem (Zhu et al., 2000). BioPortal makes available an advanced visualization module, called the Spatial Temporal Visualizer (STV) to facilitate exploration of infectious disease case data and to summarize query results. STV is a generic visualization environment that can be used to visualize a number of spatial temporal datasets simultaneously. It allows the user to load and save spatial temporal data in a dynamic manner for exploration and dissemination. STV has three integrated and synchronized views: periodic, timeline, and

GIS. The periodic view provides the user with an intuitive display to identify periodic temporal patterns. The timeline view provides a timeline along with a hierarchical display of the data elements organized as a tree. The GIS view displays cases and sightings on a map. Figure 13-8 illustrates how these three views can be used to explore infectious disease dataset: The top left panel shows the GIS view. The user can select multiple datasets to be shown on the map in a layered manner using the checkboxes (e.g., disease cases, natural land features, and land-use elements). The top right panel corresponds to the timeline view displaying the occurrences of various cases using a Gantt chart-like display. The user can also access case details easily using the tree display located left to the timeline display. Below the timeline view is the periodic view through which the user can identify periodic temporal patterns (e.g., which months have an unusually high number of cases). The bottom portion of the interface allows the user to specify subsets of data to be displayed and analyzed.

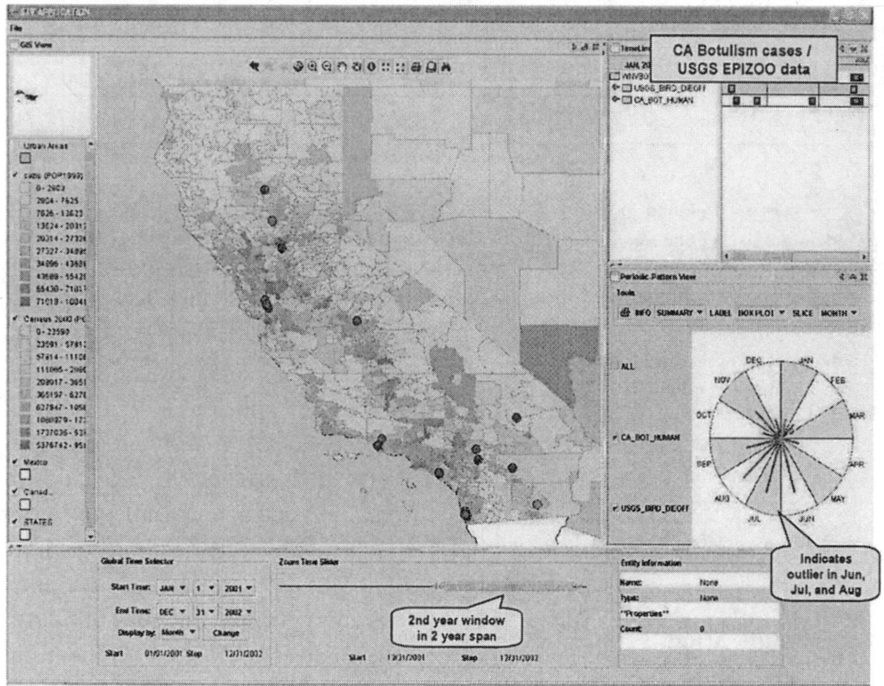

Figure 13-8. Spatial Temporal Visualization

3.3 West Nile Virus Outbreak Analysis

We now discuss a case study focusing on using various hotspot analysis techniques to detect possible WNV outbreaks. All the analyses were conducted using the BioPortal infrastructure. Pending completion of MOUs for sharing of actual data, in this case study for method development we used a test dataset containing altered dead bird sighting records in the spring and summer seasons of 2002 in New York State. Each sighting is identified with its reporting time and the corresponding geocoded location. There are 364 sightings in total.

The analysis performed using this dataset is retrospective in that we consider all the sightings of interest in one batch with the objective of identifying unusual clusters (which may be suggestive of WNV outbreaks) relative to the baseline data. For all the spatio-temporal data analysis techniques evaluated, the baseline data are defined as all the sightings reported before the day when the first dead bird was diagnosed with WNV. All the sightings after that day are the data items of interest. In our computational experiments, the baseline contains 140 sightings and the data of interest contains 224 sightings.

Three techniques have been studied: (a) spatial scan using SaTScan with the Bernoulli model option, (b) RNNH using CrimeSTAT, and (c) an SVM-based clustering method implemented in the MATLAB programming environment.

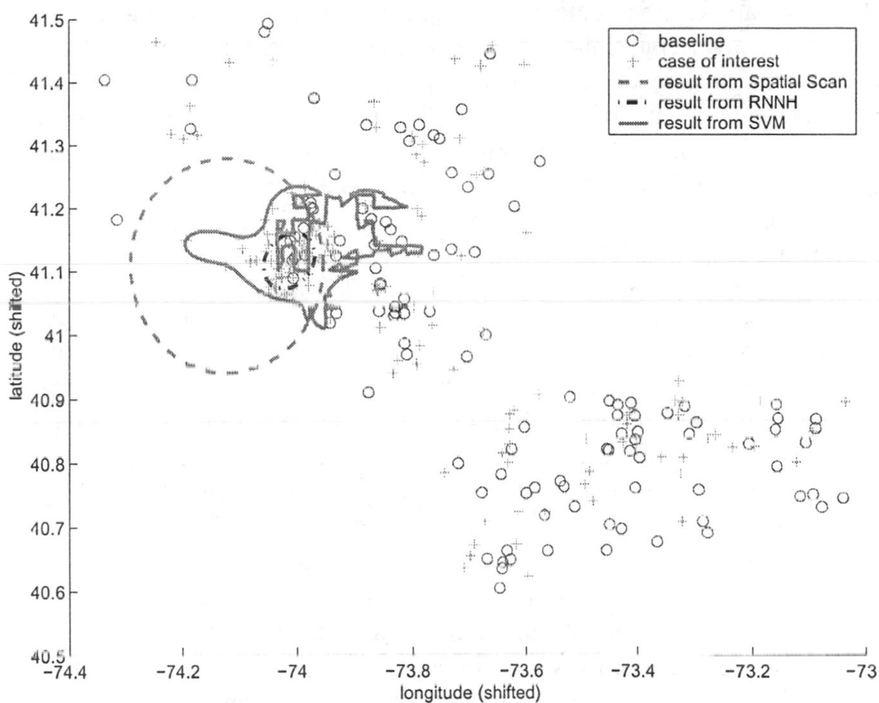

Figure 13-9. Dead Bird Hotspots Identified by Spatial Scan, RNNH, & SVM

Figure 13-9 shows the hotspots identified by each of these three techniques. As indicated above, due to data confidentiality issues pending completion of the BioPortal MOUs, the coordinates given are "shifted" versions of the real ones and the background map is not provided. Circle points represent baseline sightings and cross points sightings of interest. We observe that (a) the spatial scan approach identified a large circular area that encompasses many dead bird sightings but also includes empty regions without sightings, (b) the RNNH approach identified a hotspot that is strictly a subset of the hotspot identified with the spatial scan approach, and (c) the SVM method identified a hotspot with a nonregular shape that overlaps the hotspots identified by other methods but also includes other dead bird sightings. We also note that all these three methods are driven by the locations of the baseline and cases of interest and are not capable of incorporating relevant factors such as natural land features, land-use elements, and temperatures, into model development and outbreak detection. Models that are able to analyze such factors or data elements in addition to time and geolocation represent a promising new research direction.

4. CONCLUSIONS AND DISCUSSION

This chapter presents an overview of infectious disease informatics (IDI). IDI's broad application context and major components are reviewed. Using case studies involving real-world applications and research prototypes, this chapter highlights the key design decisions and technical challenges facing the design and implementation of infectious disease information systems. We also provide a tutorial on spatio-temporal hotspot analysis techniques which are an effective tool to facilitate identification of possible outbreaks.

We conclude this chapter by discussing issues still to be addressed for IDI research and system development and deployment.

- There is a critical need to develop cross-jurisdiction, cross-species infectious disease information systems to maximize the potential benefit and practice impact of IDI. The BioPortal system discussed in Section 3.2 represents an ongoing effort in this direction. Due to the technical and policy challenges facing the development and deployment of such systems, we envision that their development path will follow a *bottom-up, evolutionary* approach. Initially, each individual state will develop its own integrated infectious disease infrastructure for a limited number of diseases. Then, prototype systems linking several states will be developed involving a small number of diseases. Following successful deployment of such systems, regional nodes linking neighboring states can be established, leveraging both state sources and data from federal agencies. National and international infrastructures will then become a natural extension and integration of these regional nodes, covering most types of infectious diseases.

  This bottom-up approach is highly complementary to existing federal efforts such as CDC's National Notifiable Disease Surveillance System.

These federal efforts take a *top-down, centralized* approach, with reporting of cases for a limited number of diseases by local and state health departments through electronic means. Data transmitted to such federal databases are typically comprehensive in geographical coverage but time-delayed and coarse-grained (e.g., case locations are often aggregated at the county level). In addition, health agencies reporting data do not have access to the data system in order to monitor their own or neighboring area disease trends, but instead have only access to public summaries generated for the Morbidity and Mortality Weekly Report or public websites which do not provide data in real time or in sufficient detail for local disease control decisions. In the bottom-up approach, on the other hand, it is expected that information shared between states will be (near) real-time and fine-grained, enabling detailed local data analysis and timely dissemination of possible alert messages. Such capabilities in turn provide strong incentives to participating partners who voluntarily join in cross-jurisdiction collaboration. Technologically, data sharing and dissemination standards developed as part of the federal efforts can be directly leveraged in the information sharing infrastructures between states. We also expect that as the infectious disease information infrastructure matures and moves from the regional level to the national level, federal agencies such as the CDC will play an increasingly important role as both a partner and a regulatory body.

- As discussed in Section 2.3, data and messaging standards to facilitate IDI system design and development are far from being settled and universally accepted. We expect continued development and community buy-in for these technical standards in the years to come. There is also room for new middleware technologies that can serve as a bridge between competing standards to enable system interoperability and minimize the risk of being locked in standards that may become obsolete.

Growing concerns about the rapid spread of emerging diseases like WNV and the potential use of disease agents for bioterrorism have placed great emphasis on development of real-time surveillance tools, especially for detecting disease activity prior to receiving laboratory-confirmed reports of cases. Examples include syndromic surveillance systems (Bravata et al., 2004), most of which by definition rely on indicators other than laboratory-confirmed cases. Although surveillance systems that utilize laboratory-confirmed cases will continue to have a primary and critical role, more rapid indices of disease activity, including those based on hotspot analysis approaches, are expected to play an increasingly important role in some situations.

From a technical perspective, new research is critically needed to develop integrated modeling and data access/analysis/visualization support as part of the IDI infrastructure. Further research is needed to improve various outbreak detection and predictive modeling methods, and to establish a theoretical and empirical comparison base such that various methods can be evaluated in meaningful and relevant ways. One particular research direction that may lead to fruitful findings is to explore ways to integrate surveillance and outbreak detection methods with classical mathematical epidemiological models. Another important research direction is to combine surveillance and outbreak detection methods with decision-making or decision-aiding frameworks used to guide incident management decisions.

- Considerably more research and development is required for various options in system design to determine the optimal ways for early detection and monitoring of infectious disease trends. Some of the pressing issues include: (1) Scan statistics have an advantage of providing a level of statistical significance in determining whether clusters are sufficiently different from background to warrant further interest or action. However, what level of statistical significance should be used? This will be determined by the specific disease and surveillance and control actions to be taken. If the consequences of missing an outbreak are high, and the costs associated with control are reasonable, the level of significance may be set to a relatively high value such as 0.10 in order to have more statistical power and increased sensitivity for detection of outbreaks, even though this increases the chance of implementing control for "false positives"—outbreaks that are not real. If the consequences of missing an outbreak are lower, and the ramifications of triggering a response are severe, the level of significance may be set to a significantly lower value such as 0.01 in order to increase certainty that outbreaks are real before taking actions. (2) Methods that rely on comparison of data with background/baseline levels are highly dependent on the quality of surveillance during the non-outbreak periods, and if such surveillance is poor, other methods may be necessary.

- Although considerable progress is being made, systems for reporting animal diseases are still not as well developed as systems for reporting human diseases. With most bioterrorism agents being zoonotic (capable of transmission from animals to humans), animal surveillance is an essential component of an IDI system. Systems for reporting critical diseases in livestock to agriculture agencies are furthest in development; IDI systems for reporting diseases in pets or wildlife are also getting attention but need considerably more resources. Unlike for humans, disease case definitions have not been well developed for all animal species and diseases, nor are there resources for obtaining laboratory diagnosis of illness in many cases.

- IDI systems are being actively developed. In parallel with system development efforts, user evaluation and technology adoption research plays a critical role in receiving feedback to improve the IDI systems and promote their use in the workplace. There is an array of frameworks guiding such evaluation studies. For instance, the success model addresses the importance of technology dissemination within an adopting organization (Delone and Mclean, 1992) and the user community (Fichman and Kemerer, 1999). Another framework systematically examines technology impacts on inter-organizational collaboration and coordination (Clemons and Row, 1993), together with plausible causal relationships and reinforcement forces between or among key evaluation issues in public health (Chau and Hu, 2004). It is important for IDI researchers to be aware of these frameworks and apply them to conduct evaluation studies as IDI systems are being developed, deployed, and updated towards full use and maturity.

5. ACKNOWLEDGEMENTS

This work is supported in part by NSF Digital Government Grant #EIA-9983304 and NSF Information Technology Research Grant #IIS-0428241. We would like to thank the members of the NSF-sponsored BioPortal project, and a federal inter-agency advisory committee called "Infectious Disease Informatics Working Committee," in particular, Dr. Michael Ascher and Dr. James Kvach, for insightful discussions. Help from the members of the Artificial Intelligence Laboratory at the University of Arizona, in particular, Mr. Wei Chang, Mr. Chunju Tseng, and Ms. Catherine Larson, is also appreciated.

REFERENCES

Ben-Hur, A., Horn, D., Siegelmann, H.T., and Vapnik, V. (2001). "Support Vector Clustering," *Journal of Machine Learning Research*, 2, 125–137.

Bennett, K.P. and Campbell, C. (2000). "Support Vector Machines: Hype or Hallelujah?," *SIGKDD Explorations*, 2(2), 1–13.

Brauer, F. and Castillo-Chavez, C. (2001). *Mathematical Models in Population Biology and Epidemiology*, Springer.

Bravata, D., McDonald, K., Smith, W., Rydzak, C., Szeto, H., and Buckerdge, D. (2004). "Systematic review: Surveillance Systems for Early Detection of Bioterrorism-Related Diseases," *Annals of Internal Medicine*, 140, 910–922.

Buehler, J., Hopkins, R., Overhage, J., Sosin, D., and Tong, V. (2004). "Framework for Evaluating Public Health Surveillance Systems for Early Detection of Outbreaks: Recommendations from the CDC Working Group," *Morbidity and Mortality Weekly Report*, 53(RR-5), 1–13.

Chang, M., Glynn, M., and Groseclose, S. (2003). "Endemic, Notifiable Bioterrorism-related Diseases, United States, 1992-1999," *Emerging Infectious Diseases*, 9(5), 556–564.

Chau, P. and Hu, P. (2004). "Technology Implementation for Telemedicine Programs," *Communications of the ACM*, 47(2), 87–92.

Clemons, E. and Row, M. (1993). "Limits to Interfirm Coordination Through Information Technology: Results of a Field Study in Consumer Packaged Goods Distribution," *Journal of Management Information Systems*, 10(1), 73–95.

Damianos, L., Ponte, J., Wohlever, S., Reeder, F., Day, D., Wilson, G., and Hirschman, L. (2002). "MiTAP for Bio-Security: A Case Study," *AI Magazine*, 23(4), 13–29.

Delone, W. and Mclean, E. (1992). "Information Systems Success: the Quest for the Dependent Variable," *Information Systems Research*, 3(1), 60–95.

Eidson, M. (2001). "Neon Needles in A Haystack: The Advantages of Passive Surveillance for West Nile Virus," in White, D.J. and Morse, D.L., (Ed.), *West Nile Virus: Detection, Surveillance, and Control*, pages 38–53. New York Academy of Sciences, New York.

Eidson, M., Miller, J., Kramer, L., Cherry, B., and Hagiwara, Y. (2001). "Dead Crow Densities and Human Cases of West Nile Virus, New York State, 2000," *Emerging Infectious Diseases*, 7, 662–664.

Fichman, R. and Kemerer, C. (1999). "The Illusory Diffusion of Innovation: An Examination of Assimilation Gaps," *Information Systems Research*, 10(3), 255–275.

Goldsmith, J., Blumenthal, D., and Rishel, W. (2003). "Federal Health Insurance Policy: A Case of Arrested Development," *Health Affairs*, 22(4), 44–55.

Gotham, I., Eidson, M., White, D., Wallace, B., Chang, H., Johnson, G., Napoli, J., Sottolano, D., Birkhead, G., Morse, D., and Smith, P. (2001). "West Nile virus: A Case Study in How New York State Health Information Infrastructure Facilitates Preparation and Response to Disease Outbreaks," *Journal of Public Health Management and Practice*, 7(5), 75–86.

Gotham, I.J., Smith, P.S., Birkhead, G.S., and Davisson, M.C. (2002). "Policy Issues in Modern Surveillance," in O'Carroll, P., Yasnoff, W.A., Ward, M.E., Rubin, R., and Ripp, L., (Ed.), *Public Health Informatics and Information Systems*, chapter 25. Springer-Verlag.

Guptill, S.C., Julian, K.G., Campbell, G.L., Price, S.D., and Marfin, A.A. (2003). "Early-season Avian Deaths from West Nile Virus as Warnings of Human Infection," *Emerging Infectious Diseases*, 9, 483–484.

Hyman, J. and LaForce, T. (2003). "Modeling the Spread of Influenza among Cities," in Banks, H. and Castillo-Chàvez, C., (Ed.), *Bioterrorism: Mathematical Modeling Applications in Homeland Security*, chapter 10, pages 211–236. Society for Industrial and Applied Mathematics.

Jacquez, G.M., Grimson, R., Waller, L.A., and Wartenberg, D. (1996). "The Analysis of Disease Clusters, Part II: Introduction to techniques," *Infection Control and Hospital Epidemiology*, 17(6), 385–397.

Jain, A., Murty, M., and Flynn, P. (1999). "Data Clustering: A Review," *ACM Computing Surveys*, 31(3), 264–323.

Johnson, G.D., Eidson, M., Schmit, S., Kulldorff, M., and Ellis, A. (2004). "Geographic Prediction of Human Onset of West Nile Virus Using Dead Crow Clusters: A Quantitative Assessment of Year 2002 Data in New York State," Technical Report (in submission), New York State Department of Health.

Julian, K.G., Eidson, M., Kipp, A.M., Weiss, E., Petersen, L.R., and Miller, J.R. (2002). "Early Season Crow Mortality as A Sentinel for West Nile Virus Disease in Humans, Northeastern United States," *Vector Borne Zoonotic Disease*, 2, 145–155.

Kargupta, H., Liu, K., and Ryan, J. (2003). "Privacy Sensitive Distributed Data Mining from Multi-Party Data," in *Proceedings of the fi rst NSF/NIJ Symposium on Intelligence and Security Informatics, Springer LNCS 2665*, pages 336–342.

Kay, B., Timperi, R., Morse, S., Forslund, D., McGowan, J., and O'Brien, T. (1998). "Innovative Information-Sharing Strategies," *Emerging Infectious Diseases*, 4(3), 465–466.

Kulldorff, M. (1997). "A Spatial Scan Statistic," *Communications in Statistics: Theory and Methods*, 26(6), 1481–1496.

Kulldorff, M. (2001). "Prospective Time Periodic Geographical Disease Surveillance Using A Scan Statistic," *Journal of the Royal Statistical Society: Series A*, 166(1), 61–72.

Levine, N. (2002). *CrimeStat: A Spatial Statistics Program for the Analysis of Crime Incident Locations (v 2.0)*. Ned Levine & Associates, Houston, TX, and the National Institute of Justice, Washington, DC.

Mazumdar, J. (1999). *An Introduction to Mathematical Physiology and Biology*, second edition, Cambridge University Press.

McNeill, W.H. (1976). *Plagues and Peoples*, Doubleday, New York.

Mostashari, F., Kulldorff, M., Hartman, J., Miller, J., and Kulasekera, V. (2003). "Dead Bird Clusters as an Early Warning System for West Nile Virus Activity," *Emerging Infectious Diseases*, 9(6), 641–646.

Ohno-Machado, L., PS, P.S. Silveira, and Vinterbo, S. (2004). "Protecting Patient Privacy by Quantifiable Control of Disclosures in Disseminated Databases," *International Journal of Medical Informatics*, 73(7-8), 599–606.

Patil, G.P. and Tailie, C. (2004). "Upper Level Set Scan Statistics for Detecting Arbitrarily Shaped Hotspots," *Environmental and Econological Statistics*, 11(2), 183–197.

Pinner, R., Rebmann, C., Schuchat, A., and Hughes, J. (2003). "Disease Surveillance and the Academic, Clinical, and Public Health Communities," *Emerging Infectious Diseases*, 9(7), 781–787.

Rogerson, P. (1997). "Surveillance Systems for Monitoring the Development of Spatial Patterns," *Statistics in Medicine*, 16, 2081–2093.

Ruiz, M.O., Tedesco, C., McTighe, T.J., Austin, C., and Kitron, U. (2004). "Environmental and Social Determinants of Human Risk During a West Nile Virus Outbreak in the Greater Chicago Area," *International Journal of Health Geographics*, 3(8).

Siegrist, D. (1999). "The Threat of Biological Attack: Why Concern Now?," *Emerging Infectious Diseases*, 5(4), 505–508.

Sonesson, C. and Bock, D. (2003). "A Review and Discussion of Prospective Statistical Surveillance in Public Health," *Journal of the Royal Statistical Society Series A*, 166, 5–12.

Tax, D. and Duin, R. (2004). "Supper Vector Data Description," *Machine Learning*, 54, 45–66.

Vapnik, V. (1998). *Statistical Learning Theory*, John Wiley and Sons, Inc.

Wonham, M., de Camino-Beck, T., and Lewis, M. (2004). "An Epidemiological Model for West Nile Virus: Invasion Analysis and Control Applications," *Proceedings of Royal Society: Biological Sciences*, 271(1538), 501–507.

Wylie, J.E. and Mineau, G.P. (2003). "Biomedical Databases: Protecting Privacy and Promoting Research," *Trends in Biotechnology*, 21(3), 113–116.

Zeng, D., Chang, W., and Chen, H. (2004). "A Comparative Study of Spatio-Temporal Data Analysis Techniques in Security Informatics," in *Proceedings of the 7th IEEE International Conference on Intelligent Transportation Systems*, pages 106–111.

Zhu, B., Ramsey, M., and Chen, H. (2000). "Creating a Large-scale Content-based Airphoto Image Digital Library," *IEEE Transactions on Image Processing, Special Issue on Image and Video Processing for Digital Libraries*, 9(1), 163–167.

SUGGESTED READINGS

P. O'Carroll, W. Yasnoff, E. Ward, L. Ripp, and E. Martin, *eds*. 2002. *Public Health Informatics and Information Systems*. Springer. 824 pages.
This edited book covers all aspects of public health informatics and presents a strategic approach to information systems development and management. The targeted audience includes both state and local public health practitioners as well as faculty and students of public health.

H.T. Banks and C. Castillo-Chàvez, *eds*. 2003. *Bioterrorism: Mathematical Modeling Applications in Homeland Security*. The Society for Industrial and Applied Mathematics. 240 pages.
This edited volume covers recent research on bio-surveillance, agroterrorism, bioterror response logistics, and assessment of the impact of bioterror attacks. The specific emphasis of this book is on mathematical modeling and computational studies relevant to bioterrorism research.

F. Brauer and C. Castillo-Chàvez. 2001. *Mathematical Models in Population Biology and Epidemiology*. Springer. 416 pages.
This book presents an in-depth examination of Population Biology and Epidemiology from a mathematical modeling perspective. Chapter 7, titled "Basic Ideas of Mathematical Epidemiology," is particularly relevant to IDI.

ONLINE RESOURCES

CDC's NEDSS homepage at http://www.cdc.gov/nedss/index.htm

CDC's PHIN Homepage at http://www.cdc.gov/phin/

Health Level Seven standards and software implementation at http://www.hl7.org

Scan statistics-related outbreak detection software, datasets, and selected publications http://www.satscan.org

The BioPortal project's homepage at http://bioportal.eller.arizona.edu

The RODS project's homepage at http://rods.health.pitt.edu/. This project investigates methods for the real-time detection and assessment of outbreaks of infectious disease.

The ESSENCE project's homepage at http://www.geis.fhp.osd.mil/aboutGEIS.asp. This is part of the U.S. Department of Defense Global Emerging Infections Surveillance and Response System.

QUESTIONS FOR DISCUSSION

1. What patient confidentiality issues need to be considered in the context of IDI? What confidentiality issues might there be with data on owned animals, free-ranging wildlife, land use, and vectors (e.g., mosquitoes for WNV)?

2. What are the potential benefits of developing an across-jurisdiction and across-species IDI infrastructure? What are the potential policy and organizational barriers to the deployment of such an infrastructure? How can we overcome these barriers?

3. What role can visualization play in IDI data analysis? What are the types of visualizations commonly used by public health officials (not necessarily computerized)?

4. Scan statistics and hotspot analysis techniques can identify unusual clustering of events or cases in space and time. How can one interpret the findings based on these techniques in the IDI context? What are the implications of false-positives (i.e., hypothesized outbreaks that turn out to be normal background disease occurrences)? What are the implications of false-negatives (i.e., missing outbreaks)?

5. What are the potential technology adoption issues that may hinder the wide acceptance of IDI systems?

6. How should IDI material be taught? From a pedagogical standpoint, what is the right balance between traditional lecture material versus case studies versus hands-on experience with IDI systems?

UNIT III

Text Mining and Data Mining

Chapter 14

SEMANTIC INTERPRETATION FOR THE BIOMEDICAL RESEARCH LITERATURE

Thomas C. Rindflesch, Marcelo Fiszman, and Bisharah Libbus

National Library of Medicine, Bethesda, MD 20894

Chapter Overview

Natural language processing is increasingly used to support biomedical applications that manipulate information rather than documents. Examples include automatic summarization, question answering, and literature based scientific discovery. Semantic processing is a method of automatic language analysis that identifies concepts and relationships to represent document content. The identification of this information depends on structured knowledge, and in the biomedical domain, one such resource is the Unified Medical Language System. After providing some linguistic background, we discuss several semantic interpretation systems being developed in biomedicine. Finally, we briefly investigate two applications that exploit semantic information in MEDLINE citations; one focuses on automatic summarization and the other is directed at information extraction for molecular biology research.

Keywords

natural language processing; semantic interpretation; information extraction; automatic summarization; UMLS

1. INTRODUCTION

Automatic access to online information is an integral part of daily life as well as academic research. In this chapter, we explore the use of natural language processing (NLP), that is, automatic analysis of online text, as a way of supporting and enhancing professional access to the biomedical research literature. We discuss a particular approach that identifies concepts and relations through the (partial) semantic interpretation of text. For example, this processing identifies the semantic proposition (2) from (1).

(1) A randomized trial of etanercept as monotherapy for psoriasis

(2) ETANERCEPT TREATS Psoriasis

Although such an interpretation does not capture the complete meaning of (1) (*randomized trial* and *monotherapy* are not addressed), it provides the basis for systems that depend on the manipulation of information rather than documents.

Information retrieval is a mature application that provides documents relevant to a user-specified topic. The information sought is presumed to be in the documents retrieved but is not made overt. Emerging applications focus on explicit manipulation of information as the basis for decision support systems (Cimino and Barnett, 1993; Mendonça and Cimino, 2000) or for connecting patient records to bibliographic resources (Cimino, 1996), for example. Others use extracted information for literature-based scientific discovery (Srinivasan and Libbus, 2004; Fuller et al., in press).

These applications often depend on MeSH indexing terms assigned (by humans) to MEDLINE citations. However, there are important reasons for supplementing MeSH resources. Reliable indexing is not always available outside MEDLINE, and the information needed by an application may not be supplied by MeSH terms. Increasingly, NLP is used to support information manipulation applications, including, in addition to those mentioned, automatic summarization (Fiszman et al., 2004), question answering (Jacquemart and Zweigenbaum, 2003), and enhanced information retrieval (Grishman et al., 2002).

2. NATURAL LANGUAGE PROCESSING

2.1 Overview

NLP methodologies in the biomedical domain can be considered from the point of view of the text they address and the NLP technology used. Two important content subdomains are clinical medicine and molecular biology. In the clinical domain, the emphasis is on disease, anatomy, etiology, and intervention, along with the interaction among these phenomena. A second important content area is molecular biology. A major challenge is recognizing entities such as genes (and other aspects of the genome) and proteins. Important relationships refer to the way these interact among themselves, as well as with genetic diseases. Below, we briefly discuss one approach to NLP in molecular biology. More extensive coverage is provided in another chapter, in Unit III (Palakal et al., in this volume).

Another way to investigate NLP systems is to consider the genre of the text being processed. Two relevant genres in biomedicine are clinical records (such as discharge summaries and imaging reports) and the research literature. Important differences in both syntactic structure and terminology distinguish the two, and in this chapter we concentrate on the literature, particularly MEDLINE citations. Semantic processing in clinical text is discussed in another chapter in this unit (Friedman, in this volume).

Various linguistic approaches have been used to process biomedical text. These can be broadly categorized as either statistical or symbolic rule-based systems. In medicine, the latter predominate; however, Taira and Soderland (1999) and Pakhomov et al. (2002) have pursued statistical approaches, which assign an analysis to input text by matching it to training text annotated (usually by hand) with target structures. Rule-based NLP systems in medicine fall into one of three categories, based on the linguistic formalism used: phrase structure grammar (Christensen et. al., 2002), which concentrates on syntactic constituents; dependency grammar (Hahn, 2002), which emphasizes relations between words; and semantic grammar (Friedman et al., 1994), which relies on distributional patterns of semantic concepts.

Due to the complexity of language, systems often focus on one aspect of linguistic structure: words, phrases, semantic concepts, or semantic relations. Words can be identified with little (or no) linguistic processing. Phrases are normally identified on the basis of at least some syntactic analysis, using part-of-speech categories and rules for defining phrase patterns in English (Leroy et al., 2003). The identification of concepts and relations constitutes semantic processing and requires that text be mapped to a knowledge structure. In the biomedical domain, the Unified Medical Language System (UMLS) provides one such resource.

2.2 Levels of Linguistic Structure

Textual information management systems based solely on words have enjoyed considerable popularity, largely because the underlying processing is relatively easy to implement. After grammatical function words such as determiners *the* and *this* and prepositions *of* and *with* are eliminated, the remaining words are taken as a surrogate representation of semantic content. In (3), for example, *arthritis*, *children*, *hexacetonide*, and *triamcinolone* represent part of the meaning of the text.

(3) The purpose of this study was to compare the efficacy and safety of intra-articular triamcinolone hexacetonide and triamcinolone acetonide in children with oligoarticular juvenile idiopathic arthritis.

However, such a representation lacks expressiveness. It does not, for example, explicitly represent the fact that the disorder discussed is juvenile idiopathic arthritis or that there are two drugs, triamcinolone hexacetonide and triamcinolone acetonide mentioned.

Phrasal processing addresses some of these deficiencies. For example, the identification of *intra articular triamcinolone hexacetonide* and *oligoarticular juvenile idiopathic arthritis* isolates the relevant strings. However, these phrases alone do not indicate that the first is a drug and the second a disease. Nor do they provide the information that childhood arthritis is another name for this disorder.

Semantic processing enhances phrasal analysis with this kind of information. For example, the phrases in the previous paragraph can be mapped to concepts in the UMLS Metathesaurus (discussed in more detail below): the first to "triamcinolone hexacetonide" and the second to "Chronic Childhood Arthritis." From information in the Metathesaurus it is possible to determine that the first is a drug and the second a disease.

Identification of concepts provides an enriched representation of the meaning of text; however, an additional level of processing combines concepts into relationships that explicitly represent their interaction. These relationships are often called predications or propositions and are made up of arguments (concepts) and a predicate (relation). Processing to construct semantic predications (called semantic interpretation) determines in (3), for example, that "triamcinolone hexacetonide" treats (rather than causes) "Chronic Childhood Arthritis." Since the UMLS knowledge sources serve as an enabling resource for semantic interpretation in the biomedical domain, we discuss their main characteristics.

3. DOMAIN KNOWLEDGE: THE UMLS

The UMLS (Humphreys et al., 1998) consists of three components that provide structured knowledge in the biomedical domain: the SPECIALIST

Lexicon (McCray et al., 1994), the Semantic Network (McCray, 2003), and the Metathesaurus. The Lexicon supports syntactic analysis, while the Metathesaurus allows concepts to be identified in text; finally, the Semantic Network underpins the identification of semantic relationships.

3.1 SPECIALIST Lexicon

The SPECIALIST Lexicon describes syntactic characteristics of biomedical and general English terms, and this comprehensive resource provides the basis for NLP in the biomedical domain. In addition to part-of-speech labels for each entry, spelling variation when it occurs (particularly British forms) and inflection for nouns, verbs, and adjectives are included. Inflection is encoded by referring to rules for regular variants (*-s* for nouns and *-s*, *-ed*, *-ing* for verbs, for example) as well as Greco-Latin plurals. Irregular forms are listed where they apply. The variant annotation for *sarcoma* (4), for example, indicates that this form may either appear invariant (*sarcoma*), with a regular plural (*sarcomas*), or with Greco-Latin morphology (*sarcomata*).

(4) sarcoma
    cat=noun
    variants=uncount
    variants=reg
    variants=glreg

For verbs, complement patterns and nominalizations are included. The verb *manage* (5) takes regular verbal inflection and has nominalization *management*. It may occur with no object (intran), with a noun phrase object (tran=np), or with an infinitival complement, in which case the subject of *manage* is also the subject of the infinitive (tran=infcomp:subjc), as in *she managed to win the race*.

(5) manage
    cat=verb
    variants=reg
    intran
    tran=np
    tran=infcomp:subjc
    nominalization=management

3.2 Metathesaurus

The Metathesaurus is a compilation of more than 100 terminologies and controlled vocabularies in the biomedical domain, and includes those with comprehensive coverage, such as Medical Subject Headings (MeSH) and Systematized Nomenclature of Medicine (SNOMED), as well as those focused on subdomains such as dentistry (Current Dental Terminology) or nursing (Nursing Interventions Classification). Others provide specialized terms for components of the medical domain, such as anatomy (University of Washington Digital Anatomist) or medical devices (Universal Medical Device Nomenclature System).

Terms from the constituent vocabularies are organized into more than a million concepts (in the 2004 release) that reflect synonymous meaning. For example, the concept "Chronic Childhood Arthritis" contains synonymous terms "Arthritis, Juvenile Rheumatoid" (from MeSH and SNOMED) and "Rheumatoid arthritis in children" (Library of Congress Subject Headings), among others.

Hierarchical information inherent in component vocabularies is maintained in the Metathesaurus. For example, part of the structure for the concept "Juvenile Rheumatoid Arthritis" is given in (6).

(6) Immunologic Diseases
    Autoimmune Diseases
        Arthritis, Rheumatoid
            Arthritis, Juvenile Rheumatoid

Each concept in the Metathesaurus is assigned at least one semantic type, selected from 135 general categories relevant to the biomedical domain. Examples include 'Pharmacological Substance', 'Disease or Syndrome', 'Therapeutic or Preventive Procedure', and 'Amino Acid, Peptide, or Protein'.

Identical concepts with different meanings reflect word sense ambiguity in English, and such terms are distinguished in the Metathesaurus. For example, "Strains <1>" (with semantic type 'Injury or Poisoning') has synonyms "Muscle strain" and "Pulled muscle" and is distinguished from "Strains <2>" (semantic type 'Intellectual Product') with synonym "Microbiology subtype strains."

3.3 Semantic Network

The UMLS Semantic Network constitutes an upper-level ontology of medicine. Its components are the 135 semantic types assigned to Metathesaurus concepts as well as 54 relationships. The semantic types are organized into two hierarchies whose roots are 'Entity' and 'Event'. The two immediate children of 'Entity' are 'Physical Object' and 'Conceptual Entity', while 'Activity' and 'Phenomenon or Process' are immediately dominated by 'Event'. The hierarchical structure of the semantic type 'Pharmacologic Substance' is given in (7).

(7) Entity
  Physical Object
    Substance
      Chemical
        Chemical Viewed Functionally
          Pharmacologic Substance

Semantic types are also organized into higher level groups (McCray et al., 2001), which reflect semantic coherence among members. For example, the semantic group Disorders includes such semantic types as 'Acquired Abnormality', 'Disease or Syndrome', and 'Injury or Poisoning', while the group Procedures includes 'Diagnostic Procedure' and 'Therapeutic or Preventive Procedure'.

The 54 relationships in the Semantic Network are organized hierarchically under nodes that include PHYSICALLY_RELATED_TO (e.g. PART_OF and CONNECTED_TO), FUNCTIONALLY_RELATED_TO (e.g. DISRUPTS and TREATS), and CONCEPTUALLY_RELATED_TO (e.g. PROPERTY_OF and MEASURES). These relationships serve as the predicates of semantic predications whose arguments are semantic types. Some examples are given in (8).

(8) 'Therapeutic or Preventive Procedure' TREATS 'Injury or Poisoning'
'Organism Attribute' PROPERTY_OF 'Mammal'
'Body Space or Junction' CONNECTED_TO 'Tissue'
'Bacterium' CAUSES 'Pathologic Function'

The predications in the Semantic Network define a model of the medical domain and provide an important constraint on semantic interpretation.

4. SEMANTIC INTERPRETATION FOR THE BIOMEDICAL LITERATURE

4.1 Overview

Semantic interpretation relies on the identification of concepts in an outside knowledge structure and then determines relationships asserted between these concepts in text. We consider three approaches to semantic processing in the biomedical domain: AQUA (Johnson et al., 1993), PROTEUS-BIO (Grishman et al., 2002), and SemRep (Rindflesch and Fiszman, 2003). All three depend on biomedical knowledge sources and produce semantic predications as output. They differ primarily regarding the goals for which they were devised. They are based on varying linguistic formalisms and the particular knowledge sources used. Each system has specific strengths (and limitations). Given the challenges posed by natural language it is not possible for any system to produce a complete semantic analysis.

4.2 AQUA

AQUA (A QUery Analyzer) is an underspecified semantic interpreter that was originally devised for processing MEDLINE queries. The general approach is to identify salient medical concepts along with the syntactic phenomena that cue relations between them, without constructing a complete analysis. There are general principles for ignoring syntactic aspects of the input that are not directly concerned with key relations, such as *I am interested in articles about...*

The linguistic approach is based on operator grammar (Johnson and Gottfried, 1989), which provides rules for the ordering of operators and arguments in sentences. For example, the operator *with* occurs between its arguments in *patients with liver abscess*, while the operator *treatment* precedes its arguments in *the treatment of tuberculosis with rifampin.* Operator grammar supports a principled means of formulating generalizations that relate syntactic operator-argument patterns to underlying semantic predications.

The parsing formalism in AQUA is implemented as a definite clause grammar, which affords a flexible way of recognizing the argument-operator patterns defined by the operator grammar. This formalism allows both syntactic and semantic constraints to be included in parsing rules and also accommodates skipping part of the input. The parser depends on a lexicon that was derived from the UMLS (final editing was by hand). The AQUA lexicon contains semantic information (including semantic types) as well as part-of speech labels, and explicitly indicates whether an entry functions as an argument or an operator.

The combination of operator grammar, definite clause grammar, and semantic lexicon underpins AQUA's ability to map queries to semantic predications, which are represented as conceptual graphs, a more expressive form of the first-order predicate calculus (Sowa, 2000). For example the query (9) is interpreted as the proposition (10), which captures the key relations that infections and liver abscesses occur in patients who also have Hodgkin's disease.

(9) Request search for papers detailing infections, specifically liver abscesses, in patients with Hodgkin's disease

(10) [Pathologic Function: {infections, liver abscesses}] -
(occurs_in) → [Patient or Disabled Group: patients] -
(occurs_in) ← [Disease or Syndrome: Hodgkin's disease]

Semantic predications produced by AQUA have been validated against the UMLS Semantic Network. Recent work using AQUA focuses on semantic relations in clinical text and connecting that text with MEDLINE citations (Mendonça et al., 2002).

4.3 PROTEUS-BIO

PROTEUS-BIO is an information extraction system that depends on underspecified semantic interpretation as its core element. The system applies to Web documents on infectious disease outbreaks; it extracts semantic predications relevant to this domain and stores them in a database, which can be queried by users.

Semantic interpretation in PROTEUS-BIO identifies relationships pertinent to the domain, such as "outbreak of <disease> killed <victims>." Concepts in the entity classes in this domain, namely diseases, victims, and geographic locations, are stored in a hierarchical knowledge structure, which was specifically constructed for this application.

Initial processing concentrates on syntactic patterns to find the entities that can serve as arguments. In addition to noun phrases, verb groups such as *were killed* are identified. Noun phrases are labeled with semantic classes (such as <disease> or <victim>) during this phase and are then available to the next phase.

Processing to identify semantic predications is based on event patterns, which are defined in terms of the argument classes identified in the previous phase. For example, the pattern (11) matches the text (12).

(11) np(<disease>) vg(KILL) np(<victim>)

(12) Cholera killed 23 inhabitants

Additional patterns are defined to accommodate passive structures (based on the verb groups identified in the first phase). A metarule is designed to allow an event pattern to apply to text that includes adverbial constructions either before or after the components of the pattern. The metarule, for example, allows all the examples in (13) to match the event pattern (11), despite the occurrence of the adverbial expression *last week.*

(13) last week 23 inhabitants were killed by cholera
23 inhabitants were killed last week by cholera
23 inhabitants were killed by cholera last week

The accuracy of the semantic predications extracted by PROTEUS-BIO was evaluated on an annotated test collection of 32 documents. Precision was 79% and recall was 41%. As noted earlier, semantic processing in this system is meant to support information retrieval applications. Predications identified by PROTEUS-BIO are stored in a database and are linked to the documents from which they were extracted. It is thus possible to use this database to enhance the results of queries seeking documents in the disease outbreak domain. A task-oriented evaluation to measure effectiveness in achieving this goal was conducted, and initial results indicate that precision was notably increased using the PROTEUS-BIO system.

4.4 SemRep

SemRep is being developed to recover semantic propositions from the biomedical research literature (concentrating on MEDLINE citations) using underspecified syntactic analysis and structured domain knowledge. Processing begins with a lexical analysis based on the SPECIALIST Lexicon and a stochastic tagger. This serves as input to an underspecified parser, which provides the basis for semantic analysis (also underspecified). In analyzing (14), for example, after tokenization, the SPECIALIST Lexicon is consulted.

(14) Doppler echocardiography can be used to diagnose left anterior descending artery stenosis in patients with type 2 diabetes

Each lexical entry (including multiword forms like *Doppler echocardiology*) is assigned a part-of-speech label, and lexical ambiguities are assigned more than one label. For example, *used* has labels "verb" and "adj" in the lexicon, while *left* has "adj," "adv," "noun," and "verb."

A stochastic tagger (Smith et al., 2004) then resolves part-of-speech ambiguities based on common patterns seen in training data. The tagged text in (15) serves as input to the parser.

(15)

Doppler echocardiography	can	be	used	to	diagnose	left	anterior
noun	modal	aux	verb	adv	verb	noun	adj

descending	artery	stenosis	in	patients	with	type	2	diabetes
adj	noun	noun	prep	noun	prep	noun	num	noun

Note that taggers do not have 100% accuracy. For example, *left* should be tagged as an adjective in this context rather than as a noun.

The underspecified syntactic analysis is based on part-of-speech labels and segments the input into phrases that correspond to the lowest level structures in a full syntactic analysis. Segmentation is based on barrier words, which serve as boundaries between phrases. These include modals (*can* in the current example), auxiliaries (*be*), verbs (*used, diagnose*), and prepositions (*in, with*). The exploitation of these barriers in an algorithm that uses them to close one phrase and open another produces the analysis in (16). Any phrase containing a noun constitutes a (simple) noun phrase. The rightmost noun is relabeled as "head" and items to the left of the head (other than determiners and prepositions) are labeled as "mod."

(16) [[head('Doppler echocardiography')],
[modal(can)],
[aux(be)],
[verb(used)],
[adv(to)],
[verb(diagnose)],
[mod(left), mod(anterior), mod(descending), mod(artery),
head(stenosis)],
[prep(in), head(patients)],
[prep(with), head('type 2 diabetes')]]

Simple noun phrases constitute the referential vocabulary. The concepts they refer to in the domain model are computed by using MetaMap (Aronson, 2001) to match elements in each noun phrase to concepts in the UMLS Metathesaurus. MetaMap examines all the words in a phrase and then determines the best match with a term in the Metathesaurus, taking into account inflectional and derivational variation and allowing for partial and multiple mappings.

The phrases *Doppler echocardiography* and *patients*, for example, match exactly to concepts: "Echocardiography, Doppler" (with semantic type 'Diagnostic Procedure') and "Patients" ('Patient or Disabled Group'). The phrase *left anterior descending artery stenosis* maps to two concepts: "Anterior descending branch of left coronary artery" ('Body Part, Organ, or Organ Component') and "Acquired stenosis" ('Finding' and 'Pathologic Function'). When MetaMap has found a viable match between text words and a Metathesaurus term, it provides the preferred Metathesaurus name for that term, as in the case of the coronary artery mentioned here. Similarly, although the term *type 2 diabetes* occurs in the Metathesaurus, its preferred name is "Diabetes Mellitus, Non-Insulin-Dependent" ('Disease or

Syndrome'). Metathesaurus concepts for a noun phrase become a part of the representation of that phrase as semantic enhancement.

The interpretation of semantic predications asserted in the input text depends on the syntactic and semantic information contained in the underspecified parse structure enhanced with UMLS concepts and semantic types. Syntactic phenomena (including verbs, prepositions, nominalizations, and the head-modifier relation in noun phrases) "indicate" semantic predicates and are mapped to relations in the Semantic Network. The indicators in (14) are the verb *diagnose*, the prepositions *in* and *with*, and the modifier-head structure in the noun phrase whose head is *stenosis*.

Indicators are syntactic predicates that anchor the interpretation of syntactic structures as semantic predications, and two phenomena are involved in this process: argument identification and mapping to relations in the Semantic Network. Argument identification is controlled by a dependency grammar that establishes a syntactic relation between the indicator and the head of a simple noun phrase serving as its argument. Rules in this grammar are stated in very general terms for each class of indicator. For example, the argument identification rules for verbs stipulate that subjects occur to the left of the verb and objects to the right.

The syntactic constraint imposed by the dependency grammar serves as a necessary condition on the interpretation of a syntactic indicator and its arguments as a semantic predication. In (14), for example, the rules applied to *diagnose* limit the subject of this verb to the noun phrase *Doppler echocardiography*; the object, however, could be any of the three noun phrases to the right of *diagnose*: *left anterior descending artery stenosis*, *patients*, or *type 2 diabetes*. Further semantic conditions apply in determining which of these is the object of *diagnose* in (14).

All indicators are linked by rule to relations in the UMLS Semantic Network. The indicator rules needed to interpret (14) are given in (17); syntactic phenomena (part-of-speech or structure) occur to the left of the arrow and Semantic Network relations occur to the right.

(17) *diagnose* (verb) → DIAGNOSES
modifier-head (structure) → LOCATION_OF
in (preposition) → OCCURS_IN
with (preposition) → CO-OCCURS_WITH

The complete relationships, with semantic types as arguments, are given in (18) for the Semantic Network predicates in (17).

(18) 'Diagnostic Procedure' DIAGNOSES 'Pathologic Function'
'Body Part, Organ, or Organ Component' LOCATION_OF

'Pathologic Function'
'Pathologic Function' OCCURS_IN 'Patient or Disabled Group'
'Pathologic Function' CO-OCCURS_WITH 'Disease or Syndrome'

A metarule ensures that all semantic propositions identified by SemRep are sanctioned by a predication in the Semantic Network, and this restriction limits the identification of arguments. For example, the Semantic Network predication DIAGNOSES has the semantic type 'Pathologic Function' as one of its arguments. Therefore, any syntactic indicator linked to DIAGNOSES must have an argument whose head has been mapped to a Metathesaurus concept with the same semantic type. In (14), the only potential object of the verb *diagnose* that fulfills this requirement is the head of *left anterior descending artery stenosis* (whose semantic type is 'Pathologic Function'). *Doppler echocardiography* was identified syntactically as an argument of *diagnose*, and its semantic type, 'Diagnostic Procedure', matches the other argument of DIAGNOSES in the Semantic Network.

When these syntactic and semantic conditions are satisfied, a semantic predication can be constructed that is the interpretation of the syntactic indicator and its (syntactic) arguments. The predicate in this semantic proposition is the Semantic Network relation to the right of the arrow in the indicator rule; the arguments are the Metathesaurus concepts from the syntactic arguments of the indicator. In the case of the indicator *diagnose*, the predicate is DIAGNOSES and the arguments are the concepts "Echocardiography, Doppler" and "Acquired stenosis." The complete predication is

(19) Echocardiography, Doppler DIAGNOSES Acquired stenosis

When similar rules are applied to the other indicators in (14), namely the prepositions *in* (OCCURS_IN) and *with* (CO-OCCURS_WITH) and the head-modifier construction in the *stenosis* noun phrase (LOCATION_OF), the semantic propositions in (20) are produced.

(20) Acquired stenosis OCCURS_IN Patients
Acquired stenosis CO-OCCURS_WITH Diabetes Mellitus, Non-Insulin-Dependent
Anterior descending branch of left coronary artery LOCATION_OF Acquired stenosis

SemRep has recently been enhanced to address hypernymic propositions (Fiszman et al., 2003), in which a more specific concept is asserted to be in a taxonomic relation with a more general concept. For example, SemRep is able to extract the predication (22) as a representation of the relationship between *posaconazole* and *antifungal agent* in (21).

(21) Posaconazole is a potent broad-spectrum azole antifungal agent in clinical development for the treatment of invasive fungal infections.

(22) posaconazole ISA Antifungal Agents.

The interpretation of hypernymic predications depends on the arguments involved being in a hierarchical relationship in the Metathesaurus.

4.4.1 Evaluation of SemRep

Preliminary evaluation of SemRep has been conducted on a collection of 2,000 sentences from MEDLINE citations, concentrating on drug treatments for disease. Initial focus has been on a core set of semantic predicates, such as TREATS, LOCATION_OF, CO-OCCURS_WITH, and OCCURS_IN. Precision and recall on this test collection are 78% and 49% respectively. The majority of the false positive errors (contributing to diminished precision) are due to word sense ambiguity. For example, in (23), *concentration* maps to the corresponding Metathesaurus concept with semantic type 'Mental Process'.

(23) . . .the mean fluorescein concentration in the cornea of the lyophilisate group was two times higher than at baseline.

This mapping allows the incorrect predication (24) to be constructed, in which the cornea is interpreted as the location of a mental process.

(24) Cornea <1> LOCATION_OF Concentration

A significant percentage of false negative errors are due to current deficiencies in processing comparative structures. For example, SemRep retrieves the predication (26) while interpreting (25), but fails to identify that co-trimoxazole treats pneumonia, which is also asserted in the sentence.

(25) The purpose of this study was to compare the clinical effectiveness of co-trimoxazole with amoxicillin for treatment of childhood pneumonia

(26) Amoxicillin TREATS Pneumonia

4.5 Comparison of AQUA, PROTEUS-BIO, and SemRep

The three systems discussed are intended to provide useful results, without attempting a full semantic analysis. AQUA uses operator grammar to manipulate traditional syntactic constituent structure. The flexibility of this formalism allows the system to focus on grammatical structures relevant to the interpretation of users' queries to MEDLINE. Domain knowledge used by AQUA is based on the UMLS, and a wide range of semantic topics are accommodated. PROTEUS-BIO is intended to retrieve timely information from Web documents in a specific content area, namely infectious disease outbreaks. It uses partial constituent structure for noun phrases and verb groups, along with robust pattern matching in cooperation with specially constructed knowledge sources to achieve practical results in a limited domain. SemRep also relies on partial constituent structure, and in addition uses an underspecified dependency grammar for argument identification. It exploits the UMLS knowledge sources without modification. Although limited in the semantic relations it addresses, SemRep applies to a wide range of syntactic structures asserting the treatment of disease in the biomedical research literature.

5. APPLICATION OF SEMREP

Above, we briefly mentioned applications for semantic interpretation in the discussion of AQUA and PROTEUS-BIO. We now consider recent applications of SemRep. This program serves as the basis for several ongoing research initiatives in biomedical information management, including efforts directed at automatic summarization of the results of PubMed searches and extracting molecular biology information from text.

5.1 Automatic Summarization

Automatic summarization is an important emerging application in the biomedical domain. With the growing emphasis on evidence-based medicine it is important for physicians to keep abreast of the research literature. This is challenging due to the large size of the MEDLINE database. For example, a PubMed query on the treatment of diabetes, limited to articles published in 2003 and having an abstract in English, finds 3,621 items; further limitation to articles describing clinical trials still returns 390 citations.

One goal of automatic summarization in biomedicine is to provide practitioners with current, focused information on the treatment of specific diseases, including summaries with pointers to the most relevant citations.

SemRep is being used as the basis for an automatic summarization application in the abstraction paradigm (Fiszman et al., 2004), in which the semantic interpretation of text is manipulated, rather than the text itself (extraction summarization).

The system we are developing takes as input a list of semantic predications produced by SemRep from a set of documents on a specified disorder topic. The output is a conceptual condensate (in graphical format) containing just those predications that represent key information in the input documents. There are links to the original text that generated the propositions.

The core of the method is a transformation process that condenses and generalizes the input predications, guided by four principles (27) that use semantic information from the UMLS and frequency of occurrence of concepts and relations in the input predications.

(27) Relevance: Include predications on the topic of the summary
Novelty: Do not include predications that the user already knows
Connectivity: Also include "useful" additional predications
Saliency: Only include the most frequently occurring predications

Relevance processing condenses the list of predications by ensuring that they conform to a schema describing disorders (Jacquelinet et al., 2003) that contains general statements such as "{Treatment} treats {Disorders}." "Domains" such as {Treatment} and {Disorder} define sets of UMLS semantic types derived from the semantic groups. Predications conforming to the schema are called "core predications." Novelty provides further condensation by eliminating predications having generic arguments, as determined by hierarchical depth in the Metathesaurus. For example, predications containing arguments such as "Patients" and "Pharmaceutical Preparations" are eliminated by the Novelty principle.

Connectivity is a generalization process that identifies predications occurring in neighboring semantic space of the core, namely non-core predications that share an argument with a core predication. For example, from "Naproxen TREATS Osteoarthritis," non-core predications such as "Naproxen ISA NSAID" are included in the condensate. Finally, the Saliency principle calculates frequency of occurrence of arguments, predicates, and predications; those occurring less frequently than the average are eliminated from the final condensate (Hahn and Reimer, 1999).

Figure 14-1 is a conceptual condensate summarizing the 300 most recent citations retrieved by a PubMed search using the query "Diabetes Mellitus, Type II" (a MeSH term).

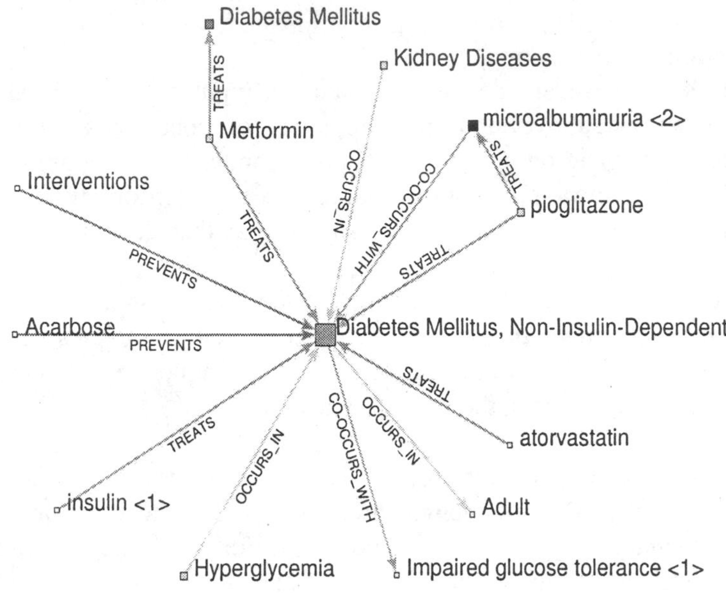

Figure 14-1. Conceptual condensate summarizing 300 citations on type 2 diabetes

SemRep generated 3,092 semantic predications from the input documents, and the transformation process reduced these to 73 predications (only the unique types are given in Figure 14-1).

The summary of type 2 diabetes given in Figure 14-1 provides an overview of the latest research on interventions for this disorder. Insulin is becoming increasingly important in this regard and is included in the summary. Traditionally, oral pharmacotherapy has been the treatment of choice, as shown by the appearance of metformin in the condensate. New drugs such as pioglitazone (thiazolinediones) and acarbose (both are included in Figure 14-1) are showing promise in either treating or preventing type 2 diabetes.

The conceptual condensate can be viewed from the perspective of citations rather than predications and doing so may have implications for improving information retrieval effectiveness. Of the 300 citations summarized, 52 contributed at least one predication to the final condensate. The three citations that contributed at least four predications are all highly relevant to the treatment of type 2 diabetes. For example, one of these has the title "Effect of antidiabetic medications on microalbuminuria in patients with type 2 diabetes." Of the citations that contributed a single predication to the conceptual condensate, only one directly discusses the treatment of type

2 diabetes; others are about related issues, for example: "Persistent remodeling of resistant arteries in type 2 diabetic patients on anti-hypertensive treatment."

5.2 Information Extraction in Molecular Genetics

A second application of SemRep currently being pursued investigates the use of NLP for studying the etiology of genetic diseases. The focus of this work is to identify semantic predications in the research literature that assert a relationship either between a gene and a disease or between two genes implicated in a disease. The underlying technology is a program called SemGen (Rindflesch et al., 2003), which is a modification of SemRep. SemGen has the same core structure as SemRep, and processing other than mapping noun phrases to semantic concepts is identical in the two programs.

While enhancing SemRep to construct SemGen, a program called ABGene (Tanabe and Wilbur, 2002) was added in order to augment MetaMap processing for genetic terminology. ABGene is based on part-of-speech tagging technology and uses several statistical and empirical methods to identify gene names that do not occur in the UMLS Metathesaurus.

The domain knowledge underpinning semantic interpretation specific to the etiology of genetic diseases that SemGen relies on was constructed by hand. This knowledge substitutes for the UMLS Semantic Network in SemRep. The allowable arguments of the semantic predications addressed by SemGen are characterized by two semantic classes: disorders and genetic phenomena. Disorders are defined as concepts having the UMLS semantic types in the Disorder semantic group. For genetic phenomena, concepts with semantic types from the semantic group Gene (including semantic type 'Gene or Genome', for example) are augmented with output from ABGene.

The relevant predicates for gene-disease relationships are ASSOCIATED_WITH, PREDISPOSE, and CAUSE. The subject of these predicates is a genetic phenomenon and the object is a disorder. Predicates defined for gene interactions are INTERACT_WITH, STIMULATE, and INHIBIT. Both arguments of these predicates are genetic phenomena. For example, SemGen extracts the gene-disease interaction predication (29) from (28) and the gene-gene predication (31) from (30).

(28) An elevated frequency of the CYP2D6*4 allele has been found in Parkinson's disease.

(29) cyp2d6*4 allele ASSOCIATED_WITH Parkinson Disease

(30) PDX-1 interacts with multiple transcription factors and coregulators, including the coactivator p300, to activate the

transcription of the insulin gene and other target genes within pancreatic beta cells.

(31) pdx-1 STIMULATE insulin

We are pursuing research on several fronts that exploits SemGen output in bioinformatics applications. One project compares a curated database to the current literature. OMIM (Online Mendelian Inheritance in Man) is an information resource on genetic diseases that has nearly 15,000 hand-curated entries describing clinical phenotypes and associated genes. We have used SemGen output as the basis for comparing OMIM entries on a particular disorder to MEDLINE citations (Libbus et al., 2004). The goal was to explore the possibility of automatically suggesting recent research to supplement OMIM information.

For example, we ran SemGen on OMIM text for Alzheimer's disease and also on the output of a PubMed query on that disorder, limited to citations that postdate the most recent OMIM entry. We then automatically compared the SemGen predications from OMIM to those from MEDLINE. We were most interested in discovering predications that occurred in MEDLINE, but not in OMIM, and the following are examples of such predications.

(32) TGFB1 ASSOCIATED_WITH Amyloid deposition
MAPT INTERACT_WITH HSPA8
CD14 STIMULATE amyloid peptide

On the basis of this kind of output, SemGen can potentially serve as an important tool for researchers in scanning a large number of citations and providing information that could promote hypothesis generation and scientific discovery.

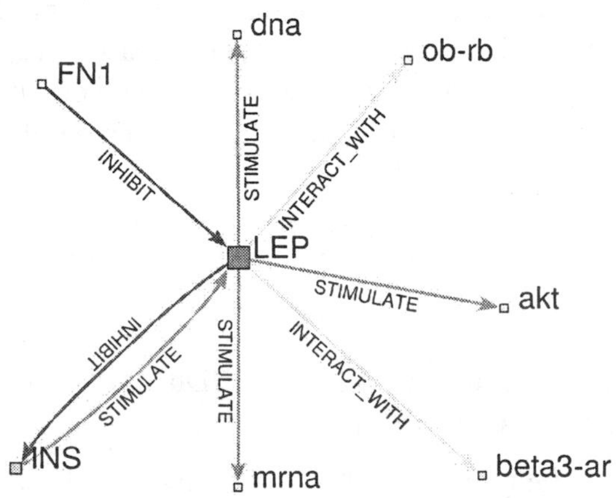

Figure 14-2. Some LEP gene interactions extracted by SemGen from text on diabetes

Finally, visualization techniques can be used to construct gene-gene interaction networks automatically from predications extracted from text by SemGen. Such networks provide an easily accessible overview of the molecular mechanisms implicated in genetic disease. As an example, Figure 14-2 is a partial network for some of the predications describing the genes that interact with the leptin gene (LEP). The relationships illustrated were extracted from documents discussing diabetes and genes and may provide insight into the genetic underpinnings of that disorder.

Figure 14-2, for example, indicates that LEP inhibits insulin (INS), while INS stimulates LEP. This feedback relationship is involved in appetite suppression and is perturbed during diabetes or obesity. Further, LEP, which is elevated in obesity, stimulates the gene AKT, which ultimately leads to the formation of new vessels underlying diabetic retinopathy.

6. CONCLUSION

The development of NLP systems for semantic interpretation in the biomedical research literature is motivated by the need to support emerging applications that focus on the manipulation of information rather than documents. Implemented systems address a range of information management tasks, including automatic summarization, connecting patient records with the research literature, question answering, literature-based scientific discovery, and the extraction of information to support molecular biology research, as well as enriched query processing and document manipulation.

A variety of linguistic formalisms are used for semantic processing. Due to the complexity of natural language, practical systems focus on biomedical subdomains as well as specific syntactic structures and semantic relations. The identification of semantic concepts and predications in the research literature relies on structured domain knowledge, such as the UMLS. This large resource includes lexical information to support NLP, and the content it contains is organized hierarchically and as an upper-level ontology of biomedicine.

Two examples of the application of semantic interpretation to the biomedical research literature include automatic summarization for the treatment of disease and extraction of molecular biology information on the etiology of genetic disorders. Visualization techniques can profitably be used to give users an overview of extracted information. Continued development of semantic processing systems in biomedicine promises to provide professionals with more powerful tools for effectively exploiting online textual resources.

REFERENCES

Aronson, A. R. (2001). "Effective Mapping of Biomedical Text to the UMLS Metathesaurus: The MetaMap Program," in *Proceedings of the AMIA Symposium*, 17-21.

Cimino, J. J. (1996). "Linking Patient Information Systems to Bibliographic Resources," *Methods of Information in Medicine,* 35, 122-6.

Cimino, J. J. and Barnett, G. O. (1993). "Automatic Knowledge Acquisition from MEDLINE," *Methods of Information in Medicine,* 32, 120-30.

Christensen, L., Haug, P. J., and Fiszman, M. (2002). "MPLUS: A Probabilistic Medical Language Understanding System," *ACL Workshop on Natural Language Processing in the Biomedical Domain,* 29-36.

Fiszman, M., Rindflesch, T. C., and Kilicoglu, H. (2003). "Integrating a Hypernymic Proposition Interpreter into a Semantic Processor for Biomedical Text," in *Proceedings of the AMIA Symposium*, 239-43.

Fiszman, M., Rindflesch, T. C., and Kilicoglu, H. (2004). "Abstraction Summarization for Managing the Biomedical Research Literature," in *Proceedings of the Workshop on Computational Lexical Semantics*, 76-83. HLT-NAACL.

Fuller, S., Revere, D., Bugni P., and Martin, G.M. "Telemakus: A Schema-based Information System to Promote Scientific Discovery," *Journal of the American Society for Information Science and Technology,* in press.

Friedman, C., Alderson, P. O., Austin, J. H., Cimino, J. J., and Johnson, S. B. (1994). "A General Natural-language Text Processor for Clinical Radiology," *Journal of the American Medical Informatics Association*, 1(2), 161-74.

Grishman, R., Huttunen, S., and Yangarger, R. (2002). "Information Extraction for Enhanced Access to Disease Outbreak Reports," *Journal of Biomedical Informatics,* 35(4), 236-46.

Hahn, U. and Reimer, U. (1999). "Knowledge-based Text Summarization: Salience and Generalization Operators for Knowledge Base Abstraction," in I. Mani (Ed.), *Advances in Automatic Summarization*, Cambridge, MA: MIT Press. 215-32.

Hahn, U., Romacker, M., and Schulz, S. (2002). "MEDSYNDIKATE—Design Considerations for an Ontology-based Medical Text Understanding System," in *Proceedings of the AMIA Symposium*, 330-4.

Humphreys, B. L., Lindberg, D.A., Schoolman, H.M., and Barnett, G.O. (1998). "The Unified Medical Language System: An Informatics Research Collaboration," *Journal of the American Medical Informatics Association,* 5(1), 1-11.

Jacquelinet, C., Burgun, A., Delamarre, D., Strang, N., Djabbour, S., Boutin, B., and Le Beux, P. (2003). "Developing the Ontological Foundations of a Terminological System for End-stage Diseases, Organ Failure, Dialysis and Transplantation," *International Journal of Medical Informatics,* 70(2-3), 317-28.

Jacquemart, P. and Zweigenbaum P. (2003). "Towards a Medical Question-answering System: A Feasibility Study," *Stud Health Technol Inform,* 95, 463-8.

Johnson, S. B., Aguirre, A., Peng, P., and Cimino, J. J. (1993). "Interpreting Natural Language Queries Using the UMLS," in *Proceedings of the AMIA Symposium,* 294-8.

Johnson, S. B. and Gottfried, M. (1989). "Sublanguage Analysis as a Basis for Controlled Medical Vocabulary," *SCAMC*, 519-23.

Leroy, G., Chen, H., and Martinez, J.D. (2003). "A Shallow Parser Based on Closed-class Words to Capture Relations in Biomedical Text," *Journal of Biomedical Informatics,* 36(3), 145:58.

Libbus, B., Kilicoglu, H., Rindflesch, T. C., Mork, J. G., and Aronson, A. R. (2004). "Using Natural Language Processing, Locus Link, and the Gene Ontology to Compare OMIM to

MEDLINE," in *Proceedings of the Workshop on Linking the Biological Literature, Ontologies and Databases: Tools for Users*, 69-76. HLT-NAACL.

McCray, A. T. (2003). "An Upper-level Ontology for the Biomedical Domain," *Comp Funct Genom*, 4, 80-4.

McCray, A. T., Burgun, A., and Bodenreider, O. (2001). "Aggregating UMLS Semantic Types for Reducing Conceptual Complexity," in *Medinfo*, 10(Pt 1), 216-20.

McCray, A. T., Srinivasan, S., and Browne, A. C. (1994). "Lexical Methods for Managing Variation in Biomedical Terminologies," *SCAMC*, 235-9.

Mendonça, E. A. and Cimino, J. J. (2000). "Automated Knowledge Extraction from MEDLINE Citations," in *Proceedings of the AMIA Symposium*, 575-9.

Mendonça, E. A., Johnson, S. B., Seol, Y., and Cimino, J. J. (2002). "Analyzing the Semantics of Patient Data to Rank Records of Literature Retrieval," *ACL Workshop on Natural Language Processing in the Biomedical Domain*, 69-76.

Pakhomov, S. V., Ruggieri, A., and Chute, C. G. (2002). "Maximum Entropy Modeling for Mining Patient Medication Status from Free Text," in *Proceedings of the AMIA Symposium*, 587-91.

Rindflesch, T. C., and Fiszman, M. (2003). "The Interaction of Domain Knowledge and Linguistic Structure in Natural Language Processing: Interpreting Hypernymic Propositions in Biomedical Text," *Journal of Biomedical Informatics*, 36(6), 462-77.

Rindflesch, T. C., Libbus, B., Hristovski, D., Aronson, A. R., and Kilicoglu, H. (2003). "Semantic Relations Asserting the Etiology of Genetic Diseases," in *Proceedings of the AMIA Symposium*, 554-8.

Smith, L., Rindflesch, T., and Wilbur, W. J. (2004). "MedPost: A Part of Speech Tagger for Biomedical Text," *Bioinformatics*, in press.

Sowa, J. F. (2000). *Knowledge Representation*, Pacific Grove: Brooks/Cole.

Srinivasan P. and Libbus, B. (2004). "Mining MEDLINE for Implicit Links Between Dietary Substances and Diseases," *Bioinformatics*, 20, i290-i296.

Taira, R. K., and Soderland, S. G. (1999). "A Statistical Natural Language Processor for Medical Reports," in *Proceedings of the AMIA Symposium*, 970-4.

Tanabe, L. and Wilbur, W. J. (2002). "Tagging Gene and Protein Names in Biomedical Text," *Bioinformatics*, 18(8), 1124-32.

SUGGESTED READINGS

Friedman, C. and Hripcsak, G. (1999). "Natural Language Processing and its Future in Medicine," *Acad Med*, 74(8), 890-5.

Reviews several current NLP methodologies in biomedicine with discussion of potential applications.

Jurafsky, D. and Martin J.H. (2000). *Speech and Language Processing: An Introduction to Natural Language Processing, Computational Linguistics, and Speech Recognition*, 1st ed., Upper Saddle River: Prentice Hall.

Provides an overview of techniques for natural language processing, both rule-based and statistical approaches.

Rindflesch, T. C. and Aronson, A.R. (2002). "Semantic Processing for Enhanced Access to Biomedical Knowledge," in *Real World Semantic Web Applications*, V. Kashyap and L. Shklar (Eds.), IOS Press, 157-72.

Gives an overview of SemRep and MetaMap with examples of their application.

ONLINE RESOURCES

UMLS documentation:
http://www.nlm.nih.gov/research/umls/documentation.html

Semantic Knowledge Representation Project at NLM:
http://skr.nlm.nih.gov

SPECIALIST Lexicon and lexical tools:
http://specialist.nlm.nih.gov

MetaMap Transfer (MMTx):
http://mmtx.nlm.nih.gov

QUESTIONS FOR DISCUSSION

1. Why is it important to pursue research on semantic processing of the biomedical literature? Discuss biomedical applications (other than those noted in this chapter) that could benefit from semantic representation.

2. What are the levels of knowledge required for semantic processing? List the steps required for semantic interpretation of "low dose aspirin for the prevention of myocardial infarction," if SemRep is used as the semantic processor.

3. Discuss strengths and limitation as well as similarities and differences of the systems designed to provide semantic interpretation of the biomedical literature (AQUA, PROTEUS-BIO, and SemRep).

4. What is automatic summarization and why is it important in the biomedical domain? What is the importance of semantic processing as the basis for automatic summarization?

5. Discuss differences between task-oriented evaluation of semantic processing and evaluation of the accuracy of semantic predications identified in text. Which one do you think is harder and why?

Chapter 15

SEMANTIC TEXT PARSING FOR PATIENT RECORDS

Carol Friedman

Department of Biomedical Informatics, Columbia University, New York, New York 10027

Chapter Overview

Accessibility to a comprehensive variety of different types of structured patient data is critical to improvement in the health care process, yet most patient information is in the form of narrative text. Semantic methods are needed to interpret and map clinical information to a structured form so that the information will be accessible to other automated applications. This chapter focuses on semantic methods that map narrative patient information to a structured coded form.

Keywords

natural language processing; data mining; electronic patient record; automated coding

"...better management of clinical information is a prerequisite for achieving patient safety and improved care ... Because health care data are often narrative, natural language processing (NLP) is another important technique for mining data for quality improvement and patient safety purposes"

Institute of Medicine Report on Patient Safety, *Achieving a New Standard of Care*, 2003

1. INTRODUCTION

Advancement of health care is dependent on integration, organization, and utilization of massive amounts of genomic, pharmacological, cellular, tissular, environmental, and clinical information. The promise of the electronic health record (EHR) and the Clinical Data Architecture (CDA) (Dolin, Alschuler, et. al., 2001), an XML-based standardized exchange model for patient records, is that they will lead to substantial improvements in health care and biological research by facilitating the harnessing and accessibility of information in the EHR. The EHR is mainly expressed using natural language (i.e. narrative text), which is the primary means for communicating patient information in the patient reports. Although the content of patient reports includes a rich source of data, it is also a major bottleneck hindering widespread deployment of effective clinical applications because textual information is difficult if not impossible to reliably access by computerized processes. Natural language processing (NLP) systems are automated methods containing some linguistic knowledge that aim to improve the management of information in text. NLP systems have been shown to be successful for realistic clinical applications, such as decision support, surveillance of infectious diseases, research studies, automated encoding, quality assurance, indexing patient records, and tools for billing. Currently, NLP systems in clinical environments process patient records to: 1) index or categorize reports, 2) extract, structure, and codify clinical information in the reports so that the information can be used by other computerized applications, 3) generate text to produce patient profiles or summaries, and 4) improve interfaces to health care systems.

In a recent report, the Institute of Medicine recognized that NLP is potentially a very powerful technology for the medical domain because it enables a new level of functionality for health care applications that would not be otherwise possible (Institute of Medicine Report, 2003). NLP provides a method whereby large volumes of text reports (i.e. all textual patient reports) can be processed and the clinical information in the reports automatically encoded in a timely manner. It is not possible to have experts manually capture and encode a broad range of clinical information in the reports because it is too costly and time consuming, although limited coding is now being done manually. For example, primary and secondary diagnoses and procedures are coded manually for billing purposes. Once clinical information is encoded, it will be possible to develop a wide range of automated high throughput clinical applications, which should become invaluable tools that assist clinicians and researchers. These applications depend on structured data, and therefore are not currently feasible on a large scale. For example, an automated system could process enormous volumes of reports to detect medical errors, whereas it would not be possible for experts to check such large volumes. Table 15-1 lists some queries that could be reliably answered using output from an NLP system. For example, it would be possible to check if a patient may have *pneumonia* based on information occurring in a radiological report of the chest. To answer this query accurately, it would be inadequate to perform only a keyword search for *pneumonia* because other contextual information in the report associated with pneumonia would also be required for a correct interpretation. For example by using a query with only the keyword *pneumonia*, false positives would be obtained if *pneumonia* 1) was in the process of being ruled out (e.g. *workup to rule out pneumonia*), 2) was ruled out (e.g. *no evidence of pneumonia*), 3) occurred in the past (e.g. *history of multiple admissions for pneumonia*), or 4) was associated with a family member (e.g. *father had pneumonia*). In addition, the query would also have to include clinical knowledge in order to associate descriptive findings, such as *consolidation,* with the condition *pneumonia*. Such expert knowledge would be necessary because descriptive findings indicative of *pneumonia* may occur in a report whereas the term *pneumonia* may not. The clinical rules that associate findings with clinical conditions typically consist of a logical combination of multiple findings. Depending on the condition and type of report, the rules could be quite complex. Because the output of an NLP system is consistent, NLP technology facilitates the development of automated alerting systems that are based on structured NLP output. An automated alerting system will consist of rules that are determined by experts, and it will be possible to incorporate the best and most current medical knowledge into the rules, promoting consistency, objectivity, and evidence-based medicine. Another significant advantage is that NLP technology can be used to standardize reports from diverse institutions and applications because the same automated system will uniformly encode clinical information that occurs in heterogeneous reports, thereby facilitating interoperability.

Table 15-1. Examples of queries that can be answered using structured output obtained by an NLP system as a result of processing textual patient reports

Does the patient have a particular clinical condition?
What medications is the patient on?
What problems does the patient currently have?
What procedures were performed on this patient?
Are there changes in a particular condition?
What problems did the patient have in the past?

This chapter will first present an overview of NLP in the clinical domain and describe challenges associated with processing clinical reports. Additionally, background information will be discussed concerning the current state of the art in this domain. The chapter will end with a case scenario, describing the processing of a clinical narrative and the utilization of the output for an alerting clinical application.

2. OVERVIEW

2.1 Challenges of Processing Clinical Reports

NLP in the clinical domain has multiple challenges because of the health care setting, which we summarize below, and it is important for an NLP system to address such challenges if it is to be deployed in a clinical setting. A more detailed discussion can be found in (Friedman and Johnson, 2005).

2.1.1 Performance

Because the output of an NLP system will be employed by a healthcare application, it must have adequate recall, precision, and specificity for the intended clinical application, but it should be possible to adjust its performance according to the needs of the application. Different applications require varying levels of performance, which means that a clinical application involving NLP will have to undergo an evaluation before actually being deployed to ensure that the performance is appropriate.

2.1.2 Availability of Clinical Text and Confidentiality

Development of an NLP system is based on analysis (manual or automated) of samples of the text (i.e. a training set) to be processed. In the clinical domain, this means that large collections of online patient records in textual form must be available to the developers. However, patient records are confidential, and in order to make them accessible for research purposes, personal identifying information must be removed to comply with laws protecting patient confidentiality. Automated detection of identifying information within the text of the clinical reports, such as names, addresses, phone numbers, unique characteristics (i.e. mayor of New York), is an extremely difficult task that often requires manual review, although even after removal of names, addresses, etc. identification may still be possible, because as discussed by Sweeney et. al., rare characteristics may occur in the report that help identify a patient (Sweeney, 1997). Additionally, even if the data were manually checked, approval to use the records must be obtained from an Institutional Review Board and from institutional administrators.

2.1.3 Intra- and inter-operability

In order to be disseminated, an NLP system has to be able to function well in different health care facilities and for different clinical applications. It also has to be seamlessly integrated into a Clinical Information System, and generate output that is in a form usable by other components of the system. This generally means that the system will have to handle different interchange formats (i.e. XML, HL7) and heterogeneous formats that are associated with the different types of reports (i.e. formats of radiology reports, discharge summaries, echocardiograms, and pathology reports are generally different). An additional problem is that the NLP system will have to generate output that can be stored in an existing clinical repository, but the output often has complex and nested relations, and it may be impossible to map the NLP output to the schema of the clinical database without substantial loss of information. One more serious challenge is that to achieve widespread deployment, the NLP output has to be comparable so that it can be used across institutions for a variety of automated applications. This means it must be mapped to a controlled vocabulary and to a standard representation for the domain. Although different clinical vocabularies exist (i.e. UMLS, ICD-9, SNOMED-CT), none are complete and there is no single standard. An equally serious problem is that although there are standard controlled vocabularies, there is no standard representational model for medical language, and a representational model is also essential in order to interpret the underlying meaning of the clinical information in the reports and relationships among the information. Such a model would include relations between separate clinical concepts. For example, *treats*, is a relation between a medication event and a disease event in *On Asmacort for asthma*. Similarly, *suggestive of* is a relation between a diagnostic event and a disease event in *Chest x-ray suggestive of pneumonia*. In addition to relations among events there is information that modifies an individual event, such as negation (e.g. *no evidence of pneumonia*), certainty (e.g. *possible pneumonia*), severity (e.g. *mild cough*), change (e.g. *worsening cough*), and temporal information (e.g. *past history of pneumonia*). In 1994, the Canon Group (Evans, Cimino, et. al. 1994) attempted to merge different representational models to create a widely used model for medical language. That effort resulted in a common model for radiological reports of the chest (Friedman, Huff, et. al. 1995), but it has not been adopted by the community.

2.1.4 Evaluation

Evaluation of an NLP system is critical but difficult in the healthcare domain because of the difficulty of obtaining a gold standard and because it is difficult to share the data across institutions. A fuller discussion on evaluation of NLP systems can be found in (Friedman and Hripcsak, 1998; Hripcsak, Austin et. al., 2002). Usually, there is no gold standard available that can be used to evaluate the performance of an NLP system. Therefore, for each evaluation, recruitment of subjects who are medical experts is generally required to obtain a gold standard for a test set. Obtaining a gold standard for this type of evaluation is very time consuming and costly.

2.1.5 Expressiveness

Language is extremely expressive in the sense that there are often different ways to describe the same medical concept and also numerous ways to express modifiers of the concept. For example, findings associated with cancer can be expressed using a very large number of terms, such as *neoplasm, tumor, lesion, growth, mass, infiltrate, metastasis, lymphoma, carcinoma, etc.* Similarly, a modifier, such as certainty information, is associated with more than 800 different phrases in the MedLEE lexicon, with terms such as *conceivable, definite, borderline, questionable, convincing evidence for, unlikely,* and *negative for.* Modifiers make it more complex to retrieve reports based on NLP structured output since they have to be accounted for, but then the query for retrieving the information will have a very fine granularity.

2.1.6 Heterogeneous Formats

There is no standardized structure for reports. Although sections and subsections of reports are important for many applications because they provide context, their names have not been standardized. For example, in New York Presbyterian Hospital (NYPH), there are many different section headers for reporting diagnostic tests (i.e. *Diagnostic Studies, Examination, Examination Type, Studies Performed*). Sometimes section headers are omitted or several sections are merged into one. For example, family and social history is occasionally reported in the History of Present Illness section. The Clinical Document Architecture (CDA) is an effort to address this problem because its aim is to establish standards for the structure of clinical reports (Dolin, Alschuler, et. al., 2001). Another problem occurs because the format for text within the reports is not standardized. Punctuation is often missing or is inappropriate, and a new line may be used instead of a period to signify the end of a sentence. An additional problem is that some reports contain tables with different configurations as well as text. Structured fields, such as those of a table, are easy for a human to interpret but are very problematic for a general NLP program because formatting characteristics rather than phenomena associated with language determine the meaning of the fields and their relations.

2.1.7 Abbreviated Text

Generally, clinical reports are very compact, contain abbreviations, and often omit information that can easily be inferred by health care professionals based on their knowledge of the context and the domain. One problem with abbreviations is that they are highly ambiguous. For example, *pe,* may mean *physical examination, pleural effusion,* or *pulmonary embolism.* A clinical note, as shown in Figure 15-3, may have numerous abbreviations, which typically will occur in many reports (e.g. *yo, F, hx, HTN, COPD, CRI*). An additional problem is that a unique abbreviation may be defined in a single report. Omitted or implicit information present another challenge because an automated system, which utilizes the structured information generated from the reports, would have to automatically capture the implicit information based on knowledge of the domain, and this is a very complex task. A simple example is that the body location is frequently missing in a report of a particular diagnostic examination. For example, when *mass* occurs in a radiological report of the chest, it means *mass in lung* whereas if it occurs in a mammography report, it means *mass in breast.*

2.1.8 Interpreting Clinical Information

Clinical information in a report is important for an application, but frequently additional medical knowledge along with knowledge of the report structure is needed in order to associate findings with possible diagnoses. Interpretations of the findings also vary depending on the type of report and section. For example, retrieving information from the Admission Diagnosis section of a discharge summary is generally more straightforward than retrieving information from the Description section of a radiological report. Radiological reports typically do not have definitive diagnoses, and contain a continuum of findings that range from patterns of light (e.g. *patchy opacity*), to descriptive findings (e.g. *focal infiltrate*) to possible diagnoses (e.g. *pneumonia*). In some radiological reports, only the descriptive findings may be present and there may be no interpretation by the reporting radiologist (e.g. a finding *pneumonia* may actually be included in a report; instead, findings consistent with *pneumonia*, such as *consolidation* or *infiltrate* may occur). Therefore, in order to use an NLP system to detect pneumonia based on chest x-ray findings, the NLP system or application using the system would have to contain medical knowledge associated with findings that are suggestive of pneumonia. For example, two systems developed by Fiszman and colleagues (Fiszman and Haug, 2000) and Hripcsak and colleagues (Hripcsak, Friedman, et. al., 1995) to detect patients with possible *pneumonia* from chest x-ray reports contain components that are used to infer *pneumonia* from the findings. However, it is also important to account for contextual information. Hripcsak and colleagues (Hripcsak, Friedman, et. al., 1995) found that the occurrence of pneumonia in the Clinical Information section was ambiguous: it could signify that 1) the patient had pneumonia based on the current examination, 2) the examination was a follow up examination for a patient with a known case, or 3) pneumonia was being ruled out. In contrast, if a finding of *pneumonia* occurs in the Impression or Description section, it is associated with findings in the current x-ray. Similarly, *history of heart disease* in the Family History section of a discharge summary does not mean the patient has heart disease. The rules needed to detect a particular condition based on output generated by an NLP system can be quite complex. In order to develop such a component, machine learning techniques can be used. This involves collecting instances of positive and negative samples, which would be used to develop rules automatically (Wilcox and Hripcsak, 1999), but this may be costly, since performance is impacted by sample size (McKnight, Wilcox, et. al., 2002) and for many conditions, a large number of instances would have to be obtained for satisfactory performance. An alternative involves having an expert manually write the rules by observing the target terms that the NLP system generates along with sample output. In that case, the rules will generally consist of combinations of Boolean operators (e.g. and, or, not) and findings. For example, a rule written by Chuang and colleagues (Chuang, Friedman, et. al., 2002), which detects a comorbidity of neoplastic disease, consisted of a Boolean combination of over 200 terms.

2.1.9 Rare Events

Natural language systems generally need a large number of training examples to train, refine, or test the system. Since some events occur rarely, it may be difficult to find a large number of reports for these events. Terminological knowledge sources, such as the UMLS (Lindberg, Humphreys, et. al., 1993) and the Specialist Lexicon (Browne, Divita et. al., 2003), may be helpful for providing lexical knowledge for rare clinical terms, but they may not include the variety of phrases that occur in natural language text.

2.2 Components of an NLP System

There are different approaches to NLP in the clinical domain. Most approaches use a combination of syntactic and semantic linguistic knowledge as well as heuristic domain knowledge, but they vary as to the extent of each type of knowledge and as to how the different types are integrated. Some use manually developed rules, and others are more statistically oriented. Figure 15-1 shows a high level overview of a generic clinical application that utilizes NLP extraction technology.

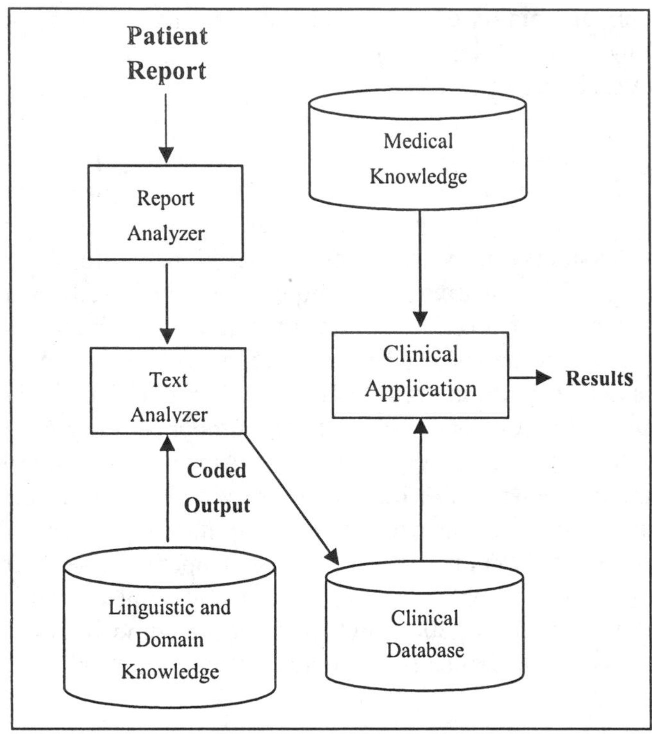

Figure 15-1. Components of a generic NLP Application in the Clinical Domain

First a **Report Analyzer** may be used to process the report in order to identify segments and to handle textual irregularities (i.e. tables, domain-specific abbreviations, missing punctuations). This process is typically straightforward to write, is heuristically-driven, and usually requires tailoring for each type of report. The next process is the **Text Analyzer**, which is the information extraction component that is the core NLP engine. It uses linguistic knowledge associated with syntactic and semantic features, and a conceptual model of the domain to structure and encode the clinical information and to generate output, which is then stored for subsequent use, generally in a structured coded clinical database. Once the data is structured, it may be used by an automated clinical application. However, use of the extracted patient information for an application typically requires additional medical knowledge, depending on the application. For example, to detect patients who may have pneumonia based on radiological reports of the chest, the findings that were extracted will have to be interpreted based on expert knowledge.

A core NLP system will usually have a number of components, which are summarized below, but many variations are possible. A fuller discussion of syntax and semantics will follow the summary:

- **Morphological Analysis** is a process that breaks up the original words in a text so that they are in their canonical forms (i.e. *hands* → hand + -s), which are then used for lexical lookup. This process reduces the number of entries needed in a lexicon, because only the canonical forms and irregular forms need be defined, but not all the forms. One alternative approach eliminates this step and uses a lexicon containing entries for all forms; another approach uses a part of speech (POS) tagger, which identifies the syntactic part of speech of the words and possibly their canonical forms. This is discussed further below.
- **Lexical Look up** is a process where the words or phrases in the text are matched against a lexicon to determine their syntactic (i.e. noun, adjective, verb) and semantic properties (i.e. body part, disease, procedure). A lexicon requires extensive knowledge engineering and effort to develop and maintain. The Specialist Lexicon (Browne, Divita, et. al., 2003) maintained by the National Library of Medicine is comprehensive and contains syntactic information associated with both medical terms and general English terms, and is a valuable biomedical resource for NLP systems. Identification of syntactic classes using lexical lookup may not be necessary if POS tagging is used, but determining semantic properties, which are clinically relevant, is critical. A knowledge source, such as the UMLS or SNOMED may be used to identify semantic properties of words and phrases; however, since the focus of these terminologies concerns the classification of concepts, they generally do not include all the variations of the terms that may be found in text.
- **Syntactic Analysis** is a process that determines the structure of a sentence so that the relationships among the words in the sentence are established. Syntactic analysis may be complete or partial, or may not be used at all. Partial analysis usually revolves around identifying noun phrases.
- **Semantic Analysis** is a process that determines what words and phrases in the text are clinically relevant, and determines their semantic relations. An important requirement of semantic analysis is a **semantic model of the domain** or **ontology**. This model may be linguistically determined based on distributional patterns observed in the text, or it may be

developed based on deep knowledge of the domain. The semantic model may be in the form of frames (Sager, Friedman, et. al., 1987) or conceptual graphs (Baud, Rassinoux, et. al., 1995; Sowa, 1984). Semantic analysis structures and encodes the information, and may or may not depend on syntactic analysis. Structured output is generated, which is used subsequently by another automated process. Alternatively, semantic analysis may include statistical or knowledge components that classify the information in the text.

- **Encoding** is a process that maps the clinically relevant terms to well-defined concepts in a controlled vocabulary, such as the UMLS or SNOMED, or a local vocabulary. Encoding is necessary to achieve widespread use of the structured information by other automated processes, and is essential for intra-operability and inter-operability.

2.2.1 Syntax

Syntax is used to delineate the structure of the sentences in order to find the structural units of information and their relations. Some systems perform a comprehensive syntactic analysis whereas others perform a partial one. A system in the clinical domain that is based on a complete syntactic analysis is the Linguistic String Project system (Sager, 1981; Sager, Friedman, et. al., 1987), a pioneering system in the clinical domain. A complete syntactic analysis or parse of the sentence *patient experienced pain in left arm,* would determine that 1) the subject is a noun *patient*, 2) the verb is *experienced,* 3) the object is a noun phrase *pain in left arm*, 4) *in left arm* is a prepositional phrase that modifies *pain*, and 5) *left* modifies *arm*. In order to obtain a syntactic analysis, linguistic knowledge components consisting of a lexicon and grammar are typically used. The lexicon enumerates the terms that occur in clinical reports, and typically specifies the parts of speech (i.e. *discharge* is a singular noun or a verb), and the canonical forms (i.e. the canonical form *experience* for *experiences, experienced, experiencing,* and *experience*), which are used when generating the output in order to reduce variety. Grammar rules delineate the components of well-defined structures. For example, one rule would specify that one type of well-formed sentence consists of a noun phrase followed by a verb phrase followed by a noun phrase, and that one type of well-formed noun phrase consists of an adjective followed by a noun and a prepositional phrase.

Frequently, words are associated with more than one POS, and in order to obtain a correct syntactic analysis, that ambiguity must first be resolved. Prior to syntactic analysis, a process called POS tagging is generally used, which incorporates syntactic knowledge and contextual features of words surrounding a syntactically ambiguous word to determine the most likely syntactic category. A statistically-based POS-tagging process is most frequently used. It is trained on a large corpus that has been correctly tagged (usually manually) for part of speech information. POS-tagging has been shown to be highly effective (> 95% accuracy) in the general English domain (http://www.coli.uni-sb.de/~thorsten/tnt/), but because it relies on a sample corpus, loss of accuracy occurs when moving from the general English domain to the clinical domain. Training on a clinical POS-tagged corpus would be highly desirable, but obtaining such a corpus is a very costly and time-consuming effort, and currently there are no publicly available clinical corpora because of the issue of patient confidentiality.

One difficulty with syntactic systems is the complexity of the grammar and parsing, and the prevalence of ambiguous structures. Often many analyses are obtained as a result of parsing, but most would not be correct. In contrast, humans have no trouble finding the correct parse(s) because they seamlessly use contextual and world-knowledge. For example, most people would be able to determine that *in joints* modifies *pain* in *patient experienced pain in joints*, and that *in the morning* does not modify *pain* but modifies *experienced* in *patient experienced pain in the morning.* A statistical parser can alleviate some of the ambiguity problems, but would require a large domain corpus that has been syntactically tagged so that it could be used as a training corpus to establish the appropriate parse probabilities. This has been accomplished for the general English domain (http://www.cis.upenn.edu/~treebank/home.html), but, as discussed above, no such corpus exists in the clinical domain. A partial syntactic analysis is a much simpler, more efficient, and more robust process, but has lower precision. Such an analysis could be used to determine that the sample sentence above has the noun phrases *patient, pain,* and *left arm*, and a verb phrase *experienced* but the relations between the phrases are not identified. Partial parsing could be rule-based or statistical-based. For the latter case, phrases such as noun phrases, could be detected using statistical methods that are trained on a large training corpus that has been correctly tagged to identify noun phrases in addition to POS tags. Since clinical information primarily occurs in noun phrases, this technique was used by Hersh and colleagues (Hersh, Mailhot, et. al., 2001) to index radiology reports for retrieval purposes. Another use of partial parsing would be to determine the verb and its arguments. Partial parsing results in a decrease in precision, especially when processing long complex sentences, and may not be adequate for clinical applications, such as an alerting application.

2.2.2 Semantics

Semantic analysis is used to determine the clinically relevant semantic components of a sentence, and to establish the semantic relationships among them. Thus, a semantic analysis establishes a real-world clinical interpretation of the sentence. It is generally based on a semantic knowledge base, which semantically classifies words and phrases that occur in the domain, and then establishes an interpretation of their relationships. For example, *pain* would have a semantic category **sign/symptom**, whereas *arm* would have a category **body part.** *Pain in arm* would be interpreted as a symptom occurring in a body part. Semantic knowledge is incorporated into systems in many different ways. In some systems the semantic components are separated from the syntactic components, although the processor may interleave their use.

The LSP system (Sager, 1987) used a semantic constraint component to specify well-formed semantic patterns for particular syntactic structures in order to obtain an improved syntactic analysis of the sentences. For example, an incorrect analysis of *patient experienced pain in the morning* would be eliminated if a phrase *in the morning* modified *pain.* Another separate semantic component was used by the LSP system to map a syntactic analysis that was normalized into an appropriate semantic template or format. In that system, templates form the model of the clinical domain, and different templates were designed to model different types of information. A template represents clinical information by associating slots with specific types of information and interprets the relations among the slots by associating predefined relations for the slots of the templates. For example, a patient state frame would be built for the sample sentence above, and it would contain slots that will be instantiated with patient state information, such as a **sign/symptom** slot (*pain)*, a **body part** slot (*arm*), and an **evidential** slot (*certainty of sign/symptom information*) by using a process that fills slots based on the semantic classes of the words.

The MPLUS system (Christensen, Haug, et. al., 2002) represents another substantial NLP effort in the clinical domain. It is similar to the LSP system in that it has a separate syntactic parsing component, lexicon, and grammar. However, the semantic component consists of Bayesian networks (BN) that are integrated into the syntactic parsing process. The BN establishes possible values of a node based on a set of training cases, which are used to learn a probability function by considering probabilities of neighboring nodes. Parsing proceeds from the bottom up, from the word level to more complex phrase levels. When a word is parsed, a BN instance is attached to it establishing a semantic interpretation to the word, and when a phrase is parsed, the BN instances of the components of the phrase are unified and attached to the phrase, resulting in a semantic interpretation of the relations between the components.

Another comprehensive NLP system in the clinical domain is the MedLEE system (Friedman, Alderson, et. al., 1994). It differs from the LSP and MPLUS systems in that it has an integrated syntactic and semantic component, which is realized in the form of a grammar. The MedLEE grammar consists of a specification of semantic (and sometimes syntactic) components and is used to interpret the semantic properties of the individual terms and of their relations with other terms, and to generate a target output form. Thus, one grammar rule could contain both syntactic and semantic components. For example, a rule specifies that a sentence containing sign/symptom information consists of a phrase associated with a patient (i.e. *patient*), followed by an evidential verb (*e.g. experienced*), followed by a phrase that contains a sign/symptom (e.g. *pain in arm*). The semantic grammar rules were developed based on co-occurrence patterns observed in clinical text.

2.2.3 Domain Knowledge

Some systems are based on a sound knowledge representation that models the underlying domain, and use the model to achieve a semantic analysis. For example, Baud and colleagues (Baud, Rassinoux, et. al., 1995) modeled a typology of concepts and relations for the domain of digestive surgery. Their NLP system uses a proximity parsing (Baud, Rassinoux, et. al., 1992) approach to obtain a semantic analysis of the text. In this approach words and sequences of words are grouped together if their concepts are semantically compatible according to the model under consideration. Syntax may be used if available, but the method does not rely on it. The method provides advantages for multi-lingual capabilities because it is based on domain concepts and is less dependent on characteristics of different languages. Another system that relies on a knowledge-rich infrastructure was developed by Hahn and colleagues (Hahn, Romacker, et. al., 2002). This system maps text to a knowledge base that contains a formal representation of the content of the text. The system also performs a syntactic analysis that is driven by lexical definitions and a syntactic dependency grammar. A domain-specific lexicon is used that is geared to the needs of the particular clinical subdomain.

2.3 Clinical Applications

Natural language systems have been employed and evaluated in the clinical domain for a variety of clinical applications. Table 15-2 provides a list of papers describing use of NLP for processing clinical textual documents in different domains for a variety of different applications. Although the list is comprehensive and represents a broad collection of different clinical applications, it is not complete. In addition, methods that use NLP for medical applications that do not involve the processing of clinical documents, such as those focusing on clinical terminology, the processing of journal articles and consumer health messages, are not included.

Table 15-2. Clinical Applications Using NLP Technology

Reference	*Clinical Domain*	*Application*
Lyman, Sager, et al., 1991	Progress notes	Quality assessment
Baud, Rassinoux, et al., 1992	Surgical notes	Encoding and improved browsing; Multilingual capabilities
Moore and Berman, 1994	Pathology	Key diagnoses for indexing
Zweigenbaum, et al., 1994	Discharge summary	ICD-9-CM for indexing Multilingual capabilities
Hripcsak, Friedman, et al., 1995	Radiology	Detecting clinical conditions
Gunderson, Haug, et al., 1996	Admission diagnoses	ICD-9 encoding
Sager, Nhan, et al., 1996, and Lussier, Y., and Shagina, et al,. 2001	Discharge summary	SNOMED encoding for granular retrieval
Knirsch, Jain, et al., 1998	Radiology	Isolating patients with tuberculosis
Fiszman, M., and Haug, et al., 1998	Ventilations/ perfusion lung scan	Interpret findings
Blanquest and Zweigenbaum, 1999	Discharge summary	ICD-10 encoding
Aronsky and Haug, 2000	Radiology	Assessing severity of pneumonia
Friedman, Knirsch, et al., 1999	Radiology and Discharge summary	Assessing severity of pneumonia
Fiszman, M., and Haug 2000	Radiology	Pneumonia guidelines
Tuttle, M.S., Olsen, N.E., et al., 1998; Aronson, 2001; Nadkarni, Chen, et al., 2001; Leroy and Chen, 2001; Zou, Chu, et al., 2003; Friedman, Shagina, et al., 2004	Biomedical text	Mapping text to UMLS codes
Heinze, Morsch, et al., 2001b	Radiology, Emergency Medicine	Coding for billing

continued

Reference	*Clinical Domain*	*Application*
Heinze, Morsch, et al., 2001a	Variety of domains	Data mining
Hripcsak, Austin, et al., 2002	Radiology	Data mining
Hahn, Romacker, et al., 2002	Histopathology of gastro-intestinal domain	Knowledge acquisition
Mamlin, Heinze, et al., 2003	Radiology	Extract cancer-related findings
Schadow and McDonald, 2003	Surgical pathology	Extract specimens, related pathological findings
Xu, Anderson, et al., 2004	Surgical pathology	Obtaining variables for clinical study
Mitchell, Becich, et al., 2004	Surgical pathology	Detecting negated concepts
Liu and Friedman, 2004; and Meng, Taira, et al., 2004	Clinical reports	Summarize patient information

3. CASE SCENARIO

The MedLEE NLP system was developed to address a need to transform narrative text in patient reports to structured and encoded data so that the data could be stored in the Clinical Repository at NYPH and accessed by other applications, such as a monitoring system, which consists of medical rules that access data in the repository. For example, an alerting application could be used for newborn infants in a neonatal intensive care unit (ICU) to screen for hospital-acquired pneumonia. There are many different types of clinical reports in a healthcare institution, such as radiology reports, resident sign out notes, discharge summaries, and pathology reports. Although each contains text, their overall formats differ.

CLINICAL INFORMATION:
3 day old male with resp dist.
IMPRESSION:
Opacities are noted in left and right lobe of lung, which is consistent with pneumonia or atelectasis.
DESCRIPTION:
a. p. portable chest radiograph is submitted. Bibasilar opacities, right greater than left. this may represent pneumonia or atelectasis; otherwise, there is no interval change from 6/19/01.

Figure 15-2. Sample Radiology Report

76 yo F hx of HTN, COPD, CRI, DM2 from NH after noted to have fever, cough, change in MS. R/O PNA. Given IV ABX in ED, R/O Staph. CXR with bilobar involvement
ALL: NKDA
Meds: Vanco 500mg, Azithro, tylenol, atenolol, alb/atrov, NPH

Figure 15-3. Sample Resident Sign Out Note

Figure 15-2 shows a radiology report of the chest taken from an infant in the ICU with respiratory distress. Radiology reports generally have three sections, which are indicated by special section headers: **Clinical Information**, which usually contains some information related to the indication for the examination, **Impression**, which lists or interprets the most relevant findings, and **Description**, which describes the findings. All sections contain text, but the sentence structures range from noun phrases to complex sentences, and include some abbreviations. The report shown has important information for a pneumonia ICU alerting application because it contains a finding *opacities*, which may indicate *pneumonia* or *atelectasis* (e.g. a collapsed lung or lobe of lung). In order to process the report, a special preprocessor will first be used to transform the report so that the overall structure (i.e. sections followed by text associated with the section) is correct for the MedLEE core NLP processor. This will be a preprocessor tailored to the specific type of report. In the case of radiology reports at NYPH, a special report preprocessor is not needed because the report structure is in a form that is appropriate for MedLEE to use directly. However, for other types of reports, special preprocessing may be necessary. For example, the resident sign out report shown in Figure 15-3 requires a special purpose preprocessor to establish appropriate sections, to add sentence endings when necessary, and to handle abbreviations, which are prevalent in this type of report. For the sign out note example, the preprocessor will add a section header, SUMMARY, to the beginning of the report, will change *ALL* to a section ALLERGIES, and *Meds* to a section MEDICATIONS. A period will be added after NKDA and NPH to indicate the end of the two sentences. Additionally, abbreviations will be expanded based on knowledge of the abbreviations used in the domain. For example, *F* will be changed to *female*, *hx* to *history,* etc.

The MedLEE core NLP engine will be used next to process the report; it first identifies each section and then processes all the sentences within the section. Thus, in Figure 15-2, the sentence *opacities are noted in left and right lobe of lung, which is consistent with pneumonia or atelectasis*, which is in the Impression section, will be processed first. The lexicon will be used to identify the semantic and or the syntactic categories of the terms and to specify the canonical form. For example, the words *opacities, pneumonia,* and *atelectasis* will each be associated with the semantic category **pathological finding.** In addition, *left* and *right* will be associated with the semantic category **region**, *lung* will be associated with **bodyloc**, and *consistent with* will be associated with a **relation** that connects findings. The parser will use the semantic grammar to identify the semantic relations in the sentence and to generate structured output, which for the sample sentence, will consist of four findings, each of which is represented by a tag **problem**. Several grammar rules will be used and satisfied in the course of parsing this sentence in order to match the semantic sequence of the words and phrases in the sentence to the rules. One rule that will be satisfied is a rule specifying that a finding (e.g. *opacities*) is connected (e.g. via the relation *consistent with*) to one of two findings (e.g. *pneumonia* or *atelectasis*). Another rule that will be satisfied is a rule specifying that a finding (e.g. *opacities*) is followed by a verbal phrase (e.g. *are noted*), which is associated with certainty information, and a phrase specifying body location information (e.g. *lobe*). Finally, another rule will be satisfied specifying that body location *lobe* is modified by a body location *lung*, and also by two regions *left* and *right.* Each rule in the grammar not only identifies the semantic components for that structure, but also interprets their relations and specifies an output structure, which is a composition of the output structures of each of the components. A simplified XML form of the output generated by processing the sample sentence is shown in Figure 15-4. The complete XML form has additional attributes representing the associated codes if coding is requested, and other contextual attributes, but these were omitted from the figure for simplicity. In Figure 15-4, the first problem tag has the value **opacity**, and modifiers **bodyloc** and **certainty**, which are nested within the **problem** tag. The **bodyloc** modifier with the value **lobe** has a nested body location whose value is **lung,** and a **region** modifier whose value is **left.** The value of the certainty modifier is **high certainty**, which is the target form of *are noted.* The second problem, whose value is also **opacity**, is almost identical to the first except the **region** modifier of **lobe** has the value **right**. This output is the result of expanding the conjunction relation in *left and right lobe of lung* to obtain separate findings. The remaining two findings have a **problem** tag with the value **pneumonia** and a **problem** tag with the value **atelectasis.** Each of the two findings has a **certainty** modifier, stemming from *consistent with* whose target output form is **moderate certainty**.

Once the output is in a structured form as shown below, it will be transformed to a form suitable for storage in a particular clinical database, and subsequently, a query may be used to reliably access the information. For example, a rule used to detect infants who may have hospital-acquired pneumonia will look for findings with certain characteristics, such as one which has a **problem** tag with a value **opacity** along with certain other modifiers, such as modifiers signifying that the opacity is in both lobes of the lung. The rule will also look for a **problem** with the value **pneumonia**. However, findings that have characteristics that match the rule may also be filtered out depending on the presence of certain modifiers that negate the event, signify that it occurred in the past, or signify that it did not actually occur. For example, if the **certainty** modifier of **problem** is **no** or **rule out**, it will be filtered out because it signifies that there was no evidence for the finding or that the patient is being evaluated for the finding. Similarly, the finding will be filtered out if it has a temporal type of modifier with a value that signifies the finding is associated with a previous event. This will happen if the modifier is **status** with a value **previous** or a temporal modifier **date** with a previous date, such as **19900502** (5/2/1990).

```
<section v="impression">
  <problem v = "opacity"><bodyloc v = "lobe"><bodyloc v = "lung"/>
                                   <region v = "left"/></bodyloc>
                               <certainty v = "high certainty"/>
  </problem>
  <problem v = "opacity"><bodyloc v = "lobe"><bodyloc v = "lung"/>
                                  <region v = "right"/></bodyloc>
                                  <certainty v = "high certainty"/>
  </problem>
  <problem v = "pneumonia"><certainty v = "moderate certainty"/>
  </problem>
  <problem v = "atelectasis"><certainty v = "moderate certainty"/>
  </problem>
</section>
```

Figure 15-4. Simplified output form generated as a result of processing the sentence shown in the Impression section of Figure 15-2.

In the sample sentence above, all the words of the sentence were known to the system and the grammar rules were completely satisfied. In some cases, it may not be possible for the system to obtain a parse for the complete sentence because there may be words that are unknown to the system or the sentence may not conform to the grammar rules. For those cases, the parsing strategy is relaxed and different strategies are attempted so that a parse is always obtained if there is relevant clinical information in the sentence. The first relaxation attempt ignores unknown words and then tries to obtain a parse. For example, if the sample sentence is *opacities are noted in lft lobe of lung,* the segment *opacities are noted in lobe of lung* will be parsed successfully. Other strategies are based on segmenting the sentence in different ways and attempting to parse the segments. However, these relaxation strategies are aimed at improving recall at the expense of some loss of precision, and may be undesirable for some clinical applications. The way this is handled is that the parse mode (i.e. the parsing strategy that was used) is saved as a special modifier **parse mode** in the structured output form and its value is the parse mode that was used to obtain the output. In this way, findings can be filtered out if the parse mode is not reliable enough for the application.

4. CONCLUSIONS AND DISCUSSION

Improved automated methods are needed to advance the quality of patient care, to reduce medical errors, and to lower costs, but these methods depend on reliable access to a comprehensive variety of patient data. However, the primary means of communication in healthcare is narrative text, and therefore natural language processing techniques are needed to structure and encode the narrative text so that the clinical information will be in a form suitable for use by automated applications. NLP is a difficult, complex, and knowledge intensive process. It involves the integration of many forms of knowledge, including syntactic, semantic, lexical, pragmatic, and domain knowledge. Successful NLP methods have been developed, and applications using NLP methods have been evaluated demonstrating their effective use in healthcare settings. However, these efforts represent individual instances, mostly at individual sites, and have not been widely disseminated. In order for NLP to become more widespread, standardization at several different levels is critical: namely a standardization of report structures, a standardization of models representing clinical information, and a standardized clinical vocabulary.

5. ACKNOWLEDGEMENTS

Work on the MedLEE NLP system was supported by grants LM07659 and LM06274 from the National Library of Medicine, by the New York State sponsored Columbia University Center for Advanced Technology, and by the Research Foundation of CUNY.

REFERENCES

Aronsky, D. and Haug, P. J. (2000). "Assessing the Quality of Clinical Data in a Computer-based Record for Calculating the Pneumonia Severity Index," *Joiurnal of the American Medical Informatics Association,* 7(1):55-65.

Aronson, A.R. (2001). "Effective Mapping of Biomedical Text to the UMLS Metathesaurus: The MetaMap Program," in *Proceedings of the 2001 AMIA Symposium*:17-21.

Baud, RH., Rassinoux, A.M. and Scherrer, J.R. (1992). "Natural Language in Processing and Semantical Representation of Medical Texts," *Methods of Information in Medicine,* 31(2):117-125.

Baud, RH., Rassinoux, A.M., Wagner, J.C. and Lovis, C. (1995). "Representing Clinical Narratives Using Conceptual Graphs," *Methods of Information in Medicine,* 1/2:176-186.

Blanquet, A. and Zweigenbaum, P. (1999). "A Lexical Method for Assisted Extraction and Coding of ICD-10 Diagnoses from Free Text Patient Discharge Summaries," in *Proceedings of the 1999 AMIA Symposium:*1029.

Browne, A. C., Divita, G., Aronson, A.R. and McCray, A.T. (2003). "UMLS Language and Vocabulary Tools," in *AMIA Annual Symposium Proceedings*:798.

Christensen L, Haug, P.J., and Fiszman, M. (2002). "MPLUS: A Probabilistic Medical Language Understanding System," in *Proceediongs of the ACL 2002 Workshop on Natural Language in Processing in Biomedicine:*29-36.

Chuang, J. H., Friedman, C., Hripcsak, G. (2002). "A Comparison of the Charlson Comorbidities Derived from Medical Language in Processing and Administrative Data," in *Proceedings of the AMIA.Symposium.:*160-164.

Dolin, R.H., et al. (2001). "The HL7 Clinical Document Architecture," *Journal of the American Medical Informatics Association,* 8(6):552-569.

Evans, D.A., et al. (1994). "Toward a Medical Concept Representation Language," *Journal of the American Medical Informatics Association,* 1(3):207-217.

Fiszman, M. and Haug, P.J. (2000). "Using Medical Language in Processing to Support Real-time Evaluation of Guidelines," in *Proc AMIA Symp 2000;* 235-239.

Fiszman, M, Haug, P.J., and Frederick, P.R. (1998). "Automatic Extraction of PIOPED Interpretation from Ventilation/perfusion Lung Scan Reports," in *Proc 1998 AMIA Symp:*860-864.

Fiszman, M. and Haug, P.J. (2000). "Using Medical Language in Processing to Support Real-time Evaluation of Pneumonia Guidelines," in *Proc 2001 AMIA Symp*:235-239.

Friedman, C. and Johnson, S.B. (2005). "Natural Language and Text in Processing in Biomedicine," in Shortliffe EH and JJ. Cimino, Eds. *Biomedical Informatics: Computer Applications in Health Care and Medicine.* Springer.

Friedman, C., et al. (1994). "A General Natural Language Text in Processor for Clinical Radiology," *Journal of the American Medical Informatics Association,* 1(2):161-174.

Friedman, C. and Hripcsak, G. (1998). "Evaluating Natural Language in Processors in the Clinical Domain," *Methods of Information in Medicine,* 37:334-344.

Friedman, C., et al. (1995). "The Canon Group's Effort: Working Toward a Merged Model," *Journal of the American Medical Informatics Association,* 2(1):4-18.

Friedman, C., Knirsch, C.A., Shagina, L., and Hripcsak, G. (1999). "Automating a Severity Score Guideline for Community-acquired Pneumonia Employing Medical Language in Processing of Discharge Summaries," in *Proc 1999 AMIA Symp:*256-260.

Friedman, C., Shagina, L., Lussier, Y. and Hripcsak, G. (2004). "Automated Encoding of Clinical Documents Based on Natural Language in Processing," *Journal of the American Medical Informatics Association,;* 11(5):392-402.

Gundersen, M.L., et al. (1996). "Development and Evaluation of a Computerized Admission Diagnoses Encoding System," *Computers and Biomedical Research*; 29:351-372.

Hahn, U., Romacker, M. and Schulz, S. (2002). "Creating Knowledge Repositories from Biomedical Reports: The MEDSYNDIKATE Text Mining System," in *Pac Symp Biocomput.*:338-349.

Heinze, D.T., Morsch, M.L., and Holbrook, J. (2001a). "Mining Free-text Medical Records," in *Proceedings AMIA.Symp.*:254-258.

Heinze, D.T., et al. (2001b). "LifeCode - A Deployed Application for Automated Medical Coding," *AI Magazine;* 22(2):76-88.

Hersh, W., Mailhot, M., Arnott-Smith, C., and Lowe, H. (2001). "Selective Automated Indexing of Findings and Diagnoses in Radiology Reports," *J.Biomed.Inform.;* 34(4):262-273.

Hripcsak, G., Austin, J.H., Alderson, P.O., and Friedman, C. (2002). "Use of Nutural Language in Processing to Translate Clinical Information from a Database of 889,921 Chest Radiographic Reports," *Radiology*; 224(1):157-163.

Hripcsak, G., et al. (1995). "Unlocking Clinical Data from Narrative Reports," *Ann.of Int.Med.;* 122(9):681-688.

Knirsch, C.A., et al. (1998). "Respiratory Isolation of Tuberculosis Patients Using Clinical Guidelines and an Automated Decision Support System," *Infection Control and Hospital Epidemiology;* 19(2):94-100.

Leroy, G. and Chen, H. (2001). "Meeting Medical Terminology Needs--The Ontology-Enhanced Medical Concept Mapper," *IEEE Trans.Inf.Technol.Biomed.;* 5(4):261-270.

Lindberg, D.A.B., Humphreys, B. and McCray, A.T. (1993). "The Unified Medical Language System," *Methods of Information in Medicine,* 32:281-291.

Liu, H. and Friedman, C. (2004). "CliniViewer: A Tool for Viewing Electronic Medical Records Based on Natural Language in Processing and XML," *Medical Informatics,* 2004:639-643.

Lussier, Y., Shagina, L., and Friedman, C. (2001). "Automating SNOMED Coding Using Medical Language Understanding: A Feasibility Study," in *Proceedings of the 2001 AMIA Symposium,* 418-422.

Lyman, M., et al. (1991). "The Application of Natural-language in Processing to Healthcare Quality Assessment," *Medical Decision Making,* 11(suppl):S65-S68.

Mamlin, B.W., Heinze, D.T., and McDonald, C.J. (2003). "Automated Extraction and Normalization of Findings from Cancer-related Free-text Radiology Reports," *AMIA Annual Symposium Proceedings:*420-424.

McKnight, L.K., Wilcox, A. and Hripcsak, G. (2002). "The Effect of Sample Size and Disease Prevalence on Supervised Machine Learning of Narrative Data," in *Proceedings of the AMIA.Symposium:*519-522.

Meng, F., et al. (2004). "Automatic Generation of Repeated Patient Information for Tailoring Clinical Notes," *Medical Informatics,* 2004:653-657.

Mitchell, K.J., et al. (2004). "Implementation and Evaluation of a Negation Tagger in a Pipeline-based System for Information Extract from Pathology Reports," *Medical Informatics,* 2004:663-667.

Moore, G.W. and Berman, J.J. (1994). "Automatic SNOMED Coding," in *Proceedings of the Eighteenth Annual Symposium on Computer Applications in Medical Care.*

Nadkarni, P., Chen, R. and Brandt, C. (2001). "UMLS Concept Indexing for Production Databases: A Feasibility Study," *Journal of the American Medical Informatics Association,* 8(1):80-91.

National Academy of Science. (2003). "Patient Safety: Achieving a New Standard for Care," Aspden, P., Corrigan, J.M., Wolcott, J., and Erickson, S.M. Washington, D.C.

Sager, N. (1981). *Natural Language in Processing: A Computer Grammar of English and Its Applications.* Mass, Addison-Wesley.

Sager, N., Friedman, C., Lyman, M. and et al. (1987). *Medical Language in Processing: Computer Management of Narrative Data.* Reading, MA: Addison-Wesley.

Sager, N., Nhan, N.T., Lyman, M., and Tick, L.J. (1996). "Medical Language in Processing with SGML Display," in *Proceedings of the 1996 AMIA Symposium*:547-551.

Schadow, G. and McDonald, C.J. (2003). "Extracting Structured Information from Free Text Pathology Reports," in *AMIA Annuual Symposium Proceedings*:584-588.

Sowa, J. (1984). *Conceptual Structures: Information in Processing in Mind and Machine.* Reading: Addison-Wesley.

Sweeney, L. (1997). "Weaving Technology and Policy Together to Maintain Confidentiality," *Journal of Law, Medincein and Ethics,* 25(2-3):98-110, 82.

Tuttle, M.S., et al. (1998). "Metaphrase: An Aid to the Clinical Conceptualization and Formalization of Patient Problems in Healthcare Enterprises," *Methods of Information in Medicine,* 37:373-383.

Wilcox, A. and Hripcsak, G. (1999). "Classification Algorithms Applied to Narrative Reports," in *Proceedings of the AMIA.Symposium*:455-459.

Xu, H., Anderson, K., Grann, V.R., and Friedman, C. (2004). "Facilitating Cancer Research Using Natural Language in Processing of Pathology Reports," *Medical Informatics* 2004:565-572.

Zou, Q., et al. (2003). "IndexFinder: A Method of Extracting Key Concepts from Clinical Texts For Indexing," in *AMIA Annual Sympoisum Proceedings* :763-767.

Zweigenbaum, P., and et al. (1994). "MENELAS: An Access System for Medical Records Using Natural Language," *Computer Methods and Programs in Biomedicine,* 45:117-120.

SUGGESTED READINGS

Friedman, C. and Johnson S.B., 2005. *Natural Language and Text in Processing in Biomedicine.* In Shortliffe EH and Cimino JJ, editors, Biomedical Informatics, third edition. Chapter 8, New York; Springer (in press).
Provides a methodological overview of natural language processing in the biomedical domain.

Jurafsky, D. and Martin, J. H., 2000. *Speech and Language in Processing: An Introduction to Natural Language in Processing, Computational Linguistics, and Speech Recognition*, Prentice-Hall.
A comprehensive textbook describing general natural language processing techniques.

Friedman, C., ed., 2002. Sublanguage - Zellig Harris Memorial, *J Biomed Inform* **35**:213-277.
A special issue of the Journal of Biomedical Informatics devoted to Zellig Harris's theory on sublanguage processing.

Manning, C. and Sch]tze H. 1999. *Foundations of Statistical Natural Language in Processing.* MIT Press.
A textbook describing statistical natural language processing techniques.

Sager, N., C. Friedman, M. Lyman, and et al. 1987. *Medical Language in Processing: Computer Management of Narrative Data.* Reading, MA: Addison-Wesley.
A book describing medical language processing techniques used by the Linguistic String Project that are based on the sublanguage theory of Zellig Harris.

ONLINE RESOURCES

This site contains a summary of the capabilities and sources of a large amount of natural language processing (NLP) software available to the NLP community:
http://registry.dfki.de/

This site contains links to general NLP software tools.
http://www-a2k.is.tokushima-u.ac.jp/member/kita/NLP/nlp_tools.html

The Penn Treebank Project annotates general narrative text for linguistic structure. It produces skeletal parses showing rough syntactic and semantic information, which also includes part of speech tagging:
http://www.cis.upenn.edu/~treebank/home.html

A demonstration of the MedLEE clinical NLP System providing examples of the parsing of discharge summaries, chest radiological reports, and mammograms:
http://lucid.cpmc.columbia.edu/medlee/

A demonstration of the MedLEE clinical NLP System providing examples of the parsing of an electrocardiogram report, a discharge summary, and a radiological report of the abdomen:
http://lucid.cpmc.columbia.edu/dli2/demo/

The Unified Medical Language System is a comprehensive terminological resource for NLP in the clinical domain:
http://www.nlm.nih.gov/research/umls/

The Specialist Lexicon and software tools are linguistic resources for NLP in the biomedical domain:
http://specialist.nlm.nih.gov/

QUESTIONS FOR DISCUSSION

1. How does NLP have the potential to change the practice of medicine?

2. What are some of the hurdles that will need to be overcome?

3. What role does machine learning have in NLP? What are some of the difficulties in using machine learning techniques?

4. Describe some components of an NLP system, and provide some examples of how they would be used.

5. What are some of the potential benefits of using NLP to process clinical reports?

6. Describe some clinical applications that use NLP. What other types of applications would be useful? What impediments are there to integrating

an NLP system with a Clinical Information System?

7. There is typically a trade off between recall (number of true positives that were found by the system/the total number of true positives in the gold standard) and precision (the number of true positives found by the system/the number of all positives found by the system). What are the advantages/disadvantages of aiming for high precision or aiming for high recall in the clinical domain.

8. In what ways do abbreviations cause problems for NLP systems? What types of contextual information can be used to help NLP systems handle abbreviations and how would the information help? Would these suggestions completely solve the problem?

9. Explain the similarities and differences between controlled vocabularies and terms occurring in clinical text.

10. Modifiers change the underlying interpretation of clinical information in text. Provide examples illustrating how modifiers associated with negation, uncertainty, time, and body locations change the meaning of clinical findings.

Chapter 16

IDENTIFICATION OF BIOLOGICAL RELATIONSHIPS FROM TEXT DOCUMENTS

Mathew Palakal, Snehasis Mukhopadhyay, and Matthew Stephens

Indiana University Purdue University Indianapolis, Indianapolis, IN 46202

Chapter Overview

Identification of relationships among different biological entities, e.g., genes, proteins, diseases, drugs and chemicals, etc, is an important problem for biological researchers. While such information can be extracted from different types of biological data (e.g., gene and protein sequences, protein structures), a significant source of such knowledge is the biological textual research literature which is increasingly being made available as large-scale public-domain electronic databases (e.g., the Medline database). Automated extraction of such relationships (e.g., gene A inhibits protein B) from textual data can significantly enhance biological research productivity by keeping researchers up-to-date with the state-of-the-art in their research domain, by helping them visualize biological pathways, and by generating likely new hypotheses concerning novel interactions some of which can be good candidates for further biological research and validation. In this chapter, we describe the computational problems and their solutions in such automated extraction of relationships, and present some recent advances made in this area.

Keywords

biological objects; associations; text mining; transitivity; flat relationships; directional relationships; hierarchical relationships

1. INTRODUCTION

The scientific literature is an important source of knowledge for the scientist during the course of study of any research problem. The huge and rapidly increasing volume of scientific literature makes finding relevant information increasingly difficult. Content level information rather than collection level information is needed for scientific research. The availability of scientific literature in electronic format, such as Medline, has made the development of automated text mining systems and hence "data-driven discovery" possible. Text mining enables analysis of large collections of unstructured documents for the purposes of extracting interesting and non-trivial patterns or knowledge (Tan, 1999). Text mining has a very high potential for knowledge discovery, as the most natural form of reporting, storing, and communicating information is text. Informatics tools can assist the traditional hypothesis-driven research (Smalheiser, 2001). Many hypotheses are formed by extrapolating the current knowledge; for example, if we know that apoptosis in breast cancer is mediated by calpain, we can ask if apoptosis in other related cancer (e.g., ovarian cancer) is also mediated by calpain.

The query "breast cancer" on Medline returned 128,171 documents on June 21, 2004. This shows the large volume of scientific information available in the form of text. It is impossible or impractical for anyone to read through all of these documents to find the relevant information. It is even more difficult to capture the knowledge in those documents. Researchers and scientists are challenged by this increasing knowledge gap. Associations among biological objects such as genes, proteins, molecules, processes, diseases, drugs and chemicals, are one such form of underlying knowledge. For example, Swanson (Swanson and Smalheiser, 1997) found an association between magnesium and migraine headaches that was *not explicitly reported in any one article*, but based on *associations extracted from different journal titles*, and later validated experimentally.

In this chapter, we describe the progress made in the development of a complete "knowledge base" of associations among biological objects, such as those mentioned above, that are important for biologists to study and understand specific biological processes. The term object refers to any biological object (e.g. protein, gene, cell cycle, etc.) and relationship refers to an action one object has on another. Biological relationships discovered from literature and experiments can be used to set up templates for biologists to model a biological process and to formulate new hypotheses for guided laboratory research.

The main goals of this chapter are to describe the progress made in: (a) developing a very large knowledge base, called BioMap, using the entire

Medline collection (over 14 million) of literature documents, and (b) developing an interactive knowledge network for users to access this secondary knowledge (BioMap) along with its primary databases such as Medline, GenBank, etc., in an integrated manner based on a specific area of problem enquiry. In order to build the BioMap and its associated access "window" (the knowledge network), various algorithms and tools need to be developed for: (i) identifying biological object names; (ii) discovering object-object relationships; (iii) creation of the knowledge base (BioMap); (iv) a hypergraph realization of the knowledge network (generating pathways and hypothesis) in response to a user query, and, (v) global access capability for the entire system.

Identification of biological objects and their relationships from free running text is a very difficult problem. This problem is compounded by several factors, specifically, when multiple objects and multiple relationships need to be detected. Typically, the extraction of object relationships involves object name identification, reference resolution, ontology and synonym discovery, and finally extracting object-object relationships. We describe a multi-level hybrid approach that incorporates statistical, connectionist, and N-Gram models along with multiple dictionaries to handle the multi-object identification and relationship extraction problem for BioMap.

The relationships thus discovered from the entire Medline collection are to be maintained in a relational database along with the specific links to literature sources, genes and protein sequence databases. A user can access this knowledge base using simple or complex queries, such as a disease name, a set of gene names, or any such combinations. Unlike in traditional databases, the outcome of a user query will be a complex set of data with multiple associations among them. Hence, the results of a query will be constructed as a *knowledge network* (knowledge view of BioMap) and presented to the user.

The knowledge network is to be constructed as a hypergraph "on the fly" based on each user query. A hypergraph is an extension of a graph in the sense that each hyperedge can connect more than two vertices, thus allowing to connect relationships among multiple objects simultaneously. A system based on such a hypergraph model has a number of advantages: the model is independent from updates to the underlying databases; it enables the formation of hypergraphs from entities in different databases; it allows the system to be accessed by multiple users simultaneously, each with an independent hypergraph; further queries can be made to a hypergraph to obtain a better "focused" view of the knowledge base; it reduces the need to access the knowledge base multiple times when a hypergraph is to be shown to a remote user; and most importantly, the interaction with the user can be made faster when a query is made on the hypergraph since it is available in the local computer memory. Furthermore, the edges and vertices of the hypergraph are "live" (made as hyperlinks) allowing the user to access primary data sources and bioinformatics tools pertinent to the information contained in the knowledge network.

The key innovative features of the proposed BioMap are that it is adaptable and scalable, and that the core knowledge base will be constructed from a very large collection of documents (Medline) to make the system robust. The adaptation feature requires that the system should have the ability to learn new problem domains without having to rebuild the system as in the case of fully rule-based or grammar-based approaches. The scalability feature allows the system to continue to develop its knowledge base as new information arrives in the literature databases or information is incorporated from other data sources (e.g. *Science*, *Nature*, etc.).

2. OVERVIEW OF THE FIELD

2.1 Background

Ever since the emergence of the field of Bioinformatics, it has been of great interest for both the informatics and the biology communities to develop automatic methods to extract embedded knowledge from literature data. Dealing with literature data in free running text is a challenging problem that has been well studied by natural language processing (NLP) and artificial intelligence (AI) communities with some success. For example, several works report relationship extraction among biological objects (Ono et al., 2001; Humphreys et al., 2000; Thomas et al., 2000; Proux et al., 2000; Marcotte et al., 2001; Oyama et al., 2002) and biological object recognition problems (Leroy and Chen, 2002; Tanabe and Wilbur, 2002; Krauthammer et al., 2000). Some work has also been reported on document clustering (Iliopoulus et al., 2001; Nobata et al., 2000) and pathway identification (Sanchez et al., 1999; Park et al., 2001; Ng and Wong, 1999). Current progress in supporting biomedical research activities through published literature can be broadly classified into two categories: (i) Biological information extraction (IE), and (ii) Development of "tools of the trade" bioinformatics tools.

2.2 Biological Information Extraction

Mining of literature databases to discover information relevant to biological relationships and pathways involves two key tasks: first, identification of biological object names, and second, identification of relationships among these objects. Several works report research on the identification of biological objects. The most successful tagging system, described in (Fukuda et al., 1998), is a rule-based system, called PROPER, that was specifically designed to extract protein names from the text using proper noun dictionaries and a pattern dictionary. The results of the PROPER tagging method were evaluated using precision and recall and yielded an accuracy of 98.84% and 94.70% respectively. These results, however, do not include distinction between gene and protein names and leave out words that may not be in the target object list. Also, it was designed to tag only one object (i.e., proteins), which is not adequate for extracting different object names.

Collier (Collier et al., 2000a) proposed a stochastic approach to tagging biological objects. Their model utilized 100 Medline abstracts using a pre-specified annotation method and used this data to train a Hidden Markov Model to tag similar objects. The results of this method, given as F-scores, combine recall and precision (Chinchor, 1995). In the case of tagging the proteins, the best F-score reported was 0.759 using the one hundred hand-tagged abstracts. For DNA it was significantly less at 0.472. The highest average F-score calculated was 0.728. It was assumed that the increased training data would improve the performance but how much more training data would be necessary to achieve performance above the desired F-score of 0.9 was not reported.

In order to overcome the limitation of hand tagging, Hatzivassilou (Hatzivassiloglou et al., 2001) trained models using an extensive dictionary of unambiguous gene terms from the GeneBank database. Using a nine million-word corpus, they managed to distinguish between three biological entities using a Bayesian classifier with accuracy of 80%. A two-way classifier jumps up to an accuracy of 85%. The fact that accuracy starts to decline with the addition of more classes indicates that such a method would not scale up when tagging multiple object types. In conclusion, the current methods for tagging biological objects fall short because they either cannot tag more than one object (Fukuda et al., 1998) or they rely heavily on hand-tagged training data (Collier et al., 2000a).

Natural language-based parsers have also been used on biological literature to extract relationships such as protein-protein interactions. For example, a full parser was used in (Yakushiji et al., 2000) to extract information from biomedical papers. One reported experiment consisted of 179 sentences from an annotated corpus of Medline abstracts. The first 97 sentences were used to determine the accuracy of the system. Out of 133 argument structures in those 97 sentences, 23% were extracted uniquely, 24% with ambiguity, and 53% were not extracted. Another NLP system reported in (Friedman et al., 2001) is called Genie, consisting of a term tagger, preprocessor and parser. The term tagger uses BLAST (Atschul et al., 1990) techniques, specialized rules, and external knowledge sources to identify and tag genes and proteins in the text articles. The Genie system had a measured sensitivity of 54% and a specificity of 96%. Although the reported accuracy of the system is good, it does not take into consideration where the interaction takes place and under what conditions.

A method to identify Gene-pair relationships from a large collection of text documents is reported in (Stephens et al., 2001). The goal is to discover pairs of genes from a collection of retrieved text documents such that the genes in each pair are related to one other in some manner. Details of this method are discussed in Section 3b(iv) under title "rFinder-I." The results of this study indicated that finding the actual nature of the relationship between proteins had a specificity of 67% in the unknown pathway and specificity of 50% in the known pathway. The potential drawback of this approach is that it finds only gene-pair relationships; relationships that occur indirectly across sentences will not be found. Another study (Craven et al., 1999) proposed a learning method to extract relationships and organize these relationships as structured representations or knowledge bases. This study, primarily focused on protein related interactions, reports 77% precision and 30% recall on a corpus of 633 sentences. EDGAR is another natural language processing system (Rindflesch et al., 2000) that extracts relationships between cancer-related drugs and genes from biomedical literature. Again, the scope is limited to few biological objects and their relationships.

Recently, support vector machine (SVM) based approaches (Kazama et al., 2002; Steffen et al., 2003) showed promising results for biological entity identification. Using SVM, a named entity task is formulated as the classification of each word with context to one of the classes that represent the region information and entity's semantic class. The best results reported so far show a precision of 71.4% and a recall of 72.8%, corresponding to an F-measure of 72.1%, for the closed division for only gene name identification.

In summary, the current methods for tagging biological objects fall short because they either cannot tag multiple objects or they rely heavily on hand-tagged training data. The techniques that are used for extracting relationships using NLP are too specific to be extended to new domains without creating a large number of new rules for new relationships. Also, most importantly, current approaches do not look into the creation of a "knowledge base" of all possible relationships for biological problem domain or domains, so that the research community can not only retrieve relevant literature but also retrieve and view the embedded knowledge.

2.3 Bioinformatics Tools

Numerous bioinformatics tools also exist that closely or loosely connect to provide literature support for biomedical research. These systems have relied on annotation of the biomedical literature, with the most successful system being the Online Mendelian Inheritance in Man (OMIM) database and its associated morbid map (OMIM). While OMIM has been very successful in the development of an annotated disease-based database, the very nature of its annotation means *OMIM only presents the well established and proven associations for a given disease*. As such, the OMIM database is not capable of finding novel associations with respect to different diseases of interest. For example, for the discovery of novel gene-disease relationships one needs a list of all the possible gene-disease relationships, even if currently unproven, such that a scientist may find a weak gene-disease relationship that is strengthened by the addition of their own research data.

Databases have been designed and tools built to search the biomedical literature as well. The best is PubMed, the searchable database related to biomedical literature present in Medline. This database, with over 14 million references, is the most comprehensive listing of biomedical literature in the world. These tools can be used to download and parse the appropriate data to a secondary database that can be examined based on the users needs. Secondary databases that perform these functions include MedMiner, which allows one to query Genecards using terms related to physiologic pathways and receive back a list of genes involved in that pathway. In addition, gene or drug names can be sent to PubMed to identify the biomedical literature by searching the abstracts, Keywords, and MeSH terms. A useful function that is not present in MedMiner is the capability to comprehensively search for all genes related to a keyword. Thus MedMiner does not allow the desired degree of flexibility in user search terms and the comprehensive search of all key biological names.

PubGene (Jenssen et al., 2001) uses a similar design to allow the user to query genes using the HUGO approved gene symbols in its database. This database contains relationships identified through searches of Medline and identifies pairs of genes that are mentioned in the same abstract or correlated by GO (Ashburner, 2000) classifiers. The PubGene query system returns a graphical representation of the gene-gene relationships mentioned in the same reference as the queried gene. However, *the focus of the PubGene process is to identify gene-gene relationships and not gene-search term relationships* ("search term" can be gene, protein, drug, etc.).

In summary, the available databases offer useful information related to genes and their interrelationships in the biomedical literature, however, there is a lack of a truly flexible user-driven data mining system for multiple biological objects, even for all genes. More importantly, most of the existing tools *provide well established and proven associations* in actively investigated areas and they do not provide information on associations that are less organized and obvious. The proposed BioMap and its associated tools described herein are designed to specifically address these issues.

3. CASE STUDIES

Different types of biological relationships can be extracted from literature documents. These include flat relationships, directional relationships, and hierarchical relationships. Flat relationships simply state there exists a relationship between two biological entities. Directional relations also indicate the direction of the relationship that actually applies, for example, "A inhibits B" or "A is inhibited by B." In this section, we present three case studies in biological association discoveries. The first two studies (described in section 3.1 and 3.2) illustrate in a comprehensive way all the problems arising in biological relationship finding and some computational approaches to their solutions. The second study deals with an important extension of the basic association discovery methods, i.e., using transitivity property to postulate implicit potentially novel associations. The third study looks into discovering directional and hierarchical associations using text mining approaches.

3.1 Identification of Flat Relationships from Text Documents

In this case study we present a Thesaurus-based text analysis approach to discover the existence and the functional nature of relationships between one single biological object (e.g. gene) relating to a problem domain of interest. The approach relies on multiple Thesauri, representing domain knowledge as gene names and terms describing gene functions. These Thesauri can be constructed using existing organizational sources (e.g., NCBI and EBI), by consulting experts in the domain of interest, or by the users themselves. Thesauri can also be constructed using automated vocabulary discovery techniques being developed by the Information Extraction (IE) or Information Retrieval (IR) communities. In its simplest form, a Thesaurus consists of a linear list of terms and associated concepts. The process involves Thesaurus-based content representation of the retrieved documents, identification of associations (relationships) and finally, detecting gene functionality from the represented retrieved document set. These primary steps are described in detail in the following sections along with some experimental results.

3.1.1 Text Document Representation

The document representation step converts text documents into structures that can be efficiently processed without the loss of vital content. At the core of this process is a thesaurus, an array *T* of atomic tokens (e.g., a single term) each identified by a unique numeric identifier culled from authoritative sources or automatically discovered. A thesaurus is an extremely valuable component in term-normalization tasks and for replacing an uncontrolled vocabulary set with a controlled set (Rothblatt et al., 1994). Beyond the use of the thesaurus, the *tf.idf* (the term frequency multiplied with inverse document frequency) algorithm (Rothblatt et al., 1994) is applied as an additional measure for achieving more accurate and refined discrimination at the term representation level. In this formula, the *idf* component acts as a weighting factor by taking into account inter-document term distribution, over the complete collection given by:

$$W_{ik} = T_{ik} \times \log(N / n_k) \quad (1)$$

Where T_{ik} is the number of occurrences of term T_k in document *i*, $I_k = \log(N/n_k)$ is the inverse document frequency of term T_k in the document base, *N* is the total number of documents in the document base, and n_k is the number of documents in the base that contain the given term T_k.

As document representation is conducted on a continuous stream, the number of documents present in the stream may be too few for the *idf* component to be usefully applied. To deal with this, a table is maintained containing total frequencies of all thesaurus terms in a sufficiently representative collection of documents as a base (randomly sampled documents from the source used as the training set). It is worth pointing out that such a table can be pre-constructed off-line before any on-line analysis of retrieved documents is attempted. The purpose of the document representation step is to convert each document to a weight vector whose dimension is the same as the number of terms in the thesaurus and whose elements are given by the above equation.

3.1.2 Gene-pair Relationship

The goal here is to discover pairs of genes from a collection of retrieved text documents such that the genes in each pair are related to one other in some manner. Whether two genes are to be related depends on somewhat subjective notion of "being related." We have investigated Gene-pair discovery from a collection of Medline abstracts using the Vector-Space *tf*idf* method and a thesaurus consisting of Gene terms. Each Gene term, in turn, contains several synonymous keywords that are gene names. Each document d_i is converted to a M dimensional vector W_i where W_{ik} denotes the weight of the kth gene term in the document and M indicates the number of terms in a Thesaurus. W_{ik} is computed by equation 1 described in Section 3.1.1.

It is clear that W_{ik} increases with term frequency T_{ik}. However, it decreases with n_k, i.e., if a gene term occurs in increasingly larger number of documents in the collection, it is treated as a common term and its weight is decreased.

Once the vector representation of all documents are computed, the association between two gene terms k and l is computed as follows:

$$association[k][l] = \sum_{i=1}^{N} W_{ik} * W_{il} \quad \text{k} = 1...\text{m}, l = 1...\text{m} \tag{2}$$

For any pair of gene terms co-occurring in even a single document, the *association*[k][l] will be non-zero and positive. However, the relative values of *association*[k][l] will indicate the product of the importance of the kth and l^{th} term in each document, summed over all documents. This computed association value is used as a measure of the degree of relationship between the kth and l^{th} gene terms. A decision can be made about the existence of a strong relationship between genes using a user-defined threshold on the elements of the Association matrix.

3.1.3 Functional Nature of Relationships Between Gene-pairs

Once a "relationship" has been found between genes, the next step is to find out what that relationship is. This requires an additional thesaurus containing terms relating to possible relationships between genes that a user may be interested in. This thesaurus is then applied to sentences, which contain co-occurring gene names. If a word in the sentence containing co-occurrences of genes matches a relationship in the thesaurus, it is counted as a score of one. The highest score over all sentences for a given relationship is then taken to be the relationship between the two genes or proteins. A score of as little as one could be significant because a relationship may be only mentioned in one abstract. A higher score, however, would be more likely to indicate that relationship because they are often reiterated in multiple abstracts. The following equation summarizes the relationship:

$$score[k][l][m] = \sum_{i=1}^{S} p_i \,;\, (p_i = 1 : Gene_k, Gene_l, Relation_m \text{ all occur in sentence } i) \tag{3}$$

where, S is the number of sentences in the retrieved document collection, p_i is a score equal to 1 or 0 depending on whether or not all terms are present, and $Gene_k$ refers to the gene in the gene thesaurus with index *k*, and $relation_m$ refers to the term in the relationship thesaurus with index *m*. The functional nature of the relationship is chosen as arg_{max} score[k][l][m].

The idea is to narrow down the search to a few relationships which the user can check. If a functional relationship cannot be found the user can still check against articles where the terms co-occurred to see if a function might have been missing from the function thesaurus containing the relationships. Overall, this will help the user to quickly develop potential pathways and speed up the process of finding genetic interactions.

3.1.4 Experimental Results

Two experiments show how this technique performs in accuracy and as a tool for discovering a legitimate pathway based on retrieved data. The list of potential relations used for both examples, determined manually using a Molecular Biology text book (Salton, 1989), is shown in Table 16-1. The first experiment uses the gene list shown in Table 16-2.

Table 16-1. The Thesaurus of Relationships

"activates, activator"	"inhibits, inhibitor"	"phosphorylates"		
"binds, binding, complexes"	"catalyst, catalyses"	"hydrolysis, hydrolyzes"		
"cleaves"	"adhesion"	"donates"	"regulates"	"induces"
"creates"	"becomes"	"transports"	"exports"	"releases"
"suppresses, suppressors"				

This list includes genes and proteins not taken from any particular pathway but is associated with cell structure and muscle cells.

Table 16-2. Thesaurus of Genes (Unknown Pathway)

"actinin"	"actn2"	"ank1, ankyrin"	"atf4"	"ca3"	"CD36"	"cd54"
"COI"	"cox1"	"CSE1"	"cst3"	"desmin"	"FKBP51"	"FKBP54"
"FUS, TLS"	"GAPDH"	"hmsh2"	"hrv"	"hsp90"	"importin"	
"lim"	"mcm4"	"myoglobin"	"nebulin"	"nfatc"	"myosin"	
"nop-30"	"NPI-1"	"p55"	"titin"	"ubiquinone"	"filamin"	

The training documents are created by taking an equal number of abstracts from the Medline database for each gene. Altogether, 5,072 abstracts were used.

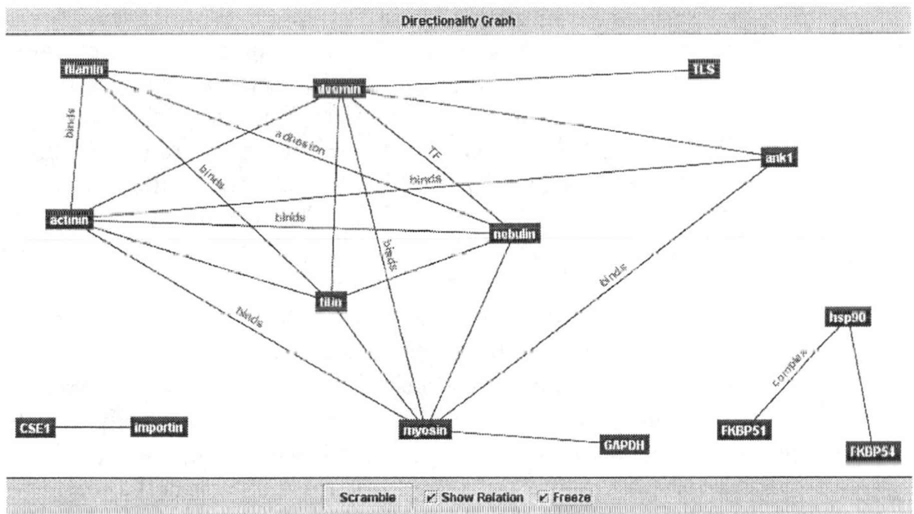

Figure 16-1. Graph showing relationships between genes in Known Pathway. The higher the Association strength the closer the genes appear on the graph. In this way the related genes are clustered together and can be picked out.

A graphical presentation of the unknown pathway (Table 16-2) is shown in Figure 16-1. The relationship discovery aspect of this method was excellent. This was verified by looking at the actual abstracts on the basis of which associations were computed. The strong central cluster includes proteins involved in construction of the cytoskeleton. The cluster containing CSE1 and *importin* are involved in the process of recycling *importin* and the other cluster contains proteins involved in making a steroid receptor complex. More details about the results and discussions can be found in (Stephens et al., 2001).

3.2 TransMiner: Formulating Novel, Implicit Associations Through Transitive Closure

An important question in biological knowledge management is whether it is possible to generate novel hypotheses concerning associations between biological objects, based on existing associations as presented in the literature. We have developed a system called TransMiner, which aims to identify transitive associations by using graph theoretic properties, in particular the transitivity property, on an underlying association graph. A strong motivation for the use of such transitivity property was provided by Swanson (Swanson and Smalheiser, 1997) and his co-workers. The idea is that if, according to existing literature, object A is related to object B, and object B is related to object C, then there is a likelihood of A being related to C, even though the last association may not have been explicitly reported. Moreover, considering such transitive (implicit) association between A and C, the likelihood of its existence is increased if more and more intermediate objects (object B) are found in the literature. Further, such transitive property can be extended through any number of intermediate nodes, as incorporated in the transitive closure of the original graph. Swanson developed a system called ARROWSMITH (Swanson and Smalheiser, 1997) that automated the one-step transitive relationship discovery process by considering only document titles and one intermediate node. TransMiner generalizes it by considering entire document abstracts (also full-text articles, if available) in addition to document titles, and also extending the transitivity property to the complete transitive closure of the original graph.

Swanson made seven medical discoveries by analyzing medical literature and applying the one-step transitivity property on titles (including the famous prediction of the magnesim-migraine association, before it was biologically verified). Smalheiser (Smalheiser, 2002) a collaborator of Swanson used ARROWSMITH to discover that genetic packaging technologies such as DEAE-dextran, cationic liposomes and cyclodextrins are plausible candidates to enhance infections caused by viruses delivered via an aerosol route – despite the fact that no studies had been reported that examined this issue directly.

Another novel feature of TransMiner is an iterative retrieval and association extraction process in an attempt to verify potential new associations from literature in an effort to overcome limited initial document set size. This prevents processing an inordinately large document set unnecessarily (possibly the entire MedLine!).

3.2.1 Transitive Association Discovery – Methods and Techniques

Relations are ways in which things can stand with regard to one another or to themselves (Honderich, 1995). Relation R is transitive if R (x, y) and R (y, z) imply R (x, z). In symbols, R is transitive if and only if $\forall x \forall y \forall z$ $((Rxy \wedge Ryz)\ Rxz)$.

Transitive Closure:

The transitive closure of a graph G is the graph G* such that there is an edge from vertex A to vertex C in G* if there is a path from A to C in G. The traditional Warshall's (Warshall, 1962) algorithm can be used to compute the transitive closure of the association graph. Given a directed graph G = (V, E) where, V is the set of vertices and E is the set of edges, represented by an adjacency matrix A[i,j], where A[i,j] = 1 if (i,j) is in E, compute the matrix P, where P[i,j] is 1 if there is a path of length greater than or equal to 1 from i to j. Thus, defining $A^0 = I$ (the Identity matrix), and $A^i = A^{i-1}.A$ for all i, where the matrix multiplication is Boolean,

$$P = \sum_{i=1}^{\infty} A^i \tag{4}$$

This algorithm extends paths by joining existing paths together. The transitive closure of a symmetric matrix (undirected graph) is also a symmetric matrix (undirected graph).

Mining Direct and Transitive Associations from Potential Transitive Associations:

The newly discovered potential transitive associations must be checked to see if those associations are indeed 'direct' (explicitly found in any of the Medline documents). We used an automated way (Algorithm 1) to find those associations that are direct and that are transitive, by submitting the two nodes (objects) of a potential transitive relationship to the Medline database with 'AND' operator in the query iteratively for all potential transitive object pairs. The documents will be retrieved only if both the objects are present in the document. For any pair of objects representing a potential transitive relationship, if the document set retrieved is non-zero, then by the principle of co-occurrence we can conclude that there exists a possibility of association between this object pair and that the association is direct. The association strength of these newly discovered 'direct' associations are given by the product of tf.idf weight of both nodes (objects) summed over all the documents retrieved for the object pair. The rest of the potential transitive associations with zero strength are implicit or transitive. These transitive associations are candidates for hypothesis generation. For these transitive associations there are no documents in Medline at present that have both the objects in their contents.

Algorithm 1: Transitive Association Discovery

1. Potential transitive associations are the difference between the transitive closure (G*) and the initial association graph (G).
2. Find the object pair for each potential transitive association and construct the Medline URL query using 'AND' operator.
3. Retrieve documents for this object pair and calculate the association strength between the object pair.
4. If the association strength is not zero, the association is direct. Keep the object pair in G*

5. If the association strength is zero, the association is transitive. Remove the object pair from G*
6. Repeat steps 2, 3, 4, and 5 for all the potential transitive associations discovered to get G' that contains the initial direct associations G and the newly discovered direct associations.

Ranking transitive associations:

Ranking of the transitive associations that are new and potentially meaningful associations will help the user to select associations (hypotheses) that can be further investigated in detail. Transitive association strength cannot be calculated directly as done in the case of direct associations, as there is no co-occurrence in any document between the nodes "A" and "C" of a transitive association. The transitive association strength is defined as the sum of weight of all words "B" that co-occur with both nodes "A" and "C" of a transitive association (intersection of words that co-occur with A and words that co-occur with B). This is based on the idea that if there is a strong link in the form of A-B-C then the possibility of AC association becoming true is more.

3.2.2 Experimental Results - Association Discovery among Breast Cancer Genes

This validation study attempted to use TransMiner to extract gene-gene associations relevant to the disease of Breast Cancer. A list of fifty-six gene symbols related to breast cancer was made from Baylor College of Medicine, Breast Cancer Gene Database (Baasiri et al., 1999) and the GeneCards database (Rebhan et al., 1997). These gene names are given in Table 16-3 and formed the dictionary for the validation study.

Table 16-1. Fifty-six breast cancer genes

APC	APS	ATM	BCL1	BCL2	BRCA1	BRCA2	CCND1
CDKN2A	COL18A1	DCC	EGF	EGFR	EMS1	ERBB2	ERBB3
MSH2	MLH1	FGF3	FGF4	FGFR1	FGFR2	FGFR4	GH1
GRB7	HRAS	IGF1R	KIT	KRAS2	MYCL1	IGF2R	MCC
MDM2	MET	MYC	NF2	NRAS	PGR	PHB	PLAT
PLG	PRL	PTH	PTPN1	RB1	SSTR1	SSTR2	SSTR3
SSTR4	SSTR5	SRC	TGFA	TP53	TSG101	VIM	WNT10B

The initial document set was 5000 Medline documents. The initial association discovery extracted 87 direct associations (i.e., association pairs with non-zero weights). This formed that initial graph G, in which the gene pair BRCA1-BRCA2 was found to have the highest association strength, which is expected. Application of Warshall's transitive closure algorithm on

G (to calculate the transitive closure G^* of G) yielded 655 potential new transitive gene pair associations were obtained in (G^* - G). The iterative retrieval and validation process identified 296 of them as direct associations (i.e., mentioned explicitly in the literature, although not mentioned in the 5000 original documents) and the remaining 359 as transitive association.

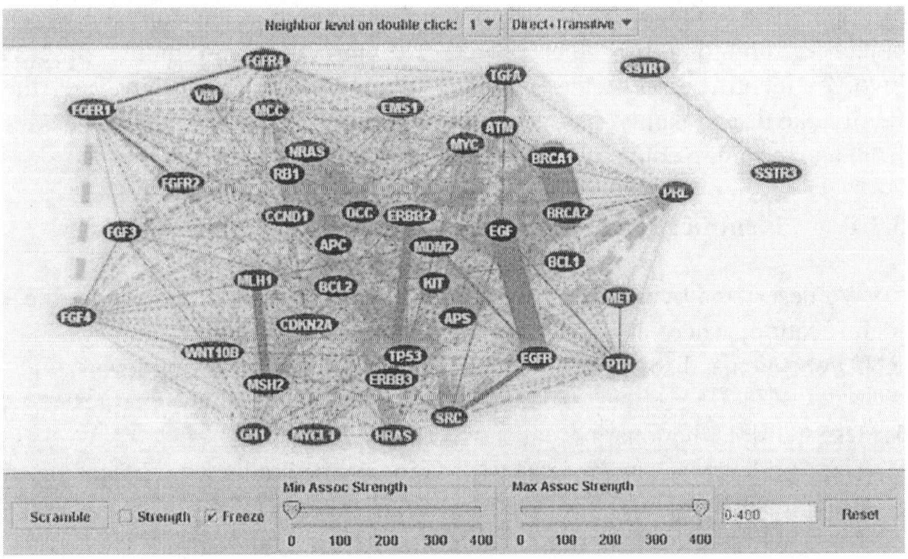

Figure 16-2. The initial direct associations among 56 gene symbols based on 5,000 Medline documents (blue edges), the direct associations discovered from potential transitive associations (blue dash edges) based on the presence of non-zero association in Medline database and the transitive associations (pink dash edges)

Figure 16-2 is a color-coded graphical display of all the associations. Based on manual evaluation of the 87 initial gene pair associations discovered by TransMiner, 75 (86.21%) gene pairs were found to have some valid biological association. Similarly, out of 296 direct gene pair associations discovered from potential transitive associations, 237 (80.06%) gene pairs were found to have biological association based on expert evaluation.

The detailed results and evaluations are available at http://sifter.cs.iupui.edu/~sifter/transMiner/TransMinerBCResults.html

3.3 Identification of Directional and Hierarchical Relationships

The association discovery methods described in Sections 3.1 and 3.2 does not take into account the directionality of the relationships. For example, if the relationship is "inhibits," then it is important to know which object is inhibiting the other object. The directionality finding process involves identification of the biological objects, the relationships, and the finally, the directionality. In this section, we describe each of these processes from a text mining context.

3.3.1 Identification of Biological Objects

We describe a hybrid method to address the specific challenges in object identification, where an object can be a gene, protein, cell type, organism, RNA, chemicals, disease, or drug. This consists of the following levels:

1. Use multiple dictionaries to identify known objects.
2. Use Hidden Markov Models (HMM) to identify unknown objects based on term suffixes, and,
3. Use N-Gram models to resolve object name ambiguity.

The tagging process begins with a Brill tagger (Brill, 1995) generating the POS. The process then continues with creating a dictionary of terms from databases (e.g. Swiss-Prot), for each class type (e.g. protein, gene, etc.) to be identified, and a dictionary such as WordNet (Brill, 1995), which contains the majority of other known nouns (e.g. lab, country, etc.) that may not be classified as a classified object. These dictionaries are to be single-token words, meaning they are very general in nature. To create these dictionaries, one would take a list of multi-token words (e.g. IkB inhibitor, RNA polymerase) which are defined in a class (e.g. protein) and then take the last word from each (e.g. inhibitor, polymerase). This can be described as $w_1w_2w_3...w_n \in$ MTD then $w_n \in$ STD where MTD is the multi-token dictionary, and STD is the single token dictionary, and w_i is the ith term in a multi-token word. The other piece of data the user needs is a set of training documents from the area for which the objects are to be tagged.

Once the training data is obtained, two important steps are involved in the tagging process. First, an N-gram model describes a class (e.g. protein, gene, etc.) using the phrases of the surrounding context. Second, an HMM model describes a class based on the internal context. If abbreviations are present, then a separate HMM model is created to describe them using a separate dictionary for each class where abbreviations commonly occur. Protein and gene classes would need this extra model to fully describe them as they often use abbreviations.

N-Gram Models:

An N-Gram model is used to disambiguate object tags. An N-Gram model is described as a simple Markov model where the probability of a word W_1 in position *n* can be given by the following equation (Jurafsky and Martin, 2000):

$$P(w_1^n) = \prod_{k=1}^{n} P(w_k \mid w_{k-N+1}^{k-1}) \tag{5}$$

where $P(w_k \mid w_{k-N+1}^{k-1})$ is the probability that w_k follows the previous *N* words. This is a simplification and assumes a word's probability is only dependent on the previous N characters. In order to calculate $P(w_k \mid w_{k-N+1}^{k-1})$ for each word in a given training corpus, the following general equation is used (Jurafsky and Martin, 2000):

$$P(w_n \mid w_{n-N+1}^{n-1}) = \frac{C(w_{n-N+1}^{n-1} w_n)}{C(w_{n-N+1}^{n-1})} \tag{6}$$

where $C(w_{n-N+1}^{n-1} w_n)$ is the number of times the previous *N* words are followed by w_n and $C(w_{n-N+1}^{n-1})$ is the total number of times the previous *N* words occur. Often times a given corpus is not sufficient to encompass all words that may be encountered in a given corpus. It is necessary to use smoothing to help describe more accurately the probability of a given word. One of the best methods used is the Good-Turing method and is described as (Jurafsky and Martin, 2000):

$$c^* = \frac{(c+1)\dfrac{N_{c+1}}{N_c} - c\dfrac{(k+1)N_{k+1}}{N_1}}{1 - \dfrac{(k+1)N_{k+1}}{N_1}} \tag{7}$$

for $1 \le c \le k$, where *c* is the original count of the word and N_C is the number of words counted *c* times.

Object Disambiguation Using N-Gram Models:

The N in N-Gram is the number of words in a given *pregram* or *postgram.* For this tagging method, the N-gram model takes the phrase data that was obtained using an N-Gram training process and uses it to define the probabilities for each class given a phrase. This probability is defined as:

$$P(c \mid phrase) = \frac{C(phrase_c)}{C(allphrases_c) + \sum_{i=0}^{M} C(phrase_i)} \qquad (8)$$

where $C(phrase_c)$ is the number of *times* the phrase appears in class c, *C(allphrasesc)is* the number of times all phrases appear in class c, *M* is the total number of classes, and $C(phrase_j)$ is the number of times the phrase appears in class *i.*

For N-Grams, that have N > 1, it becomes necessary to set up the model so that if the match for the full length N-Gram is not found, then the (N-l)-Gram can be tried, and if that does not work, the (N-2)-Gram, etc., would be needed. This stepping down can continue all the way down to the 1-Gram. If there is no match for the 1-Gram then the N-Gram fails to classify the object. The stepping down also allows the best possible N-Gram that matches to be found. Of course, a probability obtained using an N-Gram will always be greater than that obtained using an M-Gram where M < N. The following two-step process finds a class for a word having a *pregram*, *postgram*, or both:

1. Given a postgram and pregram,
2. Return the class c having the maximum $P(c_i \mid postgram) + P(c_i \mid pregram)$

If the probability for the class is zero, it shows that the word is not represented by the model and that it would require additional processing. Typically, smoothing would be done (e.g. Add-One Smoothing, Good-Turing Discounting, etc.; see (Jurafsky and Martin, 2000)).

Tagging of Abbreviations Using HMM:

An HMM is used to classify words that are abbreviations composed of less than six characters. The size of six was chosen based on the observation that most abbreviations are less than six characters long. This is done using a separate set of dictionaries specified by the user that are example abbreviations of several words known to fall into a specific class (e.g. SPF-l is a known protein abbreviation) that is known to contain abbreviations. These abbreviations are used as training data for the HMM. This abbreviated HMM will be referred to as the short HMM (SHMM). In addition, there may be longer words which are comprised of unusual symbols but represent an important object (e.g. TrpI53->Gly, which represents a specific change in a protein sequence). These longer words would have a separate HMM which will be referred to as the long HMM (LHMM).

An HMM contains states that represent a defined character type and the events in the states represent the specific characters in the word. To separate different words based on their characters, the model uses two groups of states. These states are identical in that they represent the same character types, but are different in that one represents a character type for a particular word type while the other represents the same character type for another word type. Figure 16-3 shows an example of different states of an HMM model to distinguish between words and gene names.

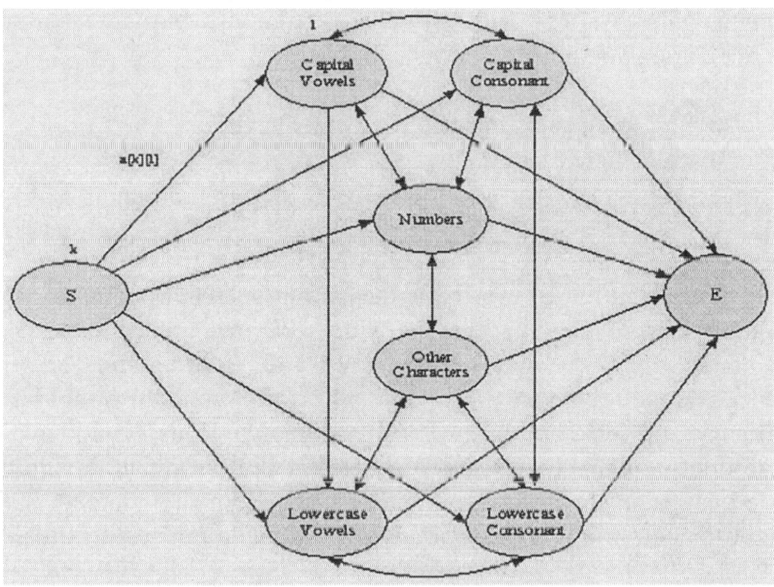

Figure 16-3. States of word tagger using HMM.

In this example, the state S represents the starting state. Within each state is the event probability of a given character occurring and is defined as $e[l](x[i])$ where *l* is the state and *x[i]* is the ith character in sequence *x*. There is also a transition probability between each state showing the probability of going from one state to another defined as $a[k][l]$ where *k* is the state that is being left and *l* is the new state being visited.

The path is the sequence of states that occur and the probability of a given path for a sequence of characters is given by the sequence of states and the corresponding characters occurring in the sequence of characters. Formally, it can be expressed as follows (Durbin et al., 1998):

$$p(x, p) = a[0][1]\prod_{i=1}^{L} e[i](x[i])a[i][i = 1] \tag{9}$$

The ending state of the most probable path is used to determine what object type the word is. In order to get the most probable path, the Viterbi algorithm (Durbin et al., 1998) is used.

3.3.2 Grouping Object Synonyms

The second stage in the whole process is resolving object synonyms correctly. The grouping of synonyms becomes complicated when (i) Synonyms share words and (ii) when Synonyms do not share words. Consider, for example, when they share words:

1. For the first time, **somatolactin (SL) cells** have ...
2. The **SL cells** were ...
3. The SL-immunoreactivity was mostly located in the granules of the **cells** ...

All three highlighted words in the above sentences refer to the same biological object. Knowing that the word *cells* in sentence three means the same thing as *SL cells* in sentence two and *somatolactin (SL) cells* in sentence one would have led to additional information that would have been specific enough for a biologist to use. Not knowing this would have caused the information extracted to be too general in the sense that the word "cell" by itself can represent more than one cell (e.g. somatolactin cell, heart cell, gonadotrope, etc.). The second case is when they do not share words:

1. Thyroid hormone receptors (T3Rs) are ...
2. T3Rs are bound by ...
3. **It** is found on ...
4.

Here the highlighted word "it" has little in common with the other two words. How to identify pronouns becomes important as information can be lost if they remain ambiguous.

In our approach, the word abbreviations are first processed through the PNAD-CSS algorithm (Yoshida et al., 2000). The abbreviations thus identified are used to group words together by merging words associated with the abbreviation with words associated with the full word from which the abbreviation was derived. In the first step in the grouping process, words are separated into their different classes (e.g. protein, gene). The next step is to build a generalized ontology and grouping of related words using a graph structure. The algorithms for this process can be broken down into two parts: the insertion of words from a group into a tree, and the extraction of word and their synonyms. The extraction process will create two categories of relationship between terms: one is a direct relationship like that of a word and its abbreviation, and the other is a hierarchical type where one word refers to several different words but those words do not refer to each other. For example, if someone talks about *proteins,* this is referring to more than one protein and not necessarily related, while if someone is referring to a *protein,* it is encompassing only one protein and a method similar to the pronoun tagging should be used.

A Grouping Example:

Consider the following text passage:

An anti-TRAP (AT) protein, a factor of previously unknown function, conveys the metabolic signal that the cellular transfer RNA for tryptophan (tTNA^{TRP}) is predominantly uncharged. Expression of the operon encoding AT is induced by uncharged tRNATRP. AT associates with TRAP, the trp operon attenuation protein, and inhibits its binding to its target RNA sequences. This relieves TRAP-mediated transcription termination and translation inhibition, increasing the rate of tryptophan biosynthesis. AT binds to TRAP primarily when it is in the tryptophan-activated state. The 53-residue AT polypeptide is homologous to the zinc-binding domain of DnaJ. The mechanisms regulating tryptophan biosynthesis in Bacillus subtilis differ from those used by Escherichia coli.

The tagging process would yield:

An **<p>anti-TRAP (AT) protein</p>**, a **<p>factor</p>** of previously unknown function, conveys the metabolic signal that the **<rna>cellular transfer RNA for tryptophan</rna>** (**<rna>tTNA^{TRP} </rna>**) is predominantly uncharged. Expression of the **<dna>operon encoding AT</dna>** is induced by **<rna>uncharged tRNA^{TRP}</rna>.** **<p>AT</p>** associates with **<p>TRAP</p>**, the **<p>trp operon attenuation protein</p>**, and inhibits its binding to its **<rna>target RNA sequences</rna>.** This relieves **<s>TRAP-mediated transcription termination</s>** and **<s>translation inhibition</s>,** increasing the rate of **<s>tryptophan biosythesis</s>. <p>AT</p>** binds to **<p>TRAP</p>** primarily when it is in the tryptophan-activated state. The **<p>53-residue AT polypeptide</p>** is homologous to the **<d>zinc-binding domain</d>** of **<p>DnaJ</p>.** The mechanisms regulating **<s>tryptophan biosynthesis</s>** in **<o>Bacillus subtilis</o>** differ from those used by **<o>Escherichia coli</o>.**

The Grouping Process will then generate:

Group 1	Group 3
anti-TRAP (AT) protein	cellular transfer RNA for tryptophan
factor	tTNATRP
AT	uncharged tRNATRP
53-residue AT polypeptide	
Group 2	**Group 4**
TRAP	operon encoding AT
trp operon attenuation protein	

It can be observed that grouping these words can greatly change the statistical nature of the terms when one word in the group is used for all other words in the same group, helping methods to achieve accurate statistical measures.

3.3.3 Extracting Object Relationships

The final step in the process is to extract relationships between the tagged entities. This process defines two types of relationships. The first is referred to as *directional relationships*. These relationships include for example, *protein A inhibits protein B*. In this case, a biologist not only needs to know what the relationship is but also the direction in which the relationship occurs. The second type of relationship is referred to as hierarchical relationships. This type of relationship would include, for example, *the brain is part of the nervous system.* The next two sections discuss the techniques used for each type of relationship.

Directional Relationships:

Directional relationships are found using a Hidden Markov Model. This is accomplished by generalizing words based on their POS tag or object tag so that the model encompasses a wide variety of relationships without having a lot of training data. The idea behind the scheme is that each relationship has a certain form. When a sentence is given to the model, its state sequence will contain the states created for a specific relationship classifying the sentence to be that relationship. The direction is detected by creating two event sequences, one where one of the objects in the relationship is classified as the subject while the other sentence has the other object classified as the subject. The model would then give the highest probability to the sentence with the correct subject, indicating the direction of the relationship. This is important as there may be instances where in one sentence the subject comes before the verb (e.g. protein A binds protein B) and in another it comes after the verb (e.g. protein B is bound by protein A).

To understand this model, a short example showing how the model is built for two sentences representing the same relationship but having a different position for the subject and object is shown. The two sentences are as follows:

1. Protein A inhibits Protein B.
2. Protein B is inhibited by Protein A.

The sentences are then Brill tagged and the objects identified as:

1. *<protein>Protein A</protein> inhibits/VBZ <protein>Protein B</protein>*
2. *<protein>Protein B</protein> is/VBZ inhibited/VMX by/IN <protein>Protein A</protein>*

All possible relationships (e.g. protein-protein) that were defined by the user (during training) as directional are then extracted from the sentence to give the following possibilities:

1. Possibilities for sentence 1: <subject>Protein A</subject> <object>Protein B</object>, <object>Protein A</object> <subject>Protein B</subject>
2. Possibilities for sentence 2: <subject>Protein B</subject> <object>Protein A</object>, <object>Protein B</object> <subject>Protein A</subject>

A user would then define the type of relationship each sentence is and the relationship with the correct labeling. The produced event and state sequences needed to train the HMM would be as follows:

1. Class for sentence 1: Inhibits
   Event sequence: [subject][inhibits][object]
   State sequence: [subject][/VBZ][object]
2. Class for sentence 2: Inhibits
   Event sequence: [object][is][inhibited][by][subject]
   State sequence:[object][/VBZ][/VMX][by/IN][subject]

Once the event and state sequences are known, the parameters of the HMM are determined using the following equations (Durbin et al., 1998):

$$a_{kl} = \frac{A_{kl}}{\sum_{l'} A_{kl'}} \qquad e_k(b) = \frac{E_k(b)}{\sum_{b'} E_k(b')} \tag{10}$$

where a_{kl} is the probability of a transition from state k to state l, $e_k(b)$ is the probability that an event b occurs in state k, A_{kl} is the number of transitions from state k to state l in the training data, and $E_k(b)$ is the number of times the event b occurs in state k in the training data. Once the model is trained, the relationship finding using the model can be carried out. For example, consider the following sample sentence:

1. ADH is inhibited by alcohol.

The sentence is tagged and the possible relationships extracted. Assuming one of the defined directional relationships is between a chemical and protein, the process produces the following two event sequences:

1. [subject] [is] [inhibited] [by] [object]
2. [object] [is] [inhibited][by][subject]

The model generates the probability of event sequence 1 as 0.0 while the probability of event sequence 2 as 0.0625 (the actual algorithm is omitted here for the sake of brevity). Taking the higher probability, this produces the relationship that *alcohol* inhibits *ADH* as opposed to *ADH* inhibits *alcohol.* The general process for producing an event sequence for any sentence which is tagged can be given by the following process.

1. If the word is a tagged object, make the event the object tag.
2. If the word is not an object, make the event the word.
3. For each possible directional relationship found, produce two event sequences where each object in the relationship is represented as the subject event, each object is also represented as an object event, and each sequence has one subject event and one object event.

One advantage of this method is that it avoids the complex issues of creating rules encompassing all possible relationships that are needed in a rule-based approach. In addition, generalizing objects allow for more flexibility in the model and enables the detection of new relationships without having to define a specific event probability for an object which is not specified by the model in its unclassified form. Another advantage for using the HMM for classification includes the ability to overcome noise by allowing default event probabilities in cases where an event may not be defined. This allows sentences to be classified to their most probable classification despite the HMM not having seen the event sequence previously in training. In addition, the verb states (e.g. /VBZ, /VMX) can be modified to include new verbs which define new directional relationships.

Hierarchical Relationships:

The Hierarchical relationships take advantage of the fact that verbs are not important to the relationship. Hence, this type of relationship can be defined in purely statistical terms using only the parent and child of the relationship. This technique is closely related to the co-occurrences of the gene extraction process described in Section 3. This is different because before the association matrix was square and considered relationships between two objects that were classified in the same class. Now the relationship is defined in such a way that it considers relationships between any objects regardless of what class they are classified in.

3.3.4 Experimental Results

Various experiments were carried out to evaluate the performance of the system at all three stages of the process, namely, tagging, grouping and relationship extraction. Results on each are considered separately below.

Tagging Performance:

The tagging method was applied to 100 abstracts from Medline obtained using the keyword "pituitary." The results were quantified using the measurements precision, recall, and F-Score defined earlier. The N-Gram model's default phrase length was three, making it a 3-Gram model. The training data used for the 3-Gram was comprised of 2,000 abstracts obtained from Medline using the keyword phrase "protein interaction."

The tagging performance using the dictionary only was only 50-60% despite using a large dictionary of words extracted from Swiss-Prot. The addition of the HMM and N-gram to the tagging process produced the results in Table 16-4 and has an average F-Score of 70%. The final step of the tagging process which made corrections for mis-tagged abbreviations greatly increased specificity and recall by eliminating false positives from the HMM tagged protein and gene objects, and increased recall for other object types as their formally mis-tagged abbreviations are tagged correctly.

Experiments were also conducted by increasing the length of the 3-Gram model to a 4-Gram model, and only the protein object tagging performance increased slightly in both precision and recall. This can be expected considering the low number of 3-Grams found in the 3-Gram model, which would indicate an even lower number of 4-grams. Add to this the fact that the majority of 3-Grams that were found were surrounding the protein object type, it would be expected that the new 4-Grams found would most likely effect protein object tagging.

Table 16-4. Results of Gene/Protein Classification

Tag Type	Correct	Missed	Recall	Precision	F-Score
Protein	533	150	78%	67%	72%
Gene	54	66	45%	57%	50%
Chemical	115	44	72%	69%	71%
Organism	305	130	70%	76%	72%
Organ	171	96	64%	81%	72%
Disease	202	93	68%	60%	64%

Performance of Grouping:

Performance of the grouping process used the same set of documents used for evaluating the tagging performance. A good way to measure the grouping performance is to see how it reduces the amount of information in terms of unique objects. The grouping showed a drop in the number of protein-protein relationships, cell-protein relationships, and organ-cell relationships. The grouping of synonyms thus greatly reduced the complexity of the data and helped objects to become more specific. The number of unique objects dropping by 24.7% indicates that an object may be written in many different ways. This is particularly true when looking at protein names. The drop of 38% in the relationships between proteins, due to grouping, directly shows the way in which the same protein takes on many different word forms. This drop is less when referring to the relationships between proteins and cells, indicating that the use of different names to refer the same cell object are much less common than that of proteins. This is further illustrated where the drop is only 2% for binary relationships between organs and cells. This would indicate that the use of different names for organs is almost non-existent. These results are expected as proteins are much more likely to take on different names in a document than a cell type or organ.

The overall performance of grouping was obtained by going through ten grouped abstracts and counting the number of terms grouped correctly and grouped incorrectly. These results are shown in Table 16-5.

Table 16-5. Performance of Grouping process

# of Correct group terms	# of Missed group terms	# of Incorrect group terms	Recall	Specificity
46	4	10	92%	82%

Results on Object-Object Relationship Extraction:

Directional Relationships: The HMM model was first trained using four directional relationships: inhibit, activate, binds, and same for the problem of protein-protein interactions and were trained for the directional HMM. When the model was used on the training set of sentences, it was found to have recorded the relationships with a recall and specificity of 100%, which would be expected given the small number of relationships.

The model was then tested against a larger corpus of text consisting of 1,000 abstracts downloaded from Medline using the key word of "protein interaction." These abstracts were then tagged and grouped, and all possible protein-protein interactions were extracted without specification to direction. The possible relationships were then passed through the trained HMM to extract directional relationships. In all, there were 53 such relationships extracted of which 43 were correct giving a specificity of 81%.

Hierarchical Relationships: To test the hierarchical relationships, the same pituitary corpus that was used to test grouping was used. Of these, 83 were specific and accurate enough to be useful while 49 were either wrong or too general to be useful giving a specificity of 65%. Having a threshold on the association value of 20 would change it to be 57 and 14, respectively, giving a specificity of 82%. More details about this work can be found in (Palakal et al., 2002c; Palakal et al., 2003).

4. BIOMAP: A KNOWLEDGE BASE OF BIOLOGICAL LITERATURE

In this section we present the progress made to develop a complete "knowledge base" of associations between biological objects that are important for biologists to study and understand specific biological processes. The main goals of this effort are (a) to develop a very large knowledge base, called BioMap, using the entire Medline collection of literature documents (over 12 million), and (b) to develop an interactive knowledge network for users to access this secondary knowledge (BioMap) along with its primary databases such as Medline, GenBank, etc., in an integrated manner based on a specific area of problem enquiry. The development of BioMap and its associated access "window" (the knowledge network), all of the text mining tasks that were discussed in the previous sections (such as identification of biological object names and discovering object-object relationships) will be utilized. The overall architecture of BioMap is shown in Figure 16-4.

The BioMap system basically consists of a set of organism-specific Knowledge Bases, a collection of intelligent algorithms for biological Object Tagging, Identification, and Relationship Discovery, System Interface, and a User Interface. A multi-level hybrid approach that incorporates statistical, stochastic, neural network and N-Gram models along with multiple dictionaries are used to handle the multi-object identification and relationship extraction problem for BioMap as described in Section 3.

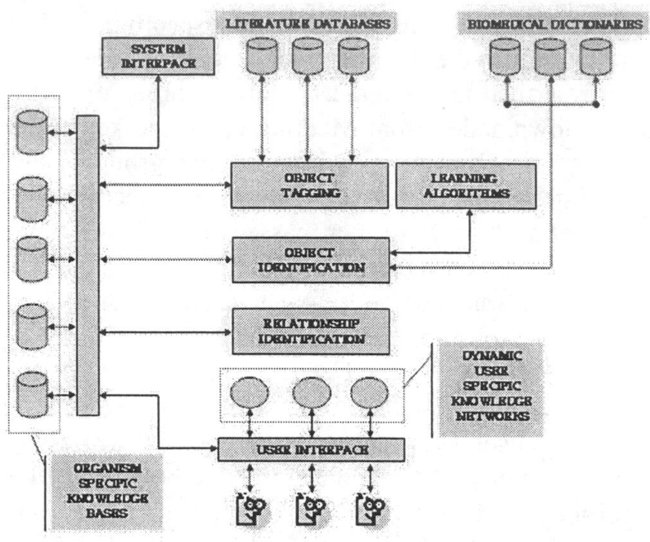

Figure 16-4. The overall organization of the BioMap System

The relationships thus discovered from the Medline collection are maintained in a relational database along with the specific links to literature sources, genes and protein sequence databases, popular bioinformatics tools such as PubGene, GO, etc., as well as links to image sources if the proof of relationships appear as images in the literature (as in the case of microarray experimental results). A user can access this knowledge base using any simple or complex queries, a disease name, a set of gene names, or any such combinations. Unlike in traditional databases, the outcome of a user query will be a complex set of data with multiple associations among them. Furthermore this network will be viewed in a hierarchical manner, allowing biologists to transcend the molecular view and see the physiological context from which a relationship is pulled. An example of the constructed knowledge network (knowledge view of BioMap) is shown in Figure 16-5. Knowledge outside of the hierarchical view can be pulled in, but the biologist will be able to specify what the context of the knowledge brought in is. For example, a biologist may start out looking at protein interactions shown to occur in the hippocampus of the human brain. The user may then choose to bring in additional interactions from the hippocampus of the rat brain. Understanding the context in which different interactions are playing a role is a key in understanding the function of a biological object. Different context often means the biological object can have a different function as described recently in (Brill, 1995). This context view is often overlooked.

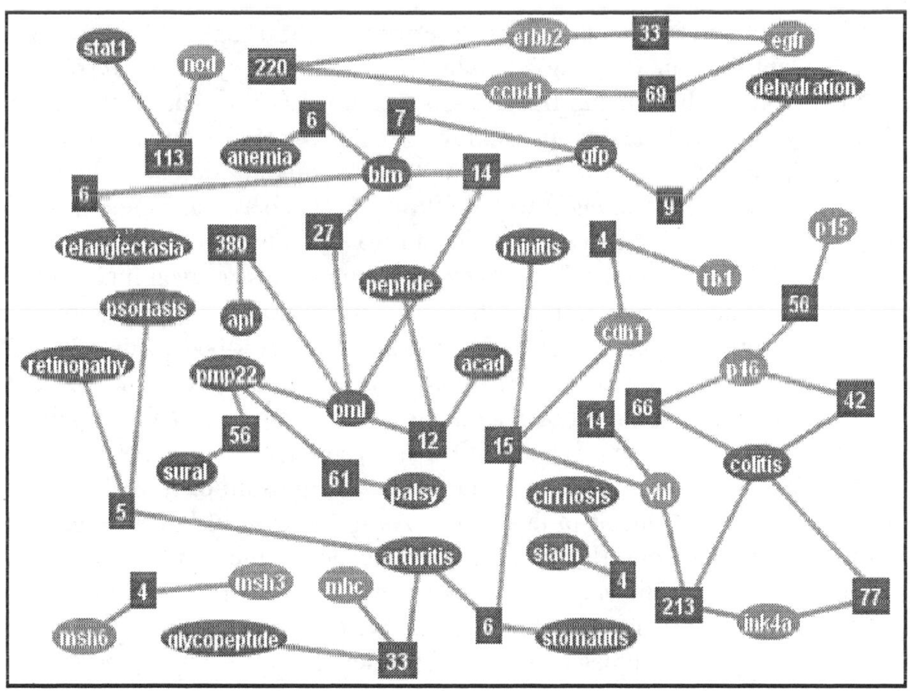

Figure 16-5. BioMap Knowledge represented as Hypergraph

The knowledge network is constructed as a hypergraph "on the fly" based on each user query. A hypergraph is an extension of a graph in the sense that each hyperedge can connect more than two vertices, thus allowing connecting relationships among multiple objects simultaneously. A system based on the hypergraph model has a number of advantages: the model is independent from updates to the underlying database; it enables the formation of hypergraphs from entities in different databases; it allows the system to be accessed by multiple users simultaneously, each with an independent hypergraph; further queries can be made to a hypergraph to obtain a better "focused" view of the knowledge base; it reduces the need to access the knowledge base multiple times when a hypergraph is to be shown to a remote user; and most importantly, the interaction with the user can be made faster when a query is made on the hypergraph since it is available in the local memory. Furthermore, the edges and vertices of the hypergraph will be "live" (made as hyperlinks) allowing the user to access primary data sources and bioinformatics tools pertinent to the information contained in the knowledge network.

In order to develop such a derived knowledge-base of associations from the primary source of biological literature databases, several research issues need to be resolved and these are described in the chapter. These include object identification (tagging), ambiguity resolution, synonym resolution, abbreviation resolution, and finally associations discovery and visualization. The knowledge network is a "window" to the BioMap's large knowledge base. Unlike traditional *binary relationships* among objects, BioMap's knowledge base has rich *multi-way relationships* such as that captured by the sentence "gene A inhibits protein B in pathway C in the context of disease D in organ E." This naturally leads to ternary, quaternary or even higher-order relationships and hence, to the notion of a hypergraph. A hypergraph is a generalization of a binary relationship graph (as described in the chapter) and is characterized by G = (V,E) where V is the set of vertices and E is the set of hyperedges. Unlike regular graphs where elements of E are pairs of vertices, denoting binary relationships, a hyperedge in a hypergraph is a subset of V and corresponds to a multi-way relationship of (possibly) more than two objects included in the subset. As in the case of binary edges, the multi-way associations (hyperedges) can be determined by co-occurrence based mining from BioMap's knowledge base. It is clear that the number of such possible hyperedges is combinatorially exponential with the number of objects, since, the number of subsets of a set A of cardinality n (i.e., the cardinality of the power set of A) is 2^n. This is in contrast to the binary graph, encoding binary relationship, where the number of possible associations (edges) is quadratic in n. Hence, any exhaustive attempt to check for all hyperedges will run into extremely high computational complexity, particularly since the total number of objects in the entire BioMap knowledge base is expected to be very large (in the hundreds of thousands). Hence, heuristic approximations are needed to limit the number of possible hyperedges.

4.1 BioMap Knowledgebase

The BioMap knowledgebase can be viewed as one large database or it can be conceptually divided by a concept such as organism to allow for greater scalability. The schema for the database is shown in the Figure 16-6. The Noun_Phrases table stores the extracted noun-phrases from text. This table is used for classifying these noun-phrases into different biological objects using various sources like UMLS, LocusLink etc. and machine learning techniques. These classified noun phrases are identified in the table Classified_Noun_Phrases that contains the type of object represented (e.g. organism) by this noun-phrase and the method used to classify it (e.g. dictionary), which are in turn stored in Categories and Methods tables. Each noun phrase can then be associated with a defined object through the Defined_Noun table. Relationships between defined objects can then be stored in the Relationship table. Within the relationship table there is a typical binary relation between two objects but also dependencies on that relationship can be described using the Relationship and Object dependency tables. The dependency can be either positive or negative. In this way, more complex relationships can be described beyond the binary relationships. The Complex_Relationships table adds to this by creating objects made up of relationships between other objects (e.g. a protein complex is made up of binding relationships between proteins and in some cases RNA)

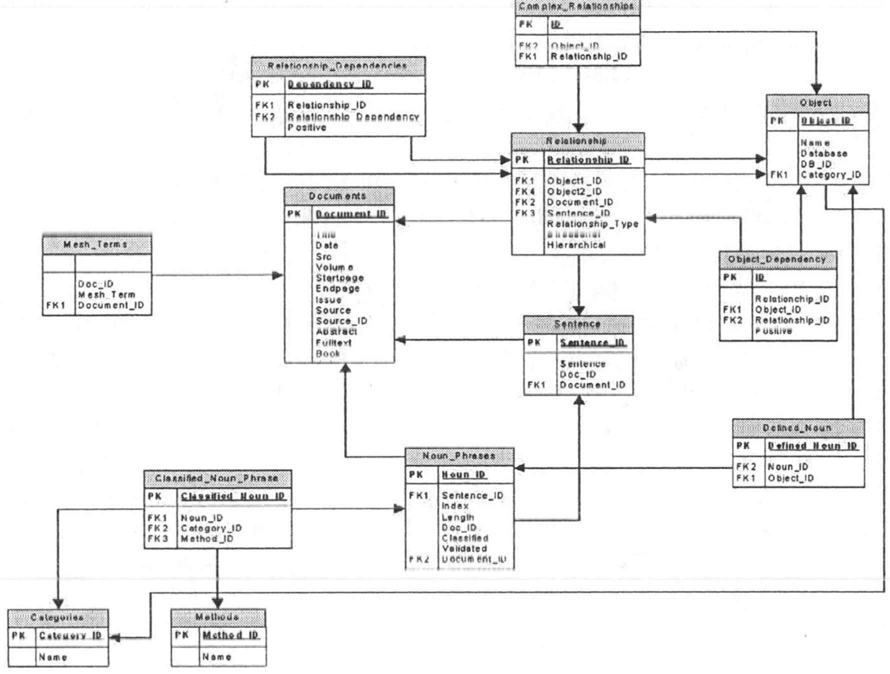

Figure 16-6. BioMap database schema

The BioMap knowledge base is implemented using Oracle 9i databases. The documents for each database are acquired from Medline. Once the database is populated and the noun-phrases are classified using different methods, they will form a basis for the knowledge network as discussed in the previous section, and they provide readily available data for testing and employing new techniques for object name resolution and other interesting text mining problems.

A major step in the creation of the knowledge base is to populate the database with relevant information from the text documents. This process involves identifying objects such as Gene, Protein, Cell Type, Organ, Organelle, Chemical/Drug, and Disease, and was carried out using the multi-level approach described in the previous section.

In our prototype study, we used a set of 30,000 documents and used UMLS and LocusLink to identify and classify objects. UMLS is a major source to resolve the noun-phrases into a number of categories. As a second step to improve on direct matches to UMLS concepts, the MetaMap Transfer (MMTx) program was used to map the noun-phrases to UMLS concepts. The MetaMap Transfer API in Java is used for this purpose. Again, only unambiguous matches are considered that also give a maximum score of 1,000. A score of 1,000 means that the mapping MetaMap came up with is the best one. The default parameters for the MetaMap are used. Overall, UMLS classifies the entities into a number of categories, which include genes, proteins, drugs/chemicals, and diseases, among others. The reason for the two steps used for UMLS is the following. Doing a direct match is much faster and the majority of entities resolved by the two steps is covered by the first step of direct comparison. MetaMap is used as an important second step to catch those noun-phrases that are similar but are not exact matches to a UMLS concept. LocusLink is then used for classifying gene names. LocusLink is a resource provided by NCBI that provides "genecentric" information for various organisms. LocusLink is particularly suited to the task as it has genetic information for multiple organisms. Currently "human," "rat" and "mouse" are being used to create BioMap. For each database for human, rat, and mouse, respective dictionaries of gene names are created from LocusLink. These dictionaries are then used to resolve the gene names in each respective organism's database. The noun-phrases that have not been resolved by UMLS are looked up in the LocusLink gene dictionary. If a match is found, then that entity is classified as "gene."

4.2 Results and Discussions

The results for entity name resolution for documents relating to human, rat and mouse are presented here. These results are based on the databases created using 30,000 documents from Medline and resolving noun-phrases using UMLS and LocusLink. The results are summarized in Table 16-6.

Table 16-6. Results of Name Resolution using Dictionaries

Noun-phrases		UMLS	LocusLink	Total
Human	Total	789,551		
	Classified	217312	9561	226873
	Percentage	27.52%	1.21%	28.73%
Rat	Total	94,212		
	Classified	21408	1261	22669
	Percentage	22.72%	1.34%	24.06%
Mouse	Total	89422		
	Classified	21385	2139	23524
	Percentage	23.91%	2.39%	26.31%

UMLS is a major contributor for resolving the object names followed by LocusLink. UMLS has resolved the entities into 132 distinct categories. LocusLink has resolved roughly 2% of the total nounphrases into genes.

As an example, from the experiments we have done, we consider some of the noun-phrases and walk through the process. In the first step of classification using UMLS, the noun "apoptosis" is classified as a "Cell Function." After that in the next step when we applied the MetaMap Transfer method, the noun-phrase "urinary infection," which was missed by the direct match method was mapped to "Urinary infection NOS (Urinary tract infection)" concept in UMLS, which belongs to a "Disease or Syndrome" category in the UMLS Metathesaurus. In the last step using LocusLink, let us consider the noun "FADD." The noun "FADD" was not classified by either UMLS methods, but LocusLink classified it as a gene.

As discussed in the previous section and evident from sample results, a multilevel approach to resolving names is quite effective in identifying important entities using a specialized dictionary for those types of entities. However, as we can see from the above results, the dictionaries can resolve only up to 30% of the nouns. This may probably be improved to 40% by using more specialized dictionaries for more types of entities. However, there is only so much that can be achieved using only a dictionary lookup approach. There is clearly a need for sophisticated algorithms to successfully classify the entities. The machine-learning techniques such as Hidden Markov Models (HMM) and N-grams to tackle the entities left unresolved by the dictionary look up approach is currently being developed.

The key innovative features of the proposed BioMap are that it is adaptable and scalable, and that the core knowledge base will be constructed from a very large collection of documents (Medline) to make the system robust. The adaptation feature requires that the system should have the ability to learn new problem domains without having to rebuild the system as in the case of fully rule-based or grammar-based approaches. The scalability feature allows the system to continue to develop its knowledge base as new information arrives in the literature databases or incorporating information from other data sources (e.g. Science, Nature, etc.). BioMap is novel in its ability to transcend typical views of data that only consider a small scope of objects and relations and allows for a global view of interactions among objects. It is hoped that this view will help biologists transcend the bottlenecks that keep them from relating findings at the molecular level to real physiological changes which characterize disease. Further discussions on BioMap can be found in (Kumar et al., 2004).

5. CONCLUSIONS

The biological literature databases continue to grow rapidly with vital information that is important for conducting sound biomedical research. The objective of the research described in this chapter is to develop, for the first time, a scalable knowledge base (BioMap) of biological relationships from vast amount of literature data. The results of this research will significantly enhance the ability of biological researchers with diverse objectives to efficiently utilize biomedical literature data. BioMap will be a new type of "secondary" knowledge resource derived from primary resources such as Medline. It will be the "window" to every biomedical researcher who will be seeking knowledge from the literature databases, however, without being overwhelmed by its large volume.

When the knowledge network is presented to the user as a rich hypergraph, it enables easier browsing of the content shown to the user. Each node in the hypergraph can be made to be rich in content. They can be clicked on to show a new hypergraph dynamically with the selected entity as the seed. They contain information such as citations from where the corresponding term is derived. The biological objects such as genes, proteins, drugs, etc., will be represented as nodes in a hypergraph. Hence, BioMap will not only be an effective aid for biomedical research, but also a teaching and learning tool for high school, undergraduate, and graduate students pursuing academic programs in biomedical sciences. Another significant contribution of this work is the ability of the system to efficiently discover associations not explicitly reported in any one document, but based on associations implicitly hidden in multiple documents.

6. ACKNOWLEDGEMENTS

The research reported in this paper was supported in part by National Science Foundation Information Technology Research grant number NSF-IIS/ITR 0081944, NIH BISTI grant, NIH-NIGMS P20 GM66402, and by an NSF/NSDL grant NSF-DUE-0333623. The computing facilities used for this research were supported in part by the Indiana Genomics Initiative (INGEN). The Indiana Genomics Initiative (INGEN) of Indiana University is supported in part by Lilly Endowment Inc.

REFERENCES

Ashburner, M. (2000). "Gene Ontology: Tool for the Unification of Biology," *Nature Genetics*, vol. 25, 25-29

Atschul, S.F., Gish, W., Miller, W., Myers, E., Lipman, D. (1990). "Basic Local Alignment Search Tool," *Journal of Molecular Biology* 215, 403-410

Baasiri, R.A., Glasser, S.R., Steffen, D.L., Wheeler, D.A. (1999). "The Breast Cancer Gene Database: A Collaborative Information Resource," *Oncogene* 18:7958-7965, http://tyrosine.biomedcomp.com/4d.acgi$tsrchname?Name=&topic=BCIR

Brill, E. (1995). "Transformation-based Error-driven Learning and Natural Language Processing: A Case Study in Part-of-Speech Tagging," *Computational Linguistics*, 21 (4):543-566

Chinchor, N. (1995). "MUC-5 Evaluation Metrices," in *Proceedings of the Fifth Message Understanding Conference (MUC-5)*, Baltimore, Maryland, USA, 69- 78

Collier, N., Nobata, C., and Tsujii, J. (2000a). "Extracting the Names of Genes and Gene Products with a Hidden Markov Model," *Coling 2000*, 201-207

Craven, M., Kumlien, J. (1999). "Constructing Biological Knowledge Bases by Extracting Information from Text Sources," *ISMB* : 10 – 20

Durbin, R., Eddy, S., Krogh, A., Mitchison, G. (1998). *Biological Sequence Analysis*. Cambridge University Press. New York, N

Friedman, C., Kra, P., Yu, H., Krauthamrner, M., Rzhetsky, A. (2001). "Genies: A Natural-Language Processing System for the Extraction of Molecular Pathways from Journal Articles," *Bioinformatics* 17 Suppl. 1, S74 -S82

Fukuda, K., Tsunoda. T., Tamura, A., and Takagi, T. (1998). "Toward Information Extraction: Identifying Protein Names from Biological Papers," in *Proceedings of the Pacific Symposium on Biocomputing*, 705-716

Genecards, http://mach1.nci.nih.gov/cards/index.html

Hatzivassiloglou, V., Duboue, P., Rzhetsky, A. (2001). "Disambiguating Proteins, Genes, and RNA in Text: A Machine Learning Approach," *Bioinformatics*, 17 Suppl. 1, S97 -S106

Honderich, T. (1995). *The Oxford Companion to Philosophy*, Oxford University Press, http://www.xrefer.com/entry/553381.

HUGO, http://www.gene.ucl.ac.uk/nomenclature/

Humphreys, K., Demetrios, G., and Gaizauskas, R. (2000). "Two Applications of Information Extraction to Biological Science Journal Articles: Enzyme Interactions and Protein Structures," in *Proceedings of the Pacific Symposium on Biocomputing*, 505-516

Iliopoulos, I., Enright, A.J., and Ouzounis, C.A. (2001). "Textquest: Document Clustering of Medline Abstracts for Concept Discovery in Molecular Biology."

Jenssen, T.K., Laegreid, A., Komorowaki, J., Hovig, E. (2001). "A Literature Network of Human Genes for High-Throughput Analysis of Gene Expression," *Nature Genetics*. May 28 (1); 21-8, http://www.pubgene.org/

Joshi, A.K. (1998). "Role of Constrained Computational Systems in Natural Language Processing," *AI Journal*, 103, 117-132

Jurafsky, D., and Martin, J. Speech and Lanuae (2000). *Processing*. Prentice-Hall, Inc. Upper Saddle River, New Jersey

Kazama, J., Makino, T., Otha, Y., Tsujii, J. (2002). "Tuning Support Vector Machines for Biomedical Named Entity Recognition," http://www.snowelm.com/~t/research/pub/./kazama_aclbio02.pdf

Krauthammer, M., Rzhetsky, A., Morozov, P., Friedman, C. (2000). "Using BLAST for Identifying Gene and Protein Names in Journal Articles," *Gene* 259: 245 – 252

Kumar, K., Palakal, M., and Mukhopadhyay, S. (2004). "BioMap: Toward the Development of a Knowledge Base of Biomedical Literature," in *2004 ACM Symposium on Applied Computing,* Nicosia, Cyprus

Leroy, G. and Chen, H. (2002). "Filling Preposition-Based Templates to Capture Information from Medical Abstracts," in *Proceedings of the Pacific Symposium on Biocomputing* 7, 350-361

LocusLink, http://www.ncbi.nlm.nih.gov/LocusLink/

Lodish, H., Berk, A., Matsudaira, P., Baltimore, D., Zipursky, S., Darnell, J. (1995). *Molecular Cell Biology.* Third Edition. Scientific Books, Inc. New York

Marcotte, E.M., Xenarios, I., and Eisenberg, D. (2001). "Mining Literature for Protein-Protein Interactions." *Bioinformatics*, 17: 359-363

MedMiner, http://discover.nci.nih.gov/textmining/filters.html

MetaMap Transfer, http://mmtx.nlm.nih.gov

Ng, S. and Wong, M. (1999). "Toward Routing Automatic Pathway Discovery from On-line Scientific Text Abstracts." *Genome Informatics*, 10:104-112

Nobata, C., Collier, N., and Tsujii, J. (2000). "Automatic Term Identification and Classification in Biology Texts," in *Proceedings of the Natural Language Pacific Rim Symposium (NLPRS '2000)*, 369- 375

OMIM, http://www3.ncbi.nlm.nih.gov/htbin-post/Omim/

Ono T., Hishigaki H., Tanigami A., and Takagi T. (2001). "Automatic Extraction of Information on Protein-Protein Interactions from the Biological Literature," *Bioinformatics*, 17(2):155-161

Oyama T., Kitano K., Satou K., and Ito T. (2002). "Extraction of Knowledge on Protein-Protein Interaction by Association Rule Ddiscovery," *Bioinformatics*, 18(5):705-714

Palakal, M., Stephens, M., Mukhopadhyay, S., Raje, R. and Rhodes, S. (2003). "Identification of Biological Relationships from text documents using efficient computational Methods," *Journal of Bioinformatics and Computational Biology*, Vol. 1(2), 1-34

Palakal, M., Stephens, M., Mukhopadhyay, S., Raje, R. (2002c). "A Multi-level Text Mining Method to Extract Biological Relationships," in *CSB2002*, Stanford, CA

Park, J.C., Kim, H.S., and Kim, J.J. (2001). "Bidirectional Incremental Paring for Automatic Pathway Identification with Combinatory Categorical Grammar," Oct., http://citeseer.nj.nec.com/384291.html

Proux, D., Rechenmann, F., and Julliard, L. (2000). "A Pragmatic Information Extraction Strategy for Gathering Data on Genetic Interactions," in *Proc Int Conf Intell Syst Mol Biol* 8, pages 279-285

PubGene, http://www.pubgene.com/

PubMed, http://www.ncbi.nlm.nih.gov/entrez

Rebhan, M., Chalifa-Caspi, V., Prilusky, J., Lancet, D. (1997). "GeneCards: Encyclopedia for Genes, Proteins and Diseases," Weizmann Institute of Science, Bioinformatics Unit and Genome Center (Rehovot, Israel), http://bioinformatics.weizmann.ac.il/cards

Rindflesch, T.C., Tanabe, L., Weinstein, J.N., and Hunter, L., (2000). "EDGAR: Extraction of Drugs, Genes and Relations from the Biomedical Literature," in *Proceedings of the Pacific Symposium on Biocomputing*; 517-28

Rothblatt, J., Novick, P., Stevens, T. (1994). *Guidebook to the Secretory Pathway.* Oxford University Press Inc., New York

Salton, G. 1989, Automatic Text Processing. Addison-Wesley

Sanchez C., Lachaize C., Janody F., Bellon B., Roder L., Euzenat J., Rechenmann F., and Jacq B. (1999). "Grasping at Molecular Interactions and Genetic Networks in Drosophila Melanogaster Using Flynets, an Internet Database," *Nucleic Acids Res,* 27(1):89-94

Smalheiser, N.R. (2002). "Informatics and Hypothesis-driven Research," *EMBO Rep* 3:702

Smalheiser, N.R. (2001). "Predicting Emerging Technologies with the Aid of Text-Based Data Mining: A Micro Approach," *Technovation* 21: 689-693

Steffen, B., Ulf, B., Faulstich, L., Hakenberg, J., Leser, U., Plake, C., Scheffer, T. (2003), http://www.pdg.cnb.uam.es/BioLink/workshop_BioCreative_04/handout/pdf/user12_1a.pdf

Stephens, M., Palakal, M., Mukhopadhyay, S., Raje, R., Mostafa, J. (2001). "Detecting Gene Relations from Medline Abstracts," in *PSB* 2001: 483-495

Swanson, D.R. and Smalheiser, N.R. (1997). "An Interactive System for Finding Complementary Literatures: A Stimulus to Scientific Discovery," *Artificial Intelligence* 91: 183-203

Tan, A-H. (1999). "Text Mining: The State of the Art and the Challenges," in *Proc of the Pacific Asia Conf on Knowledge Discovery and Data Mining PAKDD'99 workshop on Knowledge Discovery from Advanced Databases*, 65-70

Tanabe, L. and Wilbur, W.J. (2002). "Tagging Gene and Protein Names in Biomedical Text," *Bioinformatics*, 18(8): 1124-1132

Thomas, J., Milward, D., Ouzounis, C., Pulman, S., and Carroll M. (2000). "Automatic Extraction of Protein Interactions from Scientific Abstracts," in *Proceedings of the Pacific Symposium on Biocomputing*, 541-551

UMLS, http://www.nlm.nih.gov/research/umls/umlsmain.html

Warshall, S. (1962). "A Theorem on Boolean Matrices," *JACM* 9:11-12,

Yakushiji, A., Tateisi, Y., Tsujii, J., Miyao, Y. (2000). "Use of a Full Parser for Information Extraction in Molecular Biology Domain," *Genome Informatics* II: 446-447

Yoshida, M., Fukuda, K., and Takagi, T (2000) "PNAD-CSS: A Workbench for Constructing a Protein Name Abbreviation Dictionary," *Bioinformatics,* 16, 169-175

SUGGESTED READINGS

Hearst, M.A. (1998). *Automated discovery of wordnet relations.* In Christiane Fellbaum, editor, WordNet: An Electronic Lexical Database. MIT Press, Cambridge, MA
This chapter discusses methods to extract general lexico-semantic relationships (e.g., x is a kind of y) by extracting corresponding patterns from text. Some biological relationships could also be of lexico-semantic nature.

Salton, G. (1983). *Introduction to Modern Information Retreival.* McGraw-Hill, New York
A popular authentic textbook on information retrieval techniques, including discussion of vector-space (tf-idf) and other models of text retrieval.

Ashburner, M. (2001). Gene Ontology Consortium. Creating the Gene Ontology Resource: Design and Implementation. *Genome Research*, 11, pp. 1425-1433
The gene ontology consortium provides a database of a large number of gene names and their relationships to biological functions, processes, and cellular locations. While this exercise is useful in its own right, this also can be used for tagging and pathway generation in biological context.

Hirschman L, Park JC, Tsujii J, Wong L, Wu CH. (2002). Accomplishments and challenges in literature data mining for biology. *Bioinformatics*. 18(12):1553-1556
This paper provides a review of text mining in biology. While reviewing several works in object tagging, relationship extraction, and pathway generation, it discusses the requirements of benchmark information extraction task datasets for biological text mining.

Scheffer, T. (2004). *Proceedings of the Second European Workshop on Data Mining and Text Mining for Bioinformatics*, Pisa, Italy, September, Available on-line at: http://www.informatik.hu-berlin.de/Forschung_Lehre/wm/ws04/
A collection of papers describing some very recent research in text and data mining for Bioinformatics. There are papers on extracting protein-protein interactions and protein-function relationships, as well as other applications of text mining.

ONLINE RESOURCES

GENECARD: Rebhan, M., Chalifa-Caspi, V., Prilusky, J. and Lancet, D. (1997).
GeneCards: encyclopedia for genes, proteins and diseases. Weizmann Institute of Science, Bioinformatics Unit and Genome Center (Rehovot, Israel).
http://bioinformatics.weizmann.ac.il/cards

Medline
http://www.ncbi.nlm.nih.gov/entrez/query.fcgi

PIR (Protein Information Resources)
http://pir.georgetown.edu/

SUN's GRAPH VIEWING APPLET:19. Sun Microsystems, Inc. (1995)
Graph.java demonstration software. Sun Microsystems Inc.
http://java.sun.com/applets/jdk/1.0/demo/GraphLayout/index.html

TRANSMINER
http://sifter.cs.iupui.edu/~sifter/transMiner/TransMinerBCResults.html

QUESTIONS FOR DISCUSSION

1. One of the problems in association discovery is determining the direction, if any, of a particular association. What can be some of the approaches in determining such directionality through text mining?

2. Some of the approaches that can be used for object identification in text include rule (grammar) based, statistical, and connectionist or other machine learning approaches. What are the relative advantages and disadvantages of the different approaches?

3. Since there seems to be the possibility of a variety or a bank of multiple taggers (object identifiers) designed using possibly different computational techniques, a question arises as to whether it is possible to improve the tagging performance further by combining them in a judicious way. What are some of the issues involved in designing such a meta-tagger?

4. The general hypergraph construction algorithm is believed to be computationally very complex. Why? What could be some of heuristics and/or approximations that can be used to make it more tractable?

5. Information visualization: The user-specific knowledge graph (or, hypergraph) can be quite complex involving a large number of nodes and associations. What could be some approaches to visualizing such large graphs in a cognition-rich manner?

Chapter 17

CREATING, MODELING, AND VISUALIZING METABOLIC NETWORKS

FCModeler and PathBinder for Network Modeling and Creation

Julie A. Dickerson[1,2], Daniel Berleant[1,2], Pan Du[1,2], Jing Ding[1,2], Carol M. Foster[3], Ling Li[3], and Eve Syrkin Wurtele[2,3]

[1]*Electrical and Computer Engineering Dept;* [2]*Virtual Reality Applications Center;* [3]*Genetics Development and Cell Biology, Iowa State University, Ames, IA*

Chapter Overview

Metabolic networks combine metabolism and regulation. These complex networks are difficult to understand and create due to the diverse types of information that need to be represented. This chapter describes a suite of interlinked tools for developing, displaying, and modeling metabolic networks. The metabolic network interactions database, MetNetDB, contains information on regulatory and metabolic interactions derived from a combination of web databases and input from biologists in their area of expertise. PathBinderA mines the biological "literaturome" by searching for new interactions or supporting evidence for existing interactions in metabolic networks. Sentences from abstracts are ranked in terms of the likelihood that an interaction is described and combined with evidence provided by other sentences. FCModeler, a publicly available software package, enables the biologist to visualize and model metabolic and regulatory network maps. FCModeler aids in the development and evaluation of hypotheses, and provides a modeling framework for assessing the large amounts of data captured by high-throughput gene expression experiments.

Keywords

fuzzy logic; microarray analysis; gene expression networks; fuzzy cognitive maps; text mining; naïve Bayes

1. INTRODUCTION

The field of systems biology in living organisms is emerging as a consequence of publicly-available genomic, transcriptomics, proteomics, and metabolomics datasets. These data give us the hope of understanding the molecular function of the organism, and being able to predict the consequences to the entire system of a perturbation in the environment, or a change in expression of a single gene. In order to understand the significance of this data, the functional relationships between the genes, proteins, and metabolites must be put into context. This chapter describes an iterative approach to exploring the interconnections between biomolecules that shape form and function in living organisms. This work focuses on the model plant system, Arabidopsis. The systems biology approach itself can be used as a prototype for exploration of networks in any species.

The two major motivating currents in this work are the need to build systems for biologists and the need to better understand the science of knowledge extraction from biological texts. The biologist's information about the function of each RNA and protein is limited. Currently, about 50% of Arabidopsis genes are annotated in databases (e.g., TAIR (Rhee, Beavis et al., 2003) or TIGR (www.tigr.org)). In part because the process of evolution results in families of genes with similar sequences and related functions, much of the available annotation is not precise, and some annotation is inaccurate. Even more limited is our understanding of the interactions between these biomolecules. To help bridge this gap, metabolic networks are being assembled for Arabidopsis (e.g., AraCyc, KEGG). To date, these contain many derived pathways based on other organisms; consequently, they have errors and do not capture the subtleties of the Arabidopsis (or even plant) biochemistry and molecular biology that are necessary for research.

Considerable high-quality information is buried in the literature. A given pathway is known predominantly to those researchers working in the area. Such a pathway is not easily generated by curators whom are not experts in the particular field. This information is not rapidly accessible to a biologist examining large and diverse datasets and investigating changing patterns of gene expression over multiple pathways in which she/he may have little expertise. Furthermore, the interconnections between the multiple complex pathways of a eukaryotic organism cannot be envisioned without computational aid. To assist biologists in drawing connections between genes, proteins and metabolites, cumulative knowledge of the known and hypothesized metabolic and regulatory interactions of Arabidopsis must be supported by advanced computing tools integrated with the body of existing knowledge.

2. OVERVIEW

2.1 Metabolic Pathway Databases

The database used in this work is MetNetDB (Wurtele, Li et al., 2003). MetNetDB, combines knowledge from experts, Aracyc (Mueller, Zhang et al., 2003) and more specialized pathway sequence data, with experimental data from microarrays, proteomics and metabolomics and dynamically displays the results in FCModeler (Dickerson, D. Berleant et al., 2003; Wurtele, Li et al., 2003). The database is designed to include information about subcellular location, and to handle both enzymatic and regulatory interactions.

There are a few major database projects designed to capture pathways: What Is That? (WIT) Project (Overbeek, Larsen et al., 2000) (http://wit.mcs.anl.gov/WIT2/WIT), Kyoto Encyclopedia of Genes and Genomes (KEGG (Kanehisa and Goto, 2000), and EcoCyc/MetaCyc (Karp, Riley et al., 2000; Karp, Riley et al., 2002)).

WIT and KEGG contain databases of metabolic networks, which focus on prokaryotic organisms. The WIT2 Project produced static "metabolic reconstructions" for sequenced (or partially sequenced) genomes from the Metabolic Pathway Database. KEGG computerizes current knowledge of molecular and cellular biology in terms of the interacting genes or molecules and links the pathways with the gene catalogs being produced by the genome sequencing projects. EcoCyc is a pathway/genome database for *E. coli* that describes its enzymes and transport proteins. It has made significant advances in visualizing metabolic pathways using stored layouts, and linking data from microarray tests to the pathway layout (Karp, Krummenacker et al., 1999; Karp, 2001). The metabolic-pathway database, MetaCyc, describes pathways and enzymes for many different organisms (e.g. *Arabidopsis thaliana*, AraCyc), and combines information from sequences.

Other database designs emphasize data visualization. Cytoscape visualizes existing molecular interaction networks and gene expression profiles and other state data using Java (Shannon, Markiel et al., 2003). Cytoscape has facilities for constructing networks and displaying annotations from fixed files. MetNetDB is web-accessible and users can create their own custom-pathways, which can then be used to analyze expression data.

2.2 Network Modeling and Reconstruction

There have been many attempts to reconstruct gene regulatory and metabolic networks from microarray data using machine learning methods

(Weaver, Workman et al., 1999; Akutsu, Miyano et al., 2000; D'haeseleer, Liang et al., 2000; Matsuno, 2000; Genoud, Trevino Santa Cruz et al., 2001; Hartemink, Gifford et al., 2001; Wessels, Someren et al., 2001; Hanisch, Zien et al., 2002). However, these methods are based on the assumption that genes in the same pathways are co-regulated and show the same expression patterns. This assumption does not always hold for genes in pathways. Additionally, several pathways can show similar responses to a stimulus which leads to many false positive links. Data must be combined from multiple sources such as gene function data and expert information to give a complete picture of the interactions.

2.3 Extracting Biological Interactions from Text

Mining of the biological "literaturome" is an important module in a comprehensive creation, representation, and simulation system for metabolic and regulatory networks. Without it, many biomolecular interactions archived in the literature remain accessible in principle but underutilized in practice. Competitions to test the performance of automatic annotation such as the BioCreative Workshop (EMBO BioCreative Workshop, 2004), the TReC (Text Retrieval Conference) genomic track, and the KDD Cup 2002 show encouraging results, but high rates of error show that the systems are not yet accurate enough. None of these competitions directly focused on the problem of finding, and combining, evidence from sentences describing biomolecular interactions. This is a key need for a biological database system like MetNetDB, in which evidence provided by sentences must be rated to support ranking in terms of the likelihood that an interaction is described, must be combined with evidence provided by other sentences, and must support efficient human curation. Furthermore, sentence-based retrieval can be useful in and of itself to biologists, who are typically limited to retrieval based on larger text units as supported e.g. by PUBMED and Agricola in the biological domain and common Web search engines in general.

2.3.1 Empirical Facts about Biological Texts.

Although many researchers have investigated mining of biomolecular interactions from text, the reporting of empirical facts about interaction descriptions remains quite limited. Craven and Kumlien (Craven and Kumlien, 1999) investigated word stems and the ability of each to predict that a sentence describes the subcellular location of the protein if it contains a stem, a protein name, and a subcellular location. Marcotte et al., (Marcotte, Xenarios et al., 2001) gave a ranked list of 20 words found useful in identifying abstracts describing protein interactions in yeast-related abstracts. Results were derived from and therefore may be yeast-specific. Ono et al., (Ono, Hishigaki et al., 2001) quantitatively assessed the abilities of four common interaction-indicating terms, each associated with a custom set of templates, to indicate protein-protein interactions. The quantitative performances of the four are hard to interpret because each used a different template set, but it is interesting that their ranks in terms of precision were the same for both the yeast and the *E. coli* domains, suggesting domain independence for precision.

Thomas et al., (Thomas, Milward et al., 2000) proposed four categories of biological text passages using a rule-based scoring strategy, and gave the information retrieval (IR) performance of each category. Sekimizu et al., (Sekimizu, Park et al., 1998) measured the IR performances of 8 interaction-indicating verbs in the context of a shallow parser. The IR capabilities of the verbs could be meaningfully compared, but whether these results would hold across different parsers or other passage analyzers is an open question. In our lab, we have obtained similar results using passages containing two protein names. Counting passages describing interactions as hits and others as misses, sentences had slightly higher IR effectiveness than phrases despite lower precision, and considerably higher IR effectiveness than whole abstracts (Ding, Berleant et al., 2002). Ding et al., (Ding, 2003; Ding, Berleant et al., 2003) applied an untuned link grammar parser to sentences containing protein co-occurrences, finding that using the presence of a link path as an additional retrieval criterion raised the IR effectiveness by 5 percentage points (i.e. 7%). These works highlight the gap in knowledge of empirical facts about biological texts. In the future, researchers will focus increasing attention on this important gap.

2.3.2 Combining Evidence

Combining different items of evidence can result in a single composite likelihood that a sentence describes a biomolecular or other interaction. This can enable putative interactions in automatically generated biomolecular interaction network simulators to be rated, or sentences to be ranked for human curation. The following paragraphs compare two methods for evidence combination: the Naïve Bayes model and semi-naïve evidence combination.

Naïve Bayes and semi-naïve evidence combination both have a similar scalability advantage over full Bayesian analysis using Bayes Theorem to account for whatever dependencies may exist. That scalability is why they are useful. However, when used to estimate probabilities that an item (e.g. a sentence) is in some category (e.g. describes a biomolecular interaction), semi-naïve evidence combination makes fewer assumptions (Berleant, 2004).

Evidence combination with the Naïve Bayes model. This standard method produces probability estimates that can used for categorization (Lewis, 1998). The formula is:

$$\begin{aligned} p(h \mid f_1, \ldots f_n) &= \frac{p(h)p(f_1, \ldots f_n \mid h)}{p(f_1)p(f_2)p(f_3) \ldots p(f_n)} \\ &\approx \frac{p(h)p(f_1 \mid h)p(f_2 \mid h)p(f_3 \mid h) \ldots p(f_n \mid h)]}{p(f_1)p(f_2)p(f_3) \ldots p(f_n)} \end{aligned} \quad (1)$$

where h is the probability that a sentence is a "hit" (has a description of the expected interaction), and f_i is feature i. The approximation provides a computationally tractable way to estimate the desired probability given the assumption that the features occur independently of one another. A readable derivation is provided by Wikipedia (Wikipedia: The Free Encyclopedia, 2004).

Semi-naïve evidence combination. This method is scalable in the number of features, like Naïve Bayes, but has the advantage of making fewer independence assumptions. Unlike the Naïve Bayes model, it does not assume that the features are independent regardless of whether sentences are hits or not.

The parsimonious formula for semi-naïve evidence combination, in terms of odds (the ratios of hits to misses) is (Berleant, 2004):

$$O(h|f_1,\ldots,f_n)=O_1\ldots O_n/(O_0)^{n-1} \quad (2)$$

where the odds that a sentence describes an interaction if it has features $f_1,\ldots,f_n$ are $O(h|f_1,\ldots,f_n)$. The odds that a sentence with feature k is a hit are O_k, and the prior odds (i.e. over all sentences in the test set irrespective of their features) that a sentence is a hit are O_0. The odds of flipping a head are 1/1=1 (1 expected success per failure), while the odds of rolling a six are 1/5 (one success expected per five failures). Odds are easily converted to the more familiar probabilities by applying $p=O/(O+1)$. Similarly, $O=p/(1-p)$.

Comparison of the Naïve Bayes and semi-naïve evidence combination models. Naïve Bayes is often used for category assignment. The item to be classified is put into the category for which Naïve Bayes gives the highest likelihood. In the present context there are two categories, one of hits and one of non-hits, but in general there can be N categories. In either case, the denominator of the Naïve Bayes formula is the same for each category, so it can be ignored. However, when the Naïve Bayes formula is used for estimating the *probability* that a sentence is in a particular category, the denominator must be evaluated. This is problematic because the assumption of unconditional independence is not only unsupported, but most likely *wrong*. The reason is that the features that provide evidence that the sentence belongs in a particular category are probably correlated.

For the problem of estimating the *probability* that a particular sentence is a hit (or, more generally, belongs to a particular category), semi-naïve evidence combination appears more suitable because it estimates odds (which are easily converted to probabilities) without requiring the problematic assumption that features occur unconditionally independently (i.e. independently regardless of whether the sentence is a hit or not).

3. METNET

MetNet is designed to provide a framework for the formulation of testable hypotheses regarding the function of specific genes, proteins, and metabolites, and in the long term provide the basis for identification of genetic regulatory networks that control plant composition and development. Our approach to reveal complex biological networks is to extract information from gene expression data sets and combine it with what is already know about metabolic and regulatory pathways to achieve a better understanding of how metabolism is regulated in a eukaryotic cell.

3.1 Metabolic Networking Data Base (MetNetDB)

A critical factor both in establishing an efficient system for mining the literaturome and in modeling network interactions is the network database itself. MetNetDB is a searchable database with a user-friendly interface for creating and searching the *Arabidopsis* network map (Wurtele, Li et al., 2003). ***MetNetDB contains a growing metabolic and regulatory map of Arabidopsis.*** Entities (represented visually as nodes) in the database include metabolites, genes, RNAs, polypeptides, protein complexes, and 37 hierachically-organized interaction types, including catalysis, conversion, transport, and various types of regulation. MetNetDB currently contains more than 50,000 entities (from KEGG, TAIR and BRENDA), 1000 expert-user-added entity definitions, and 2785 expert-user-added interactions, including transport, together with associated information fields. In addition, it contains interactions from *Arabidopsis* Lipid Gene Database, and partially curated interactions from AraCyc. Synonyms for each term in MetNetDB are obtained from sources including expert users, TAIR, and BRENDA; an adequate library of synonyms is particularly important in text mining. Database nomenclature is modeled after the *Arabidopsis* Gene Ontology (http://arabidopsis.org/info/ontologies/), for ease of information transfer between MetNetDB and other biological databases.

3.2 FCModeler: Network Visualization and Modeling

FCModeler is a Java program that dynamically displays complex biological networks and analyzes their structure using graph theoretic methods. Data from experiments (i.e., microarray, proteomics, or metabolomics) can be overlayed on the network map.

3.2.1 Network Visualization and Graph Theoretic Analysis

Visual methods allow the curator to investigate the pathway one step at a time and to compare different proposed pathways. Graph union and intersection functions assist curators in highlighting these differences. FCModeler uses graph theoretic methods to find cycles and alternative paths in the network. Alternative path visualizations help curators search for redundant information in pathways. For example, a sketchy pathway may need to be replaced with more details as they become available. Cycles in the metabolic network show repeated patterns. These cycles range from simple loops, for example, a gene causing a protein to be expressed, and accumulation of the protein inhibiting the gene's transcription. More complex cycles encompass entire metabolic pathways. The interactions or overlaps between the cycles show how these control paths interact. FCModeler searches for elementary cycles in the network. Many of the cycles in a pathway map are similar, and several similarity measures and pattern recognition models are available for grouping or clustering the cycles (Cox, Fulmer et al., 2002; Dickerson and Cox, 2003).

3.2.2 Multi-Scale Fuzzy K-Means Clustering

The analysis and creation of gene regulatory networks involves first clustering the data at different levels, then searching for weighted time correlations between the cluster center time profiles. Link validity and strength is then evaluated using a fuzzy metric based on evidence strength and co-occurrence of similar gene functions within a cluster. The Fuzzy K-means algorithm minimizes the objective function (Bezdek, 1981):

$$J(F,V)=\sum_{i=1}^{N}\sum_{j=1}^{K}m_{ij}^{2}d_{ij}^{2} \quad (3)$$

$F = \{X_i,\ i = 1,...,N\}$ are the N data samples; $V = \{V_j,\ j = 1,...,K\}$ represents the K cluster centers. m_{ij} is the membership of X_i in cluster j, and d_{ij} is the Euclidean distance between and V_j. This work uses a windowed membership function:

$$m_{ij} = \frac{1/d_{ij}^2}{\sum_{k=1}^{K} 1/d_{ik}^2} W\left(d_{ij}\right) \tag{4}$$

Adding a window function $W(d)$ to the membership function limits the size of clusters. This work uses truncated Gaussian windows with values outside the range of 3σ set to zero:

$$W\left(d_{ij}\right) = \begin{cases} e^{-\left(d_{ij}\right)^2/\left(2\sigma^2\right)} & d_{ij} < 3\sigma \\ 0 & elsewhere \end{cases} \tag{5}$$

The window function, $W(d)$, insures that genes with distances larger than 3σ from the cluster center will have no effect on the new cluster center estimates.

3.2.3 Multi-Scale Algorithm

The multi-scale algorithm is similar to the ISODATA algorithm with cluster splitting and merging (Ball, 1965; Ball and Hall, 1965). There are four parameters: K (initial cluster number), σ (scale of the window $W(d)$), T_{split} (split threshold), $T_{combine}$ (combine threshold). Whenever the genes are further away from the cluster center than T_{split}, the cluster is split and faraway genes form new clusters. Also, if two cluster centers are separated by less than $T_{combine}$, then the clusters are combined. Usually $T_{combine} \le \sigma$ and $2\sigma \le T_{split} \le 3\sigma$. The algorithm is given in Table 17-1. ε_1 and ε_2 are small numbers to determine whether the clustering converged, and $\varepsilon_1 > \varepsilon_2$. If one cluster has elements far away from the cluster centers then the cluster is split. The advantage of this algorithm is that it dynamically adjusts the number of clusters based on the splitting and merging heuristics.

Table 17-1. Multi-Scale Fuzzy K-Means Algorithm

Step	Description
1	Initialize parameters: K, σ, T_{split} and $T_{combine}$
2	Iterate using Fuzzy K-means until convergence to a given threshold ε_1
3	Split process: do split if there are elements farther away from cluster center than T_{split}.

continued

Step	Description
4	Iterate using Fuzzy K-means until convergence to a given threshold ε_1
5	Combine Process: combine the clusters whose distance between cluster centers is less than $T_{combine}$. If the cluster after combining has elements far away from cluster center (distance larger than 3σ), stop combining.
6	Iterate steps 1-5 until no splits or combination occur. Converging to a given threshold ε_2.

3.2.4 Effects of Window Size

Changing the window size can affect the level of detail captured in the clusters. If σ<<1, then clusters are individual elements. As σ increases, the window gets larger. The result is a hierarchical tree that shows how the clusters interact at different levels of detail. This work uses three level of multi-scale fuzzy K-mean clustering (σ=0.1, 0.2 and 0.3). The initial number of clusters is $K = N$, the total number of data points, $T_{combine} = \sigma$, and $T_{split} = 3\sigma$. Clustering results with different window sizes provide different levels of information. At σ=0.1, the cluster sizes are very small. These clusters represent very highly correlated profiles (correlation coefficients between gene profiles within 1- σ window size are larger than 0.9) or just the individual gene profiles because many clusters only contain a single element. At σ=0.2, smaller clusters are combined with nearby clusters. Highly correlated profiles are detected. The σ=0.3 level is the coarsest level.

3.2.5 Construction of Rene Regulatory Networks

Clustering provides sets of genes with similar RNA profiles. The next step is finding the relationships among these coregulated genes. If gene *A* and gene *B* have similar expression profiles, there are several possible relationships: 1. *A* and *B* are coregulated by other genes; 2. *A* regulates *B* or vice versa; 3. There is no causal relationship, just coincidence. Here the regulation may be indirect, i.e., interact through intermediates. These cases cannot be differentiated solely by clustering. We use cubic spline interpolation for simplicity and get equally sampled profiles.

The gene regulatory model can be simplified as a linear model (D'Haeseleer, Liang et al., 1999):

$$x_A(t+\tau_A) = \sum_B w_{BA} x_B + b_A \tag{6}$$

x_A is the expression level of gene *A* at time *t*, τ_A is the gene regulation time delay of gene *A*, w_{BA} is the weight indicating the inference of gene *B* to *A*, b_A is a bias indicating the default expression level of gene *A* without regulation.

Standardizing gene expression profiles to 0 mean and 1 standard deviation removes the bias term, b_A. The goal is to find out if genes *A* and *B* have a regulatory relationship so the weight is $w_{AB} = [0,1]$ (0 means no regulatory relation, 1 means strongly regulated). The time correlation between genes *A* and *B* can be expressed in discrete form as

$$R_{AB}(\tau) = \sum_n x_A(n) x_B(n - \tau) \tag{7}$$

Where x_A and x_B are the standardized (zero mean, standard deviation of unity) expression profiles of genes *A* and *B*. τ is the time shift. For the periodic time profile, we can use circular time correlation, i.e., the time points at the end of the time series will be rewound to the beginning of series after time shifting. For multiple data sets, the time correlation results of each data set are combined as:

$$R_{AB}^C(\tau) = \sum_k w_k R_{AB}^k(\tau) \tag{8}$$

Where $R_{AB}^C(\tau)$ is the combined time correlation result, $R_{AB}^k(\tau)$ is the time correlation result of the kth data set, w_k is the weight of kth data set that depends on the experiment reliability and the length of the expression profile.

The value $\max|R_{AB}^C(\tau)|$ can be used to estimate the time delay τ' between expression profiles of genes *A* and *B*. Given a correlation threshold T_R, if $\max|R_{AB}^C(\tau)| > T_R$ there is significant regulation between genes or clusters. By defining the clusters as nodes and significant links as edges, we can get the gene regulation network of these clusters. Assuming that the time delays are caused by regulation, we can define four types of regulation:

$R_{AB}^C(\tau') > 0,\ \tau' \neq 0$, positive regulation between genes *A* and *B*;
$R_{AB}^C(\tau') < 0,\ \tau' \neq 0$, negative regulation between genes *A* and *B*;
$R_{AB}^C(\tau') > 0,\ \tau' = 0$, genes *A* and *B* are positively coregulated;
$R_{AB}^C(\tau') < 0,\ \tau' = 0$, genes *A* and *B* are negatively coregulated.

The sign of τ' determines the direction of regulation. $\tau' > 0$ means gene *B* regulates gene *A* with time delay τ'; $\tau' < 0$ means gene *A* regulates gene *B* with time delay τ'.

3.3 Network Validation Using Fuzzy Metrics

The available gene ontology (GO) annotation information can estimate a fuzzy measure for the types or functions of genes in a cluster. The GO terms in each cluster are weighted according to the strength of the supporting evidence information for the annotation of each gene and the distance to cluster center. An additive fuzzy system is used to combine this information(Kosko, 1992). Every GO annotation indicates the type of evidence that support it. Among these types of evidence, several are more reliable and several are weaker. This evidence is used to set up a bank of fuzzy rules for each annotated data point. Different fuzzy membership values are given to each evidence code. For example, evidence inferred by direct assays (IDA) or from a traceable author statement (TAS) in a refereed journal has a value of one. The least reliable evidence is electronic annotation since it is known to have high rates of false positives.

Table 17-2. Evidence Codes and Their Weights

Evidence Code	Meaning of the Evidence Code	Membership Value, w_{evi}
IDA	Inferred from direct assay	1.0
TAS	Traceable author statement	1.0
IMP	Inferred from mutant phenotype	0.9
IGI	Inferred from genetic interaction	0.9
IPI	Inferred from physical interaction	0.9
IEP	Inferred from expression pattern	0.8
ISS	Inferred from sequence, structural similarity	0.8
NAS	Non-traceable author statement	0.7
IEA	Inferred from electronic annotation	0.6
	Other	0.5

Each gene in a cluster is weighted by the Gaussian window function in equation (5). This term weights the certainty of the gene's GO annotation using product weighting. Each gene and its associated GO term are combined to find the possibility distribution for each single GO term that occurs in the GO annotations in one cluster. One gene may be annotated by several GO terms, and each GO term has one evidence code. Each GO term may occur *K* times in one cluster, but with a different evidence code and in different genes. For the *n*th unique GO term in the *j*th cluster, the fuzzy weight is the sum of the weights for each occurrence of the term:

$$W_{GO}(j,n) = \sum_{i=1}^{K} w_{GO,j}(i,n) \tag{9}$$

Where $w_{GO,j}(i,n) = w_{evi}(i,n) \cdot W\left(d_{ij}\right)$, w_{evi} is shown in Table 17-2, and $W(d_{ij})$ is the same as equation (5).

This provides a method of pooling uncertain information about gene function for a cluster of genes. This gives an additive fuzzy system that assesses the credibility of any GO terms associated to a cluster (Kosko 1992). The results can be left as a weighted fuzzy set or be defuzzified by selecting the most likely annotation. For each cluster, the weight is normalized by the maximum weight and the amount of unknown genes. This is the weighted percentage of each GO term p_{weight} :

$$p_{weight}(j,n) = \frac{W_{GO}(j,n)}{W_{root}(j) - W_{unknown}(j)} * 100\% \quad (10)$$

Where $W_{GO}(j,n)$ represents the weight of the nth GO term in the jth cluster. $W_{unknown}(j)$ is the weight of GO term in cluster j: xxx unknown, e.g., GO: 0005554 (molecular_function unknown). $W_{root}(j)$ is the weight of root in cluster j. GO terms are related using directed acyclic graphs. The root of the graph is the most general term. Terms further from the root provide more specific detail about the gene function and are more useful for a researcher. The weight of each node is computed by summing up the weights of its children (summing the weights of each of the N GO terms in a cluster):

$$W_{root}(j) = \sum_{n=1}^{N} W_{GO}(j,n) \quad (11)$$

The higher weighted nodes further from the root are the most interesting since those nodes refer to specific biological processes.

3.4 PathBinderA: Finding Sentences with Biomolecular Interactions

The objective of the PathBinderA component of the system is to mine sentences describing biomolecular interactions from the literature. This functionality forms a potentially valuable component of a range of systems, by supporting systems for automatic network construction, systems for annotation of high-throughput experimental results, and systems that minimize the high costs of human curation. Such a component should typically mine all of MEDLINE, the *de facto* standard corpus for bioscience text mining. For the plant domain, full texts in the plant science domain should also be addressed, requiring cooperative agreements with publishers in general. The feasibility of different PathBinder components is illustrated by the system at http://metnetdb.gdcb.iastate.edu/pathbinder/. The PathBinderA system, used in this work, has been prototyped and is undergoing further development.

Attaining the desired results requires a well-motivated and tested method for processing biological texts. The design includes a two-stage algorithm. Each stage is based on probability theory. In stage 1, evidence for interaction residing in sentence features is combined to compute the sentence's credibility as an interaction description. In stage 2, the credibilities of the "bag" of sentences that mention two given biomolecules are combined to rate the likelihood that the literature describes those biomolecules as interacting. The practical rationale for this process is that an important resource is being created for use by the scientific community, as well as an important module of the overall MetNetDB system. This resource is aimed at effective curation support, which in turn is aimed at feeding the construction of interaction networks.

3.4.1 PathBinder Component Design Issues

There are three major phases of a PathBinder component such as the PathBinderA component of METNET. The mining process comprises stages 1 and 2, and using the results of the mining constitutes the third phase.

Text mining, stage 1. This involves assessing the credibility of a given sentence as a description of an interaction between two biomolecule names in it. To do this the evidence provided by different features of a sentence must be combined. Semi-naive evidence combination, described earlier in this chapter, is one such method. The Naïve Bayes model provides another possibility. Syntactic parsing to analyze sentences in depth is an alternative approach.

The abilities of various features of sentences to predict whether they describe an interaction can be determined empirically in order to enable those features to be used as input to a method for assessing sentences. One such feature is whether a sentence with two biomolecule names has those names in the same phrase or, instead, the names occur in different phrases within the sentence. We have investigated this feature using the IEPA corpus (Ding, Berleant et al., 2002). Table 17-3 (rightmost column) shows the results. Another feature is whether or not an interaction term intervenes between the co-occurring names. An interaction term is a word that can indicate that an interaction between biomolecules takes place, like "activates," "block," "controlled," etc. Such a term can appear between two co-occurring names, can appear in the same sentence or phrase but not between them, or can be absent entirely. The table below (middle two columns) shows the data we have collected, also using the IEPA corpus.

Table 17-3. Analysis of the recall and precision of co-occurrence categories with respect to mining interaction descriptions.

	Interactor intervening		Interactor elsewhere		Interactor anywhere	
Phrase co-occurrences	r=0.55	p=0.63	r=0.18	p=0.24	r=0.72	p=0.45
Sentence co-occurrences	r=0.22	p=0.30	r=0.058	p=.09	r=0.28	p=0.21
All co-occurrences	r=0.77	p=0.48	r=0.23	p=0.17	r=1	p=0.34

Text mining, stage 2. In this stage, the evidence for an interaction provided by multiple relevant sentences is combined to get a composite probability estimate for the interaction. This becomes possible after stage 1 has given a probability for each sentence. The basic concept underlying stage 2 is that if even one sentence in a "bag" containing two given names describes an interaction between them, then the interaction is present in the literature (Skounakis and Craven, 2003). The need for as little as a single example to establish an interaction leads directly to a probability calculation for combining the evidence provided by the sentences in a bag. The reasoning goes as follows.

Let notation $p(x)$ describe the probability of x. Assume the evidence provided by each sentence s_i in a bag is independent of the evidence provided by the other sentences, allowing us to multiply the probabilities of independent events to get the probability of their simultaneous occurrence.

p(one or more sentence in bag b describes an interaction between n_1 and n_2)
$=1-p$(zero sentences in bag b describe an interaction between n_1 and n_2),
$=1-p(s_1$ does not describe an interaction AND s_2 does not describe an interaction AND s_3...)
$=1-p(s_1$ does not describe an interaction)$\cdot p(s_2$ does not describe an interaction)$\cdot p(s_3$ does not describe…
$=1-[1-p(s_1$ describes an interaction)]$\cdot$[1-$p(s_2$ describes an interaction)]$\cdot$ [1-$p(s_3$ does not describe…
$=1-\prod_i 1-p(s_i$ describes an interaction).

This equation is not only mathematically reasonable but considerably simpler than the more complex formulas given by Skounakis and Craven (Skounakis and Craven, 2003).

Using the mining results, stage 3. This phase integrates the extraction capability into the larger MetNetDB system. The integration supports the following functionalities.

1. *Support for curation.* Because networks of interactions are built from individual interactions, it is important not only to mine potential interactions from the literature but to present these to curators so that they can be efficiently verified. Curation is a serious bottleneck because it requires expert humans, a scarce resource. Therefore efficient support for curation is an important need. PathBinderA supports curation by presenting mined potential interactions to curators starting from the best, most likely interactions. The goal of this design is to minimize the labor required by the curation process.
2. *Generating interaction hypotheses.* When mining the literature produces a strong hypothesis of an interaction, that interaction may be tentatively

added to the interaction database without curation. Interactions whose probabilities are assessed at 90% or better are likely to fall into this category, although this threshold is adjustable. Such likely interactions are made available pending curation.

3. *More efficient literature access.* High-volume information resources can benefit from providing convenient access to the literature relevant to particular items in the resource. Such functionality is clearly useful to non-expert users, and even expert users can benefit since no individual can be intimately familiar with the full range of the literature on biomolecular interactions even in one species. The system design provides for integrating literature access with an easy-to-use community curation functionality. In this design, users anywhere can click a button associated with the display of any sentence they retrieve from the system. This brings up a form with two other buttons. One of these registers an opinion that the sentence describes an interaction, and one registers an opinion to the contrary. Comments may be typed into an optional comment area. Submitted forms will then be used by the official curators.

Users can choose *species and other taxa* to focus their search. Users may, for example, specify *viridiplantae* (green plants) to see sentences related to *Arabidopsis* as well as any other green plant species. The current prototype of PathBinderA allows users to specify two biomolecules, an interaction-relevant verb, and a subcellular location. Sentences with the two biomolecules and the verb which are associated with the specified subcellular location can then be retrieved (see Figure 17-1).

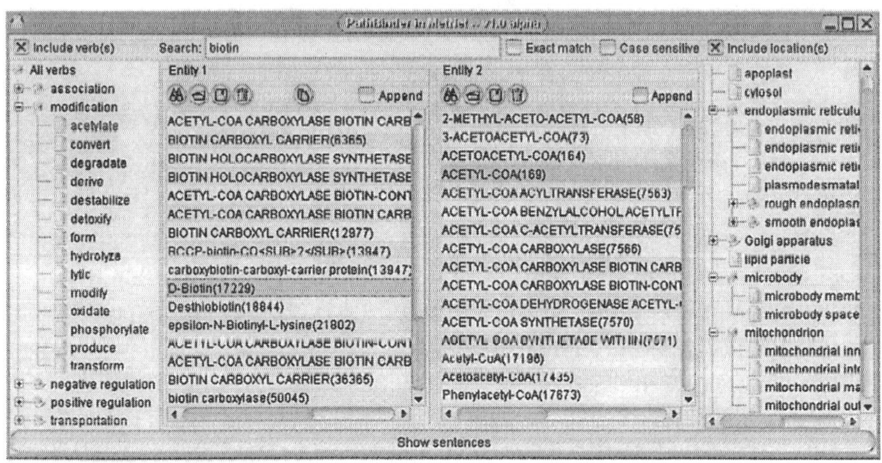

Figure 17-1. PathBinderA interface, showing the four choices a user can make. These choices include two biomolecules, a subcellular location, and an interaction-relevant verb.

4. BUILDING ON METABOLIC NETWORKS: USING METNET

Regulatory networks from Arabidopsis can be built using a combination of expert knowledge from MetNetDB, fuzzy clustering and correlation from FCModeler. The constructed networks can be validated using PathBinderA to access the literaturome and the weighted GO scores derived in FCModeler.

We illustrate with a test data set comparing wild-type (WT) *Arabidopsis thaliana* plants with those containing antisense *ACLA-1* behind the constitutive CaMV 35S promoter (referred to as *aACLA-1*) (Fatland, Nikolau et al., 2004). Plants were grown under a short day cycle of 8 hours light and 16 hours dark. Affymetrix GeneChip microarrays were used. The data consisted of two replicates; each with eleven time points (0, 0.5, 1, 4, 8, 8.5, 9, 12, 14, 16, 20 hours), harvested during the light (0 to 8 hours) and dark (8 to 20 hours) (Foster, Ling et al., 2003). Only *ACLA-1* seedlings exhibiting features characteristic of the antisense phenotype were used. Total RNA was extracted from leaves and used for microarray analyses.

The Affymetrix microarray data were normalized with the Robust Multichip Average (RMA) method (Gautier, Cope et al., 2004). Both replicates of each gene expression profile are standardized to zero mean, one standard deviation. The data was filtered by comparing the expression values of each gene between the WT and *ACLA-1* plants at 1, 8.5 and 12 hours; differentially expressed genes having larger than 2 fold changes at any time point were retained for further analysis. There are 484 such genes in total. The expression patterns of these 484 genes in the wild-type plants at all 11 time points were clustered.

4.1 Construct the Genetic Network Using Time Correlation

The genetic networks among the clusters of coregulated genes can be constructed based on their cluster center profiles. Since the data used were unequally sampled with 0.5h as minimum interval, we interpolated the gene expression profiles as equally sampled 41 time points with 0.5h intervals using cubic spline interpolation. The time correlation of each replicate was computed using equation (7), then combined using equation (8). The time period was limited to the range of [-4h, 4h] because the light period only lasted 8 hours in this data set. The genetic networks were constructed with a correlation threshold of $T_R = 0.65$. The strength of correlation was mapped into three categories: [0.65, 0.75), [0.75, 0.85), and [0.85, 1]. Three types of line thickness from thin to thick represent the strength of the correlation.

Black dashed lines represent positive coregulation; green dashed lines represent negative coregulation; solid red lines with bar head represent negative regulation; solid blue lines with arrowheads represent positive regulation.

Figure 17-2 shows the constructed gene regulatory networks based on the cluster center with window sigma equals 0.3. The networks indicate that cluster 1 and 5 are highly coregulated (0 time delay), cluster 1 and 5 positively regulate cluster 4 with a time delay of 2.5 hours and 3h, and both negatively regulated cluster 3 with a time delay of 1.5 hours; cluster 4 is negatively regulated by cluster 3 with a delay of 1 hour, the correlation between cluster 2 and cluster 4, and cluster 1 and 3 is not strong. All of these relations are coincident with the cluster center profiles.

Figure 17-3 shows the constructed regulatory networks of the 28 cluster centers at the σ= 0.2 level. The graph notations are the same as Figure 17-2. The graph shows that there is one highly connected group of clusters. The other clusters at the upper right corner are less connected. The relations between clusters may become complex with a large number of edges. Simplification of the networks is necessary when there are many highly connected clusters.

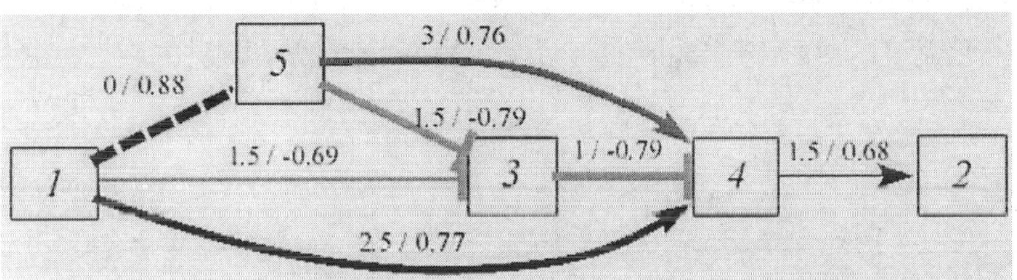

Figure 17-2. Gene regulatory networks inferred from the case with sigma equal to 0.3. The numbers on each link show the time delay for the interaction on top and the correlation coefficient of the interaction on the bottom.

Figure 17-3 shows possible duplicate relationships. This can be analyzed using the path search function in FCModeler. From cluster 15 to 19, there are two paths: one is directly from cluster 15 → 19 with time delay 1h and correlation coefficient, $\rho = -0.85$; another path is cluster 15 → 7 with time delay 0.5 h and correlation coefficient, $\rho = -0.89$, and then from 7→ 19 with time delay 0.5h and $\rho = 0.81$. The total time delays of both paths are the same. So it is very possible one of the paths is redundant. Figure 17-4 shows part of the simplified graph.

4.2 Cluster and Network Validation

Cluster validation can use the available literature and the GO annotation for each gene to find out what kind of functions or processes occur within each cluster and to search for potential interactions between genes in different clusters. In Figure 17-3, the clusters in the portion of the graph at the upper right corner are less connected both to each other and to the main graph. Most of the genes in these "less-connected" clusters are not annotated in GO. This means these genes have no biological evidence of a direct relation with the highly connected group. It also shows how the fuzzy hierarchical algorithm successfully separates these genes.

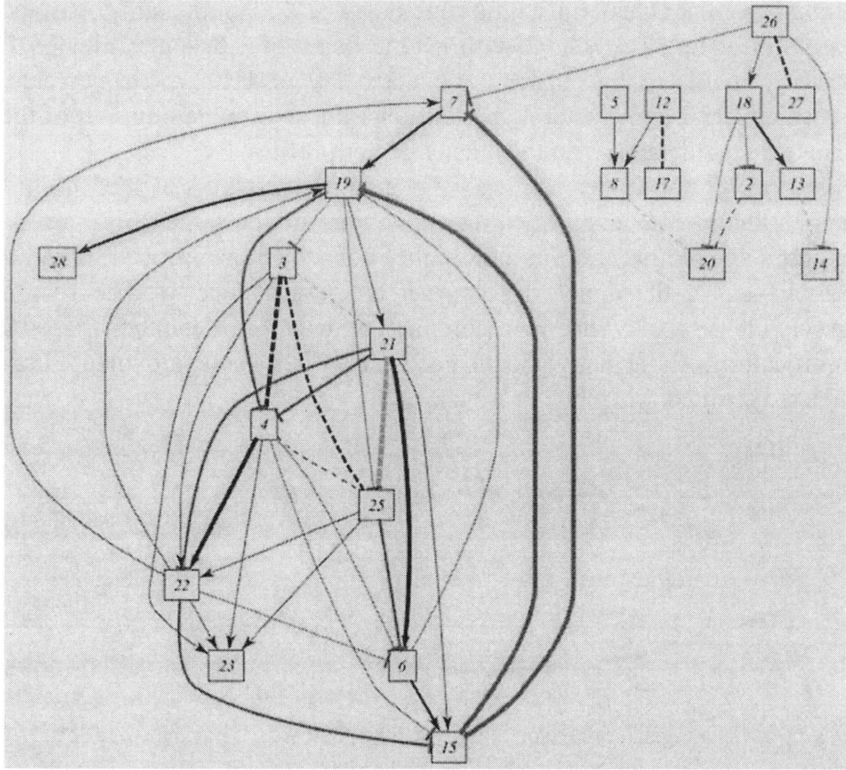

Figure 17-3.. Regulatory networks among cluster centers at the window size sigma = 0.2 level.

Figure 17-4 shows that clusters 3 and 4 are highly coregulated (correlation coefficient between cluster centers is 0.91). The cluster is split because the combined cluster 3 and 4 has a cluster diameter larger than 3σ. Table 17-4 shows the fuzzy weights for the GO terms for the genes in each cluster. The BP (Biological Process) GO annotations show that the genes in clusters 3 and 4 function in several similar biological processes. For example, both clusters contain genes of "Carboxylic acid metabolism", "Regulation of transcription, DNA-dependent", and "Protein amino acid phosphorylation". Cluster 3 has more "Regulation of transcription, DNA-dependent" genes, while cluster 4 emphasizes "Protein amino acid phosphorylation" genes. Clusters 3 and 4 provide an example of the overlapping of fuzzy clusters in which the separation of two clusters may make sense, and suggests additional biological analyses.

Clusters 21 and 25 are two highly negatively coregulated clusters. Cluster 21 contains genes of "Photosynthesis, dark reaction" and hormone response genes, while cluster 25 mainly contains genes of catabolism and stress-associated genes. Cluster 21 contains genes for "Trehalose biosynthesis". Trehalose plays a role in the regulation of sugar metabolism, which has recently been identified for Arabidopsis (Eastmond and Graham 2003). Clusters 6 and 21 involve sugar metabolism (carbohydrate metabolism GO term). This provides an interesting biological implication for understanding regulation in this experiment.

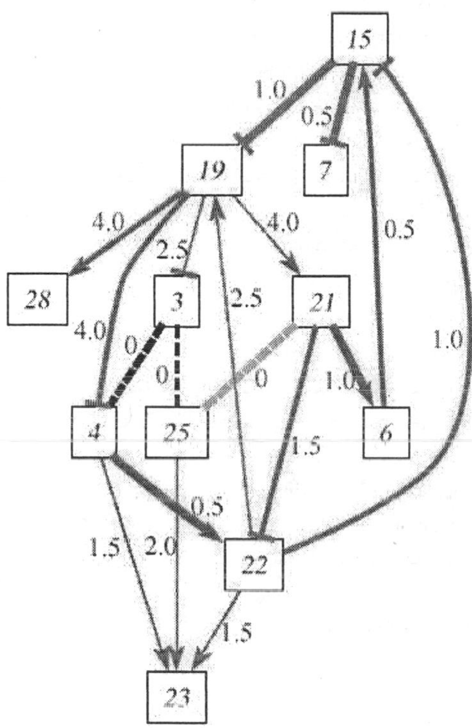

Figure 17-4. Simplified regulatory network with redundant edges removed for the window size $\sigma = 0.2$ level. The number on each link represents the estimated time delay.

Figures 17-3 and 17-4 show that cluster 19 regulates clusters 3, 4, 21, 22, 25 and 28. After checking the BP GO annotations, we found many of the annotated genes in cluster 19 fall in three categories: "Protein Metabolism" ("N-terminal protein myristoylation", and "Protein folding"), "Response to auxin stimulus", and "Cell-cell signaling". "N-terminal protein myristoylation," and "Protein folding" modulate protein activity, while

"Response to auxin stimulus" and "Cell-cell signaling" involve the processes of receiving stimulus or signals from others. Therefore these BP GO annotations are consistent with our network structures. Clusters 23 and 28 have no out-going edges. This implies that they may not be involved in regulatory activity. The genes in these clusters are metabolic or have no known function.

Table 17-4. Cluster annotation of Biological Process GO with the weights as defined in equations 9-11.

CLUSTER INDEX (W_{ROOT})	MAJOR GO TERM	$W_{GO}(J,N)$	$P_{WEIGHT}(J,N)$
Cluster 3	Response to water deprivation	4.11	16.6
(24.81)	Regulation of transcription, DNA-dependent	3.16	12.7
	Carboxylic acid metabolism	2.82	11.4
	Protein amino acid phosphorylation	2.63	10.6
Cluster 4	Protein amino acid phosphorylation	8.34	**23.1**
(36.03)	Carboxylic acid metabolism	3.58	9.9
	Response to abiotic stimulus	3.35	9.3
	Regulation of transcription, DNA-dependent	2.44	6.8
Cluster 6	Regulation of transcription, DNA-dependent	1.99	**23.5**
(8.48)	myo-inositol biosynthesis	0.95	11.2
	Abscisic acid mediated signaling	0.83	9.8
	Protein amino acid phosphorylation	0.57	6.7
Cluster 7	Carbohydrate metabolism	3.02	**22.2**
(13.58)	Cell surface receptor linked signal transduction	1.71	12.6
	Nucleobase, nucleotide, nucleic acid metabolism	1.62	11.9
	Protein amino acid phosphorylation	1.59	11.7
Cluster 15	Regulation of transcription, DNA-dependent	1.32	**52.4**
(2.52)	Electron transport	0.7	27.8
Cluster 19	Cell-cell signaling	0.78	23.5
(3.32)	Response to auxin stimulus	0.68	20.5
	Protein folding	0.65	19.6
	N-terminal protein myristoylation	0.61	18.4
Cluster 21	Carbohydrate metabolism	2.93	**29.1**
(9.71)	Response to gibberellic acid stimulus	1.86	19.2
	Photosynthesis, dark reaction	0.91	9.4
Cluster 22	Protein amino acid phosphorylation	6.74	**28.4**
(23.76)	Macromolecule biosynthesis	3.38	14.2
	Regulation of transcription DNA-dependent	2.50	10.5
	Signal transduction	2.30	9.7
Cluster 23	Response to endogenous stimulus	2.79	**60.5**
(4.61)	Response to biotic stimulus	1.83	39.7

continued

Cluster Index (W_{ROOT})	Major GO Term	$W_{GO}(J,N)$	$P_{WEIGHT}(J,N)$
Cluster 25	Carboxylic acid metabolism	8.19	**20.9**
(39.16)	Response to pest/pathogen/parasite	5.66	14.5
	Lipid biosynthesis	3.55	9.1
	Transport	3.52	9.0
Cluster 28 (0.95)	Carbohydrate metabolism	0.95	**100**

Using PathbinderA to explore the relationship between genes in clusters is illustrated in a comparison between two genes in clusters 19 and 4. Cluster 19 contains the ethylene response gene "ethylene-induced esterase". Cluster 4 contains jasmonic acid response and several jasmonate biosynthesis genes. A search encompassed both these terms together with all of the synonyms for these terms in the MetNetDB database. We used "ethylene" and "jasmonate" to search in Pathbinder and retrieved 18 sentences (Figure 17-5 shows a subset of these sentences). Clicking on each sentence gives the entire abstract. Many of the sentences provided useful connections between these two nodes. For example, the abstract for the highlighted sentence delineates a relationship between the ethylene and jasmonate signaling pathways, as shown in Figure 17-6.

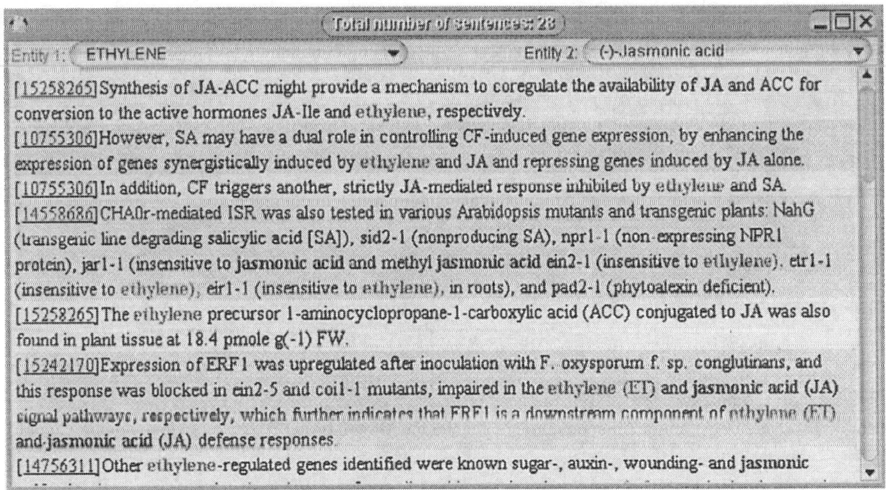

Figure 17-5. PathBinderA output for the terms ethylene and jasomonic acid. The relevant sentences and their Medline identification numbers are given.

5. DISCUSSION

The MetNet software focuses on understanding the complex molecular network in the model plant eukaryotic species, Arabidopsis. This work enables biologists to capture relationships at different levels of detail, to integrate gene expression data, and to model these relationships. Text mining in the PathBinderA system can help confirm relationships discovered by machine learning algorithms and will eventually be used to discover new relationships as methods of evidence combination are improved. Because of our absence of knowledge about many biological interactions, the software is designed to model at many levels of detail.

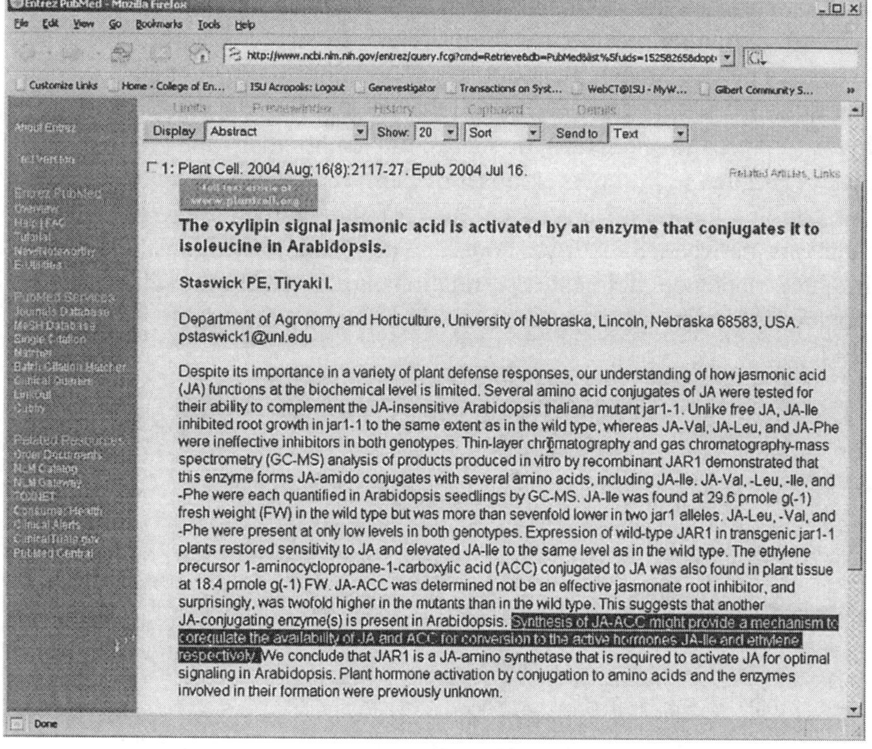

Figure 17-6.. The complete abstract for the selection shown above gives more details on the relationship between the ethylene and jasmonate signaling pathways.

6. ACKNOWLEDGEMENTS

This work is supported by grants from NSF (MCB-9998292, Arabidopsis 2010 DBI-0209809 and ITR-0219366), and the Plant Sciences Institute at Iowa State University. The network visualization was performed using the facilities at the Virtual Reality Application Center at Iowa State University. We thank Lucas Mueller and TAIR for helpful advice and Aracyc data.

REFERENCES

Akutsu, T., S. Miyano and S. Kuhara (2000). "Algorithms for Inferring Qualitative Models of Biological Networks," in *Pacific Symposium on Biocomputing* 5, Hawaii.

Ball, G. H. (1965). "Data Analysis in the Social Sciences: What About the Details?" in *AFIPS Proc. Cong. Fall Joint Comp*, 27(1): 533-559.

Ball, G. H. and D. J. Hall (1965). *Isodata, a Novel Method of Data Analysis and Pattern Classification*, Stanford Research Institute.

Berleant, D. (2004). *Combining Evidence: The Naïve Bayes Model Vs. Semi-Naïve Evidence Combination*. Ames, IA, Software Artifact Research and Development Laboratory, Iowa State University.

Bezdek, J. C. (1981). *Pattern Recognition with Fuzzy Objective Function Algorithms*. New York, Plenum Press

Cox, Z., A. Fulmer and J. A. Dickerson (2002). *Interactive Graphs for Exploring Metabolic Pathways*. ISMB, 2002, Edmonton, CA.

Craven, M. and J. Kumlien (1999). "Constructing Biological Knowledge Bases by Extracting Information from Text Sources," in *Proc Int Conf Intell Syst Mol Biol*: 77-86.

D'Haeseleer, P., S. Liang and R. Somogyi (1999). "Gene Expression Analysis and Modeling," in *Pacific Symposium of Biocomputing* (Tutorial).

D'haeseleer, P., S. Liang and R. Somogyi (2000). "Genetic Network Inference: From Co-Expression Clustering to Reverse Engineering," *Bioinformatics* 16(8): 707-26.

Dickerson, J. A. and Z. Cox (2003). "Using Fuzzy Measures to Group Cycles in Metabolic Networks," in *North American Fuzzy Information Processing Society (NAFIPS) Annual Meeting*, Chicago, IL.

Dickerson, J. A., D. Berleant, Z. Cox, W. Qi, D. Ashlock, E. S. Wurtele and A. W. Fulmer (2003)." Creating and Modeling Metabolic and Regulatory Networks Using Text Mining and Fuzzy Expert Systems" in *Computational Biology and Genome Informatics*. J. T. L. Wang, C. H. Wu and P. Wang. Singapore, World Scientific Publishing: 207-238.

Ding, J., D. Berleant, D. Nettleton and E. Wurtele (2002). "Mining Medline: Abstracts, Sentences, or Phrases?" in *Pacific Symposium on Biocomputing* (PSB 2002), Kaua'i, Hawaii.

Ding, J. (2003). *Pathbinder: A Sentence Repository of Biochemical Interactions Extracted from Medline*. Dept. of Electrical and Computer Engineering. Ames, IA, Iowa State University.

Ding, J., D. Berleant, J. Xu and A. W. Fulmer (2003). "Extracting Biochemical Interactions from Medline Using a Link Grammar Parser," in *Proceedings of the Fifteenth IEEE Conference on Tools with Artificial Intelligence (ICTAI 2003)*, Sacramento, CA, USA.

Eastmond, P. J. and I. A. Graham (2003). "Trehalose Metabolism: A Regulatory Role for Trehalose-6-Phosphate?" *Curr Opin Plant Biol* 6(3): 231-5.

EMBO BioCreative Workshop (2004). "A Critical Assessment for Information Extraction in Biology," (Biocreative), At. Granada, ES. 2004.

Fatland, B., B. J. Nikolau and E. S. Wurtele (2004). "Reverse Genetic Characterization of Cytosolic Acetyl-Coa Generation by Atp-Citrate Lyase in Arabidopsis," *Plant Cell*, in press.

Foster, C. M., L. Ling, A. M. Myers, M. G. James, B. J. Nikolau and E. S. Wurtele (2003). "Expression of Genes in the Starch Metabolic Network of Arabidopsis During Starch Sythesis and Degradation, " In Preparation.

Gautier, L., L. Cope, B. Bolstad and R. Irizarry (2004). "Affy--Analysis of Affymetrix Genechip Data at the Probe Level, " Bioinformatics 20(3): 307-315.

Genoud, T., M. B. Trevino Santa Cruz and J. P. Metraux (2001). "Numeric Simulation of Plant Signaling Networks, " Plant Physiol. 126: 1430-1437.

Hanisch, D., A. Zien, R. Zimmer and T. Lengauer (2002). "Co-Clustering of Biological Networks and Gene Expression Data," in *Intelligent Systems for Molecular Biology (ISMB), 10th International Conference*, Edmonton, Canada, International Society for Computational Biology.

Hartemink, A. J., D. K. Gifford, T. S. Jaakkola and R. A. Young (2001). "Using Graphical Models and Genomic Expression Data to Statistically Validate Models of Genetic Regulatory Networks," in *Pacific Symposium on Biocomputing*, Hawaii.

Kanehisa, M. and S. Goto (2000). "Kegg: Kyoto Encyclopedia of Genes and Genomes, " *Nucleic Acids Research* 28(1): 27-30.

Karp, P. D., M. Krummenacker, S. Paley and J. Wagg (1999). "Integrated Pathway/Genome Databases and Their Role in Drug Discovery, " *Trends in Biotechnology* 17(7): 275-281.

Karp, P. D., M. Riley, M. Saier, I. T. Paulsen, S. M. Paley and A. Pellegrini-Toole (2000). "The Ecocyc and Metacyc Databases, " *Nucleic Acids Research* 28(1): 56-59.

Karp, P. D. (2001). "Pathway Databases: A Case Study in Computational Symbolic Theories, " *Science* 293(5537): 2040-4.

Karp, P. D., M. Riley, S. M. Paley and A. Pellegrini-Toole (2002). "The Metacyc Database, " *Nucl. Acids. Res.* 30: 59-61.

Kosko, B. (1992). *Neural Networks and Fuzzy Systems.* Englewood Cliffs, Prentice Hall.

Lewis, D. (1998). "Naïve Bayes at Forty: The Independence Assumption in Information Retrieval," in *Conf. Proc. European Conference on Machine Learning*, Chemnitz, Germany.

Marcotte, E. M., I. Xenarios and D. Eisenberg (2001). "Mining Literature for Protein-Protein Interactions, " *Bioinformatics* 17(4): 359-63.

Matsuno, H., Doi, A., Nagasaki, M. and Miyano, S. (2000). "Hybrid Petri Net Representation of Gene Regulatory Network," in *Pacific Symposium on Biocomputing* 5, Hawaii.

Mueller, L. A., P. Zhang and S. Y. Rhee (2003). "Aracyc: A Biochemical Pathway Database for Arabidopsis, " *Plant Physiol.* 132(2): 453–460.

Ono, T., H. Hishigaki, A. Tanigami and T. Takagi (2001). "Automated Extraction of Information on Protein-Protein Interactions from the Biological Literature, " *Bioinformatics* 17(2): 155-61.

Overbeek, R., N. Larsen, G. D. Pusch, M. D'Souza, E. S. Jr, N. Kyrpides, M. Fonstein, N. Maltsev and E. Selkov (2000). "Wit: Integrated System for High-Throughput Genome Sequence Analysis and Metabolic Reconstruction, " *Nucl. Acids. Res.* 28: 123-125.

Rhee, S. Y., W. Beavis, et al., (2003). "The Arabidopsis Information Resource (Tair): A Model Organism Database Providing a Centralized, Curated Gateway to Arabidopsis Biology, Research Materials and Community, " *Nucl. Acids. Res.* 31(1): 224-228.

Sekimizu, T., H. S. Park and J. Tsujii (1998). "Identifying the Interaction between Genes and Gene Products Based on Frequently Seen Verbs in Medline Abstracts, " *Genome Inform Ser Workshop Genome Inform* 9: 62-71.

Shannon, P., A. Markiel, O. Ozier, N. Baliga, J. Wang, D. Ramage, N. Amin, B. Schwikowski and T. Ideker (2003). "Cytoscape: A Software Environment for Integrated Models of Biomolecular Interaction Networks, " *Genome Research* 13(11): 2498-504.

Skounakis, M. and M. Craven (2003). "Evidence Combination in Biomedical Natural-Language Processing," in *3rd ACM SIGKDD Workshop on Data Mining in Bioinformatics.*

Thomas, J., D. Milward, C. Ouzounis, S. Pulman and M. Carroll (2000). "Automatic Extraction of Protein Interactions from Scientific Abstracts," *in Pacific Symposium of Biocomputing*: 541-52.

Weaver, D. C., C. T. Workman and G. D. Stormo (1999). "Modeling Regulatory Networks with Weight Matrices," in *Pacific Symposium on Biocomputing* 4, Hawaii.

Wessels, L. F. A., E. P. V. Someren and M. J. T. Reinders (2001). "A Comparison of Genetic Network Models," in *Pacific Symposium on Biocomputing,* Hawaii.

Wikipedia: The Free Encyclopedia (2004). Naive Bayesian Classification. Wikipedia: the free encyclopedia (http://en.wikipedia.org/). 2004.

Wurtele, E. S., J. Li, et al., (2003). "Metnet: Software to Build and Model the Biogenetic Lattice of Arabidopsis," *Comparative and Functional Genomics* 4: 239-245.

SUGGESTED READINGS

On Information Retrieval: Modern Information Retrieval, by Ricardo Baeza-Yates, Berthier Ribiero-Neto, Addison-Wesley Pub Co; 1st edition, 1999, ISBN: 020139829X.
An excellent survey of the key issues surrounding IR, from algorithms to presentation of IR results. This book contains clear explanations of all major algorithms along with quantitative analyses of the relative effectiveness of each algorithm, including the methodology used to arrive at results.

Data Analysis Tools for DNA Microarray, by Sorin Draghici, Chapman & Hall/CRC, 2003, ISBN: 1584883154.
This text on microarray analysis describes complex data analysis techniques, emphasizing specific data analysis issues characteristic of microarray data.

ON-LINE RESOURCES

MetNet, (http://www.public.iastate.edu/~mash/MetNet/) contains links to the websites for the MetNetDB, FCModeler and PathBinder tools mentioned in this chapter.

PathBinderH (www.plantgenomics.iastate.edu/PathBinderH) is a large database of sentences drawn from MEDLINE containing co-occurring terms from a large dictionary. It allows queries to be qualified by biological taxa. It is provided by the Center for Plant Genomics at Iowa State University.

The Arabidopsis Information Resource, TAIR (http://www.arabidopsis.org) is a central clearinghouse for the model organism, Arabidopsis.

Aracyc (Mueller, Zhang et al., 2003) (http://www.arabidopsis.org/tools/aracyc) AraCyc is a database containing biochemical pathways of Arabidopsis, developed at The Arabidopsis Information Resource. The aim of AraCyc is to represent Arabidopsis metabolism as completely as possible. It presently features more than 170 pathways that include information on compounds, intermediates, cofactors, reactions, genes, proteins, and protein subcellular locations.

KEGG: Kyoto Encyclopedia of Genes and Genomes (http://www.genome.jp/kegg/) KEGG is a comprehensive bioinformatics resource developed by the Kanehisa Laboratory of Kyoto University Bioinformatics Center. It contains information about genes and gene products, chemical compounds and pathway information.

Brenda: (http://www.brenda.uni-koeln.de/) is a repository for enzyme information.

R (http://www.r-project.org) is an Open Source language and environment for statistical computing and graphics. R provides a wide variety of statistical (linear and nonlinear modelling, classical statistical tests, time-series analysis, classification, clustering) and graphical techniques, and is highly extensible.

Bioconductor (http://www.bioconductor.org) is an open source and open development software project for the analysis and comprehension of genomic data. The project was started in the Fall of 2001. The Bioconductor core team is based primarily at the Biostatistics Unit of the Dana Farber Cancer Institute at the Harvard Medical School/Harvard School of Public Health. Other members come from various US and international institutions.

PubMed (http://www.ncbi.nlm.nih.gov/entrez/query.fcgi) is a search interface provided by the U.S. National Library of Medicine to a large database of biological texts, mostly but not exclusively from the MEDLINE database.

Agricola (http://agricola.nal.usda.gov/) is a database of article citations and abstracts in the agriculture field, provided by the U.S. National Agricultural Library.

Arrowsmith (kiwi.uchicago.edu) is a system for generating hypotheses about interactions from texts in MEDLINE. (The name is from Sinclair Lewis' novel Martin Arrowsmith.) Provided by the University of Chicago.

MedMiner (http://discover.nci.nih.gov/textmining/main.jsp) is a sentence retrieval system provided by the U.S. National Library of Medicine. It integrates GeneCards and PubMed.

PreBind (http://www.blueprint.org/products/prebind/prebind_about.html) is a database of sentences potentially describing biomolecular interactions. Uncurated, it feeds the Bind database, which is curated. Provided in affiliation with the University of Toronto.

QUESTIONS FOR DISCUSSION

Suppose there is a set of 8 sentences, 4 of which are hits and 4 of which are not. Feature 1 is present in all 4 hits and in 2 non-hits. Feature 2 also occurs in 4 hits and 2 non-hits. There is 1 non-hit with both features. What is the probability estimated by the Naïve Bayes formula that a sentence with both features is a hit? What are the odds for this estimated by the formula for semi-naïve evidence combination? What is the probability implied by these odds? What is the true probability? Repeat this process for the non-hit category. Discuss the results.

Chapter 18

GENE PATHWAY TEXT MINING AND VISUALIZATION

Daniel M. McDonald[1], Hua Su[1], Jennifer Xu[1], Chun-Ju Tseng[1], Hsinchun Chen[1], and Gondy Leroy[2]

[1]*University of Arizona, Management Information Systems Department, Tucson, AZ 85721;*
[2]*Claremont Graduate University, School of Information Science, Claremont, CA 91711*

Chapter Overview

Automatically extracting gene pathway relations from medical research texts gives researchers access to the latest findings in a structured format. Such relations must be precise to be useful. We present two case studies of approaches used to automatically extract gene-pathway relations from text. Each technique has performed at or near the 90 percent precision level making them good candidates to perform the extraction task. In addition, we present a visualization system that uses XML to interface with the extracted gene-pathway relations. The user-selected relations are automatically presented in a network display, inspired by the pathway maps created by gene researchers manually. Future research involves identification of equivalent relations expressed differently by authors and identification of relations that contradict each other along with the inquiry of how this information is useful to researchers.

Keywords

text mining; visualization; gene pathway; PubMed; MEDLINE; information extraction; linguistic analysis; co-occurrence

1. INTRODUCTION

The PubMed database is a valuable source of biomedical research findings. The collection contains information for over 12 million articles and continues to grow at a rate of 2,000 articles per week. The rapid introduction of new research makes staying up-to-date a serious challenge. In addition, because the abstracts are in natural language, significant findings are more difficult to automatically extract than findings that appear in databases such as SwisProt, InterPro, and GenBank. To help alleviate this problem, several tools have been developed and tested for their ability to extract biomedical findings, represented by semantic relations from PubMed or other biomedical research texts. Such tools have the potential to assist researchers in processing useful information, formulating biological models, and developing new hypotheses. The success of such tools, however, relies on the accuracy of the relations extracted from text and utility of the visualizers that display such relations. We will review existing techniques for generating biomedical relations and the tools used to visualize the relations. We then present two case studies of relation extraction tools, the Arizona Relation Parser and the Genescene Parser along with an example of a visualizer used to display results. Finally, we conclude with discussion and observations.

2. LITERATURE REVIEW/OVERVIEW

In this section we review some of the current contributions in the field of gene-pathway text mining and visualization.

2.1 Text Mining

Published systems that extract biomedical relations vary in the amount and type of syntax and semantic information they utilize. Syntax information for our purposes consists of part-of-speech (POS) tags and/or other information described in a syntax theory, such as Combinatory Categorical Grammars (CCG) or Government and Binding Theory. Syntactic information is usually incorporated via a parser that creates a syntactic tree. Semantic information, however, consists of specific domain words and patterns. Semantic information is usually incorporated via a template or frame that includes slots for certain words or entities. The focus of this review is on the syntax and semantic information used by various published approaches. We first review systems that use predominantly either syntax or semantic information in relation parsing. We then review systems that more equally utilize both syntax and semantic information via pipelined analysis. In our review, we will also draw connections between the amount of syntax and semantic information used and the size and diversity of the evaluation.

2.1.1 Syntax Parsing

Tools that use syntax parsing seek to relate semantically relevant phrases via the syntactic structure of the sentence. In this sense, syntax serves as a "bridge" to semantics (Buchholz, 2002). However, syntax parsers have reported problems of poor grammar coverage and over generation of candidate sentence parses. In addition, problems arise because important semantic elements are sometimes widely distributed across the parse of a sentence and parses often contain many syntactically motivated components that serve no semantic function (Jurafsky and Martin, 2000). To handle these challenges, some filtering is performed to eliminate non-relevant parses. Also, sentence relevance is judged before parsing to avoid parsing irrelevant sentences. As reported in the literature, however, systems relying primarily on syntax parsing generally achieve lower precision numbers as compared to relations extracted from systems using full semantic templates.

In the following predominantly syntactic systems, key substances or verbs are used to identify relevant sentences to parse. Park et al. used a combinatory categorical grammar to syntactically parse complete sentence structure around occurrences of proteins (Park et al., 2001). Sekimizu et al. used partial parsing techniques to identify simple grammatical verb relations involving seven different verbs (Sekimisu et al., 1998). Yakushiji et al. used full syntax parsing techniques to identify not just relations between substances, but the sequence of the relations as they occurred in events (Yakushiji et al., 2001). Others, while still predominantly syntactic, have incorporated different types of semantic information. Leroy et al. used shallow syntax parsing around three key prepositions to locate relevant relations (Leroy et al., 2003). Thomas et al. reported on their system Highlight that used partial parsing techniques to recognize certain syntactic structures (Thomas et al., 2000). Semantic analysis was then incorporated afterwards by requiring certain syntactic slots to contain a certain type of semantic entity. In addition, the system extracted only relations that used one of the verb phrases *interact with*, *associate with*, or *bind to*.

With the exception of Leroy et al. that reported a 90 percent precision, the highest precision reported from the syntax approaches did not exceed 83 percent. Park et al. reported 80 percent precision. Sekimizu et al. reported 83 percent precision. Yakushiji et al. reported a recall of 47 percent. The Highlight system reported a high of 77 percent precision. Semantic approaches on the other hand have achieved precision as high as 91 and 96 percent (Friedman et al., 2001; Pustejovsky, Castano et al., 2002).

Despite the lower performance numbers, the evaluations of syntax approaches typically involved more documents than used in semantic parser evaluations. Park et al. evaluated their parser on 492 sentences, while Leroy et al. used 26 abstracts, and Thomas et al. used 2,565 abstracts. Semantic parsers have been evaluated on single articles or more heavily constrained to extract only inhibit relations. Using a greater number of documents in an evaluation requires a parser to deal with more varied writing styles and topic content and thus performance tends to be lower.

2.1.2 Semantic Templates

Other systems rely more on semantic information than on syntax. Semantic parsing techniques are designed to directly analyze the content of a document. Rules from semantic grammars correspond to the entities and relations identified in the domain. Semantic rules connect the relevant entities together in domain-specific ways. Rindflesch et al. incorporate a greater amount of semantic information in their system EDGAR (Rindflesch et al. 2000). Documents are first shallow parsed and important entities are identified using the Unified Medical Language System (UMLS). Biomedical terms are then related together using semantic and pragmatic information. Performance was described as "moderate". GENIES (Friedman et al. 2001) and a system reported by Hafner and colleagues (Hafner et al., 1994) rely primarily on a semantic grammar. GENIES starts by recognizing genes and proteins in journal articles using a term tagger. The terms are then combined in relations using a semantic grammar with both syntactic and semantic constraints. The system was tested on one journal article with the reported precision of 96 percent and a recall of 63 percent. In the system developed by Hafner and colleagues, a semantic grammar was developed to handle sentences with the verbs *measure*, *determine*, *compute*, and *estimate*. The grammar contained sample phrases acceptable for the defined relations. The system was in an early state of development when reported. Pustejovsky et al. used a semantic automaton that focused on certain verbal and nominal forms. Precision of 91 percent was reported along with a recall of 59 percent. The evaluation, however, only extracted relations that used the verb *inhibit*.

Semantic approaches, while more precise, are subject to poorer coverage than syntax approaches. As a result, semantic systems are often evaluated using a smaller sample of documents or a smaller sample of relevant sentences. GENIES was evaluated using one full text article. Pustejovsky et al. limited their relations of interest to *inhibit* relations, and Hafner et al. and Rindflesch et al. did not submit precision or recall numbers.

2.1.3 Balanced Approaches

Balanced approaches utilize more equal amounts of syntax and semantic processing. Syntax processing takes place first, often resulting in an ambiguous parse. More than 100 parses can be generated for a single sentence (Novichkova, 2003). Semantic analysis is then applied to eliminate the incorrect syntactic parse trees and further identify domain words such as proteins and genes. In this fashion, systems combine the flexibility of syntax parsing with the precision of semantic analysis. Such combination has resulted in systems that have been evaluated over a large numbers of documents. Despite the use of both syntactic and semantic processing, however, problems specific to syntactic and semantic analysis persist in part because the analyses are still separate. Syntax grammars remain subject to poor coverage. Because semantic analysis only occurs after syntactic processing, a syntax grammar with poor coverage cannot be improved by the semantic analysis. At the same time, a syntax grammar with good coverage can still generate more parses than can be effectively disambiguated using semantic analysis.

Gaizauskas et al. reported on PASTA, a system that included complete syntax and semantic modules (Gaizauskas et al., 2003). The relation extraction component of PASTA was evaluated using 30 unseen abstracts. Recall was reported at 68 percent, among the highest recall number published, and a precision of 65 percent. The high recall and larger number of documents in the experiment suggest relatively good coverage of their syntax grammar. The relatively lower precision number reflects the sparser coverage of the semantic module given the incoming syntactic parses and their task of extracting protein interactions.

Novichkova et al. reported on their system MedScan which involved both syntax and semantic components (Novichkova et al., 2003). Their first evaluation focused on the coverage of their syntax module, which was tested on 4.6 million relevant sentences from PubMed. Their syntax grammar produced parses for 1.56 million sentences out of the 4.6 million tested resulting in 34 percent coverage. In a more recent study, Daraselia et al. reported a precision of 91 percent and a recall of 21 percent when extracting human protein interactions from PUBMED using MedScan (Daraselia et al., 2004). Such a high precision supports the robustness of their semantic analysis given their task. However, the recall of 21 percent still shows the problem of a syntax grammar with relatively poor coverage. Balancing the use of syntax and semantic analysis contributed to MedScan's ability to perform well on a large sample size. Adding semantic analysis to the pipe, however, did not improve the coverage of the syntax grammar.

2.2 Visualization

Visualization plays an important role in helping researchers explore, comprehend, and analyze gene pathways. Gene pathways often take the form of a graph, in which nodes represent gene and gene products such as proteins and RNAs and edges represent interactions and reactions. Many gene databases and analytical packages provide pathway visualization functionality. KEGG Kanehisa and Goto, 2000; Kanehisa et al., 2002), for instance, provides manually drawn graph representations of protein interaction networks to help users understand the functions of the cell or an organism. However, because manually drawn pathways reflect only the knowledge at the time of drawing and often are difficult to query and update, many analytical packages have employed automatic techniques to visualize gene pathway graphs.

Automatic graph visualization is a well-studied topic in the information visualization area. Graph layout, visual cues, navigation, and interactivity are the major issues associated with graph visualization. Different pathway analysis packages employ different approaches to address these issues.

- *Graph layout*. The key of graph layout is to calculate the position of each node for a given graph according to certain aesthetic criteria or constraints (Herman et al., 2000). Aesthetically pleasing graphs may be generated if nodes and edges are distributed evenly and the number of crossing edges are kept to a minimum (Purchase, 2000). Layout techniques, such as tree layout, circular layout, and spring models have been employed in several gene pathway packages. The tree layout arranges a graph in a hierarchical structure by positioning children nodes beneath parent nodes. In the circular layout, all nodes are placed on a circle and edges are drawn between these nodes. The spring algorithm models a graph as a force-directed energy system with parcels connected by springs. The nodes attract and repulse one another and the system settles down when the energy of the system is minimized (Eades, 1984). G.NET, for example, presents gene pathways using both tree layout and a spring model. Osprey (Breitkreutz et al., 2003) and the Pathway Tools (Karp et al., 2002) offer multiple layout options including linear, tree, and circular to present gene pathways.
- *Visual cues*. Because data about components in gene pathways often are of multiple dimensions, various visual cues have been used to represent different attributes of gene pathway components. For example, in GeneNet (Ananko et al., 2002) the multimerisation state of a protein is

represented by the shape of the protein icon, while the functional state of a protein is represented by colors. Another example is Osprey (Breitkreutz et al., 2003) in which both gene and gene interactions are color-coded to indicate the biological process of the gene, or the experimental system and data source of an interaction.

- *Navigation and interactivity.* As the size of gene pathways increases, the graph may become fairly cluttered and difficult to read. The most commonly used method to address this problem is the zooming operation (Herman et al., 2000). By zooming into a selected area in a graph a user can focus on its details. Most existing pathway systems allow zooming in graphs. However, a user may be disoriented while navigating and zooming between different areas of a graph because of the loss of contextual information. One solution to the focus+context problem is to present the local view and global view of a graph simultaneously. In the G.Net system, the main window presents the selected local area of a pathway and the global view of the graph is shown in a smaller window. The user can select a focal area in the global view and the main window shows the details of the focal area. Another solution is the fisheye view (Furnas, 1986) and hyperbolic tree (Lamping and Rao, 1986). Although on the basis of different mechanisms, they both maintain the focus and context by enlarging the focal area and displaying other parts of a graph with less detail to help user navigate on a complex graph. Fisheye view technology is employed in the GSCope system to visualize highly complicated biomolecular networks (Toyoda et al., 2003).

In addition to these fundamental issues associated with graphs, graph query, editing, filtering, and pathway comparison are also addressed in many pathway analysis systems. For example, Osprey (Breitkreutz et al., 2003) and GeneNet (Ananko et al., 2002) allow users to search for specific genes and filter out unneeded information based on data source, function, etc. BioMiner (Sirava et al., 2002) and ToPLign (Kuffner et al., 2000) support the comparison of metabolic pathways among different organisms.

3. CASE STUDIES/EXAMPLES

We now present three case studies, which include two implementations of parsers that extract gene-pathway information and one system for visualizing the pathways extracted.

3.1 Arizona Relation Parser

In the Arizona Relation Parser (ARP), syntax and semantic analysis are applied together in one parsing process as opposed to the pipelined approach that applies syntax and then semantic analysis in sequence. Others have shown such a combination to be effective for information extraction (Ciravegna and Lavelli,, 1999), but we have not seen such a combination in the biomedical domain. We propose that the benefits of combining syntax and semantic analysis can be realized by using a greater number of word classes or tags that reflect the relevant properties of words. Constraints limiting the type of combinations that occur are thus implicit by the absence of such parsing rules. With a greater number of tags, parsing rules must be explicitly written for each word class. When a rule is created and added to the system, it may be correct based on syntax, semantics, or some combination. The theory behind the rules is only implicit. While many rules have to be written to support the numerous tags, semantic constraints do not have to be specified in the system's lexicon. Such an approach differs from most semantic parsing in that we attempt to parse the entire sentence into relations, regardless of the verbs used. Even if a triple is not relevant to the task, ARP still extracts it. Semantic parsing approaches use templates that center around and are specific to key verbs. Such parsing approaches do not parse non-relevant structures. Semantic approaches are thus more tailored to a specific domain and require relations to be anticipated to be extracted. Our hybrid approach can extract a relation that has not been so highly specified, but requires an accurate filtering mechanism to remove non-relevant relations.

ARP combines syntax and semantic analysis together by introducing over 150 new word classes to separate words with different properties. In comparison, the PENN TREE BANK has approximately 36 common word classes. The majority of the new word classes are semantically or lexically oriented, while we also carry over a subset of the syntax tags from the PENN TREE BANK tag set. A sample of the tags is shown in Table 18-1. The set of tags was chosen using three primary methods. First, we started with a complete lexicon extracted from the PENN TREE BANK and BROWN corpora. Then the most common prepositions and verbs from our 40-abstract training set that had been assigned multiple part-of-speech tags from the PENN TREE BANK lexicon were then assigned a unique tag in our lexicon. Second, domain relevant nouns were sub classed into groups of relevant substances or entities. Third, many of the common 36 PENN TREE BANK syntax tags were included in the new tag set. The role of new tags is determined by the way they can be parsed. As tags take semantic and syntactic properties, rules that apply to the tags reflect semantic and syntactic phenomena. Using over 150 new tags with a regular grammar eliminates the problem of over generating parses. Two different parsing rules are not allowed to act on the same input token sequence. Only one parse tree is generated for each sentence, with only two levels being analyzed for relation extraction. The particulars of the parsing process now follow. Output of the parsing steps is shown in Figure 18-1 below. The boxed numbers on the left in the figure correspond to the subsection numbers listed beside the text below.

Table 18-1. A sample of tags used by the Arizona Relation Parser

Selection Method	Tags
Unique tags in our lexicon obtained by observing ambiguous tags from the PENN TREE BANK	BE, GET, DO, KEEP, MAKE, INCD (include), COV (cover), HAVE, INF (infinitive), ABT (about), ABOV, ACROS, AFT (after), AGNST, AL (although), AMG (among), ARD (around), AS, AT, BEC, BEF (before) , BEL (below), BTN (between), DUR (during), TO, OF, ON, OPP (neg/opposite), OVR (over), UNT (until), UPN (upon), VI (via), WAS (whereas), WHL (while), WI (with), WOT
Domain relevant noun classes	DATE, PRCT, TIME, GENE, LOCATION, PERSON, ORGANIZATION
PENN TREE BANK syntax tags	IN, NP, VBD, VBN, VBG, VP, NN, NNP, NNS, NNPS, PRP, PRP$, RB, RBR

3.1.1 Sentence and Word Boundaries

The parsing process begins with tokenization, where word and sentence boundaries are recognized. The sentence splitting relies on a lexicon of 210 common abbreviations and rules to recognize new abbreviations. Documents are tokenized generally according to the PENN TREE BANK tokenizing rules. In addition, words are also split on hyphens, a practice commonly performed in the bioinformatics domain (Gaizauskas et al., 2003). A phrase tagger based on a finite state automaton (FSA) is also run during this step to recognize words that are best tokenized as phrases. Common idiomatic and discourse phrases (i.e. "for example" and "on the other hand") are grouped together in this step along with other compound lexemes, such as compound gene names. Such phrases receive a single initial tag, despite being made up of multiple words.

3.1.2 Arizona Part-of-Speech (POS)/Semantic Tagger

We developed a Brill-style tagger (Brill, 1993) written in Java and trained on the Brown and Wall Street Journal corpora. The tagger was also trained using 100 PubMed abstracts and its lexicon was augmented by the words and tags from the GENIA corpus (Ohta et al., 2002). The tags used to mark tokens include the 150 new tags (generated by methods described earlier) along with the original tags from the PENN TREE BANK tag set.

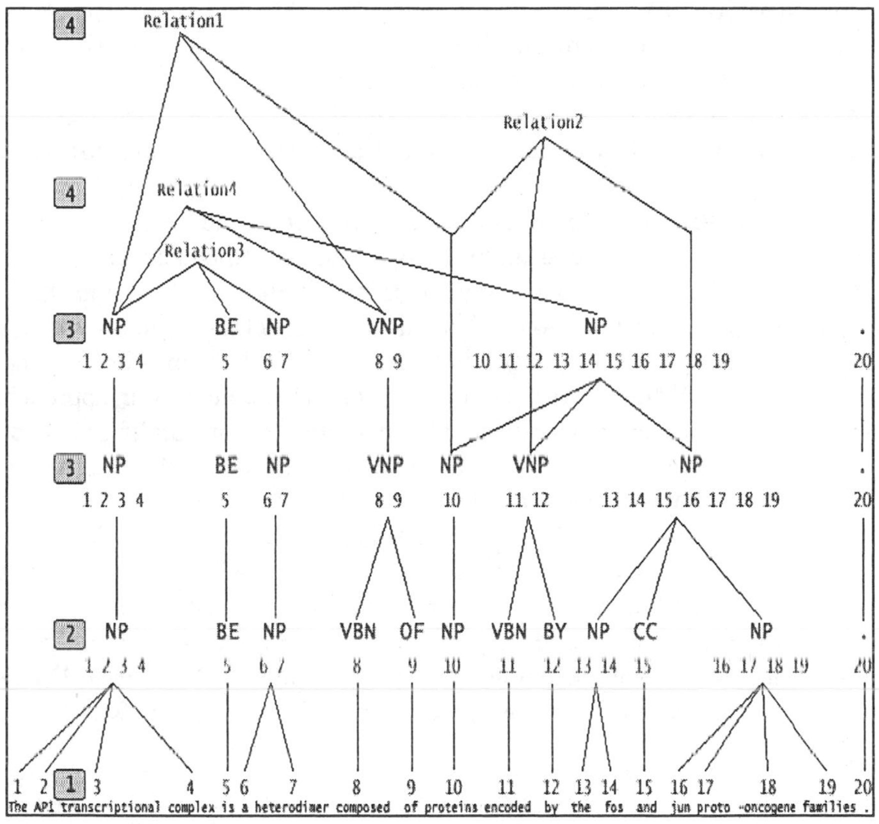

Figure 18-1. An example of a hybrid parse tree.

3.1.3 Hybrid Parsing

In this step word and phrases are combined into larger phrase classes consistent with the role the phrases play in the sentence. The parser output appears in boxed-number 3 shown in Figure 18-1. We attempt to address the poor grammar coverage problem by relaxing some of the parsing assumptions made by full-sentence parsers. For example, full sentence parsing up to a root node of a binary branching tree is not required because we extract semantic triples. Thus, ARP uses a shallow parse structure with n-ary branching, such that any number of tokens up to 24 can be combined into a new node on the tree. Internally, a cascade of five finite state automata attempts to match adjacent nodes from the parse chart to rules found in the grammar. Each FSA handles specific grammatical constructs:

- Level 0: Pre-parsing step where simple noun and verb groups are recognized.
- Level 1: Conjunctions are recognized and combined as noun phrases or treated as discourse units.
- Level 2: Prepositions are attached to verb phrases where possible and made into prepositional phrases elsewhere.
- Level 3: This level catches parsing that should have taken place at level 1 or 2, but did not because of embedded clauses or other preprocessing requirement.
- Level 4: Relative and subordinate clauses are recognized.

The ARP utilizes a regular grammar that handles dependencies up to 24 tags or phrase tags away. Most sentences end before reaching this limit. Regular grammars have previously been used to model the context-sensitive nature of the English language, notably in FASTUS and its medical counterpart HIGHLIGHT (Hobbs et al., 1996). Different in our approach, however, is that the grammar rules are constrained by surrounding tags and thus are only fired when rule core and rule context tags are filled. Figure 18-2 gives an example of a grammar rule, shown on the left, with several abbreviated rules shown on the right. The rule pattern in Figure 18-2 consists of the string "BY NP CC NP .", the rule core equal to "NP CC NP". The rule core is transformed to a "NP" when the entire rule pattern is matched. Therefore, the rule core has to be preceded by a "BY" tag and be followed by a "." tag to be combined into a new noun phrase. The GRAMMAR LEVEL designation, in Figure 18-2, refers to which cascade of the five uses this rule.

```
<GRAMMAR LEVEL= "1">
<RULE NUM=1>
<RULEPATTERN>
<PREVIOUSCONTEXT TAG="BY" />
<RULECORE>NP CC NP</RULECORE>
<FUTURECONTEXT TAG="." />
</RULEPATTERN>
<TRANSFORMATION>
NP
</TRANSFORMATION>
</RULE>
</GRAMMAR>
```

Figure 18-2. A single parsing rule with a rule pattern and transformation

Figure 18-3 below shows an abbreviated notation for grammar rules, showing nine in total. The rule core is in bold and the rule context is italicized. Some of the rules listed have empty rule context slots.

INP VDP NP . transforms to>>INP
IT VP **NP CC NP** *:* transforms to>>NP
AFT NP transforms to>>WHENP
AGNST NP transforms to>>AGPP
AMG NP CC NP II transforms to>>AMGP
ARD NP , transforms to>>ARDP
BE **NP VDP NP** transforms to>>NP
BE **RB JJ** transforms to>>JJ
WI NP CC NP , transforms to>>MOD

Figure 18-3. Abbreviated notation for grammar rules

3.1.4 Relation Identification

The top two levels of the parse chart are passed to the relation identification step. Relations can be loosely compared to subject, verb, object constructs and are extracted using knowledge patterns. Knowledge patterns refer to the different syntactic/semantic patterns used by authors to convey knowledge. Like the parsing rules, knowledge patterns consist of rule patterns with rule context and rule core and transformations that are applied only on the rule core. Different from the parsing rules, however, are the actions that take place on the rule core. First, the rule core does not get transformed into a single new tag, but rather each tag in the rule core is assigned a role, from a finite set of roles *R*. Currently, there are 10 different roles defined in the set *R*. Roles 0 - 3 account for the more directly expressed knowledge patterns. Roles 4 - 9 identify nominalizations and relations in agentive form. Sentences may contain multiple overlapping knowledge patterns, with tags playing multiple roles. The relation parsing output is shown in Figure 18-1 next to the boxed number 4. Conjunctions in the relations are split after the relation extraction step.

Once potential relations are identified, each relation has to meet a number of semantic constraints in order to be extracted. In the current system, at least one word from the first argument and at least one word from the second argument had to exist in a gene/gene products lexicon, such as those from the Gene Ontology and HUGO. In addition, relations were limited by filtering predicates with 147 verb stems. At least one of the words in the predicate had to contain a verb stem. Examples of verb stems from the lexicon include *activat, inhibit, increas, suppress, bind, catalyz, block, augment, elicit, promot, revers, control, coregulat, encod, downregulat, destabiliz, express, hydrolye, inactivat, interfer, interact, mimic, neutraliz, phosphorylat, repress, trigger, and induc.* Our biology expert generated the list of relevant verb stems by examining verbs appearing in PubMed.

3.1.5 RESULTS

The performance of the hybrid grammar together with the semantic filtering was tested in an experiment involving 100 unseen abstracts. Fifty abstracts were used to test precision and 50 for recall. The 100 abstracts were randomly selected from a collection of 23,000 abstracts related to the AP-1 family of transcription factors extracted from PubMed. For the precision experiment, an expert with a Ph.D. in biology separated the parser-generated relations into four different categories. Typically, substance-only relationships are extracted from PubMed texts. Substance-only relationships represent our category A and category B relations. Category A relations were genetic regulatory pathway relations between two recognizable substances. An example of a category A relation is "inhibit(MDM2, SMAD3)." Category B relations were gene-pathway relations between at least one substance and a process or function. An example of a category B relation is "regulates(counterbalance of protein-tyrosine kinases, activation of T lymphocytes to produce cytokines)." This group of relations contained the substances that appeared in gene-pathway maps. In addition to substance relations, we had our expert identify which of the non-substance relations would be "relevant" for pathway map creation. This type of relations, termed category C, is a more open-ended class of relations that is not typically measured in evaluations. Category C relations consisted of more general biologically relevant relations, an example being "mediate(target genes, effects of AP-1 proteins)." ARP should more effectively extract category C relations than purely semantic approaches because it parses every sentence, instead of just sentences with certain verbs or semantic constructions. Category D relations were incorrect or only partially relevant.

The primary results from the experiment are listed in Table 18-2. The parser extracted 130 relations from 50 abstracts, with 79 of those relations belonging to categories A or B. Thus the precision of the parser in extracting pathway relations was 61 percent. When we widened the pool of correct relations to include pathway-relevant relations (category C), correct relations

Table 18-2. Parser performance results

Precision (categories A & B) after filtering	79/130	61%
Recall (categories A & B) after filtering	43/125	35%

jumped to 116 producing an 89 percent precision number, as shown in Table 18-3. The ability to capture category C relations, which are more open-ended, shows the strength of the hybrid approach. By adding category C to the experiment, we affirm the parsing component is performing well, while the semantic filtering function lacks precision. Our current filtering approach did not distinguish well between the A/B group and the C group.

Table 18-3. Parser performance less filtering errors

Precision (A, B & C) after filtering	116/130	89%
Recall (categories A & B) before filtering	76/125	61%

To perform the recall experiment, an expert in biology manually identified all gene pathway relations from categories A and B from 50 randomly selected unseen abstracts. She identified a total of 125 pathway relations between substances from 36 of the 50 abstracts. Fourteen of the abstracts produced no relations. Recall equaled the ratio of system-identified relations to the expert identified relations. Table 18-2 shows the system's recall score of 35 percent. In addition to the standard recall score, we wanted to show the recall of the parsing component unaffected by semantic filtering. Table 18-3 shows the parser extracted 61 percent of the relations.

Including the errors introduced by the semantic filtering component, Table 18-4 lists the reasons why relations were not extracted. The largest number of relations was missed due to imprecise filtering. An incomplete lexicon caused the majority of filtering errors. We are in the process of replacing our semantic filtering step with a biological named entity extraction module to overcome this problem. The next largest group of relations was missed due to incomplete extraction rules. Since this evaluation, the number of extraction rules has more than doubled (rules totaled 210 at the time of the experiment). The third largest group of relations was missed due to the parser's inability to handle co-reference. The expert identified relations that required co-reference resolution 12.5 percent of the time. The co-reference usually occurred between sentences. Along with the entity identification module, a co-reference module is being developed to address this problem. Finally, the parser missed 2.7 percent of the expert identified relations due to parsing errors.

Table 18-4. Why the parser did not recall relations

Reason	% Missed
Removed at semantic filtering stage	26.3%
Incomplete extraction rules	23.6%
Required co-reference information	12.5%
Parsing error	2.7%

3.2 GeneScene Parser

This case study discusses the Genescene parser, which extracts relations between noun phrases and is tuned for biomedical text.

3.2.1 Extracting Semantic Elements

The parser begins by formatting, tokenizing, and tagging the PUBMED abstract with part-of-speech and noun phrase tags. The abstracts are prepared by removing phrases referring to publisher and copyright information. Then the sentence splitter is run followed by the AZ Noun Phraser (Tolle and Chen, 2000) to extract noun phrases. Verbs and adverbs are tagged with their part-of-speech (POS) based on a rule set and lexical lookup in the UMLS Specialist Lexicon. Closed class words such as prepositions, negation, conjunctions, and punctuation are also tagged. Nouns and noun phrases are checked for nominalizations. When a nominalization is discovered, e.g., "activation," then both the infinitive and the original nominalization are retained. Nominalizations can be replaced by the infinitive to facilitate text mining and visualization.

3.2.2 Extracting Structural Elements

Relations have a syntactic basis: they are built around basic sentence structures and prepositions. Prepositions were chosen because they form a closed class and can help capture the structure of a sentence. The closed classes' membership does not change and allows us to build very specific but semantically generic relation templates. In addition, prepositions often head phrases (Pullum and Huddleston, 2002) and indicate different types of relations, such as time or spatial relations (Manning and Schütze, 2001). Although prepositional attachment ambiguity may become a problem, we believe that researchers in biomedicine use a common writing style and so the attachment structures will not vary much for a specific structure.

This case study describes relations built around three prepositions, "by," "of," and "in," which occur frequently in text and lead to interesting, diverse biomedical relations. "By" is often used to head complements in passive sentences, for example in "Mdm2 is not increased by the Ala20 mutation." "Of" is one of the most highly grammaticised prepositions (Pullum and Huddleston, 2002) and is often used as a complement, such as for example in "the inhibition of the activity of the tumor suppressor protein p53." "In" is usually an indication of location. It forms interesting relations when combined with verbs, for example in "Bcl-2 expression is inhibited in precancerous B cells."

Negation is captured as part of these relations when there are specific nonaffixal negation words present. This is the case for *Not*-negation, e.g., not, and *No*-negation, e.g., never. We do not deal with inherent negatives (Tottie, 991), e.g., deny, which have a negative meaning but a positive form. Neither do we deal with affixal negation. These are words ending in –less, e.g., childless, or starting with non–, e.g., noncommittal.

3.2.3 Combining Semantic and Structural Elements

Cascaded, deterministic finite state automata (FSA) describe the relations: basic sentences (BS-FSA) and relations around the three prepositions (OF-FSA, BY-FSA, IN-FSA). All extracted relations are stored in a database. They can contain up to 5 elements but require minimally 2 elements. The *left-hand side* (*LHS*) of a relation is often the active component and the *right-hand side* (*RHS*) the receiving component. The *connector* connects the LHS with the RHS and is a verb, preposition, or verb-preposition combination. The relation can also be *negated* or augmented with a *modifier*. For example from "Thus hsp90 does not inhibit receptor function solely by steric interference; rather …" the following relation is extracted "NOT(negation): Hsp90 (LHS) – inhibit (connector) – receptor function (RHS)." Passive relations based on "by" are stored in active format. In some cases the connector is a preposition, e.g., the relation "single cell clone – of – AK-5 cells." In other cases, the preposition "in" and the verb are combined, e.g., the relation "NOT: RNA Expression – detect in – small intestine".

The parser also recognizes coordinating conjunctions with "and" and "or." A conjunction is extracted when the POS and a UMLS Semantic Types of the constituents fit. Duplicate relations are stored for each constituent. For example, from the sentence "Immunohistochemical stains included Ber-EP4, PCNA, Ki-67, Bcl-2, p53, SM-Actin, CD31, factor XIIIa, KP-1, and CD34," ten relations were extracted based on the same underlying pattern: "Immunohistochemical stains - include - Ber-EP4," "Immunohistochemical stains - include - PCNA," etc.

3.2.4 Genescene Parser Evaluation

1. Overall Results

In a first study, three cancer researchers from the Arizona Cancer Center submitted 26 abstracts of interest to them. Each evaluated the relations from his or her abstracts. A relation was only considered correct if each component was correct and the combined relation represented the information correctly. On average, there were 13 relations extracted per abstract and 90 percent were correct.

2. FSA-specific Results

We also report whether the relations were extracted by the appropriate FSA and calculate precision and recall (see Table 18-5) and coverage. All details of this study can be found in (Leroy et al., 2003).

Table 18-5. Genescene: Precision and Recall

FSA	Total Correct	Total Extracted	Total in Text	Precision (%)	Recall (%)
Relationships:					
BS-FSA:	8	15	23	53	35
OF-FSA:	145	157	203	92	71
BY-FSA:	15	17	24	88	63
IN-FSA:	11	13	37	85	30
All:	**179**	**202**	**287**	**89**	**62**
Conjunctions:					
BS-FSA:	1	1	1	100	100
OF-FSA:	10	10	22	100	45
BY-FSA:	0	0	1	-	0
IN-FSA:	1	1	6	100	16
All:	**12**	**12**	**30**	**100**	**40**

Precision was calculated by dividing the number of correct relations by the total number of extracted relations. The correct relations are those relations considered correct by the researchers, as described above, but with the additional restriction that they need to be extracted by the appropriate FSA. This is a more strict evaluation. Recall was calculated as the number of correct relations divided by the total number of relations available in the text. Only those relations that could have been captured with the described FSA were considered.

There were 267 relations (excludes the conjunctional copies) extracted from the abstracts. Overall, we achieved 62 percent recall of the described patterns and 89 percent precision. The numbers varied by FSA. The highest recall was found for the OF-FSA and the lowest for the IN-FSA where a relation was considered missing when any noun phrase introduced by "in" was missing. These relations were often extracted by another FSA but considered incorrect here. Many of the errors were due to incomplete noun phrases, e.g., a missing adjective.

To evaluate conjunctions, we counted all relations where a conjunction was part of the FSA. Conjunctions where the elements needed recombination, e.g., "breast and ovarian cancer," were not counted since we explicitly avoid them. A conjunction was considered correct if each constituent is correctly placed in the FSA. The conjunctions were either correctly extracted (100 percent precision) or ignored. This adds a few selective relations without introducing any errors. To learn the coverage of the FSA, we counted all occurrences of "by," "of," and "in," with a few exceptions such as "in addition," which are explicitly disregarded by the parser because they result in irrelevant relations. Seventy-seven percent of all "of" prepositions, 29 percent of all "by" prepositions, and 14 percent of all "in" prepositions were correctly captured. This indicates that the OF-FSA is relatively complete for biomedical text. The BY-FSA and IN-FSA cover a smaller portion of the available structures.

3.2.5 Ontology and Concept Space Integration

1. Additional Genescene Components

We parsed more than 100,000 PUBMED abstracts related to p53, ap1, and yeast. The parser processes 15 abstracts per second on a regular desktop computer. We stored all relations and combined them with Concept Space (Chen and Lynch 1992), a co-occurrence based semantic network, in Genescene. Both techniques extract complementary biomedical relations: the parser extracts precise, semantically rich relations and Concept Space extracts co-occurrence relations. The Gene Ontology (Ashburner et al., 2000), the Human Genome Nomenclature (Wain et al., 2002), and the UMLS were used to tag terms. More than half of the terms received a tag. The UMLS provided most tags (57 percent), and GO (1 percent) and HUGO (0.5 percent) fewer.

2. Results of Ontology Integration

In an additional user study, two researchers evaluated terms and relations from abstracts of interest to them. The results showed very high precision of the terms (93 percent) and parser relations (95 percent). Concept Space relations with terms found in the ontologies were more precise (78 percent) than without (60 percent). Terms with more specific tags, e.g., from GO versus the UMLS, were evaluated as more relevant. Parser relations were more relevant than Concept Space relations. Details of this system and study can be found in (Leroy and Chen, in press).

3.2.6 Conclusion

This study described an efficient parser based on closed-class English words to efficiently capture relations between noun phrases in biomedical text. Relations are specified with syntactic constraints and described in FSA but may contain any verb, noun, or noun phrase. On average, the extracted relations are more than 90 percent correct. The parser is very efficient and larger collections have been parsed and combined with the UMLS, GO,

HUGO and a semantic network called Concept Space. This facilitates integration.

3.3 GeneScene Visualizer

The GeneScene Visualizer is a visualization tool designed to support the searching and browsing of the pathway relations extracted from PubMed abstracts by relation parsers. These relations when viewed in large quantities resemble a comprehensive knowledge map. When less common relations are sought, the interface displays genetic networks as reported in the literature. Several screenshots of the visualizer are shown in Figure 18-4. We had previously met with gene researchers and observed the manner in which they constructed pathway maps from PubMed abstracts. These pathway maps inspired the functionality of the GeneScene Visualizer. The visualizer presents the extracted relations as a gene and gene-product interaction network or map.

Architecturally, the GeneScene Visualizer is composed of two main modules: a) a relation repository and b) a visualization interface. These two modules are loosely coupled via custom-designed XML messages on top of HTTP protocol. The GeneScene Visualizer could thus interface any backend repository that presented appropriately formatted XML. The visualizer has already been used with different pathway extraction tools.

The relation repository is implemented in Microsoft SQL Server, relying on over 40 tables and thousands of lines of stored procedure code. The relation repository serves three main purposes: 1) to provide storage for over 500,000 relations extracted from PubMed abstracts 2) to provide retrieval, searching and sorting functionality for the stored relations based on user-input keywords and various ranking strategies and 3) to provide storage for the PubMed abstracts themselves, which are loaded in the visual interface when requested by the user.

The visualization interface also has three primary functions: 1) to provide a search interface for the users to retrieve relations of interest using boolean operators and partial and exact matching 2) to automatically layout the retrieved relations in a meaningful and intuitive manner via a spring embedded algorithm and 3) to show the original PubMed abstract content as reference, when the user clicks on links of interest.

The visualizer is written in Java and is launched from a web browser using Java WebStart™. The application is thus cross-platform and can be run on Windows, Apple, and Linux client machines. The application communicates with the relation repository via the HTTP protocol and XML messages in order to perform searching and fetching of the PubMed abstracts.

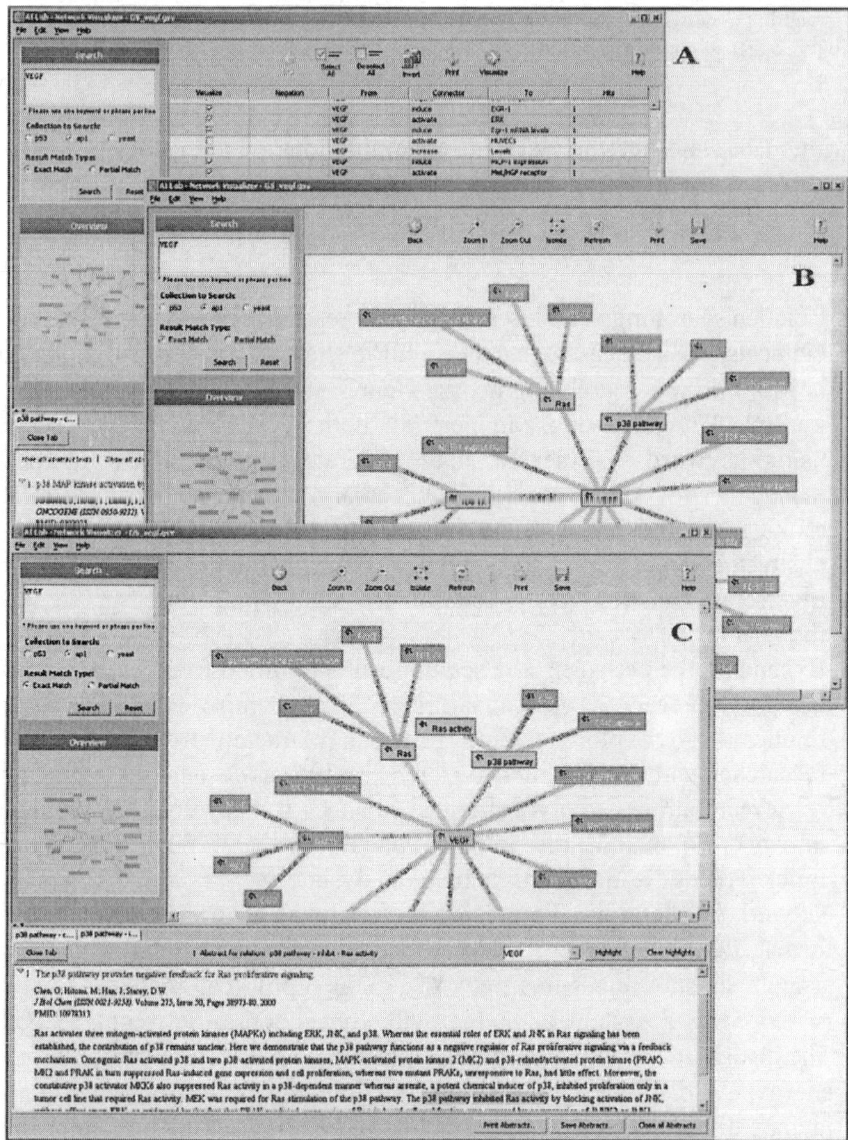

Figure 18-4. GeneScene interface. A. A user wants to search regulatory relations for VEGF (vascular endothelial growth factor), a growth factor that can stimulate tumor angiogenesis. A search in the AP1 collection by exact match results in a number of relations listed in the table in the right panel; the user then selects interesting relations (highlighted), e.g., "VEGF activate Ras", and "p38 pathway convey VEGF (signal)". B. The selected relations are visualized in a network; selected nodes ("Ras", "p38 pathway" and "RAFTK") are expanded to bring additional relations. C. Expanding the network helps the user identify more interesting relations; in this example, expanding the "p38 pathway" brings up a relation ("p38 pathway inhibit Ras activity") that seems controversial to other relations in the network ("p38 pathway convey VEGF (signal)" and "VEGF activate Ras"); by retrieving the abstracts corresponding to the relations (e.g., "The p38 pathway provides negative feedback for Ras

proliferative signaling" (Chen, Hitomi et al. 2000)), the user is able to explain the relations observed in the network (p38 provides negative feedback for Ras signaling by inhibiting Ras activity so that an equilibrium is established among VEGF, Ras and p38 pathway).

various table and network viewing manipulations of the retrieved relations. Examples include filter, sort, zoom, highlight, isolate, expand, and print. These functionalities are grouped into six categories and explained in detail below.

1. Relation searching: The interface allows researchers to search for specific elements, e.g., diseases or genes, and to view all retrieved relations as a network. Like a search engine, the GeneScene Visualizer system can take multiple keywords and can perform both "AND" and "OR" searches using keywords. A search may generate a tremendous number of matches. Thus, result ranking is necessary. We are exploring a relation ranking scheme based on various elements such as the source of the abstract, the location of the sentence in the abstract, keyword frequency, inverse abstract frequency, relation frequency, and the completeness of the relation.
2. Expanding the network: The search results returned from keyword search usually represent a very small part of the entire PubMed extracted relations. To explore related relations extracted from PubMed, the researcher can double-click on any network node and the system will retrieve relations involving that substance. By keyword searching and network expansion, researchers can find related network paths not quickly recognized through manual analysis.
3. Network presentation: To present the relation network in an easy to read format, the GeneScene Visualizer uses a spring-embedded algorithm to arrange the substances in a network. Using a physical analogy, a network is viewed as a system of bodies with forces acting between the bodies. The algorithm seeks a configuration of the bodies with locally minimized energy. With minor modification, our implementation of this algorithm layouts disconnected networks nicely. The drawback of this algorithm is that being node focused, it does not optimize the placement of links. In a complex relation network, links tend to cross and sometimes are not easy to follow.
4. Network navigation: Network navigation is similar to that found in Geographic Information Systems (GIS). The GeneScene Visualizer presents the relation network like a map. Researchers can zoom, and pan the view area of the network to gain the birds-eye view of the network and thus not miss the details. The GeneScene Visualizer also provides

researchers an overview of the network with which they can always keep a global view, providing a focus | context interface.

5. Network manipulation: The GeneScene Visualizer allows not just network navigation, but also provides the ability to manipulate the relation networks in both content and presentation. The researchers can move the elements in the network to change network layout. In addition to the network interface, relations are displayed in rows as part of a table view. From the table view, relations can be sorted, selected, and deleted. Only selected relations are included in the gene/gene products network visualization. An iterative approach to network refinement is supported.
6. Displaying relation source: By clicking on a link in the network, the GeneScene Visualizer loads the original PubMed abstract from which that particular relation was extracted. The source sentence producing the relation is highlighted to speed identification. The abstract metadata is also displayed for reference.

A demo system has been created using abstracts from the PubMed database, including relations from three test collections, which include p53 (23,243 abstracts), yeast (56,246 abstracts), and AP1 (30,820 abstracts). The demo system can be accessed at: http://genescene.arizona.edu:8080/NetVis/index.html. Preliminary user studies by cancer researchers in the Arizona Cancer Center have shown the system to effectively provide access to biomedical literature and research findings. We are currently evaluating the user interface and the visual network exploration component for their utility in exploring genetic pathway-related biomedical relations.

4. CONCLUSIONS AND DISCUSSION

We have presented two text mining systems for extracting gene-pathway relations from PubMed abstracts. The Arizona Relation Parser uses a hybrid grammar and parses sentences into their grammatical sentence structure to extract relations. The Genescene Parser recognizes noun phrases and then uses the semantics of key prepositions to anchor templates that recognize pathway relations. Both systems have achieved performance of or near 90 percent precision. Such a precision score is among the higher performing published systems for gene-pathway extraction. With a large number of precise pathway relations extracted, presenting these relations to researchers in a meaningful way becomes an important focus. We also presented a visualization tool for displaying the relations, the Genescene Visualizer. The visualizer automatically lays out selected relations in a network display for analysis and future exploration. The network display was inspired by the manually created pathway maps researchers often make when reading PubMed abstracts and articles. We are currently evaluating the utility of the visualizer to enhance cancer researchers' access to research findings.

Having extracting and analyzed large numbers of gene-pathway relations, our future research direction has focused on consolidating or aggregating the relations. Aggregating relations combines equivalent relations, even though they may have been expressed with different words (i.e., non-mutant genes are often called wild-type genes). Such a task requires a deeper understanding of the relations. Knowing which relations say the same thing or express contradictory findings allows the visualizer to display only unique relations as well as point out possibly interesting contradictions for researchers to pursue. Such information has the potential to improve the ranking of relations returned from a search as well as to present a less cluttered gene network for visual analysis. In addition, as relations are more fully understood, they can be automatically linked to other existing knowledge sources such as the UMLS semantic network and the REFSEQ database. Expert formed relations from the database could then be merged with the automatically extracted relations to increase the accuracy and coverage of the network.

5. ACKNOWLEDGEMENTS

This research was sponsored by the following grant: NIH/NLM, 1 R33 LM07299-01, 2002-2005, "Genescene: a Toolkit for Gene Pathway Analysis."

REFERENCES

Ananko, E. A., N. L. Podkolodny, et al. (2002). "GeneNet: a Database on Structure And Functional Organisation of Gene Networks," *Nucleic Acid Research* 30(1): 398-401.

Ashburner, M., C. A. Ball, et al. (2000). "Gene Ontology: Tool For the Unification of Biology," *Nature Genetics* 25: 25-29.

Breitkreutz, B.-J., C. Stark, et al. (2003). "Osprey: A Network Visualization System," *Genome Biology* 4(3): R22.

Brill, E. (1993). A *Corpus-Based Approach To Language Learning.* Computer Science. Philadelphia, University of Pennsylvania.

Buchholz, S. N. (2002). *Memory-Based Grammatical Relation Finding.* Computer Science. Tilburg, University of Tilburg: 217.

Chen, H. And K. J. Lynch (1992). "Automatic Construction of Networks of Concepts Characterizing Document Databases," *IEEE Transactions on Systems, Man And Cybernetics* 22(5): 885-902.

Ciravegna, F. And A. Lavelli (1999). *Full Text Parsing Using Cascades of Rules: An Information Extraction Perspective.* EACL.

Daraselia, N., A. Yuryev, et al. (2004). "Extracting Human Protein Interactions from MEDLINE Using a Full-sentence Parser," *Journal of Bioinformatics* 20(5): 604-611.

Eades, P. (1984). "A Heuristic For Graph Drawing," *Congressus Numerantium* 42: 19-160.

Friedman, C., P. Kra, et al. (2001). "GENIES: a Natural-language Processing System For theExtraction of Molecular Pathways From Journal Articles," *Journal of Bioinformatics* 17(1): S74-S82.

Furnas, G. W. (1986). "Generalized Fisheye Views," in *Proceedings of the Human Factors in Computing Systems Conference* (CHI '86).

Gaizauskas, R., G. Demetriou, et al. (2003). "Protein Structures And Information Extraction From Biological Texts: thePASTA System," *Journal of Bioinformatics* 19(1): 135-143.

Hafner, C. D., K. Baclawski, et al. (1994). "Creating a Knowledge Base of Biological Research Papers," *ISMB* 2: 147-155.

Herman, I., G. Melancon, et al. (2000). "Graph Visualization And Navigation in Information Visualization: A Survey," *IEEE Transactions on Visualization And Computer Graphics* 6(1): 24-43.

Hobbs, J., D. Appelt, et al., Eds. (1996). *FASTUS: Extracting Information From Natural Language Texts. Finite State Devices For Natural Language Processing*, MIT Press.

Jurafsky, D. And J. H. Martin (2000). *Speech and Language Processing: An Introduction to Natural Language Processing, Computational Linguistics, And Speech Recognition.* Upper Saddle River, Prentice Hall.

Kanehisa, M. And S. Goto (2000). "KEGG: Kyoto Encyclopedia of Genes And Genomes," *Nucleic Acid Research* 28(1): 27-30.

Kanehisa, M., S. Goto, et al. (2002). "The KEGG Databases At GenomeNet," *Nucleic Acid Research* 30(1): 42-46.

Karp, P. D., S. Paley, et al. (2002). "The Pathway Tools Software," *Bioinformatics* 18(1): S225-S232.

Kuffner, R., R. Zimmer, et al. (2000). "Pathway Analysis in Metabolic Databases Via Differential Metabolic Display," *Bioinformatics* 16(9): 825-836.

Lamping, J. And R. Rao (1986). "The Hyperbolic Browser: a Focus+Context Technique for Visualizing Large Hierarchies," *Journal of Visual Language And Computing* 7(1): 33-55.

Leroy, G. And H. Chen, in Press, "Genescene: An Ontology-enhanced Integration of Linguistic And Co-occurrence Based Relations in Biomedical Texts," *Journal of the American Society For Information Science And Technology* (Special Issue).

Leroy, G., J. D. Martinez, et al. (2003). "A Shallow Parser Based on Closed-class Words to Capture Relations in Biomedical Text," *Journal of Biomedical Informatics* 36: 145-158.

Manning, C. D. And H. Schütze (2001). *Foundations of Statistical Natural Language Processing.* Cambridge, Massachusetts, MIT Press.

Novichkova, S., S. Egorov, et al. (2003). "MedScan, a Natural Language Processing Engine For MEDLINE Abstracts," *Journal of Bioinformatics* 19(13): 1699-1706.

Ohta, T., Y. Tateisi, et al. (2002). "The Genia Corpus: An Annotated Research Abstract Corpus in Molecular Biology Domain," Human Language Technology Conference, San Diego, CA, USA.

Park, J. C., H. S. Kim, et al. (2001). "Bidirectional Incremental Parsing For Automatic Pathway Identification with Combinatory Categorical Grammar," Pacific Symposium on Biocomputing. 6: 396-407.

Pullum, G. K. And R. Huddleston (2002). *Prepositions And Preposition Phrases: theCambridge Grammar of the English Language.* R. Huddleston And G. K. Pullum. Cambridge, UK, Cambridge University Press.

Purchase, H. C. (2000). "Effective Information Visualization: A Study of Graph Drawing Aesthetic And Algorithms," *Interacting with Computers* 13: 147-162.

Pustejovsky, J., J. Castano, et al. (2002). "Robust Relational Parsing over Biomedical Literature: Extracting Inhibit Relations," in *Pacific Symposium on Biocomputing*, Hawaii.

Rindflesch, T. C., L. Tanabe, et al. (2000). "EDGAR: Extraction of Drugs, Genes and Relations From the Biomedical Literature," in *Pacific Symposium on Biocomputing*: 517-528.

Sekimisu, T., H. Park, et al. (1998). "Identifying the Interaction Between Genes and Gene Products Based on Frequently Seen Verbs in Medline Abstracts," *Genome Inform*: 62-71.

Sirava, M., T. Schafer, et al. (2002). "BioMiner: Modeling, Analyzing, and Visualizing Biochemical Pathways And Networks," *Bioinformatics* 18(2): 219-230.

Thomas, J., D. Milward, et al. (2000). "Automatic Extraction of Protein Interactions from Scientific Abstracts," in *Pacific Symposium on Biocomputing*: 510-52.

Tolle, K. M. And H. Chen (2000). "Comparing Noun Phrasing Techniques for Use with Medical Digital Library Tools," *Journal of the American Society of Information Systems* 51(4): 352-370.

Tottie, G. (1991). *Negation in English Speech And Writing: A Study in Variation*, Academic Press, Inc.

Toyoda, T., Y. Mochizuki, et al. (2003). "GSCope: a Clipped Fisheye Viewer Effective For Highly Complicated Biomolecular Network Graphs," *Bioinformatics* 19(3): 437-438.

Wain, H. M., L. M, et al. (2002). "The Human Gene Nomenclature Database," *Nucleic Acids Research* 30(1): 169-171.

Yakushiji, A., Y. Tateisi, et al. (2001). "Event Extraction From Biomedical Papers Using a Full Parser," in *Pacific Symposium on Biocomputing* 6: 408-419.

SUGGESTED READINGS

Battista, G. D., et al., 1999, *Graph drawing: Algorithms for the visualization of graphs.* Prentice Hall.

This book presents a set of algorithms and approaches for displaying the structure of a graph, including planar orientations, flow and orthogonal drawing, incremental construction, layered drawings of digraphs, and spring embedder methods.

Becker, R.A., Eick, S.G., and Wilks, A.R., 1995, Visualizing network data. *IEEE Transactions on Visualization and Computer Graphics* 1 (1), 16-28.

The focus of this article is not on displaying the structure of a network but on effectively presenting the data associated with nodes and links in the network.

Jackson, P. and I. Moulinier (2002, *Natural Language Processing for Online Applications.* Amsterdam / Philadelphia, John Benjamins Publishing Company.

This book describes the use of natural language processing for various tasks from named entity extraction to ad hoc query tasks. The book describes some of the current work from the Message Understanding Conferences (MUC).

Jurafsky, D. and J. H. Martin (2000, *Speech and Language Processing: An Introduction to Natural Language Processing, Computational Linguistics, and Speech Recognition.* Upper Saddle River, Prentice Hall.

This book presents a comprehensive treatment of natural language processing from a linguistic perspective. Many algorithms are presented for various levels of linguistic analysis.

Manning, C. D. and H. Schütze (2001, *Foundations of Statistical Natural Language Processing*. Cambridge, Massachusetts, MIT Press.
This book presents a statistical approach to natural language processing. The book includes both the statistics used in the models presented, but also examples where the statistical models are used.

ONLINE RESOURCES

A comprehensive list of major gene pathway analysis systems
http://ihome.cuhk.edu.hk/~b400559/arraysoft_pathway.html

An NLM funded resource for lexical information from PUBMED
http://www.medstract.org/

André Moreau and Associates Inc., English Language Resources
http://www.ajmoreau.com/english.me.html

Gene Ontology Consortium
http://www.geneontology.org/

Genia Project Home Page
http://www-tsujii.is.s.u-tokyo.ac.jp/~genia/index.html

Human Gene Nomenclature Committee
http://www.gene.ucl.ac.uk/nomenclature/

National Library of Medicine, Entrez PubMed
http://www.ncbi.nlm.nih.gov/entrez/query.fcgi

NIH, NCI, CCR, DHHS Genomics and Bioinformatics Group
http://discover.nci.nih.gov/

The Unfied Medical Language System (UMLS)
http://www.nlm.nih.gov/research/umls/

University of Arizona Artificial Intelligence Lab, NLM funded research
http://ai.arizona.edu/go/GeneScene/index.html

QUESTIONS FOR DISCUSSION

1. What are some common approaches used to extract gene-pathway relations from text?

2. How do syntactic parsing approaches differ from semantic parsing approaches?

3. How can biomedical relational triples be used once they have been extracted from text?

4. Why do prepositions play such a key role in medical abstracts?

5. What are major issues associated with automated graph drawing?

6. What domain-specific difficulties does gene pathway visualization face?

7. What tools help meet the focus+context need of end users?

Chapter 19

THE GENOMIC DATA MINE

Lorraine Tanabe

National Center for Biotechnology Information, Computational Biology Branch, National Library of Medicine, Bethesda, MD 20894

Chapter Overview

The genomic data mine represents a fundamental shift from genetics to genomics, essentially from the study of one gene at a time to the study of entire genetic metabolic networks and whole genomes. Experimental laboratory data are deposited into large public repositories and a wealth of computational data mining algorithms and tools are applied to mine the data. The integration of different types of data in the genomic data mine will contribute towards an understanding of the systems biology of living organisms, contributing to improved diagnoses and individualized medicine. This chapter focuses on the genomic data mine consisting of text data, map data, sequence data, and expression data, and concludes with a case study of the Gene Expression Omnibus (GEO).

Keywords

genomics; text mining; data mining; gene expression data

> "... Medical schools, slow to recognize the profound implications of genomics for clinical medicine, have been lurching, if not stumbling, forward to embrace the genomification of medicine..."
>
> *Canadian Medical Association Journal* editorial, 2003

1. INTRODUCTION

The field of genomics began in the late 20th century with the physical and genetic mapping of genes, followed by the application of DNA sequencing technology to the genetic material of entire organisms to elucidate the blueprints of life. The main branches of genomics are distinguished as 1) structural genomics, including mapping and sequencing, 2) comparative genomics, including genetic diversity and evolutionary studies, and 3) functional genomics, the study of the roles of genes in biological systems. In addition to DNA sequencing, one of the most important technologies for genomics is DNA microarrays, which can measure the expression of thousands of genes simultaneously. Largely due to the generation of voluminous gene expression data from microarrays, genomics in the 21st century is evolving from its sequence-based origins towards a systems biology perspective which encompasses the molecular mechanisms as well as the emergent properties of a biological system.

Systems Biology is not a new research area, but it has been revitalized by genomic data. In January 2003, the Massachusetts Institute of Technology (MIT) started a Computational and Systems Biology Initiative, and Harvard and MIT's Broad Institute was specifically designed to bridge genomics and medicine. The NASA Ames Research Center currently funds the Computational Systems Biology Group, an association of statisticians, computer scientists, and biologists at Carnegie Mellon University, the University of Pittsburgh, and the University of West Florida. The Systems Biology Markup Language (SBML) is a computer-readable format for representing models of biochemical reaction networks (Hucka et al., 2003). Because human patients are biological systems, the systems biology approach has enormous potential to ease the transition of genomics knowledge from the laboratory to the clinical setting. Before this transfer of knowledge can happen, the large-scale genomics data need to be interpreted with a combination of hypothesis-driven research and data mining. This chapter will present some data mining techniques for genomic data.

Data Mining is the exploration of large datasets from many perspectives, under the assumption that there are relationships and patterns in the data that can be revealed. This can be a multi-step procedure with an automatic component followed by human investigation. It is a data-driven approach, exploratory in nature, which complements a more traditional hypothesis-driven methodology. Large-scale genetic sequence and expression data generated from high-throughput experimental techniques constitute a huge data mine from which new patterns can be discovered, contributing to a greater understanding of biological systems and their perturbations, leading to new therapeutics in medicine. The genomic data mine represents a fundamental shift from genetics to genomics, essentially from the study of one gene at a time to the study of entire genetic and metabolic networks and whole genomes. Experimental laboratory data are deposited into large public repositories, and a wealth of computational data mining algorithms and tools are applied to mine the data.

Genomic databases are continually growing in depth and breadth. The Molecular Biology Database Collection lists many of these resources at the *Nucleic Acids Research* web site http://nar.oupjournals.org/. Each year, *Nucleic Acids Research* publishes a special database issue, including updates on many broad genomics databases like GenBank (Benson et al., 2004), the EMBL Nucleotide Sequence Database (Kulikova et al., 2004), the Gene Ontology (GO) database (Gene Ontology Consortium, 2004), the KEGG resource (Kanehisa et al, 2004), MetaCyc (Krieger et al., 2004), and UniProt(Apweiler et al., 2004), as well as more specialized databases like WormBase (Harris et al., 2004), the Database of Interacting Proteins (Salwinski et al., 2004), and the Mouse Genome Database (Bult et al., 2004).

In this chapter, the focus will be on genomic text data, map data, sequence data, and expression data. Protein 3-D structural data will not be covered. For brevity, the genomic data freely available at the National Center for Biotechnology Information (NCBI) at the National Library of Medicine (NLM) will be highlighted. The chapter will conclude with a case study of NCBI's Gene Expression Omnibus (GEO) data mining tool.

2. OVERVIEW

The genomics data mine contains text data, map data, sequence data, and expression data.

Table 19-1. Genomics questions can be answered using different types of data

	Text Data	Map Data	Sequence Data	Expression Data
Where on the chromosome is this gene located?	X	X		
Is there a model organism with a related gene?	X	X	X	
How has this gene evolved?	X	X	X	X
What tissues is this gene expressed in?	X			X
How does a drug affect gene expression?	X			X
What is the function of this gene?	X	X	X	X

NCBI's Entrez system is a starting point for exploring these rich datasets (Schuler et al., 1996). LocusLink is a gene-centered interface to sequence and curation data. RefSeq is a database supplying citations for transcripts, proteins, and entire genomic regions for 2000 organisms. RefSeq and LocusLink provide a non-redundant view of genes, to support research on genes and gene families, variation, gene expression, and genome annotation (Pruitt and Maglott, 2000). Unigene classifies GenBank sequences into about 108,000 gene-related groups (Schuler, 1997). Some basic questions that can be answered by mining genomics data types are summarized in Table 19-1.

2.1 Genomic Text Data

Text mining is an emerging field without a clear definition in the genomics community. Text mining can refer to automated searching of a sizeable set of text for specific facts. A more rigid definition of text mining requires the discovery of new or implicit knowledge hidden in a large text collection. In genomics, text mining can also refer to the creation of literature networks of related bimolecular entities. Text mining, like data mining, involves a data-driven approach and a search for patterns.

Scientific abstracts, full-text articles, and the internet all contain text data that can be mined for specific information or new facts. Because it contains the collective facts known about nearly all genes that have ever been studied, the genomics text data mine represents the entire genomics knowledge base. This knowledge is encoded in natural language and is a meta-level representation of the information gleaned from hypothesis-driven and numerical-data-driven experimentation.

Text mining research in genomics is a growing field of research comprising: 1) relationship mining, 2) literature networks, and 3) knowledge discovery in databases (KDD). Relationship mining refers to the extraction of facts regarding two or more biomedical entities. Literature networks are meaningful subsets of MEDLINE based on co-occurring gene names and/or functional keywords. Literature networks based on co-occurrence are motivated by the fact that functionally related genes are likely to occur in the same documents. Stapley and Benoit define *biobibliometric distance* as the reciprocal of the Dice coefficient of two genes i and j:

$$d_{ij} = \frac{|i| + |j|}{|i \cap j|} \tag{1}$$

The distance between all pairs of genes can be calculated for an entire genome and the results can be visualized as edges linking co-occurring genes (Stapley and Benoit, 2000). PubGene (Jenssen et al., 2001) adds annotation to pairs of genes using functional terminologies from MeSH and

GO. MedMiner (Tanabe et al., 1999) uses functional keywords like *inhibit, upregulate, activate,* etc. to filter the documents containing a pair of genes into subsets based on the co-occurrence of gene names in the same sentence as a functional keyword. Thematic analysis (Shatkay et al., 2000, Wilbur, 2002) finds themes in the literature, sets of related documents based on co-occurring terms. Table 17-2 summarizes some of the genomics research performed in these areas since 1998. KDD genomics tasks include prediction of gene function and location (Cheng et al., 2002) and automatic analysis of scientific papers (Yeh et al., 2003).

Table 19-2. A sample of genomic text mining. M = a MEDLINE corpus, J = biomedical journal articles, T = any biomedical text

	Date	Relation Mining	Literature Networks	Given	Returns
Sekimizu et al.	1998	X		M, verb list	Subjects, objects of verbs
BioNLP/BioJAKE Ng and Wong	1999	X		T, query term	Graphical pathways
ARBITER Rindflesch et al.	1999	X		M	Binding relationships
Blaschke et al.	1999	X		M, genes, verbs	Gene-gene relationships
MedMiner Tanabe et al.	1999	X	X	M query	Keyword summaries
Craven & Kumlien	1999	X		T, classes	Classes/ relationships
Thomas et al.	2000	X		M, verbs, frames	Filled frames
EDGAR Rindflesch et al.	2000	X		M	Gene/drug/cell relations
Stapley, Benoit	2000	X	X	M, gene list	Gene/gene networks
Shatkay et al.	2000		X	M, query term	Literature themes
Stephens et al.	2001	X	X	M, thesauri	Gene pair relationships
XplorMed Perez-Iratxeta et al.	2001	X	X	M query or T	Literature topics
MeSHmap Srinivasan P.	2001		X	M query	Searchable MeSH terms
PubGene Jenssen et al.	2001	X	X	M, gene list	Literature networks
Yakushiji et al.	2001	X		T, verb list	Predicate/arguments
PIES Wong, L.	2001	X	X	M, action verbs	Generated pathways

continued

	Date	Relation Mining	Literature Networks	Given	Returns
GENIES Friedman et al.	2001	X		T, grammar, lexicon	Gene/protein interactions
SUISEKI Blaschke, Valencia	2001	X		T, frames	Filled frames
MEDSTRACT Chang et al.	2002	X		M, relationships	Extracted relationships
Palakal et al.	2002	X		M, E/R model	Entities and relationships
Temkin et al.	2003	X		T, grammar, lexicon	Gene/protein interactions
PreBIND Donaldson et al.	2003	X		J, gene list	Protein/protein relations
MedGene Hu et al.	2003	X	X	M, disease/gene list	Gene/disease summaries
PASTA Gaizauskas et al.	2003	X		M, templates	Filled templates
MeKE Chiang, Yu	2003	X		M, gene list	Protein roles
MedScan Novichkova et al.	2004	X		M, protein list	Protein interactions
MedBlast Tu et al.	2004		X	Sequence	MEDLINE summary

2.1.1 Text Mining Methods

Methods for genomics text mining vary and can be classified into three main approaches: statistical, linguistic, and heuristic. Statistical and linguistic methods both require natural language processing (NLP), a broad expression covering computerized techniques to process human language. Statistical NLP ignores the syntactic structure of a sentence, hence it is often referred to as a "bag-of-words" approach, although it can also involve non-word features like co-occurrence, frequency, and ngrams to determine sentence and document relatedness. Often terms or other features are used in machine learning algorithms including Bayesian classification, decision trees, support vector machines (SVMs), and hidden markov models (HMMs) (Baldi and Brunak, 1998).

More linguistically-motivated approaches utilize part-of-speech (POS) tagging and/or partial or full parsing. One difference between statistical and linguistic methods is that statistical processing often discards common words like *and, or, become, when, where*, etc. (called a stop list), while linguistic techniques rely on these terms to help identify parts of speech and/or sentence syntax. In statistical NLP the resulting words or ngrams are isolated from the larger discourse, making anaphoric reference resolution impossible. For example, in the following text, the *A2780 cells* mentioned in the first sentence are later referred to as *cells*, *oblimersen-pretreated cells*, and *these cells*. The relationships between *A2780 cells* and *temozolomide*, *PaTrin-2,* and *oblimerson* in the second sentence cannot be extracted unless *cells* are resolved to *A2780 cells*:

> Using a human ovarian cancer cell line (A2780) that expresses both Bcl-2 and MGMT, we show that cells treated with active dose levels of either oblimersen (but not control reverse sequence or mismatch oligonucleotides) or PaTrin-2 are substantially sensitized to temozolomide. Furthermore, the exposure of oblimersen-pretreated cells to PaTrin-2 leads to an even greater sensitization of these cells to temozolomide. Thus, growth of cells treated only with temozolomide (5 Î¼g/mL) was 91% of control growth, whereas additional exposure to PaTrin-2 alone (10 Î¼mol/L) or oblimersen alone (33 nmol/L) reduced this to 81% and 66%, respectively, and the combination of PaTrin-2 (10 Î¼mol/L) and oblimersen (33 nmol/L) reduced growth to 25% of control.

Linguistically-motivated methodologies adapt syntactic theory and semantic/discourse analysis to the biomedical domain. Although this is a difficult task due to the complexity of biomedical text, it is necessary for capturing the full meaning of the text.

Heuristic methods make use of biomedical domain knowledge. The manual effort required to translate expert knowledge into rules and patterns is often not prohibitive, and systems using this approach have been successful at extracting pertinent facts from text collections. However, heuristic methods often miss facts that appear in unpredicted contexts, are subject to human bias, and can have problems scaling up to large full text corpora.

Many text mining systems for genomics involve a combination of statistical, linguistic, and/or heuristic methods; for example, the PubMiner system (Eom and Zhang, 2004) uses an HMM-based POS tagger, an SVM-based named-entity tagger, a syntactic analyzer, and an event extractor that uses syntactic information, co-occurrence statistics, and verb patterns. More detail on text mining methods can be found in other chapters in Unit III of this book.

2.1.2 Knowledge Discovery

Text mining is an essential component of map, sequence, and expression data mining efforts in genomics, since no experimental results can be interpreted without reference to pre-existing knowledge. The multitude of facts stored in natural language text databases, like MEDLINE, constitute a rich source of potential new discoveries.

New information can be assembled from separate texts by *literature synthesis*, which involves finding implicit connections between facts. For example, Swanson found articles showing that fish oils cause blood and vascular changes, and connected these to separate articles revealing certain blood and vascular changes that might help patients with Raynaud's syndrome (Swanson, 1990). Two years later, a clinical trial reported the benefit of fish oil for Raynaud patients. Swanson found further examples of productive literature synthesis including connections between magnesium deficiency and migraine headaches and arginine intake and somatomedins in the blood, leading him to suppose that such connections are not rare. Weeber et al. simulated Swanson's fish oil discovery using the drug-ADR-disease (DAD) -system, a concept-based NLP system for processing PubMed documents (Weeber et al., 2000). The DAD-system uses the UMLS Metathesaurus (NLM, 2000) as a basis for text mining PubMed. Query terms are mapped to UMLS concepts using MetaMap (Aronson, 1996), and then the relevant PubMed abstracts are retrieved to a local database. The UMLS concepts contained in these abstracts are presented to the user, who selects concepts for further document retrievals. The ranking of concepts depends on their interconnection and the user formulates and checks hypotheses based on this ranking. The DAD-system has been used to mine biomedical literature on side effects and adverse drug reactions (ADR).

As an alternative to documents, words, or UMLS concepts, gene/protein relations can be used as the basic analytical unit for text mining. *Relational chaining* is the linkage of entities through their relations across multiple documents, facilitating the discovery of interesting combinations of relations that would be impossible to find in a single document. Blaschke and Valencia compared the interactions of yeast cell cycle genes/proteins before and after the year 2000 and found that recent discoveries often originated from entities near each other in previously networked relations, suggesting that initially extracted gene interaction data can be combined into a plan for knowledge discovery (Blaschke and Valencia, 2001). A different strategy for text mining with gene/protein relations involved: 1) establishment of a database of gene/protein relations extracted from MEDLINE and 2) a query mechanism to mine the database for implicit knowledge based on relational chaining. In a prototype system implementing this approach, typical gene/protein queries resulted in PubMed documents automatically linked by gene/protein relations (*decreased_levels_of*, *associated_with, etc.*) (Tanabe, 2003).

2.2 Genomic Map Data

Genomic map data identify the position of a gene on a chromosome or on the DNA itself, vital information for identifying human disease genes and mutations. Chromosome maps are created by cytogenetic analysis (also known as karyotyping), linkage, or *in situ hybridization,* where a DNA probe is used to visualize the chromosomal position. Many disease-related genes are found by linkage to chromosomal regions. For example, chromosomal aberrations have been found to be associated with cancer (Mitelman et al., 1997). Physical maps show the location of a gene on the DNA itself, measured in basepairs, kilobasepairs, or megabasepairs. The Entrez Map Viewer presents genomic map data using sets of aligned chromosomal maps that can be explored at various levels of detail, including UniGene clusters. Map Viewer offers maps for a variety of organisms including mammals, plants, fungi, and protozoa (Wheeler et al., 2004). Graphical views show genes, markers, and disease phenotypes along each chromosome of an organism, as well as the genomic locations showing hits on all chromosomes. Cytogenetic map location is also available through LocusLink.

2.2.1 Finding Candidate Disease Genes

Cytogenetic map data was used for data mining by Perez-Iratxeta et al. to associate genes with genetically inherited diseases using a scoring system based on fuzzy set theory (Perez-Iratxeta, 2002). First, the system used MEDLINE to find disease and chemical terms with frequent co-occurrence in the literature. Next, the RefSeq database was mined for associations between function and chemical terms for annotated genes. Finally, the function terms, chemical terms, and disease terms were combined to get relations between diseases and protein functions. For 455 diseases with chromosomal maps, a score was assigned based on the relation of the RefSeq sequences to the disease given their functional annotation. The disease gene candidates on relevant chromosomal regions were sequence compared to the scored RefSeq sequences. Hits were scored based on the scores of RefSeq homologous sequences. In a test involving 100 known disease genes, the disease gene was among the best-scoring candidate genes with a 25% chance, and among the best 30 candidate genes with a 50% chance.

2.3 Genomic Sequence Data

DNA sequence data are made publicly available through GenBank. Nucleotide Basic Local Alignment Search Tool (BLAST) searches allow one to input nucleotide sequences and compare these against other sequences. Pairwise BLAST performs a comparison between two sequences using the BLAST algorithm. MegaBLAST allows for a sequence to be searched against a specific genome. Position-Specific Iterated (PSI)-BLAST is useful for finding very distantly related proteins. The basic BLAST algorithm looks for areas of high similarity to a query sequence in the sequence database, returning hits that are statistically significant. Non-gapped segments with maximal scores that cannot be extended or trimmed (high scoring segment pairs, HSP) represent local optimal alignments. HSPs above a score threshold are subject to gapped extensions and the best alignment is chosen. If the score of the chosen alignment is statistically significant, it is returned as a hit (Altschul et al., 1990). NCBI tools for sequence data mining include HomoloGene and TaxPlot. HomoloGene is an automated system for finding homologs among eukaryotic gene sets by comparing nucleotide sequences between pairs of organisms. Curated orthologs are incorporated from a variety of sources via LocusLink. TaxPlot is a tool for 3-way comparisons of genomes on the basis of the protein sequences they encode. A reference genome is compared to two additional genomes, resulting in a graphical display of BLAST results where each point for each predicted protein in the reference genome is based on the best alignment with proteins in each of the two genomes being compared. Generally, sequence similarity is associated with similar biological function (although this is not always the case), so mining sequence databases can lead to the discovery of new genes, regulatory elements, and retroviruses.

2.3.1 Predicting Protein Function

Sequence data can also be used to predict protein functional class. Using sequence data and other relevant features like annotation keywords (words used to describe protein function, for example, *apoptosis*), species, and molecular weight, King et al. predicted protein functional classes in *M. tuberculosis* using a combination of Inductive Logic Programming (ILP) and decision tree learning (King et al., 2000). ILP is a machine learning strategy that uses a set of positive and negative training examples to induce a theory that covers the positive but not the negative examples (Muggleton, 1991). ILP requires a set of features that can be used to construct the theory. Decision tree algorithms partition training data into a tree structure where each node denotes a feature in the training data that can be used to partition the positive and negative examples. King et al. retrieved the sequence data with a PSI-BLAST search for homologous proteins to *M. tuberculosis* genes with known function. Relevant features were extracted including percent amino acid composition, PSI-BLAST similarity score, number of iterations, and amino acid pair frequency. ILP was used to mine for patterns in the sequence descriptions and the decision tree algorithm C4.5 (Quinlan, 1993) was used to learn rules predicting function. A simple rule example is: If the percentage composition of lysine in the gene is > 6.6%, then its functional class is "Macromolecule metabolism." This rule was 85% accurate on a test set, predicting proteins involved in protein translation. Overall, the system predicted the function of 65% of *M. tuberculosis* genes with unknown function with 60-80% accuracy.

Protein function can also be predicted using the Clusters of Orthologous Groups of proteins (COGs) database at NCBI (Tatusov et al., 1997, Tatusov et al., 2000). Orthologs are genes in different species that evolved from a common ancestor, as opposed to paralogs, which are genes within the same species that have diverged by gene duplication. COGs are determined by all-against-all sequence comparisons of genes from complete genomes using gapped BLAST. For each protein, the best hit in each of the other genomes is found and patterns of best hits determine the COGs. Each COG is assumed to have evolved from one common ancestral gene. Short stretches of DNA sequences called expressed sequence tags (ESTs) of unknown function can be mined for sequences likely to have protein function using COG information. Using more than 10,000 ESTs from dbEST (Boguski et al., 1993) and 77,114 protein sequences from COG, Faria-Campos et al. mined 4,093 ESTs for protein characterization based on homology to COG groups (Faria-Campos et al., 2003).

In addition to the global view of full genomes represented by COGs, a complementary approach detecting protein families by clustering smaller pieces of sequence space is possible. In a fully automated method, BLAST-scored sequences are clustered around a query protein using pairwise similarities, and then adjacent clusters are pooled to generate potential protein families that are similar to COGs based on a sample of 21 complete genomes (Abascal and Valencia, 2002). The clustering algorithm is a derivation of the minimum cut algorithm (Wu and Leahy, 1993). The merging algorithm pools two clusters if the relative entropy of the merged clusters decreases. Like COGs, the resulting groups can be used to predict the protein function of uncharacterized genes or ESTs.

2.4 Genomic Expression Data

Gene expression data generated from DNA microarrays, oligonucleotide chips, Digital Differential Display (DDD), and Serial Analysis of Gene Expression (SAGE) enable researchers to study genetic and metabolic networks and whole genomes in a parallel manner (Shalon et al., 1996, Spellman et al., 1998, Weinstein et al., 1997, Ross et al., 2000). These technologies can generate data for thousands of genes per experiment, creating a need for data mining strategies to interpret and understand experimental results. DNA microarrays contain probe DNA of known sequence attached to a slide, which is exposed to target samples that have been differentially labeled (Schena et al., 1995). The expression of genes in the target samples can be detected and quantified by their level of competitive hybridization to the probe DNA. Affymetrix, Inc. developed oligonucleotide chips, which use a single probe followed by exposure to target samples. DDD is a method for comparing sequence-based cDNA pools, using UniGene clusters to narrow sequences to genes expressed in humans. Serial analysis of gene expression (SAGE) is a methodology using sequence tags representing specific transcripts assembled into long molecules which are cloned and sequenced, allowing for the measurement of each transcript by the detection of its sequence tags (Velculescu et al., 1995). Microarrays and SAGE can be used complementarily, for example, microarrays can be used to identify cell-specific transcripts and SAGE can be used to determine the percentage of these that are mitochondrial (Gnatenko et al., 2003). SAGEmap at NCBI provides a mapping between SAGE tags and UniGene clusters (Lash et al., 2000).

Microarrays and Affymetrix chips have the advantage of being fast and comprehensive; however, they are expensive and are subject to hybridization and image analysis artifacts. DDD and SAGE have a cost advantage, but there are fewer data analysis tools for them and the data are not comprehensive (SAGE data are available for a limited number of organs).

Gene expression data can be combined with protein data to find key patterns involving gene expression and protein function (Nishizuka et al., 2003). Data mining for relationships between gene expression and protein function is vital, because protein function can be uncorrelated with gene expression and proteins, not genes, are usually the targets for therapeutic intervention. Since proteins are often valuable drug targets, gene and protein expression data are crucial components of what has been termed the "genomification of medicine" (CMAJ, 2003).

2.4.1 Cancer Gene Discovery

The Cancer Genome Anatomy Project (CGAP) at the National Cancer Institute was established in 1996 to uncover the "molecular anatomy" of cancer cells (Strausberg et al., 1997). Digital differential display (DDD) can be used to mine CGAP's EST databases for cancer genes. For example, Scheurle et al. used DDD to identify genes of interest in breast, colon, lung, ovary, pancreas, and prostate solid tumor tissues (Scheurle et al., 2000). DDD allows for normal and tumor cDNA libraries to be compared by generating transcript fingerprints using statistical analysis. Combined with hits from the UniGene database, DDD predicted 12 genes up- or down-regulated in colon tumor tissue, and 3 genes were verified by laboratory experimentation to fit the DDD prediction, making them potential diagnostic and therapeutic targets for colon cancer.

2.4.2 Prognosis Prediction

Gene expression data help cancer researchers depict the states of several genes at a time under varying experimental conditions. The resulting molecular profiles are useful for predicting the prognosis of some types of cancer. For example, van de Vijver et al. established a 70-gene expression profile for breast cancer metastases (van de Vijver et al., 2002). At 10 years, the probability of remaining free of distant metastases was 50.6±4.5 percent in the group with a poor-prognosis signature and 85.2±4.3 percent in the group with a good-prognosis signature. The authors concluded that the gene expression profile was a more effective predictor of disease outcome than clinical/histological criteria. A similar study by Shipp et al. analyzed the expression of 6,817 genes in tumor specimens from Diffuse large B-cell lymphoma (DLBCL) patients who received chemotherapy and applied a weighted-voting classification algorithm to identify cured versus fatal or refractory disease (Shipp et al., 2002). The weighted-voting algorithm is a supervised machine learning algorithm that distinguishes two classes of inputs. An idealized expression profile is created for each class where the expression is high in one class and low in the other class. The voting scheme involves one vote per gene for class A or class B, depending on the similarity of its expression profile to each class's idealized profile. The winner of the votes, either class A or class B, is the predicted class. The weighted-voting algorithm classified two categories of patients with very different five-year overall survival rates (70% versus 12%), indicating that supervised learning classification techniques can predict outcome in DLBCL and identify rational targets for intervention. These results suggest that the integration of a patient's particular constellation of gene expression patterns with other diagnostic tools and resources would pave the way towards personalized medicine.

2.4.3 Tumor Classification

Different classes of tumors are associated with variations in therapeutic response to anti-cancer agents. Golub et al. determined that gene expression profiles can be used to assign tumors to established classes, using human acute leukemias as a test case (Golub et al., 1999). Distinction between acute myeloid leukemia (AML) and acute lymphoblastic leukemia (ALL) is imperative for effective treatment of the disease. Using an expression profile of 6,817 genes, Golub et al. applied a self-organizing map (SOM) to cluster 38 leukemia samples into AML (24/25 samples correctly classified) and ALL (10/13 samples correctly classified). An SOM is learned from an unsupervised algorithm applied to a network of nodes that tunes the inputs to pattern classes (Kohonen, 1981). These results suggest that individualized treatment would be possible using a tumor classification based on gene expression profiles.

3. CASE STUDY: THE GENE EXPRESSION OMNIBUS

The Gene Expression Omnibus (GEO) was developed to address a need in the genomics community for a standardized data repository for gene expression data, including microarray, oligonucleotide chip, hybridization filter, SAGE, and protein data (Wheeler et al., 2004). GEO data are assembled into comparable sets (GDS) which can be searched using Entrez GDS or Entrez GEO. GEO retrieves pre-computed graphical representations of experimental data, as well as gene name, GenBank accession, clone ID, ORF, mapping information, the dataset title, and additional flags regarding outliers and detection calls. Retrievals are listed in order of most-interesting-first, based on a scoring scheme which considers flagged effects, expression level, outliers, and variability.

The following examples from GEO's online tutorial illustrate the types of queries that can be posed to GEO.

To identify all dual channel nucleotide microarray experimental datasets exploring metastasis in humans, enter:

```
"dual channel"[Experiment Type] AND metastasis AND
human[Organism]
```

To view profiles of kallikrein family genes across all datasets, enter:

```
kallikrein
```

To limit these kallikrein retrievals to datasets investigating progesterone, enter:

```
kallikrein AND progesterone
```

To view profiles that fall into the top 1% abundance rank bracket in at least one sample in dataset GDS186, enter:

```
GDS186 AND 100[Max Value Rank]
```

A range can also be specified. To view the top 5%, enter:

```
GDS186 AND 96:100[Max Value Rank]
```

To view profiles that fall into the top 1% variable molecular abundance profiles in dataset GDS186, enter:

```
GDS186 AND 100[Ranked Standard Deviation]
```

GEO BLAST can be used to query Entrez GEO for expression profiles based on sequence similarity. GEO datasets can be browsed and mined using online tools including hierarchical clustering. Hierarchical clustering is an unsupervised method for detecting similar sets of data based on shared features, which can be visualized as a dendrogram. The connectivity of the dendrogram depends upon the clustering algorithm and similarity measure used. Details on hierarchical clustering can be found in a separate chapter in this book.

The following example begins with selecting the "Mammary epithelial cells and breast cancer" dataset GDS90, which returns a GEO record showing a summary of the data, including description of experiment, organism, type of experiment, number of probes, and date. Twenty six unordered samples are indicated, which can be selected or deselected using checkboxes. A cluster analysis can be performed using user-defined distance metrics and hierarchical clustering methods (see Figure 19-1).

4. CONCLUSIONS AND DISCUSSION

The genomics data mine includes text data, map data, sequence data, and expression data. Public repositories house much of the genomic data mine, along with computational tools to organize, integrate, and understand the data. Text data represent the genomics knowledge base and can be mined for relationships, literature networks, and new discoveries by literature synthesis and relational chaining. Map data are crucial for the discovery of new disease genes. Sequence data can be mined to gain insight about protein function and evolution. Microarray experiments generate expression data on thousands of genes at a time, requiring data mining tools that help to visualize, analyze, and interpret the data. Gene and protein expression data are crucial components of what has been termed the "genomification of medicine" (CMAJ, 2003), and careful and effective mining of genomic data has huge potential for future clinical applications. The integration of text, map, sequence, and expression data will contribute towards an understanding of the systems biology of living organisms, contributing to improved diagnoses and individualized medicine.

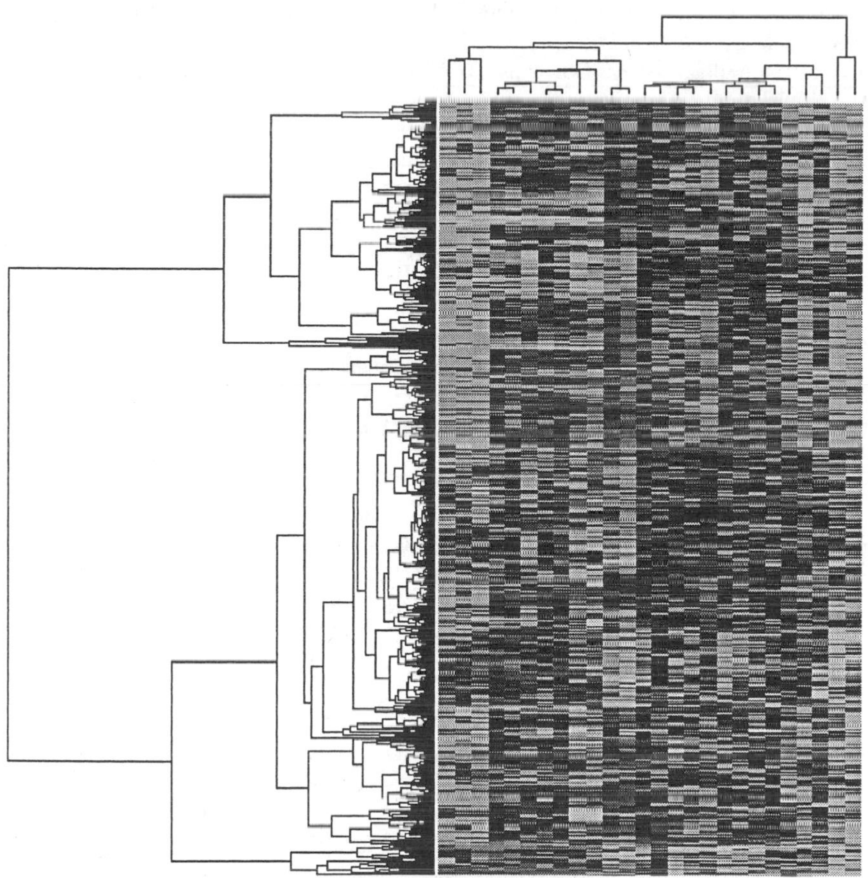

Figure 19-1. GDS90 dataset. Dark = high expression levels, light = low expression levels. Rows = genes, columns = cell/tissue type. Genes are clustered according to cell/tissue type (left hand side) and the 26 samples according to gene expression (upper area). Users can zoom in on areas of interest and download the selected data.

REFERENCES

Abascal, F. and Valencia, A. (2002). "Clustering of Proximal Sequence Space for the Identification of Protein Families," *Bioinformatics,* 18:908-21.

Altschul, S.F., Gish, W., Miller, W., Myers, E.W., Lipman, D.J. (1990). "Basic Local Alignment Search Tool," *J Mol Biol.,* 215:403-10.

Apweiler, R., Bairoch, A., Wu, C.H., Barker, W.C., Boeckmann, B., Ferro, S., Gasteiger, E., Huang, H., Lopez, R., Magrane, M., Martin, M.J., Natale, D.A., O'Donovan, C., Redaschi, N., and Yeh, L.-S.L. (2004). "UniProt: The Universal Protein Knowledgebase," *Nucl. Acids. Res.,* 32:D115-D119.

Aronson, A.R. (1996). "The Effect of Textual Variation on Concept Based Information Retrieval," *Proc AMIA Annu Fall Symp.*:373-7.

Baldi, P. and Brunak, S. (1998). *Bioinformatics: The Machine Learning Approach (Adaptive Computation & Machine Learning)*, MIT Press.

Benson, D.A., Karsch-Mizrachi, I., Lipman, D.J., Ostell, J. and Wheeler, D.L. (2004). "GenBank: Update," *Nucl. Acids. Res.,* 32:D23-D26.

Blaschke, C., Andrade, M.A., Ouzounis, C., Valencia, A. (1999). "Automatic Extraction of Biological Information from Scientific Text: Protein-protein Interactions," in *Proc Int Conf Intell Syst Mol Biol.,* 60-7.

Blaschke, C. and Valencia, A. (2001). "The Potential Use of SUISEKI as a Protein Interaction Discovery Tool," *Genome Inform Ser Workshop Genome Inform.,* 12:123-34.

Blaschke, C. and Valencia, A. (2002). "The Frame-based Module of the SUISEKI Information Extraction System," *IEEE Intelligent Systems,* 17(2):14-20.

Boguski, M.S., Lowe, T.M., Tolstoshev, C.M. (1993). "DbEST--Database for 'Expressed Sequence Tags'," *Nat Genet.,* 4:332-3.

Bult, C.J., Blake, J.A., Richardson, J.E., Kadin, J.A., Eppig, J.T. and the Mouse Genome Database Group (2004). "The Mouse Genome Database (MGD): Integrating Biology with the Genome," *Nucl. Acids. Res.,* 32:D476-D481.

Chang, J.T., Schutze, H. and Altman, R.B. (2002). "Creating an Online Dictionary of Abbreviations from MEDLINE," *J Am Med Inform Assoc.,* 9(6):612-20.

Cheng, J., Hatzis, C., Hayashi, H., Krogel, M.A., Morishita, S., Page, D. and Sese, J. (2002). "KDD Cup 2001 Report," *SIGKDD Explorations*, 3(2):47--64.

Chiang, J.H. and Yu, H.C. (2003). "MeKE: Discovering the Functions of Gene Products from Biomedical Literature via Sentence Alignment," *Bioinformatics,* 19(11):1417-22.

Chiang, J.H., Yu, H.C., and Hsu, H.J. (2004). "GIS: A Biomedical Text-mining System for Gene Information Discovery," *Bioinformatics,* 20(1):120-1.

CMAJ. (2003). "The Genomification of Medicine," *CMAJ*, 168(8):949-951.

Craven, M. and Kumlien, J. (1999). "Constructing Biological Knowledge Bases by Extracting Information from Text Sources," *Proc Int Conf Intell Syst Mol Biol.*:77-86.

Daraselia, N., Yuryev, A., Egorov, S., Novichkova, S., Nikitin, A., and Mazo, I. (2004). "Extracting Human Protein Interactions from MEDLINE Using a Full-sentence Parser," *Bioinformatics,* 20(5):604-11.

Donaldson, I., Martin, J., De Bruijn, B., Wolting, C., Lay, V., Tuekam, B., Zhang, S., Baskin, B., Bader, G.D., Michalickova, K., Pawson, T., and Hogue, C.W. (2003). "PreBIND and Textomy--Mining the Biomedical Literature for Protein-protein Interactions Using a Support Vector Machine," *BMC Bioinformatics,* 4(1):11.

Edgar, R., Domrachev, M., and Lash, A.E. (2002). "Gene Expression Omnibus: NCBI Gene Expression and Hybridization Array Data Repository," *Nucleic Acids Res.,* 30(1):207-210.

Eom, J.-H. and Zhang, B.-T. (2004). "PubMiner: Machine Learning-based Text Mining Systems for Biomedical Information Mining," *Lecture Notes in Artificial Intelligence*, 3192:216-225.

Faria-Campos, A.C., Cerqueira, G.C., Anacleto, C., De Carvalho, C.M., Ortega, J.M. (2003). "Mining Microorganism EST Databases in the Quest for New Proteins," *Genet Mol Res.*, 2:169-77.

Friedman, C., Kra, P., Yu, H., Krauthammer, M. and Rzhetsky, A. (2001). "GENIES: A Natural-language Processing System for the Extraction of Molecular Pathways from Journal Articles," *Bioinformatics,* 17 Suppl 1:S74-82.

Gaizauskas, R., Demetriou, G., Artymiuk, P.J., and Willett, P. (2003). "Protein Structures and Information Extraction from Biological Texts: The PASTA System," *Bioinformatics,* 19(1):135-43.

Gene Ontology Consortium. (2004). "The Gene Ontology (GO) Database and Informatics Resource," *Nucl. Acids. Res.,* 32:D258-D261.

Gnatenko, D.V., Dunn, J.J., McCorkle, S.R., Weissmann, D., Perrotta, P.L., Bahou, W.F. (2003). "Transcript Profiling of Human Platelets Using Microarray and Serial Analysis of Gene Expression," *Blood,* 101:2285-93.

Golub, T.R., Slonim, D.K., Tamayo, P., Huard, C., Gaasenbeek, M., Mesirov, J.P., Coller, H., Loh, M.L., Downing, J.R., Caligiuri, M.A., Bloomfield, C.D., and Lander, E.S. (1999) "Molecular Classification of Cancer: Class Discovery and Class Prediction by Gene Expression Monitoring," *Science,* 286:531537.

Hahn, U., Romacker, M., and Schulz, S. (2002). "Creating Knowledge Repositories from Biomedical Reports: The MEDSYNDIKATE Text Mining System," in *Pacific Symposium on Biocomputing*:338-49.

Harris, T.W., Chen, N., Cunningham, F., Tello-Ruiz, M., Antoshechkin, I., Bastiani, C., Bieri, T., Blasiar, D., Bradnam, K., Chan, J., Chen, C.-K., Chen, W.J., Davis, P., Kenny, E., Kishore, R., Lawson, D., Lee, R., Muller, H.-M., Nakamura, C., Ozersky, P., Petcherski, A., Rogers, A., Sabo, A., Schwarz, E.M., Van Auken, K., Wang, Q., Durbin, R., Spieth, J., Sternberg, P.W. and Stein, L.D. (2004). "WormBase: A Multi-species Resource for Nematode Biology and Genomics," *Nucl. Acids. Res.* 32:D411-D417.

Hu, Y., Hines, L.M., Weng, H., Zuo, D., Rivera, M., Richardson, A., and LaBaer, J. (2003). "Analysis of Genomic and Proteomic Data Using Advanced Literature Mining," *J Proteome Res.,* 2(4):405-12.

Hucka, M., Finney, A., Sauro, H.M., Bolouri, H., Doyle, J.C., Kitano, H., Arkin, A.P., Bornstein, B.J., Bray, D., Cornish-Bowden, A., Cuellar, A.A., Dronov, S., Gilles, E.D., Ginkel, M., Gor, V., Goryanin, I.I., Hedley, W.J., Hodgman, T.C., Hofmeyr, J. H., Hunter, P.J., Juty, N. S., Kasberger, J. L., Kremling, A., Kummer, U., Le Novere, N., Loew, L. M., Lucio, D., Mendes, P., Minch, E., Mjolsness, E.D., Nakayama, Y., Nelson, M. R., Nielsen, P.F., Sakurada, T., Schaff, J.C., Shapiro, B.E., Shimizu, T.S., Spence, H.D., Stelling, J., Takahashi, K., Tomita, M., Wagner, J., and Wang, J. (2003). "The Systems Biology Markup Language (SBML): A Medium for Representation and Exchange of Biochemical Network Models," *Bioinformatics,* 19(4):524-531.

Jenssen, T.K., Laegreid, A., Komorowski, J., and Hovig, E. (2001). "A Literature Network of Human Genes for High-throughput Analysis of Gene Expression," *Nat Genet.,* 28(1):21-8.

Kanehisa, M., Goto, S., Kawashima, S., Okuno, Y. and Hattori, M. (2004). "The KEGG Resource for Deciphering the Genome," *Nucl. Acids. Res.,* 32:D277-D280.

King, R.D., Karwath, A., Clare, A., Dehaspe, L. (2000). Accurate Prediction of Protein Functional Class from Sequence in the Mycobacterium Tuberculosis and Escherichia Coli Genomes Using Data Mining," *Yeast.,* 17:283-93.

Krieger, C.J., Zhang, P., Mueller, L.A., Wang, A., Paley, S., Arnaud, M., Pick, J., Rhee, S.Y. and Karp, P.D., 2004). "MetaCyc: A Multiorganism Database of Metabolic Pathways and Enzymes," *Nucl. Acids. Res.,* 32:D438-D442.

Kulikova, T., Aldebert, P., Althorpe, N., Baker, W., Bates, K., Browne, P., Van Den Broek, A., Cochrane, G., Duggan, K., Eberhardt, R., Faruque, N., Garcia-Pastor, M., Harte, N., Kanz, C., Leinonen, R., Lin, Q., Lombard, V., Lopez, R., Mancuso, R., McHale, M., Nardone, F., Silventoinen, V., Stoehr, P., Stoesser, G., Tuli, M. A., Tzouvara, K., Vaughan, R., Wu, D., Zhu, W. and Apweiler, R. (2004). "The EMBL Nucleotide Sequence Database," *Nucl. Acids. Res.,* 32:D27-D30.

Lash, A.E., Tolstoschev, C.M., Wagner, L., Schuler, G.D., Strausberg, R.L., Riggins, G.J., and Altschul, S.F. (2000). "SAGEmap: A Public Gene Expression Resource," *Genome Res.,* 10(7):1051-60.

Kohonen, T. (1981a). "Automatic Formation of Topological Maps of Patterns in a Self-organizing System," in *Proceedings of 2SCIA, Scand. Conference on Image Analysis*, Helsinki, Finland:214-220.

Mitelman, F., Mertens, F., Johansson, B. (1997). "A Breakpoint Map of Recurrent Chromosomal Rearrangements in Human Neoplasia," *Nature Genet.,* 15:417-474.

Muggleton, S. (1991). "Inductive Logic Programming," *New Generation Computing*, 8:295-318.

Ng, S.K. and Wong, M. (1999). "Toward Routine Automatic Pathway Discovery from On-line Scientific Text Abstracts," *Genome Inform Ser Workshop Genome Inform.*, 10:104-112.

Nishizuka, S., Charboneau, L., Young, L., Major, S., Reinhold, W.C., Waltham, M., Kouros-Mehr, H., Bussey, K.J., Lee, J.K., Espina, V., Munson, P.J., Petricoin, E. 3rd, Liotta, L.A., Weinstein, J.N. (2003). "Proteomic Profiling of the NCI-60 Cancer Cell Lines Using New High-density Reverse-phase Lysate Microarrays," *Proc Natl Acad Sci U S A,* 100:14229-34.

NLM. (2000). Unified Medical Language System Knowledge Sources.

Novichkova, S., Egorov, S. and Daraselia, N. (2003). "MedScan, A Natural Language Processing Engine for MEDLINE Abstracts," *Bioinformatics,* 19(13):1699-706.

Ono, T., Hishigaki, H., Tanigami, A., and Takagi T. (2001). "Automated Extraction of Information on Protein-protein Interactions from the Biological Literature," *Bioinformatics,* 17(2):155-61.

Palakal, M., Stephens, M., Mukhopadhyay, S., Raje, R., and Rhodes, S.J. (2002). "A Multi-level Text Mining Method to Extract Biological Relationships," *IEEE CSB:*97-108.

Perez-Iratxeta, C., Bork, P. and Andrade, M. A. (2001). "XplorMed: A Tool for Exploring MEDLINE Abstracts," *Trends Biochem Sci.,* 26(9):573-5.

Perez-Iratxeta, C., Bork, P. and Andrade, M.A. (2002). "Association of Genes to Genetically Inherited Diseases Using Data Mining," *Nat Genet.,* 31(3):316-9.

Pruitt, K.D. and Maglott, D.R. (2001). "RefSeq and LocusLink: NCBI Gene-centered Resources," *Nucleic Acids Res.,* 29(1):137-140.

Pustejovsky, J., Castano, J., Zhang, J., Kotecki, M., and Cochran, B. (2002). "Robust Relational Parsing over Biomedical Literature: Extracting Inhibit Relations," *Pac Symp Biocomput.:*362-73.

Quinlan, R. (1993). *C4.5: Programs for Machine Learning*, Morgan Kaufmann, San Mateo, CA.

Rindflesch, T.C., Hunter, L., and Aronson, A.R. (1999). "Mining Molecular Binding Terminology from Biomedical Text," *Proc AMIA Symp.:*127-31.

Ross, D.T., Scherf, U., Eisen, M.B., Perou, C.M., Rees, C., Spellman, P., Iyer, V., Jeffrey, S. S., Van De Rijn, M., Waltham, M., Pergamenschikov, A., Lee, J.C., Lashkari, D., Shalon,

D., Myers, T.G., Weinstein, J.N., Botstein, D., Brown, P.O. (2000). "Systematic Variation in Gene Expression Patterns in Human Cancer Cell Lines," *Nat Genet.*, 24(3):227-35.

Salwinski, L, Miller, C.S., Smith, A.J., Pettit, F.K., Bowie, J.U., and Eisenberg, D. (2004). "The Database of Interacting Proteins: 2004 Update," *Nucl. Acids. Res.*, 32:D449-D451.

Schena, M., Shalon, D., Davis, R.W., Brown, P.O. (1995). "Quantitative Monitoring of Gene Expression Patterns with a Complementary DNA Microarray," *Science*, 270:467-70.

Scheurle, D., DeYoung, M.P., Binninger, D.M., Page, H., Jahanzeb, M., Narayanan, R. (2000). "Cancer Gene Discovery Using Digital Differential Display," *Cancer Res.*, 60:4037-43.

Schuler, G.D., Epstein, J.A., Ohkawa, H., and Kans, J.A. (1996). "Entrez: Molecular Biology Database and Retrieval System," *Methods Enzymol.*, 266:141-62.

Schuler, G.D. (1997). "Pieces of the Puzzle: Expressed Sequence Tags and the Catalog of Human Genes," *J. Mol Med.*, 75:694-698.

Sekimizu, T., Park, H.S., Tsujii, J. (1998). "Identifying the Interaction between Genes and Gene Products Based on Frequently Seen Verbs in Medline Abstracts," in *Proc. 9th Workshop Genome Informatics*, Universal Academy Press, Tokyo:62-71.

Shalon, D., Smith, S.J., and Brown, P.O. (1996). "A DNA Microarray System for Analyzing Complex DNA Samples Using Two-Color Fluorescent Probe Hybridization," *Genome Res.*, 6:639-645.

Shatkay, H., Edwards, S., Wilbur, W.J., and Boguski, M. (2000). "Genes, Themes and Microarrays: Using Information Retrieval for Large-scale Gene Analysis," in *Proc Int Conf Intell Syst Mol Biol.* 8:317-28.

Shipp, M.A., Ross, K.N., Tamayo, P., Weng, A.P., Kutok, J.L., Aguiar, R.C., Gaasenbeek, M., Angelo, M., Reich, M., Pinkus, G.S., Ray, T.S., Koval, M.A., Last, K.W., Norton, A., Lister, T.A., Mesirov, J., Neuberg, D.S., Lander, E.S., Aster, J.C., and Golub, T.R. (2002). "Diffuse Large B-cell Lymphoma Outcome Prediction by Gene-expression Profiling and Supervised Machine Learning," *Nat Med.*, 8(1):68-74.

Spellman, P.T., Sherlock, G., Zhang, M.Q., Iyer, V.R., Anders, K., Eisen, M.B., Brown, P.O., Botstein, D., Futcher, B. (1998). "Comprehensive Identification of Cell Cycle-regulated Genes of the Yeast Saccharomyces Cerevisiae By Microarray Hybridization," *Mol Biol Cell.*, 9:3273-3297.

Srinivasan, P. (2001). "MeSHmap: A Text Mining Tool for MEDLINE," in *Proc AMIA Symp.*:642-6.

Stapley, B.J. and Benoit, G. (2000). "Biobibliometrics: Information Retrieval and Visualization from Co-occurrences of Gene Names in Medline Abstracts," in *Pac Symp Biocomput.*:529-40.

Stephens, M., Palakal, M., Mukhopadhyay, S., Raje, R., Mostafa, J. (2001). "Detecting Gene Relations from Medline Abstracts," *Pac Symp Biocomput.*:483-95.

Strausberg, R.L., Dahl, C.A., and Klausner, R.D. (1997). "New Opportunities for Uncovering the Molecular Basis of Cancer," *Nat. Genet.*, Spec No 17:415–416.

Swanson, D.R. (1990). "Medical Literature as a Potential Source of New Knowledge," *Bull Med Libr Assoc.*, 78:29-37.

Tanabe, L., Scherf, U., Smith, L.H., Lee, J.K., Hunter, L., and Weinstein, J.N. (1999). "MedMiner: An Internet Text-mining Tool for Biomedical Information, with Application to Gene Expression Profiling," *Biotechniques*, 27(6):1210-4, 1216-7.

Tanabe, L. (2003). "Text Mining the Biomedical Literature for Genetic Knowledge [dissertation]," George Mason University, *AAT 3079362*.

Tatusov, R.L., Koonin, E.V., Lipman, D.J. (1997). "A Genomic Perspective on Protein Families," *Science*, 278:631-7. Genet Mol Res. 2003 Mar 31;2(1):169-77.

Tatusov, R.L., Galperin, M.Y., Natale, D.A., Koonin, E.V. (2000). "The COG Database: A Tool for Genome-scale Analysis of Protein Functions and Evolution," *Nucleic Acids Res.*, 28:33-6.

Temkin, J.M. and Gilder, M.R. (2003). "Extraction of Protein Interaction Information from Unstructured Text Using a Context-free Grammar," *Bioinformatics,* 19(16):2046-53.

Thomas, J., Milward, D., Ouzounis, C., Pulman, S., and Carroll, M. (2000). "Automatic Extraction of Protein Interactions from Scientific Abstracts," in *Pac Symp Biocomput.* 5:538-549.

Tu, Q., Tang, H. and Ding, D. (2004). "MedBlast: Searching Articles Related to a Biological Sequence," *Bioinformatics,* 20(1):75-7.

van De Vijver, M.J., He, Y.D., Van't Veer, L.J., Dai, H., Hart, A.A., Voskuil, D.W., Schreiber, G.J., Peterse, J.L., Roberts, C., Marton, M.J., Parrish, M., Atsma, D., Witteveen, A., Glas, A., Delahaye, L., Van Der Velde, T., Bartelink, H., Rodenhuis, S., Rutgers, E.T., Friend, S.H., and Bernards, R. (2002). "A Gene-expression Signature as a Predictor of Survival in Breast Cancer," *N Engl J Med.*, 347(25):1999-2009.

Velculescu, V.E., Zhang, L., Vogelstein, B., Kinzler, K.W. (1995). "Serial Analysis of Gene Expression," *Science*, 270:484-7.

Weeber, M., Klein, H., Aronson, A.R., Mork, J.G., De Jong-van Den Berg, L.T., and Vos, R., (2000). "Text-based Discovery in Biomedicine: The Architecture of the DAD-system," in *Proc AMIA Symp.:*903-7.

Weinstein, J.N., Myers, T.G., O'Connor, P.M., Friend, S.H., Fornace, A.J., Kohn, K.W., Fojo, T., Bates, S.E., Rubinstein, L.V., Anderson, N.L., Buolamwini, J.K., Van Osdol, W.W., Monks, A.P., Scudiero, D.A., Sausville, E.A., Zaharevitz, D.W., Bunow, B., Viswanadhan, V.N., Johnson, G.S., Wittes, R.E., and Paull, K.D. (1997). "An Information-intensive Approach to the Molecular Pharmacology of Cancer," *Science,* 275:343-349.

Wheeler, D.L., Church, D.M., Federhen, S., Lash, A.E., Madden, T.L., Pontius, J.U., Schuler, G.D., Schriml, L.M., Sequeira, E., Tatusova, T.A., and Wagner, L. (2003). "Database Resources of the National Center for Biotechnology Information," *Nucleic Acids Res.,* 31(1):28-33.

Wheeler, D.L., Church, D.M., Edgar, R., Federhen, S., Helmberg, W., Madden, T.L., Pontius, J.U., Schuler, G.D., Schriml, L.M., Sequeira, E., Suzek, T.O., Tatusova, T.A., and Wagner, L. (2004). "Database Resources of the National Center for Biotechnology Information: Update," *Nucleic Acids Res.,* 32:D35-40.

Wilbur, W.J. (2002). "A Thematic Analysis of the AIDS Literature," in *Pac Symp Biocomput.*:386-97.

Wong, L. (2001). "PIES, a Protein Interaction Extraction System," in *Pac Symp Biocomput.:*520-31.

Wu, Z. and Leahy, R. (1993). "An Optimal Graph Theoretic Approach to Data Clustering: Theory and Its Application to Image Segmentation," *IEEE Transactions on Pattern Analysis and Machine Intelligence*, 15:1101-13.

Yakushiji, A., Tateisi, Y., Miyao, Y., and Tsujii, J. (2001). "Event Extraction from Biomedical Papers Using a Full Parser," in *Pac Symp Biocomput.*:408-19.

Yeh, A.S., Hirschman, L., Morgan, A.A. (2003). "Evaluation of Text Data Mining for Database Curation: Lessons Learned from the KDD Challenge Cup," *Bioinformatics,* 19 Suppl 1:i331-9.

SUGGESTED READINGS

Baldi, P. and Brunak, S., 1998, Bioinformatics: The Machine Learning Approach (Adaptive Computation & Machine Learning), MIT Press.
Machine learning Methods are covered (including neural networks, hidden Markov Models, and belief networks), aimed at biologists and biochemists. It also allows physicists/mathematicians/computer scientists to explore applications of Machine learning in Molecular biology.

Causton, H., Quackenbush, J. and Brazma, A.. (2003, *Microarray Gene Expression Data Analysis: A Beginner's Guide*, Blackwell Publishers.
Microarray experimental design and analysis, geared towards graduate students and researchers in bioinformatics, with emphasis on underlying concepts.

Jurafsky, D. and Martin, J. H.. (2000, *Speech and Language Processing: An Introduction to Natural Language Processing, Computational Linguistics, and Speech Recognition*, Prentice-Hall.
Large reference on foundations of natural language and speech processing.

Korf, I., Yandell, M., and Bedell, J., Eds.. (2003, *BLAST. The Definitive Guide. Basic Local Alignment Search Tool*, O'Reilly & Associates, 1st ed., Sebastopol, CA.
A detailed look at the BLAST suite of tools. It enables users to experiment with parameters and analyze their results, and includes tutorial and reference sections.

Shatkay H., and Feldman R.. (2003, Mining the biomedical literature in the genomic era: an overview, *J Comput Biol.* **10**:821-55.
A comprehensive review of the state of the art in biomedical text Mining.

ONLINE RESOURCES

Abascal and Valencia's Protein Family Annotation
http://www.pdg.cnb.uam.es/funcut.html

Broad Institute's datasets
http://www.broad.mit.edu/cgi-bin/cancer/datasets.cgi
http://www.broad.mit.edu/mpr/lymphoma/

Cancer Genome Anatomy Project, NCI
http://cgap.nci.nih.gov/

Jeffrey Chang's BioNLP Server, Stanford University
http://bionlp.stanford.edu/

Critical Assessment of Information Extraction Systems in Biology, CNB Protein Design Group and MITRE Corp.
http://www.pdg.cnb.uam.es/BioLINK/BioCreative.eval.html

European Molecular Biology Laboratory (EMBL), Links to XplorMed and other genomic data Mining tools http://www-db.embl-heidelberg.de/jss/SearchEMBL?services=x

GENIA corpus for biomedical NLP, University of Tokyo
http://www-tsujii.is.s.u-tokyo.ac.jp/~genia/topics/Corpus/

Genomics and Bioinformatics Group, Molecular Pharmacology, NCI
http://discover.nci.nih.gov/index.jsp

GEO tutorial, NCBI
http://www.ncbi.nlm.nih.gov/geo/info/qqtutorial.html

Kyoto Encyclopedia of Genes and Genomes (KEGG)
http://www.genome.ad.jp/kegg/kegg.html

Lymphoma/Leukemia Molecular Profiling Project data
http://llmpp.nih.gov/lymphoma/

Mitelman Database of Chromosome Aberrations in Cancer
http://cgap.nci.nih.gov/Chromosomes/Mitelman

NCBI Data Mining Tools including BLAST, COGs, Map Viewer, LocusLink, and UniGene
http://www.ncbi.nlm.nih.gov/Tools/

Oxford University, ILP applications and datasets
http://web.comlab.ox.ac.uk/oucl/research/areas/machlearn/applications.html

Predicting Protein Function from Sequence using Machine Learning
http://www.aber.ac.uk/~dcswww/Research/bio/ProteinFunction/

PubGene literature and sequence networks
http://www.pubgene.org/tools/Network/Browser.cgi

SAGEmap, NCBI
http://www.ncbi.nlm.nih.gov/SAGE/

Leming Shi's site on DNA Microarrays
http://www.gene-chips.com/The Institute for Genomic Research (TIGR)
http://www.tigr.org/

Stanford University, Yeast cell cycle data
http://genome-www.stanford.edu/cellcycle/data/rawdata/

UMLS Knowledge Source Server, NLM
http://umlsks4.nlm.nih.gov/

QUESTIONS FOR DISCUSSION

1. How will genomics change the practice of medicine? How long will it take until personalized medicine is possible? What are some of the obstacles that will need to be overcome?

2. What role does the genomic data mine have in systems biology? How can data from different perspectives be integrated into a complete picture of a biological system?

3. Describe the genomics knowledge base – what is it? What are its themes? What is it missing? How can the missing knowledge be discovered?

4. What are the potential benefits of text mining the biomedical literature? How can this potential be realized?

5. Perez-Iratxeta et al. used cytogenetic data and functional annotation to score candidate disease genes using fuzzy set theory. Describe an alternative approach to finding candidate disease genes.

6. NCBI's TaxPlot is a tool for 3-way comparisons of genomes on the basis of the protein sequences they encode. What are some applications of TaxPlot? What are some of the questions that can be explored using this tool?

7. Use GEO to explore a dataset of interest to you. Interpret the hierarchical cluster tree. What does PCA tell you about this dataset?

8. How do data-driven and hypothesis-driven genomics methodologies complement each other? What are the advantages/disadvantages of each?

9. How much genomics should medical schools require? What would be the best way to teach medical students genomics? Do bioinformaticians and computational biologists need to know biology and/or medicine? If so, how much?

10. Describe an ideal genomics data mining tool that would help researchers understand human diseases. What would the program need to do? What data would it take as input? What output would it return? How accurate would it need to be? How fast would it need to run? Would the system require human reasoning or be fully automatic? How could its performance be tested?

Chapter 20

EXPLORATORY GENOMIC DATA ANALYSIS

Larry Smith

National Center for Biotechnology Information, National Library of Medicine, National Institutes of Health,Bethesda, MD 20894

Chapter Overview

In this chapter, an introductory description of the exploration of genomic data is given. Rather than attempt an exhaustive overview of the types of genomic data and methods of analysis, the chapter focuses on one type of data, gene expression profiling by microarray technology, and one method of analysis, cluster analysis for discovering and sorting mixed populations. This type of data and method of analysis is very common in bioinformatics. It illustrates recurring problems and solutions. And a major portion of bioinformatics dealing with exploratory genomic data analysis can be viewed as a refinement and extension of this basic analysis.

Keywords

gene expression profiling; microarray technology; cluster analysis; mixed populations

1. INTRODUCTION

Exploratory genomic data analysis draws on mathematical, statistical, and computational methods to discover meaningful genetic relationships from large-scale measurements of genes. It is a continuously growing area that is constantly being seeded with new approaches and interpretations. Most of this new material is easily accessible given a familiarity with basic genetics and multivariate statistics. It is my hope that this introduction to prototypical methods and case studies will be enough to get started.

Genomics researchers are interested in the genetics of specific, controllable, cellular phenomena. Microarray assay technology makes it possible for them to measure genomic variables, such as relative mRNA expression level, associated with thousands of genes in multiple controlled situations. Normally, the usefulness of experimental data ends with the experiment, when the results are evaluated in relation to a stated hypothesis. But with its vast scope, microarray data provides a unique view of the complex and largely unknown genetic organization of cells, a view into a world beyond all preconceived hypotheses. It is not surprising that large-scale microarray assays are undertaken for the purpose of scrutinizing the results and searching for potentially new genetic knowledge. This is where exploratory data analysis enters.

Data mining, also called knowledge discovery and descriptive statistics, is a process of reformulating a large quantity of unintelligible data with a view to suggest hypothetical relationships. With luck, some of these hypotheses, when validated experimentally, will lead to real discoveries. But experimentation is usually expensive and time consuming, and it is not wise to pursue arbitrarily generated hypotheses. Therefore, it is not merely enough to generate hypotheses; they must be evaluated and selected according to their significance and plausibility. Significance means that a generated hypothesis is not likely to be based on experimental errors. To establish significance requires a statistical test based on a model of the experimental errors, also called a null hypothesis. Plausibility means that a generated hypothesis is reasonable in light of current scientific research and knowledge. To establish plausibility requires a theoretical justification and a connection of the new hypothesis to previous research. This is where literature mining proves to be useful.

In this chapter we give an introduction to data mining of genomic data with an intuitive explanation of the fundamental technique of cluster analysis of mixed population data. Three published studies are described, each with a different analysis of the same gene expression profile data set.

The material found here on data mining and genomic data is by no means complete. Rather, the goal of this chapter is to introduce the field to new students, who might afterwards follow their own interests and inclinations into any of the numerous directions of current research involving genomic data analysis.

2. OVERVIEW

Gene expression profiling is any laboratory procedure that measures the expression of genes in differing controlled cellular conditions. The controlled conditions might be selected phenotypes (e.g. cancer types), stages of development or differentiation, or responses to external stimuli (e.g. drug dosages). Gene expression is the quantitative level of messenger RNA found in cells. Expression levels of genes are highly regulated in a cell and depend on many factors. The up- or down-regulation of a particular gene in a particular situation is an important clue to its function. Genes whose expressions are correlated may actually be co-regulated, and so they are potentially participating in coordinated functions. Therefore, knowledge about one gene might be applicable to the other. And genes whose regulation correlates with an externally controlled condition are potentially involved somehow in that condition. This is a modest motivation for applying data mining to expression profile data.

2.1 Gene Expression Data

Gene expression profile data is collected in large quantities using microarray technology. The basic science behind this technology is not difficult to understand. In brief, a cDNA (complementary DNA) microarray is a glass slide containing thousands of robotically miniaturized hybridization experiments. To conduct a microarray assay a sample is prepared and exposed to the slide under controlled hybridization conditions. A sample is actually a solution containing DNA that has been collected from prepared cellular samples, concentrated and purified, and infused with radioactive or fluorescent nucleotide markers. After the hybridization, each spot on the slide emits a photon signal whose strength is determined by the amount (or amounts) of sample DNA binding at that spot. The photon signals from the slide are measured with an optical scanner, and the pixelized image is loaded into a computer for data analysis.

Microarray manufacturers typically provide slides and scanning equipment as well as computer software to analyze image data. Image analysis software must assess the quality of the slide globally, identify anomalies such as dust specs, identify the precise areas of each spot and its image quality, and ultimately reduce each spot to a single numeric quantity.

There are also additional data manipulations that can be performed after image analysis, such as quality filtering, background subtraction, and channel equalization.

In addition to data quality depending on the image, quality also depends on the manufacture of the microarray slide itself. Manufacturing flaws are possible, for example. And because so many different kinds of cDNA are used to make the slide, it is not rare for several spots on a slide to contain cDNA that is not properly identified. In any exploratory data analysis, it is important to keep control of data quality and, somehow, to prevent poor quality data from influencing the results of the analysis.

It is possible to give a general and sufficient description of expression profiling data without referring to the details of the technology. Several assays are conducted using the same microarray layout. Each assay is associated with controlled experimental conditions, such as cell phenotype, stage of development or differentiation, exposure to drug or radiation, *etc.* Each microarray spot has an associated DNA sequence, which is related to one or more genes in the genome map of the species. The data can therefore be pictured as a matrix of numeric data whose columns correspond to experimental conditions and whose rows correspond to DNA sequences. Each value in the matrix is a relative measure of the expression of that row's DNA in that column's experimental condition. Alternatively, each row of the data matrix is a vector of numbers representing the expression of a particular DNA sequence in all of the experimental conditions. Simultaneously, each column is a vector of numbers representing the gene expression in a particular condition.

The concepts of data mining are independent of the categories of the underlying data. Any matrix of data, or any list of values associated to members of some population, can be used with data mining. For this reason, in most of this section I will refer to data points, individuals, or members of a population instead of genes.

2.2 Mixed Populations

One of the fundamental goals of data mining is to determine if the data arose from a mixture of several distinct populations. If it did, then each population might have a distinctive distribution of data points. If data were two or three dimensional, we could visually inspect a graph of the data for different populations. Figure 20-1 shows some examples of how two populations might appear in two dimensions. In 1a, the two populations appear to have identical distributions. In this case, there is no amount of mathematical or statistical manipulation able to find a meaningful difference. In 1b and 1c, the populations are located in non-overlapping regions. In these examples we say there are mean differences. In 1b the distribution of the two populations has the same shape around their respective means. But in 1c, the shape of the distributions are different, we would say that these have variance differences. Not shown is an example where the subpopulations have the same mean but different variance (which is a case, similar to 1a, which is difficult to detect). Of course, distributions in reality need not be shaped into spheres or ellipses, as shown in 1d they might assume arbitrary shapes while still falling into separate regions.

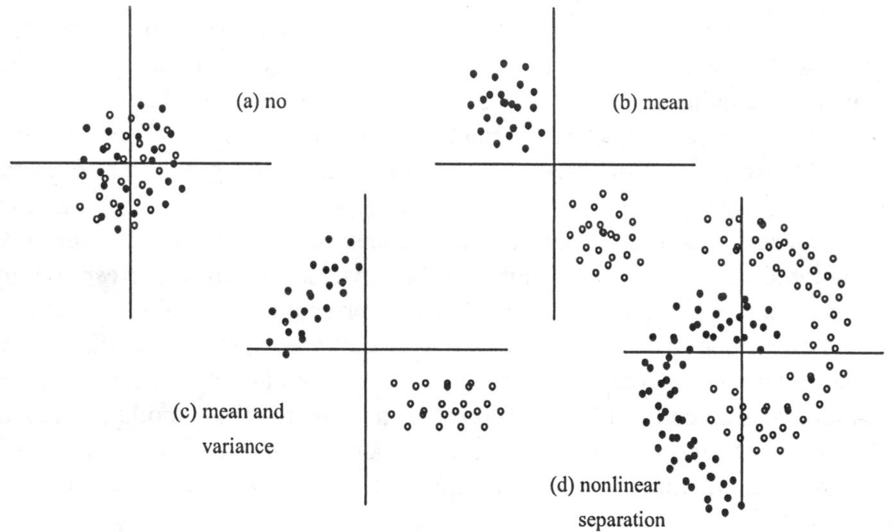

Figure 20-1. Different ways that a mixture of populations might be reflected in 2 dimensional data

In higher dimensions, visual inspection is impossible. There are well known techniques able to reduce high dimensional data to 2 or 3 dimensions for visual inspection. These include principal components analysis (PCA), multidimensional scaling, and factor analysis. These methods should preserve the appearance of subpopulations with significant mean differences. The K-means algorithm is able to detect mixed populations with less significant mean differences, or with significant variance differences. And nonlinear differences, like that shown in 1d, can sometimes be detected using (connected) clustering. The number of dimensions not only confounds our ability to visualize, but a phenomenon called the "curse of dimensionality" makes differences in the data less obvious. It is difficult to give a brief explanation of why this happens, but it has to do with the fact that small random variations in each of many dimensions tend to make points appear to be more uniformly spaced – a mean difference "signal" is drowned out by the high dimensional "noise."

If a scientific theory predicts subpopulations, there might be some expectation as to what they are, their number, and even how the distributions of the data points may differ. This would be an enormous help in analyzing the data. But without prior knowledge, unanalyzed data is a territory in complete darkness, and the various analysis techniques may seem to provide little more than a dim light. It is impossible to predict in advance whether it will be easy, difficult, or impossible to identify subpopulations. As easy as it might be to define the objective and the approach, in practice, data mining is a challenge to be respected.

And even if subpopulations are found, it is not enough to report the fact. Each subpopulation must be identified and described. The simplest approach, when it is logically justified, is to find exemplars in each subpopulation, or members (*e.g.* genes) that are well known, then to generalize what is known about them to the entire subpopulation. If there are no exemplars, or if the exemplars do not generalize, the distribution of data might be correlated in distinctive ways with the experimental conditions (this is the implication of significant mean differences), and this might suggest the function of members of the population. If no clear identification of any subpopulation can be made, the validity of the analysis must be questioned.

2.3 Methods for Mixed Populations

Encyclopedic knowledge may be available for some genes, but for most of them at the present time, little is known beyond some portions of their nucleotide sequence. Even less is known, relatively, of the ways in which genes are related. Microarray experiments generate data for thousands of genes, some known and some not. A fundamental goal in analyzing microarray data is to find new relationships between genes.

Scientists begin to understand diverse populations of things by categorizing individuals based on their similarities and differences. Data mining is able to uncover categories in a large population provided that they are reflected in the data.

K-means analysis is a statistical modeling technique (and several variations) that is able to find subpopulations of data that are distributed differently. To apply it, however, requires an estimate of the number, K, of subpopulations. It works well when data conforms to the model, as the number K can be found by trial and error. But in less ideal situations, the results are difficult to interpret.

Principal components analysis (PCA) and factor analysis, on the other hand, apply simpler linear models to the data. An ideal data set consists of points (in some high dimension) that are compactly distributed in an ellipsoidal shape. For this type of data, the first principal component is the measurement along the longest diameter of the ellipse, the second component along the next longest diameter, perpendicular to the first, *etc.* Factor analysis is different from PCA in that it allows more flexibility in the components (after the first one), but this requires subjective choices. The usual way of using PCA, or factor analysis, is to represent each data point with the first 2 or 3 components and graph them for visual examination. Data sets reflecting distinct subpopulations are not ideal, but one can imagine nevertheless that the data is distributed within an ellipsoid with gaps. Whatever the cause, PCA, or factor analysis, is frequently effective with mixed population and non-ideal data, and subpopulations can be seen in the component graph in lower dimensions. Sometimes this dimension reduction is carried out to make subsequent analysis (for example, cluster analysis to be discussed next) simpler or more robust.

Some analysis methods, however, are not based on the distribution of data but rather on the "distances" between all pairs of points. Multidimensional scaling (MDS), for instance, finds a configuration of points in lower dimension whose pairwise distances are the "closest" possible to the distances in the original higher dimensional space (variations of MDA depend on how closeness is measured). The graph that results from MDA is similar to the graph obtained from PCA, and usually there is no discernable advantage to either one, though the PCA algorithm is simpler (in fact, there is a variant of MDS that produces graphs identical to PCA).

Cluster analysis is very popular because it is both simple and intuitive. It produces a tree structure suggesting an intuitive organization of the data. Think of the cluster tree as a collection of data points, forming the leaves, and many branches that join the leaves together. Each branch has a number of leaves and may also have many branches below it. The "height" of a branch is a composite measure of the distance (sometimes called the diameter) between the leaves of that branch. The root branch is the unique branch containing all of the data. The graphical representation of a cluster tree is called a dendrogram, an example is shown in Figure 20-2. The vertical scale represents the height, and the horizontal axis lists all of the data points in the order dictated by the tree. A horizontal segment is used to represent each branch, located at the height of the branch. Vertical segments connect a branch with the data or branches that are immediately below it. The data and sub-branches of a branch can be put in any order but, by convention, they are usually sorted left to right by increasing height of the sub-branches.

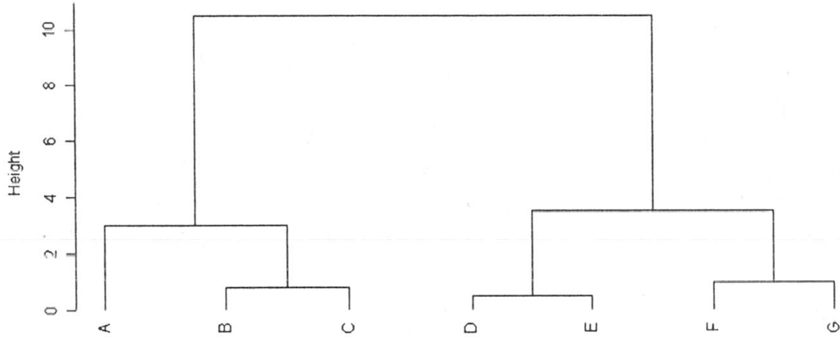

Figure 20-2. Example of a cluster tree dendrogram

There are several different clustering algorithms, and most of them are fast and efficient and use only the distances between points to be given. If the data is a mixture of well-separated subpopulations (such as the illustrations in Figure 20-1b and 20-1c), then virtually any cluster tree will have two or more distinct branches joined by a root branch at a significantly greater height. For separated but intertwined populations (such as the illustration in Figure 20-1d) a kind of clustering called *connected clustering* is sometimes able to show them, but only if it is generally the case that for each point there is at least one point in the same subpopulation that is closer to that point than any point in the other subpopulations.

One problem with cluster trees is that they must be interpreted. Cluster analysis always produces a result, even when there are no clearly delineated subpopulations. A branch that is contained in a branch at a significantly greater height is a candidate for a subpopulation. This conclusion is not automatic because it depends on interpreting the height scale, which can be arbitrary. Even when there are no obvious subpopulations, it may happen that the population members with known similarities cluster together, making it possible to postulate an "organizing principal" for the cluster tree. More often, however, cluster trees will appear to have no obvious "organization." The decision to search for meaning in a random cluster tree or to give up and try a different approach can be a vexing one, compounded by psychological, social, and even political pressures. But there are a great many variations to try; the key is in the calculation of distances, which I will discuss in the next section.

Finally, if a cluster analysis reveals subpopulations, they cannot merely be reported, they must be explained. At least some of the subpopulations must be identified by some common characteristics, as mentioned in Section 2.2. I will have more to say about this in Section 2.5.

2.4 Distance

The result of cluster analysis, regardless of the algorithm used, depends primarily on the calculated distances between individuals. After defining and collecting the raw data, deciding how to calculate distances is the next crucial step towards successful analysis, and there are many options. In this section, I will describe and illustrate some basic distance calculation methods.

The distance between two vectors of data is given by a formula that combines the differences in each component into a single value. There are many choices to make when calculating the distance between points. Each distance formula is sensitive to particular kinds of differences, and clustering results reflect those differences.

The obvious distance formula is Euclidean distance (the sum of squared differences of each component). With the Euclidean distance, each coordinate is considered equally important, and this might not be desired. But it is important to be aware of some undesirable features that are both subtle and common.

Since the components of a data vector come from separate processes (*e.g.* different experimental conditions), the distribution of component values can be very different from one another. Consider, for example, what would happen if one experiment has a high variance compared with the others. This is illustrated in Figure 20-3a for hypothetical data with two distinct subpopulations. The distances between two points in the one population are sometimes greater than the distances between points in two different populations. The problem here is that the variance of the second component is very large compared with the variance of the first. It can be corrected by "normalizing" each component. That is, each data value is modified by dividing by the standard deviation (square root of variance) computed over all values of that component. An alternative solution in this example is to note that the populations are distinguished by their first component, so the distance could be computed by disregarding the second component (this would be an example of subjective factor analysis).

Another problem is illustrated in Figure 20-3b. The data of 3a has been rotated by 45 degrees, and it is still the case that the distances between populations are sometimes less than the distances within populations. In fact, the cluster trees are identical. One approach would be to rotate the data so that it looks like 3a and then normalize (this too would be an example of subjective factor analysis). But in 3b the problem can be understood more

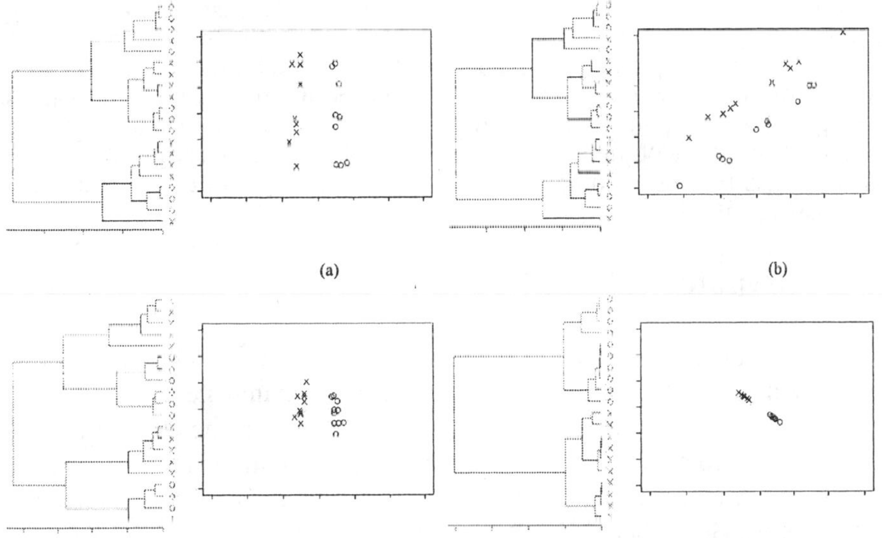

Figure 20-3. Two examples of mixed populations with unequal variances before (above) and after (below) modification of the data. In (a) each component is divided by the standard deviation, and in (b) each point is mean-centered.

simply. Each point in the sample has a random constant added to all components. It is as if each point has randomly reset the "zero-point" for its measurements. The solution is to "mean center" the data. That is, each data value is modified by subtracting the average of all of the components of that data point. This solution has the same effect as the approach to 3a of calculating the distances using only the first coordinate.

One way to account for unequal importance of different components is to multiply each component by a constant weight reflecting its relative importance. The decision to do this cannot be motivated by the data, but must be based on knowledgeable assessment of the different experimental conditions. For example, if there are 10 experimental conditions, but 5 of them are duplicates of the same condition, it would be sensible to weight each of the 5 duplicate components with a weight of 1/5.

Other distance measures use different formulas but are similar in spirit to Euclidean distance. The Manhattan distance, for example, is the sum of the absolute values (instead of square) of the differences in each component. Correlation distance is one minus the Pearson correlation coefficient of the paired components of two data points. Interestingly, correlation distance is equivalent to Euclidean distance if the data is first mean-centered and normalized.

Finally, a problem that frequently arises is missing data. In any experiment, it is to be expected that some data is usable and some is not. In the case of microarray data, this depends on quality assessments coming from the manufacture of the array, the certainty of identification of the cDNA in each spot, and the quality of the scanned image. There is no obvious way of computing distances between points that have some values missing. If the components are disregarded, the calculated distance will be grossly distorted. A common approach is to fill in the missing values with their averages or to use some estimate of the difference in each component with missing values.

2.5 Hypothesis Selection

Let us summarize the data mining process we have discussed up to now. First, raw data is obtained as vectors of data for many members of some population (*i.e.* genes). Pairwise distances are computed between members and this is used to cluster the data. The distance calculation is refined and the data is re-clustered a few times until it looks as if some interesting subpopulations have been found. The next thing to do is to shore up the analysis with evidence that the discovery is not a mirage and that it is worth the time to consider for further experimentation.

Gene expression profile data reduced to pairwise distances and visualized by clustering suggest relationships or subpopulations defined by similarities between genes. Clustering analysis does not, however, say what those similarities are based on. For a potential relationship to be worthy of further experimentation, it must satisfy three conditions: 1) it must be statistically significant, 2) it must be strong, and 3) it must be theoretically plausible.

Statistical significance. The idea of statistical significance of a conclusion is to show that it could not have arisen by accident. To do this requires a probabilistic model of how the conclusion might have been made by accident, which is then used to compute the probability that it could have happened that way. If that probability, or p-value, is sufficiently small (usually less than 0.05), the conclusion is said to be statistically significant. A detailed model for the "null hypothesis" is usually impractical to construct. And simplified models are likely to be doubted by the scientific community anyway. An alternative to this is to apply the idea of bootstrapping.

Suppose we observe a distance between two population members that appears to be very small compared to other distances, and we want to conclude from this that the members are somehow related, perhaps even duplicates. The statistical significance of the fact is the probability of a certain "random" model producing such a small distance. A reasonable alternative to constructing a model, called bootstrapping, is to assume a model in which the data values arise randomly. To estimate significance using this model, the data is shuffled a large number of times, each time calculating the distance between the populations in question and noting the percentage of times that the distance is less than or equal to the distance in question. This percentage is the p-value. Data is shuffled by randomly permuting the values in the matrix. The model can be refined by allowing only certain types of permutations. For example, depending on the experiment, it might be logical to assume that data within a particular row or column arise randomly, and to satisfy this model the random permutations used in shuffling the data could be chosen to satisfy such constraints.

A similar idea could be applied to a cluster analysis. For example, suppose that a number of genes are found in a branch of the cluster tree, and the difference in heights between that branch and the branch immediately containing it is large enough to suggest that they are in a distinct subpopulation. By shuffling the data many times, and clustering each time, the percentage of times that this height difference (or larger) is observed for the same set of genes can be used as a p-value.

Strength of finding. Independent of p-value, that is to say, even if the p-value is very small, a discovery is probably not worth discovering if it is not very strong. This happens frequently, for example, with correlation coefficients. When many independent measurements of co-occurring phenomena are taken, a very small non-zero correlation coefficient can be very significant, *i.e.* the p-value can be very small. It is correct to conclude from a small p-value that something "real" is causing the two phenomena to be related. The problem is, because the effect is so weak, it will be difficult to find the common cause. Or once it is found it will be difficult to prove that it is indeed the cause. Or even if the cause is discovered and proven, it may be so slight that it is of little value to science.

The same is true with data mining of genomic data. If the height difference of a subpopulation is determined to be statistically significant, yet it is apparently a very small difference, the distance formula is so sensitively refined that it is able to reveal similarities between members that are probably very remote. Granted, if the subpopulation is known to exist, then it can be automatically concluded that the analysis has detected it. But this is hardly a breakthrough. Data mining is looking for relationships that were not previously known.

Plausibility. Now suppose that a subpopulation has been found, it is statistically significant and strongly delineated. It is not interesting if the subpopulation is already known in the scientific literature, except as a demonstration of the approach, so suppose also that it is new. To be taken seriously, it is necessary to give some plausible explanation of what it is that characterizes this subpopulation. It is the job of the scientist to know the field and to know how to make plausible arguments. But in data mining, the field covers a very large collection of genes, and the literature bearing on all of these genes is too massive for any human scientist to grasp. In this situation, scientific research means library research: searching for and assessing research articles, exploiting electronic and print resources to expand the search, and then taking the time to carefully read and understand the articles that seem to be relevant.

Search tools are being developed and refined today to help scientists find articles focused to highly technical needs. The need is so great that some research teams include computational linguists who can develop new search tools tailored to their specific research. The ideal search tool should not only be able to retrieve articles, it should be able to summarize and rank the relevance of each article to the stated goal. Future NLP applications should be able to take a given relationship, say between two or more genes, and determine if there is a common thread in the literature that might connect them. The retrieved articles should be summarized and grouped according to the underlying explanation that is being suggested. What is more, they should be able to take as input many different possible relationships and search all of them, returning the relationships with supporting literature. All of this has been done manually in the past and is done increasingly with the help of computers.

It is important to understand that a strong relationship that is statistically significant might have an uninteresting cause. But the cause may still be worth identifying, even if it is not scientifically important. An uninteresting cause of relationships in the data may reveal a deficiency in the design of the experiment or ways in which the data could be modified to correct for the uninteresting effects.

3. CASE STUDIES

In this section, I will discuss two papers that introduced the gene expression profiles of 60 previously studied cancer cell lines. The two groups that collaborated to produce this data published separate analyses. The paper "Systematic variation in gene expression patterns in human cancer cell lines" (Ross, et al., 2000) describes the cluster analysis of the genes and cells. And "A gene expression database for the molecular pharmacology of cancer" (Scherf, et al., 2000) correlates this data with anti-cancer drug response data for the same 60 cell lines that has been collected at the NCI since the 1960s. I will also describe a third study that was done independently, some time later, for a data mining competition based on this data.

NCI60. The 60 cell lines of NCI60 have been extensively studied for many years. They represent a variety of human tissues, organs, and cancer types. The gene expression profile study measures the expression of 9,703 genes for each cell line which is a significant addition to the body of scientific data, making it possible to explore genetic correlations with virtually any of the previously studied phenomena. The data for each gene and cell line is the ratio of the expression of the gene in the cell line to the average expression of the gene in a pooled mixture of the cell lines. For example, a value of 2 for a gene and cell line indicates that the gene is expressed in the cell line twice as much as it is on average in the pool. The NCI60 data set is publicly available and might still be used to discover new relationships in the genetics of cancer.

Ross. The study of Ross, et al. explored the gene expression profiles with regard to the cell types. For this analysis, each cell line is represented by the vector of data formed from the expression values of all of the genes. Distances between cell lines were calculated using the correlation distance (this is equivalent to using Euclidean distance after mean-centering and normalizing the data). They observed that cells with the same histologic origin clustered together, indicating that they had similar gene expression patterns. Also a larger organization of cell lines was seen in the cluster tree, grouping epithelial cells and stromal cells. To explain why this occurred, they turned to the cluster tree of the genes. There they noted several large branches of genes including genes expressed characteristically in specific cell types. For example, one branch of genes included many genes related to melanin synthesis and were expressed mostly in the melanoma cell lines. These cell-specific genes probably drive the organization of the cluster tree of the cell lines along histological type. Similarly, the presence of cluster branches dominated by genes characteristically expressed in epithelial cells or stromal cells could explain the large-scale organization of the cell line cluster tree.

They explored the cluster of genes further and found several branches containing genes with common functions. For example, there was a cluster of cell-cycle-related genes, RNA processing genes, genes regulated by interferons, and genes with possible roles in drug metabolism. The expression pattern of genes in a branch containing ribosomal genes correlated with the cell doubling time of the cell lines, supporting the interpretation that the branch contained genes supporting cell proliferation rate. It is interesting to consider what the result would be if the cell lines had been clustered using only the genes from one of these functional branches. A picture would emerge, not necessarily organized around histological type, reflecting the status of the particular function in the cell lines.

They also used cluster analysis to test the feasibility of using this same microarray assay to identify the tissue of origin of actual tumor biopsies. They obtained biopsies from two breast cancer patients and performed microarray assays on samples of cell extracts. Cluster analysis was used with the new data together with some of the breast cancer and leukemia cell lines and it was found that the biopsy samples appeared in a branch independent of those cell lines. They hypothesized that the biopsy samples contained cells of several different types, unlike the pure origin of each of the 60 cell lines. To support the hypothesis, they found that genes characteristic of different cell lines, including leukocytes, epithelial cells, and stromal cells, were expressed in the biopsy samples. They also confirmed the hypothesis by staining the biopsy specimens and demonstrating different cell types.

Scherf. The study of Scherf, et al. used the same NCI60 microarray data together with drug response data collected at the Developmental Therapeutics Program of the National Cancer Institute. This drug response data is the IC50, or the dosage at which cell growth is inhibited by 50% in culture, measured for thousands of drugs that have been considered candidates for anticancer treatment. The cluster tree of cell lines obtained using the gene expression profiles was compared with the cluster tree obtained using the drug response data. The differences reiterate the conclusion that the predominant gene expression pattern is tied to the histological origin of the cell. But the cluster tree of cells based on drug response has a different pattern, reflecting different modes of drug susceptibility that is not necessarily related to tissue of origin. For example, the ovarian and colon cancer cell lines were scattered into disparate branches in the cluster tree based on drug response. The two sets of pairwise differences between cell lines, obtained from the gene expression data and from drug response data, were compared directly with the Pearson correlation. The correlation of the distances was 0.21, which is statistically significant. This number is an indicator of the positive but small degree to which gene expression could theoretically be used to predict drug response in a cell line.

They also looked for possible relationships between specific genes and specific drugs. They did this by computing the correlation between the 60 expression values for 1,376 genes and the 60 drug response values for 118 drugs with established modes of action. This resulted in 162,368 correlations, which had to be searched for significant, strong, and meaningful correlations. Three examples were chosen for illustration having strong, statistically significant correlations (all negative). Supporting arguments were made for the plausibility of these associations and references were given for the required background. It was not feasible to consider all of the significant and strong correlations carefully, because each correlation required a literature search, assessment of relevance, and a creative leap to a possible underlying mechanism. A key search tool was developed, called MedMiner, which performed MEDLINE literature searches for user selected gene-drug pairs and presented summarizing sentences for evaluation. The lead scientist estimated that MedMiner decreased the time involved in literature search by a factor of 10. But many correlations, including some potential breakthroughs and other unexpectedly low correlations, could not be pursued because of the infeasibility of finding literature to suggest or support an underlying commonality.

Coombes. The NCI60 data set is publicly available, and has been the object of further analysis since the twin studies of Ross and Scherf. The data analysis competition, "Critical Assessment of Microarray Data Analysis" or CAMDA (http://www.camda.duke.edu), is an annual open competition where new approaches to analyzing a given data set are judged. In 2001 the NCI60 data set was used as the data set for the competition, and the winner that year was K.R. Coombes, et al. for "Biology-Driven Clustering of Microarray Data: Applications to the NCI60 Data Set." By performing cluster analysis of cell lines, similar to Ross, they showed that some cancer types were easily distinguished by gene expression, *e.g.* colon or leukemia, but not breast and lung. Then by restricting the expression data to various subsets of genes, *i.e.* by calculating the distance between cells based on specific subsets of genes, they were able to show that some meaningful subsets were able to distinguish cancer types while others were not. For example, genes on chromosome 2 were, but genes on chromosome 16 were not. Also, genes associated with certain functions, such as signal transduction and cell proliferation were, but genes associated with apoptosis and energy metabolism were not able to distinguish cancer types. They reached these conclusions by examining the cluster tree in each case to see if the established cell type branches appeared.

4. CONCLUSIONS

This chapter described approaches to exploratory genomic data analysis, stressing cluster analysis. It is not necessary to have a deep understanding of genomic data in order to carry the analysis out, but a deep understanding of some facet of molecular biology is required to complete the analysis. It is likely that a large-scale microarray study will be carried out in one or more labs and with several collaborators. In this circumstance, specialists can carry out data mining, producing a list of possible conclusions (chosen for statistical significance and strength) that can be handed over to others trained in molecular biology to make the final selections and provide supporting arguments. But even with this division of labor, a team of molecular biologists does not have the expertise to judge a very large list of possible hypotheses. Sophisticated computer tools for retrieving relevant scientific literature are indispensable. Future advances in the ability of computers to make human-like judgments may relieve the human bottleneck.

REFERENCES

Efron, B. and Tibshirani, R.J. (1993*) An Introduction to the Bootstrap*. Chapman & Hall, New York.

Manly, B. (1986). *Multivariate Statistical Methods, A Primer*. Chapman & Hall, London.

Quackenbush, J. (2001). "Computational Analysis of Microarray Data," *Nature*, 2, 418-427.

Ross, et al. (2000). "Systematic Variation in Gene Expression Patterns in Human Cancer Cell Lines," *Nature Genetics*, 24, 227-235.

Scherf, et al. (2000). "A Gene Expression Database for the Molecular Pharmacology of Cancer,." *Nature Genetics*, 24, 236-244.

Tanabe, L., et al. (1999) "MedMiner: An Internet Text-mining Tool for Biomedical Information, with Application to Gene Expression Profiling," *Biotechniques*, 27(6), 1210-4, 1216-7.

Voorhees, E.M. and L.P. Buckland, eds. (2003). *Text Retrieval Conference (TREC 2003).* National Institute of Standards and Technology, Gaithersburg.

Weinstein, et al. (2002). "The Bioinformatics of Microarray Gene Expression Profiling," *Cytometry*, 47, 46-49.

Weinstein, et al. (1997). "An Information-Intensive Approach to the Molecular Pharmacology of Cancer," *Science*, 275, 343-349.

SUGGESTED READINGS

There are many books on bioinformatics that deal specifically with the analysis of microarray data, and more appear every year. There are too many to name or single out. Most provide a good introduction to the subject with programming guidelines and exercises.

For a description of microarray technology, see the January, 1999 issue of Nature Genetics (volume 21) "The Chipping Forecast."

A comprehensive overview of data mining techniques for microarray data can be found in F. Valfar "Pattern Recognition Techniques in Microarray Data Analysis." Ann. N. Y. Acad. Sci. 980, 41-64, which can be obtained in full text through PUBMED.

ONLINE RESOURCES

The R statistical package, available for free download under the GNU public license. http://www.gnu.org/software/r/R.html.

Data and analysis tools for the NCI60 data set (and others) can be found at http://discover.nci.nih.gov/.

The NCBI web site, http://www.ncbi.nlm.nih.gov, is the entry point to the PUBMED database and sequence databases. You can also find tools for data analysis of sequence data (under

"Tools") and a bioinformatics primer (in "About NCBI"). In addition there is a database of microarray assay data, called GEO, at http://www.ncbi.nlm.nih.gov/geo/.

The annual competition Critical Assessment of Microarray Data Analysis (CAMDA) can be found at http://www.camda.duke.edu.

QUESTIONS FOR DISCUSSION

The following questions are open ended.

1. When exploring data, testing all of your ideas with simulated data is an invaluable habit. The statistical package R is ideal for this. Obtain the R package and write a script that generates random microarray data and perform a cluster analysis. Here is a simplified R script to do this:

```
data <- rnorm(100, 0, 1)
data.dist <- dist(data)
data.clust <- hclust(data.dist)
plclust(data.clust)
```

   Generate random data for mixed populations and cluster them.

2. If you are familiar with any field of compiled data, scientific or other, try to find an application of cluster analysis. For example if you were interested in baseball, what data would be useful to explore different kinds of batters and pitchers? Can you draw an analogy between your approach and exploratory genomic data analysis?

3. Microarray studies generate data for thousands of genes. How many data points are you able to cluster using the R package before running out of memory? One approach to cluster analysis of huge data sets is to select a subset, either randomly or by careful selection. What kind of conclusions could be drawn from cluster analysis of random subsets of genes? How could you improve on this?

4. In the Scherf, et al. study, both gene expression data and drug response data was available for all 60 cell lines. An approach called "co-clustering" treats the drug response data as if it were gene expression data and analyzes it together with the genes. What new kind of result is possible with co-clustering? Most often, the results are disappointing. What goes wrong?

5. Suppose some of the data in a microarray study is missing or flagged as unreliable. How could you calculate distance when there are missing values? Predict the circumstances where your method would lead to

misleading results. When should genes be discarded for having too many missing values?

6. Suppose a data mining system were able to generate a list of possible hypotheses and provide each one with a p-value, a measure of strength, and an indication of the amount of accompanying literature. Could these three factors be combined in a way to rank the hypotheses from "best to "worst"?

7. Formulate some features of an ideal literature search tool to aid in data mining.

Chapter 21

JOINT LEARNING USING MULTIPLE TYPES OF DATA AND KNOWLEDGE

Zan Huang, Hua Su, and Hsinchun Chen

Artificial Intelligence Lab, Department of Management Information Systems, Eller College of Management, University of Arizona, Tucson, AZ 85721

Chapter Overview

This chapter discusses joint learning research in biomedical domains. A brief review of the field of joint learning research is given, with emphases on the large-scale data and knowledge resources used for learning and the central biological questions involved. Two representative joint learning case studies are presented with algorithmic details. The two case studies involved two representative joint learning tasks, protein function classification and regulatory network learning, and two important algorithmic frameworks for joint learning, the kernel-based framework and probabilistic graphical models. A wide range of biological data and existing knowledge was also involved in these two studies.

Keywords

data mining; machine learning; joint learning; data and knowledge; algorithm

1. INTRODUCTION

The past decade has witnessed an explosive growth of biomedical research attributed to technological advances and the completion of genome projects for a variety of species. Genomic and proteomic technologies such as DNA microarray, genome-wide two-hybrid screening, and high-throughput mass spectrometry have generated an enormous amount of data. These data contain information that may provide a global picture of cellular composition, structure, functionality, and responses to environmental changes. Thus, data-driven discovery of functional features of genes and proteins and the complex network of cellular processes becomes a critical task for functional genomics research. On the other hand, advances in information technologies and their applications in biomedical research also resulted in a boom of biomedical ontologies that store meta-level and background biomedical knowledge as well as repositories of digitized literature text and human-curated biomedical findings. We use the word "knowledge" to refer to these resources in this chapter where there is no ambiguous interpretation based on the context. The common feature of these resources is that they are based on the biomedical literature and the common understanding of the biomedical domain, while biomedical data are obtained directly from large-scale experiments or measurements. Traditional data analysis methods are typically not sufficient for learning sophisticated data patterns from complex, large-scale biological data and knowledge with a variety of representations and granularity levels. The general lack of appropriate data analysis tools for exploiting these rich biological data resources has resulted in a boom in application and development of data mining and machine learning technologies in biomedical research in recent years.

Biomedical data mining and machine learning research is largely driven by the biological data and knowledge that is available. Researchers develop new or adapt existing machine learning algorithms to exploit special characteristics of the biological data to answer particular biological questions. Many special characteristics of biological data are associated with the inherent complexity of biological systems. A biological process such as gene regulation involves a complex network of genes, RNAs, proteins, and small molecules. Such a regulation process involves multiple layers of molecular interactions. A single type of experimental data typically reveals only a certain aspect of the underlying biological phenomena. Microarray data, for instance, delineates the global gene expression patterns in a cell and has been largely used to infer genetic regulatory networks (De Jong, 2002; Friedman et al., 2000). However, since microarray data only provides measurements at the transcriptional level and cannot reflect other aspects of gene regulation (Jansen et al., 2002), microarray-based network learning models may infer incorrect gene regulatory relations. In addition to this inherent data deficiency, high-throughput datasets also often contain errors and noise arising from imperfections of the experimental technology which further impedes learning effectiveness.

To overcome the limitations of learning from a single type of biological data, many recent studies have been focusing on learning methods from multiple types of data. Results learned from a combination of different types of data are likely to lead to a more coherent model by consolidating information on various aspects of the biological process. Moreover, the effects of data noise in learning results will be dramatically reduced, assuming that technological errors across different datasets are largely independent and the probability that an error is supported by more than one type of data is small (Friedman, 2004).

A natural approach to exploit multiple types of data that are related to a common biological question is to combine the results learned from different types of data. We may combine the results in different ways such as accepting the learning results only when learned from all data types. More interestingly, and supposedly with promises of delivering better learning results, multiple types of data can be analyzed simultaneously under a unified framework. In the bioinformatics community such a general learning problem is called a *joint learning* problem, defined as the process of combining different types of data and learning knowledge from them in a single framework or algorithm (Hartemink and Segal, 2004).

This chapter familiarizes readers with the field of joint learning by providing a comprehensive review of the literature and two case studies. In Section 2 we review existing joint learning research using multiple types of biological data and knowledge, with a focus on the data and knowledge employed in these studies and the target biological questions. In Sections 3 and 4 we present two case studies on joint learning from multiple types of data and knowledge, with discussions on algorithmic details. The first case study describes a kernel-based computation framework to achieve optimal combination of various types of data ranging from protein-protein interactions to protein sequences to gene expression data for the problem of predicting protein functional classifications. The second study deals with learning regulatory networks from experimental data and existing biological knowledge, particularly the gene expression data and known genetic interactions. Both studies showcase the promises and challenges of joint learning research for the biomedical domain. We conclude the chapter in Section 5 by summarizing key insights, challenges, and future directions.

2. OVERVIEW OF THE FIELD

Joint learning has attracted much attention in the bioinformatics community because biologists need to learn knowledge from different types of data. Research in this area has been largely driven by data and knowledge that are made available by a wide range of recently developed technologies. Accordingly, our review first summarizes the data and knowledge resources available for joint learning and the biological questions that joint learning approaches try to address. We then review the major modeling techniques and their findings for joint learning from multiple types of data and from data and knowledge, respectively.

2.1 Large-scale Biological Data and Knowledge Resources

The data resources used in previous joint learning research can be classified into experimental data and biological knowledge. Here we refer to *data* as original experimental measurements or results, such as DNA and protein sequences, gene expression profiles, and protein-protein interaction measured by yeast two-hybrid screening, among others. In contrast, *knowledge* refers to human-curated research findings recorded in well-structured databases or documented in biomedical literature.

Biological data used by previous joint learning research include, but are not limited to, the following major categories:

(1) *Nucleotide and protein sequences*: Sequences of genes and proteins are typically represented by a string of characters written in a certain alphabet. Other sequence-related information includes length of genes or operons, the intergene distances, organization of introns and exons, and functional motif of proteins, etc. GenBank (Benson et al., 2004), as a part of the international nucleotide sequence database collaboration, is a comprehensive sequence database that contains publicly available DNA sequences for different organisms. It also provides information on taxonomy, genome, mapping, protein structure and domain, as well as the biomedical literature associated with a particular sequence. Swiss-Prot, TrEMBL (Boeckmann et al., 2003), and PIR (Wu et al., 2004) are the three major protein sequence databases, independently operating until the end of 2003. Recently they have been merged to form UniProt (Apweiler et al., 2004), a universal protein sequence and annotation database.

(2) *Protein structural information*: The 3-D protein structure is important because it determines the function of a protein. The 3-D structure of a protein can be defined by the coordinates of its crystal structure, which may suggest the location of catalytic sites of an enzyme or interaction sites by

which a protein interacts with other molecules. The Protein Data Bank (PDB) is the primary source of information on the 3D structure of proteins and other macromolecules (Bourne et al., 2004).

(3) *Gene and protein expression profiles*: Large-scale expression profiling of genes and proteins benefits from the development of DNA microarray and proteomic technologies. Microarray technology makes use of the sequence resources created by the genome projects, allowing exploration of expression patterns of thousands of genes at one time. The typical output of a microarray experiment is a matrix of gene expression measurements with rows designating genes and columns designating samples, conditions or patients, etc. There are dozens of publicly accessible repositories of microarray data, including the Stanford Microarray Database (Gollub et al., 2003) and ArrayExpress at EBI (European Bioinformatics Institute) (Brazma et al., 2003). Protein expression data has also been generated due to technological advances in biochemistry such as high resolution 2-D polyacrylamide gel electrophoresis (PAGE), mass spectrometry, and more recently, protein array technology. OPD (Open Proteomic Database) provides mass spectrometry-based data.

(4) *Protein-DNA binding data*: The interaction between protein and DNA molecules plays an essential role in gene regulation, hence is an important source of information on genetic regulatory networks. Protein-DNA binding data can be represented as binary relationships between a protein and a DNA molecule. BIND (Bader et al., 2003) and aMAZE (Lemer et al., 2004), among others, provide this type of information.

(5) *Protein-protein interaction data*: Protein-protein interactions provide information on cellular communication, signal transduction, and gene regulation. These interactions can also be represented by binary relationships between two proteins. The Database of Interacting Proteins (DIP) (Salwinski et al., 2004) is a database that documents experimentally determined protein-protein interactions. The MINT database (Zanzoni et al., 2002) stores experimentally verified protein interactions for mammalian proteomes. BIND also contains protein-protein interaction data.

Human-curated research findings are also a useful resource for joint learning. These knowledge sources may include human-encoded databases, ontologies, and biomedical literature text. Human-encoded databases and ontologies, sometimes called knowledge bases, have been created with pathway relations manually extracted from the literature to support keyword search, pathway visualization, and other data analysis and concept exploration tasks. These include databases for regulatory pathways, e.g., KEGG (Kanehisa et al., 2002) and TRANSPATH (Krull et al., 2003), and databases for metabolic pathways, e.g. KEGG (Kanehisa et al., 2002 and BioCyc {Karp, 2002 #61}). There are several large repositories of literature focused specifically on biomedicine. Hosted by the U.S. National Library of Medicine (NLM), PubMed is one of the most comprehensive repositories of biomedical literature, containing about 14 million references on life sciences or biomedicine. PubMed has been used in many previous studies that analyze textual documents to support efficient searching and browsing and to discover new knowledge.

The URLs of all the resources mentioned here are listed in the Online Resources section of this chapter. For a more complete list and updates we refer readers to the journal of *Nucleic Acids Research*, which dedicates the first issue of each year to biological databases.

The reviewed biological data and knowledge contain rich information about cellular composition and organization. Most joint learning research focuses on tasks that help elucidate cellular processes using data from multiple information resources. One example is identification of motifs, a short sequence of amino acids or nucleotides that form the contact interfaces between two interacting protein or DNA molecules, using gene expression, transcription binding site, or protein interaction data, etc. (Li et al., 2004; Prakash et al., 2004; Takusagawa and Gifford, 2004). Another important task is classification of proteins and prediction of their function using protein sequence, protein interaction, and gene expression data (De Hoon et al., 2004; Eskin and Agichtein, 2004; Lanckriet et al., 2004). Gene regulatory networks analysis is also a popular area where joint learning approaches have been used, in which experimental data and existing biological knowledge are both involved (Chrisman et al., 2003; Hartemink et al., 2002; Imoto et al., 2003; Nariai et al., 2004; Segal et al., 2002; Segal et al., 2003; Tamada et al., 2003; Yoo et al., 2002). In the next two sub-sections, we review the previous studies on learning from multiple types of data and knowledge.

2.2 Joint Learning Using Multiple Types of Data

In this section we review previous joint learning studies using multiple types of biological data in three major learning tasks: learning regulatory networks, functional classification of genes and proteins, and motif identification. Most previous research in joint learning has been focused on these three learning tasks. As the field of joint learning is a fast-growing field that is still in its early stages of development, we can expect to see more studies targeting a broad range of biological questions in the next several years.

2.2.1 Learning Regulatory Networks

In recent years there has been great interest in the analysis of gene expression reflected in microarray data, especially in learning gene regulatory networks (De Jong, 2002; Friedman, 2004; Friedman et al., 2000; Pe'er et al., 2001; Yoo et al., 2002). However, regulatory networks learned from gene expression data alone have several limitations. Practically, the repeatability of microarray measurements is often impeded by the instability of mRNA molecules and the fact that expression changes occur within minutes following certain triggers (Emmert-Buck et al., 2000; Model et al., 2001). Biologically, microarray mainly reflects gene regulation at the transcriptional level, while other types of regulation such as protein interaction do not necessarily correlate with it (Jansen et al., 2002). RNA samples used in microarray are pooled from a large number of cells and variation among cells is usually ignored (Chu et al., 2003). Last but not least, since microarray data have the problem of high dimensionality of input space (genes) compared to the small number of available samples, the statistical power of the learned network is thus often questionable (Husmeier, 2003; Somorjai et al., 2003).

Joint learning research in this area is largely motivated by the aforementioned deficiencies of the gene expression data. Several recent studies have shown the benefit of utilizing multiple types of data for inference of more accurate regulatory networks. Hartemink et al. (Hartemink et al., 2002) used genomic location and expression data to infer genetic regulatory networks in yeast. The genomic location of the binding site of a particular transcription factor was used as prior knowledge for network inference from microarray data using a Bayesian framework. They found that the location data was complementary to expression data and hence improved network learning. Both Segal (Segal et al., 2002) and Nariai (Nariai et al., 2004) used protein interaction and gene expression data jointly to learn genetic networks in yeast. In Segal et al.'s study, the two types of data were integrated in a unified probabilistic model where genes were partitioned into clusters based on their co-expression pattern and existence of interaction between their protein products. They found that the jointly learned model matched much better with known functional gene groups and protein complexes. Nariai et al. used a Bayesian framework to integrate the two types of data by incorporating protein-protein interaction into network learning. If two proteins interact or form a complex, the genes encoding these proteins were combined as one node in the network. They claimed that by including protein complexes, the gene network was more accurately estimated from microarray data. Transcription factor binding motif has also been used in combination with the expression data. Tamada et al. (Tamada et al., 2003) reported more accurate learning results using transcription factor binding motifs found in the promoters of genes to refine networks learned from expression data.

2.2.2 Functional Classification of Genes and Proteins

Annotating or predicting the functional class of genes and proteins is another crucial task of functional genomics. Prediction of protein function is traditionally conducted based on sequence homology between the unknown protein and proteins with known functions. Since the function of proteins is also reflected by other factors including cellular localization, expression levels under certain conditions, and interaction with other proteins, joint learning from all these types of data should provide a more complete and detailed picture of their function or functional classes.

Lanckriet et al. (Lanckriet et al., 2004) proposed a kernel-based framework to combine multiple types of data and successfully predicted the functional classes of yeast protein using an extended support vector machine (SVM) algorithm. They used five types of data including the domain structure of each protein, protein-protein interactions, genetic interactions, co-participation of a protein complex, and cell-cycle gene expression measurements. Tsuda and Noble (Tsuda and Noble, 2004) went one step further, refining the kernel representations of heterogeneous data types to improve the classification performance. Their study was used to represent protein-protein interaction and metabolic network data and demonstrated improved classification accuracy over previously used kernels. The kernel-based approach has attracted much research interest recently. Many studies have focused on developing more appropriate kernels that can represent the most relevant information of particular biological data types. Kernel-based joint learning approaches are discussed in detail in the case studies in Section 3.

Joint learning techniques have also been applied to classification of other features of genomes. For example, De Hoon et al. (De Hoon et al., 2004) used operon length, intergene distance, and gene expression information to predict the operon structure of the bacterium *Bacillus subtilis* genome. Expressing the operon prediction by each type of data as a Bayesian probability, they combined them into a Bayesian classifier, which achieved higher accuracy than those classifiers based on a single type of data.

2.2.3 Motif Discovery

Motifs are short segments of nucleotide or amino acid sequence on a DNA or protein molecule that have a particular function in molecular interaction or catalysis. Identifying motifs in DNA and proteins helps elucidate the function of macromolecules. Motifs can be inferred from sequence data, molecular interactions, and expression data. The rationale for the association between motif discovery and expression data is that if two genes are co-regulated as detected by expression measurements, they are more likely to share the same TF binding motif. Joint learning from these types of data has shown promising results. Takusagawa & Gifford (Takusagawa and Gifford, 2004) conducted TF binding motif discovery in yeast with a unified probabilistic model. Information of TF binding from chromatin immunoprecipitation microarray experiments, together with yeast genome sequence and a statistical measure of the intergenic length, were used. They chose to learn from negative intergenic sequence, i.e. sequences that are not bound to TFs, to avoid the false positive noise, which increased the accuracy of motif prediction.

To identify motifs of protein-protein interaction, Li et al. (Li et al., 2004) used a sequential approach to learn binding motif pairs from the 3-D structure of protein complexes and sequences of interacting proteins. A binding motif pair was defined as a pair of motifs each derived from one side of the binding protein sequences. They first extracted maximal contact sequence segment pairs from the complex structural data, and then grouped the sequence data of interacting proteins using the segment pairs as templates. An iterative refinement process was also taken to enable the derivation of significant binding motif pairs.

2.3 Joint Learning Using Data and Knowledge

Besides joint learning from different types of data, another type of joint learning study is the inclusion of knowledge into the learning process. Knowledge from either documented literature or human experts provides guidance for knowledge discovery in very high dimensional data and has the potential to significantly speed up the mining process and deliver more accurate learning results. Only a few previous studies fall into this category. These studies mainly exploited domain knowledge to perform joint learning with different types of biological data. Recently, learning jointly from biomedical literature text and genomic data also started to draw significant attention. In this section we review the limited literature on joint learning with domain knowledge and text mining results. In the next section we present a detailed case study on joint learning from domain knowledge, text mining results, and experimental data.

In learning regulatory networks, Chrisman et al. (Chrisman et al., 2003) incorporated biological knowledge and expression data using a Bayesian framework. Background knowledge, i.e., known regulatory relations between genes, was encoded as pre-specified regulatory relations in the regulatory network model. The remaining part of the network is inferred with experimental data given such partially fixed network structure. Including knowledge dramatically increased the statistical power of the learned network models, especially with a small sample size, which is typical for expression data. In a similar study, Imoto et al. (Imoto et al., 2003) incorporated protein-protein interaction, protein-DNA interactions, transcriptional factor binding site information, and knowledge from existing literature into a Bayesian framework to learn a regulatory network from microarray data. Their approach involves combining two regulatory network models, one based on the microarray expression data, and the other based on prior knowledge. They tuned the combination of the two models to find the optimal balance between knowledge and microarray data. Monte Carlo simulation and experimental data both showed the effectiveness of this approach.

One example study that incorporates text mining results and experimental data was Iossifov et al.'s study of protein interaction network inference (Iossifov et al., 2004). They used a unified probabilistic model to integrate the interactions observed from a yeast two-hybrid experiment and interactions documented in literature, which were automatically extracted by an information extraction system. Inference of the protein-protein interaction networks was performed with a Markov Chain Monte Carlo technique. However, this study was largely based on simulated data. Validity of this approach with real data is still to be tested.

Our review emphasizes the data sources and learning tasks addressed in previous work but only briefly describes specific learning techniques. We recommend readers look up the specific studies mentioned in our review for algorithmic details. With the two case studies we present in the next section, we cover two important algorithmic frameworks that have been the foundation for a large portion of the previous joint learning studies: the kernel-based framework and the probabilistic graphic models. In addition, we list several important studies in the suggested readings section.

In the next two sections we present two carefully selected joint learning cases that represent the major research areas and analytical techniques in the field. The two studies represent two major learning tasks (predicting functional classification of proteins and learning gene regulatory networks), two major types of joint learning study (learning from multiple types of data and learning from data and knowledge), and two important algorithmic frameworks for joint learning (the kernel-based framework and probabilistic graphical models).

3. KERNEL-BASED DATA FUSION OF MULTIPLE TYPES OF DATA

We discuss a principled framework based on kernel-based approaches for combining multiple types of biological data for classification problems proposed by Lanckriet et al. (Lanckriet et al., 2004). This study exemplifies how to achieve synergies among multiple types of biological data for an important learning problem, protein function prediction. Their proposed framework is also important in that it provides a mechanism for optimal combination of data with heterogeneous representations under a generic computational framework.

3.1 Protein Function Prediction

The function of an unannotated protein can be predicted based on multiple sources of information given the set of proteins with known functions. For example, it may be predicted based on an observed similarity between the sequences of the unannotated protein and proteins of known functions. The unknown protein may have functional relationships with other proteins similar to those of an annotated protein. The functional relationship between two proteins can also be inferred if they occur in fused form in some other organism, if they co-occur in multiple species, if their corresponding mRNAs share similar expression patterns, or if the proteins interact with one another.

It was pointed out by Lanckriet et al. (Lanckriet et al., 2004) that the comparison and fusion of these different data should produce a more detailed and useful representation for protein interactions. Machine learning algorithms working on such a combined data representation have the potential to provide better protein function prediction results.

3.2 Kernel-based Protein Function Prediction

The computational framework we discuss here relies on the use of kernel-based statistical learning methods that have proven very useful in bioinformatics. Under such methods data are represented by means of a kernel function, which defines similarities between pairs of genes, proteins, etc. Defined specifically for different types of data, the kernel functions implicitly capture various aspects of the underlying biological machinery. At the same time, these kernel functions provide the mapping between heterogeneous biological data and a common similarity representation. This common similarity representation then provides the foundation for principled approaches for optimal combination of data with originally different representations.

- Kernel Methods

Under a kernel-based method, data items are embedded into a vector/feature space (typically different from the natural data presentation) and are analyzed in this space to identify data patterns. Typically nonlinear projections are performed such that nonlinear data patterns under the original data representations will appear as linear patterns in the projected feature space. Figure 1 shows an example of such a projection *f* with which the originally nonlinear separation between the two types of data items becomes a linear one in the projected space.

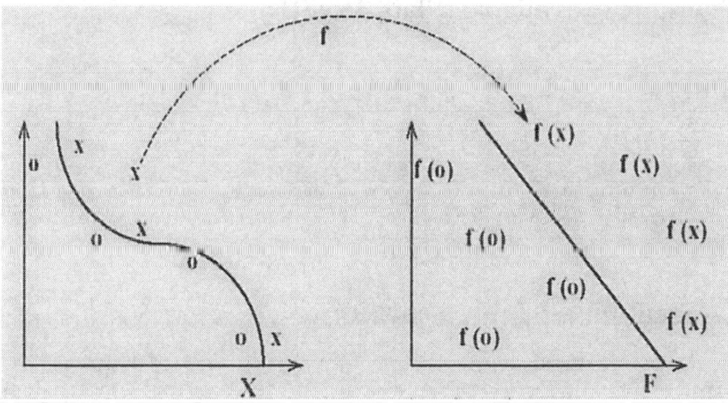

Figure 21-1. An example of data projection

Under the kernel-based approach, the projection $f(\mathbf{x})$ is specified implicitly using a kernel function $K(\mathbf{x}_i, \mathbf{x}_j) = <f(\mathbf{x}_i), f(\mathbf{x}_j)>$. The benefit of such an implicit specification of projection is that to identify data patterns that only involve inner products of the data items, such as the task of similarity-based clustering, one does not need to have the explicit representation of the mapping *f*. It suffices to be able to evaluate the kernel function, which is often much easier than computing the coordinates of the points explicitly. Thus, quite flexible kernel functions can be applied to search for the nonlinear patterns among data items without even knowing the nature of the projected feature space. For a finite set of data items we do not even require the exact specification of the kernel function itself. All we need is a square *kernel matrix* $K = (k_{ij})$, each entry $k_{ij} = <f(\mathbf{x}_i), f(\mathbf{x}_j)>$ is the inner products of the projected data points. This kernel matrix can be interpreted as one that describes a particular notion of similarity between data items. We will see examples of these kernel matrices later in this section.

Support vector machine (SVM) (Vapnik, 1995) is an important application of kernel-based methods to binary classification problems. In this study we employ this algorithm to perform protein function prediction under the kernel method framework. The *1-norm soft margin support vector machine* is used in this study, which forms a linear discriminant boundary in the feature space, $g(\mathbf{x}) = \mathbf{w}^T f(\mathbf{x}) + b$. Given a labeled sample $S_n = \{(\mathbf{x}_1, y_1),...,(\mathbf{x}_n, y_n)\}$, $\mathbf{w}$ and b are optimized to maximize the distance between the positive and negative class, allowing misclassifications:

$$\begin{aligned} &\min_{\mathbf{w},b,\xi} <\mathbf{w},\mathbf{w}> +C\sum_{i=1}^{n}\xi_i \\ &y_i(<\mathbf{w}, f(\mathbf{x}_i)> +b) \geq 1-\xi_i, \xi_i \geq 0, i=1,...,l \end{aligned} \quad (1)$$

where C is a regularization parameter, trading off error against distance. This formulation leads to the dual problem described below, for which an efficient algorithm is available (Platt, 1998).

$$\begin{aligned} &\max W(\boldsymbol{\alpha}) = \sum_{i=1}^{l}\alpha_i - \frac{1}{2}\sum_{i,j=1}^{l} y_i y_j \alpha_i \alpha_j K(\mathbf{x}_i, \mathbf{x}_j) \\ &s.t. \sum_{i=1}^{l} y_i \alpha_i = 0, 0 \leq \alpha_i \leq C, i = 1,...,l \end{aligned} \quad (2)$$

3.2.1 Kernel-based Joint Learning

Given multiple related data sets (e.g., gene expression, protein sequence, and protein-protein interaction data) one can specify a particular meaningful notion of similarity between proteins for each data set to form the corresponding kernel matrix. For example, for the gene expression data, one can use the vector similarity function to compute the similarity between each pair of genes (proteins) based on their expression profiles (expression level measurements across all experimental samples). For protein-protein interaction data we can form a protein-protein interaction network representing proteins as vertices and interactions among proteins as edges. We can then use, for example, the shortest path length between two proteins (the number of edges on the shortest path connecting the two proteins) as the kernel matrix entries. Using this approach heterogeneous data are cast into the common format of kernel matrices. With appropriate notions of similarities founded on domain knowledge, these kernel matrices could contain a significant portion of the relevant data patterns.

The common representation of kernel matrices of different data types enables us to develop principled approaches to search for the optimal combination of these data for learning tasks. An intuitive approach is linear combination. Given a set of kernels or kernel matrices $\boldsymbol{K} = \{K_1, \ldots, K_m\}$ derived from multiple types of data, one can form a linear combination:

$$K = \sum_{i=1}^{m} \mu_i K_i \qquad (3)$$

where μ_i's are the combination coefficients. This combined kernel can be used to replace the single kernel in (2). Now the optimization problem is extended to not only find the optimal linear discriminant boundaries (α_i's in (2)) but also to find the optimal combination of multiple datasets (μ_i's in (3)). It is shown in (Lanckriet et al., 2002) that the problem of finding optimal α_i's and μ_i's reduces to a convex optimization problem known as a semidefinite program (SDP). Details on efficient algorithms for solving this optimization problem can be found in (Lanckriet et al., 2002; Vandenberghe and Boyd, 1996), while general search algorithms for optimization problems such as the genetic algorithms can also be employed.

3.2.2 Experimental Study

In Lanckriet et al.'s study, various types of data were combined for predicting functional classification associated with yeast proteins. They used the functional catalogue provided by the MIPS Comprehensive Yeast Genome Database (CYGD, http://mips.gsf.de). Using the top-level categories in the functional hierarchy a set of 3,588 proteins of 13 classes was identified. The prediction problem was then cast as 13 binary classification tasks, one for each functional class. Six types of data were used to produce input kernel matrices as shown in Table 1.

Table 21-1. Six types of protein data used in Lanckriet et al.'s study

#	Data	Similarity kernel definition $K(\mathbf{x}_i, \mathbf{x}_j)$
1	Domain structure of each protein summarized using the mapping provided by SwissPort v7.5 (http://us.expasy.org/sprot) from protein sequences to Pfam domains (http://pfam.wustl.edu). Each protein is characterized by a 4950-bit vector, in which each bit represents the presence or absence of one Pfam domain.	Inner products applied on the 4950-bit protein vectors.
2	Protein-protein interactions from CYGD	Random walk measure on a network of proteins linked by interactions (Kondor and Lafferty., 2002)
3	Genetic interactions from CYGD	Same as above
4	Co-participation in a protein complex as determined by tandem affinity purification (TAP) from CYDG	Same as above

continued

#	Data	Similarity kernel definition $K(\mathbf{x}_i, \mathbf{x}_j)$
5	77 cell cycle gene expression measurements per gene (Spellman et al., 10998)	Gaussian kernel defined on the expression profiles: $\exp(-\|\mathbf{x}_i - \mathbf{x}_j\|^2/2\sigma$
6	Protein sequences	Inner products derived from the Smith-Waterman pairwise sequence comparison algorithm (Smith and Waterman, 1981)

The reported results in Lanckriet et al.'s study showed that the combined kernel achieved the best performance in protein function classification. The data types that were assigned the largest combination weights in the optimal combined kernel include: protein sequence similarity, Pfam domain structure, and protein-protein interaction.

4. LEARNING REGULATORY NETWORKS USING MICROARRAY AND EXISTING KNOWLEDGE

In this section we focus on discussing joint learning from experimental data and existing knowledge. We present a case study on learning regulatory networks. This important learning problem has attracted substantial interest in bioinformatics research. We first provide some background knowledge on learning regulatory networks from microarray data and review the relevant literature in Section 3.2.1. We discuss in detail a computational framework for learning regulatory networks jointly from microarray data and known regulatory interactions in Section 3.2.2.

4.1 Learning Regulatory Networks Using Microarray

Recent advances in microarray technologies have made possible large-scale gene expression analyses based on simultaneous measurements of thousands of genes. Such high-throughput experimental data have initiated much recent research on large-scale gene expression data analysis. Various data mining techniques (e.g., clustering and classification) have been employed to uncover the biological functions of genes from microarray data. Recently, these techniques have included a reverse engineering approach to extracting gene regulatory networks from microarray data.

The general objective of gene regulatory network analysis is to extract pronounced gene regulatory features (e.g., activation and inhibition) and to reveal the structure of the transcriptional gene regulation processes by examining gene expression patterns. A simple thought experiment illustrates the essence of the approach (Friedman, 2004). If the expression level of gene *A* is regulated by proteins *B* and *C*, then *A*'s expression level is a function of the joint activity levels of *B* and *C*. In most current biological datasets, however, protein activity levels are not available. Thus we resort to using expression levels of genes as a proxy for the activity levels of the proteins they encode. This is a problematic assumption, as the expression level of a gene does not always correlate with the activity level of the protein it encodes. Because of this fundamental data problem, the objective of regulatory network reconstruction is not to find the exact complete regulatory network corresponding to the underlying regulatory processes. We are rather looking for approximate networks and potential regulatory relations or features as hypotheses for further experiments which could lead to new discoveries.

Treating the expression levels of genes as functions of expression levels of other genes, we can rely on the variations of gene expression levels across different samples to apply reverse engineering techniques for constructing the network of regulatory relations among those genes. Taking as input a matrix of gene expression measurements, with genes and samples as the two dimensions, previous studies have proposed various network learning approaches: pair-wise comparison, differential equation estimation, and Bayesian network learning, among others.

Bayesian network learning has been the most commonly used approach for learning regulatory networks in recent years. In a Bayesian network framework, the expression level of each gene is modeled as a random variable whose value depends stochastically on other random variables that correspond to expression levels of other genes. It fits the regulatory network learning problem particularly well because of the stochastic nature of the gene expression dependencies resulted from the variability in underlying biology and measurement noise. We provide a brief description of the Bayesian network learning framework, as it is the basic learning framework adopted in the case study to be presented in this section.

Many studies have shown that regulatory networks learned from microarray data have the potential to help researchers propose and evaluate new hypotheses (De Jong, 2002; Friedman et al., 2000; Pe'er et al., 2001; Yoo et al., 2002). However, regulatory networks learned from gene expression data alone have inherent deficiencies. Biologically, the learned gene networks only provide a partial picture of the complex signaling and regulation processes. Many other important factors, including proteins and small molecules, are hidden from observation in microarray data. Technically, the small number of samples available in microarray data makes it difficult to infer statistically robust network structures. Many researchers have proposed the incorporation of other types of experimental data and existing biological knowledge to improve the statistical power of regulatory network learning. Many previous studies included in our review in Section 2 have shown that combining microarray data with a wide range of genomic and proteomic data, such as genomic location data (Gerber et al., 2003; Hartemink et al., 2002; Segal et al., 2002), DNA sequence data (Imoto et al., 2004; Segal et al., 2002; Tamada et al., 2003), known genetic interactions (Chrisman et al., 2003; Imoto et al., 2003), and protein-protein interactions (Imoto et al., 2004; Nariai et al., 2004; Segal et al., 2003), resulted in more accurate network models.

Bayesian networks are a type of graphical models that represent probabilistic relationships among variables of interest. For a finite set $X = \{X_1, \ldots, X_n\}$ of random variables, a Bayesian network $B = <G, \Theta>$ contains a qualitative component G, which is an acyclic graph that encodes the Markov assumption that each variable (represented as a vertice in the graph) is independent of its non-descendants, given its parents and a quantitative component Θ that represents the set of parameters characterizing the conditional probability distribution. When using Bayesian networks to represent gene networks, a gene is regarded as a random variable X_i, and a relationship between a gene and its parents is represented by the conditional probability.

The problem of learning a Bayesian network (gene network) can be stated as follows. Given a microarray dataset $D = \{\mathbf{x}^1, \ldots, \mathbf{x}^N\}$ of N independent instances of X, find a network $B = <G, \Theta>$ that best matches D. The common approach to this problem is to introduce a statistically motivated scoring function that evaluates each network with respect to the training data and to search for the optimal network according to this score. A commonly used scoring function is the Bayesian scoring metric:

$$Score(G : D) = \log P(G \mid D) = \log P(D \mid G) + \log P(G) + C \qquad (1)$$

where C is a constant independent of G and $P(D|G)$ is the marginal likelihood which averages the probability of the data over all possible parameter assignments to G:

$$P(D \mid G) = \int P(D \mid G, \Theta) P(\Theta \mid G) d\Theta \qquad (2)$$

The Bayesian scoring metric has been commonly used in the gene network learning literature (Friedman, et al., 2000; Hartemink, et al., 2002). The particular choice of priors $P(G)$ and $P(G \mid \Theta)$ for each G determines the exact Bayesian metric score. We followed the standard Bayesian network learning procedure using *BDe* priors (Friedman, et al., 2000; Heckerman, et al., 1995) and greedy hill-climbing search (Friedman, et al., 2000; Nariai, et al., 2004).

Figure 21-2. Bayesian networks for learning regulatory networks

4.2 Joint Learning Using Known Genetic Interactions

Among the various types of information that have been used by researchers to enhance regulatory network learning, known genetic interactions, either prepared by domain experts or collected from biological knowledge bases, probably provide the most reliable information for learning. In this section we present one of our previous studies as an example of computational approaches for learning regulatory networks using known genetic interactions and microarray data.

4.2.1 Microarray Data and Genetic Interactions

Before getting into the details of the computational approach, we first describe the data and knowledge involved in our study.

Microarray data: We used a *Homo sapiens* dataset provided by the Arizona Cancer Center. It contains expression measurements of 33 samples (11 cell lines, 2 with wild-type p53 and 9 with mutated p53, under 3 treatments with respect to introduction of exogenous p53) on a platform of 5,306 human genes. Based on discussions with the domain scientists who conducted the experiments, we selected a threshold value of 1 in logarithmic (base 2) scale and used the 200 genes with greatest variations for the network analysis. We refer to these 200 genes as *informative genes* as the specific microarray data only provides information involving regulatory relations among these genes.

Known genetic interactions: Known regulatory relations from two knowledge sources are used: gene and protein interactions found in the CSNDB database (Takai-Igarashi and Kaminuma, 1999) and regulatory relations automatically extracted from biomedical abstracts using a biological relation parser (Leroy and Chen, 2002) (refer to previous Chapters for details on biological relation parsing from literature). From CSNDB, 1,188 gene expression and cell signaling relations and 3,511 entities have been included. From 23,234 MedLine p53-related abstracts, our biological relation parser automatically extracted 1,903 relations with at least one side matched to the 200 genes in the microarray data. We analyzed the activation (1,398) or inhibition (387) relations in this study.

4.2.2 Joint Learning Approaches

The known genetic interactions were typically used to provide the ground truth regarding a (typically small) portion of the gene network to be inferred. Having part of the network guaranteed to be correct would ensure a network model of generally improved accuracy (Chrisman et al., 2003; Imoto et al.,

2003). However, this simple approach of joint learning is largely limited by the small number of known genetic interactions among a particular set of genes that show pronounced gene expression variations in a microarray dataset (we refer to this problem as the *limited overlap problem*).

The following statistics further illustrate this limited overlap problem. After gene name normalization we formed a set of known genetic interactions (contains only activation and inhibition relations) from the CSNDB database and automatically extracted from MedLine abstracts. This set consisted of 1,586 relations involving 1,212 biological entities. Only 20 of the 200 informative genes appeared in this set (these 20 genes are referred to as *overlapping* genes thereafter). In addition, there was only 1 known interaction with both ends matching the 200 informative genes. This data clearly demonstrates the limited overlap problem which limits the usefulness of approaches that directly leverage existing knowledge.

Under the direct approach of utilizing the existing knowledge for learning, the known interactions are treated as separate relations rather than a connected regulatory network that is used by biologists to make assessments and hypotheses. To be consistent with such human usages of known genetic interactions, the computational approach needs to place individual known interactions into a connected network as well. The computational approaches we present below are based on this intuition to better leverage the set of known genetic interactions for regulatory network learning.

We formed a regulatory network based on the set of known interactions and refer to this network as the *knowledge network*. We focused on analyzing the portion of the knowledge network that was reachable from the 20 overlapping genes. The resulting network consisted of 308 biological entities and 576 relations. This network serves as the basis for the following discussions.

A. Reasoning on the Knowledge Network: Qualitative Probabilistic Networks

The knowledge network is qualitative in nature, as contrasted with quantitative models such as kinetics and differential equations that provide detailed exact biological knowledge. In the biological domain there is a large amount of qualitative information emerging from functional genomic and proteomic studies. To be able to exploit such knowledge, qualitative reasoning (Forbus, 1984; Kuipers, 1986) has been applied to support simulation of qualitative information.

Qualitative reasoning has been shown to be a powerful method in the domains of medicine (Kuipers and Kassirer, 1987) and qualitative physics (Weld and de Kleer, 1990). Several studies in the bioinformatics literature have investigated qualitative reasoning and simulation of biological systems

(Heidtke and Schulze-Kremer, 1998; Karp and Mavrovouniotis, 1994; Kazic, 1993; McAdams and Shapiro, 1995; Meyers and Friedland, 1984). A rich set of qualitative reasoning tools has been adopted, including qualitative differential equations (Heidtke and Schulze-Kremer, 1998), frame-based process representation and execution (Karp, 1993), and deductive technologies (Prolog) (Kazic, 1993).

To be consistent with the probabilistic nature of gene regulatory processes and to provide reasoning capabilities that infer transitive relations and simulate gene expression levels, we employ a well-developed formalism, *qualitative probabilistic networks.* Qualitative probabilistic networks (QPNs) were designed by M. P. Wellman (Wellman, 1990) as qualitative abstractions of probabilistic networks. They have the same graphical structure as their quantitative counterparts, but instead of quantifying the probabilistic relationships between variables using conditional probabilities, they summarize the relationships in qualitative signs. The variables represent the genes and proteins while links with qualitative signs represent the inhibition/activation regulatory relations.

Given a set of biological entities $Y = \{Y_1, ..., Y_J\}$ in the knowledge network, a QPN model contains an acyclic graph structure $G_{qual} = (V(G_{qual}), A(G_{qual}))$. The nodes, $V(G_{qual})$, represent the biological entities. The arcs, $A(G_{qual})$, represent the activation/inhibition relationships. A QPN uses a set of signs assigned to the arcs, $S^p(Y_i, Y_j)$, $p \in \{+, -, 0, ?\}$, to represent qualitative influences between two nodes. A *positive qualitative influence*, for example, of a node Y_i on its (immediate) successor Y_j, denoted $S^+(Y_i, Y_j)$, expresses that observing higher values of Y_i makes higher values of Y_j more likely, regardless of any other direct influence on Y_j. A *negative qualitative influence*, denoted by S^-, and a *zero qualitative influence*, denoted by S^0, is analogously defined. If the influence of node Y_i on node Y_j is not monotonic or if it is unknown, we say that it is ambiguous, denoted $S^?(Y_i, Y_j)$.

The set of influences represented in a QPN exhibits various convenient and useful properties (Wellman, 1990). The *transitivity* and *composition* properties are central to the QPN reasoning used in this study. The property of transitivity asserts that qualitative influences along a chain that specifies at most one incoming arc for each node combine into a single net influence whose sign is given by the $\otimes$-operator from Table 2. The property of composition asserts that multiple qualitative influences between two nodes along parallel chains combine into a single net influence whose sign is given by the $\oplus$-operator.

For probabilistic inference with a QPN, an elegant algorithm is available, designed by Druzdzel and Henrion (Druzdzel and Henrion, 1993; Henrion and Druzdzel, 1991). The basic idea of the algorithm is to trace the effect of observing a node's value upon the probabilities of the values of all other nodes in the network by message passing between neighboring nodes. In essence, this sign-propagation algorithm computes the sign of influence along the active trails between the observed node and all other nodes.

Table 21-2. The ⊗-operator and ⊕-operator for combining signs

⊗	+	–	0	?
+	+	–	0	?
–	–	+	0	?
0	0	0	0	0
?	?	?	0	?

⊕	+	–	0	?
+	+	?	+	?
–	?	–	–	?
0	+	–	0	?
?	?	?	?	?

```
Given: A qualitative probabilistic network and an evidence node e.
Output: Sign of the influence of e on each node in the network.
Data structures:
{ In each of the nodes
        sign ch; // sign of change
        sign evs; // sign of evidential support }
Main program:
        for each node n in the network do ch := '0;
        Propagate-Sign (φ, e, e, '+);
Recursive procedure for sign propagation:
{ trail  // visited nodes,
  from   // sender of the message,
  to     // recipient of the message,
  sign   // sign of influence from from to to }
 Propagate-Sign (trail, from, to, sign)
 begin
        if to.ch = sign ⊕ to.ch then exit;        // exit if already made the update
        to.ch := sign ⊕ to.ch;  // update the sign of to
        trail := trail ∪ to;      // add to to the set of visited nodes
        for each n in the Markov blanket* of to do
        begin
                s := sign of the arc;      // direct or intercausal
                sn := n.ch;        // current sign of n
                if the arc to n is active and n ∉ trail and sn ≠ to.ch ⊗ s then
                Propagate-Sign (trail, to, n, to.ch ⊗ s)
        end
  end

*The Markov blanket of a variable A includes the parents of A, the children of A and the variables sharing a child with A.
```

Figure 21-3. The qualitative sign propagation algorithm

For each node a sign is determined indicating the direction of change in the node's probabilities occasioned by the new observation given all previous ones. Initially, all node signs equal '0'. For the newly observed node, an appropriate sign is entered, that is, either a '+' for the observed value 'over-expressed' or a '–' for the value 'under-expressed'. The node updates its sign and subsequently sends a message to each neighbor that is not independent of the observed node. The sign of each message becomes the sign product (⊗) of its previous sign and the sign of the arc it traverses. Each message keeps a list of the nodes it has visited and its origin so it can avoid visiting any node more than once. Each message travels on one evidential trail. Each node, on receiving a message, updates its own sign with the sign sum (⊕) of itself and the sign of the message. Then it passes a copy of the message to all unvisited neighbors that need to update their signs. A sketch of the sign propagation algorithm is presented in Figure 3.

With this sign propagation algorithm, we can achieve the following two basic reasoning capabilities:

(1) Deriving transitive relations: to assess the transitive influence of entity *a* to entity *b*, we assign a positive sign to node *a* (over-expression in *a*) and propagate the sign using the presented algorithm. The resulting sign of entity *b* specifies the transitive regulatory relation between *a* and *b* (+ for activation, – for inhibition, 0 for no effect, and ? for an uncertain relation). Because of the symmetric property of the sign product (⊗) and sign sum (⊕) operations, if we assigned a negative sign to node *a*, the sign that *b* will receive after propagation will be – for activation, + for inhibition, 0 for no effect, and ? for an uncertain relation. Thus the transitive relation between *a* and *b* is not dependent on what sign we assign to *a*.

(2) Simulating the expression levels of hidden biological entities based on expression levels of observed genes: given the signs of a set of nodes in the network, we can propagate the sign of each node throughout the entire network and aggregate the multiple effects on the nodes with unknown signs to obtain the simulated signs of these nodes under this condition.

The knowledge network obtained from the literature typically needs refinement before it can be represented as a QPN. We might get multiple (and contradictory) relations between two biological entities from the literature. In our current study our network representation does not include a relation between these two biological entities. For such relations to be included, expert validation is required to resolve the contradiction. Cycles could also appear in the knowledge network from the literature, which violate the acyclic characteristics of a QPN. Although such cycles rarely occurred in our experiments, principled approaches need to be developed to resolve this issue in future research.

B. Joint Learning Algorithms Based on Bayesian Networks and Qualitative Probabilistic Networks

Leveraging the two reasoning capabilities of the QPN formalism, we propose two joint learning approaches to combining known genetic interactions with microarray experiment data for learning regulatory networks.

***Transitive Effect Approach*:** The transitive effect approach derives transitive influences among the overlapping genes based on the knowledge network. The derived transitive influences are then incorporated into the Bayesian network learning process as known relations. In the current study we have employed a simple approach to incorporate the derived transitive relations. These relations were treated as if they were direct relations documented in the literature. More advanced incorporation approaches could be developed to better accommodate the stochastic and unreliable nature of the derived transitive relations.

***Synthesized Expression Approach*:** The synthesized expression approach simulates the "expression" data of the biological entities in the knowledge network based on the observed expression levels of the genes in the overlap set. The sign propagation algorithm is used to derive the synthesized observations. For each observation, the overlapping genes are assigned the values corresponding to their observed expression levels in the microarray data (+ for over-expression, – for under-expression, and 0 for no change in expression level). The sign propagation algorithm is used to determine the sign of the remaining nodes in the knowledge network by aggregating the influences from each of the nodes in the overlap set. The derived signs of these nodes give their synthesized expression data for that specific sample. Since the synthesized expression data are derived directly from the propagated signs based on the QPN model, the derived synthesized expression data conform exactly to the QPN model from the existing knowledge. This set of synthesized expression data is then combined with the original expression data to form an enhanced gene (biological entity) expression dataset. Bayesian network learning is then applied on this enhanced data set with the relations in the knowledge network as given.

4.2.3 Experimental Results

In the experimental study we compared the learning results obtained using only the observed expression data to results obtained through learning with the transitive effect and synthesized expression approaches. The approach of directly incorporating known interactions was not considered because of the limited overlap problem exhibited in our data. Evaluation of the learned genetic regulatory relations is always a weak point of this type of research due to the lack of a set of good new hypotheses as the benchmark for comparison. In our study we employed domain experts to assess the interestingness of the learned regulatory relations as potential hypotheses. The evaluation results showed that both joint learning approaches had improved the quality of learned regulatory networks as compared to learning without any prior knowledge. In particular, compared with learning without prior knowledge, the transitive effect approach generated a small number of additional relations surrounding the overlapping genes with high percentages of interesting relations, while the synthesized expression approach enabled the incorporation of large-scale existing knowledge into the learning process and generation of large numbers of additional, reasonably accurate regulatory relations.

5. CONCLUSIONS AND DISCUSSION

In this chapter we reviewed the field of joint learning research in biomedical domains. We presented two case studies on two representative joint learning tasks, protein function classification and regulatory network learning. These two studies involved a wide range of biological data and existing knowledge and two important algorithmic frameworks for joint learning, the kernel-based framework and probabilistic graphical models. Throughout our review and the two case studies we have been emphasizing the importance of data representation for joint learning research. To a certain extent, identifying the appropriate computable data representation is the most critical step in conducting joint learning research. With properly chosen data representations, joint learning problems may be transformed into formulations that can be solved by existing algorithms.

One critical challenge for joint learning research is to develop meaningful and scalable evaluation methods. Many learning tasks in joint learning are difficult to evaluate due to the lack of appropriate comparison benchmarks. Biological experiments that are specifically designed for data mining and machine learning purposes may need to be conducted to provide biologically-founded validation. The joint learning community also needs to address the general critiques of the biological foundation for combining various types of data and knowledge for learning. Typical machine learning validation approaches do not provide explanations for the learning performance improvements that result from the use of different types of data. This kind of explanation is crucial if biologists are to understand and utilize the learning results. Close collaboration with domain experts and formal studies justifying the underlying joint learning processes are needed.

6. ACKNOWLEDGEMENTS

The authors are supported by the grant: NIH/NLM, 1 R33 LM07299-01, 2002-2005, "Genescene: A Toolkit for Gene Pathway Analysis." We thank the National Library of Medicine, the Gene Ontology Consortium, and the Hugo Nomenclature Committee for making the ontologies available to researchers.

REFERENCES

Apweiler, R., Bairoch, A., Wu, C. H., Barker, W. C., and Boeckmann, B. (2004). "UniProt: The Universal Protein Knowledgebase," *Nucleic Acids Research* 32, D115-D119.

Bader, G. D., Betel, D. and Hogue, C. W. V. (2003). "BIND: The Biomolecular Interaction Network Database," *Nucleic Acids Research* 31, 248-250.

Benson, D. A., Karsch-Mizrachi, I., Lipman, D. J., Ostell, J. and Wheeler, D. L. (2004). ""GenBank: Update," *Nucleic Acids Research* 32, D23-D26.

Boeckmann, B., Bairoch, A., Apweiler, R., Blatter, M.-C. and Estreicher, A. (2003). "The SWISS-PROT Protein Knowledgebase and Its Supplement TrEMBL in 2003," *Nucleic Acids Research* 31, 365-370.

Bourne, P. E., Addess, K. J., Bluhm, W. F. and Chen, L. (2004). "The Distribution and Query Systems of the RCSB Protein Data Bank," *Nucleic Acids Research* 32, D223-D225.

Brazma, A., Parkinson, H., Sarkans, U., Shojatalab, M. and Al., E. (2003). "ArrayExpress: Public Repository For Microarray Gene Expression Data at the EBI," *Nucleic Acids Research* 31, 68-71.

Chrisman, L., Langley, P., Bay, S. and Pohorille, A. (2003). "Incorporating Biological Knowledge into Evaluation of Causal Regulatory Hypotheses," in *Pacific Symposium on Biocomputing,* Pp. 128-139.

Chu, T., Glymour, C., Scheines, R. and Spirtes, P. (2003). "A Statistical Problem for Inference to Regulatory Structure from Associations of Gene Expression Measurements with Microarrays," *Bioinformatics* 19, 1147-52.

De Hoon, M. J. L., Imoto, S., Kobayashi, K., Ogasawara, N. and Miyano, S. (2004). "Predicting the Operon Structure of Bacillus Subtilis Using Operon Length, Intergene Distance, and Gene Expression Information," in *Pacific Symposium on Biocomputing,* Pp. 276-287.

De Jong, H. (2002). "Modeling and Simulation of Genetic Regulatory Systems: A Literature Review," *Journal of Computational Biology* 9, 67-103.

Druzdzel, M. J. and Henrion, M. (1993). "Efficient Reasoning in Qualitative Probabilistic Networks," in *Eleventh National Conference on Artificial Intelligence,* 548-553.

Emmert-Buck, M. R., Strausberg, R. L., Krizman, D. B., Bonaldo, M. F. and Al., E. (2000). "Molecular Profiling of Clinical Tissue Specimens: Feasibility and Applications," *American Journal of Pathology,* 156, 1109-1115.

Eskin, E. and Agichtein, E. (2004). "Combining Text Mining and Sequence Analysis to Discover Protein Functional Regions," in *Pacific Symposium on Biocomputing,* Pp. 288-299.

Forbus, K. D. (1984). "Qualitative Process Theory," *Artificial Intelligence* 24, 85-168.

Friedman, N. (2004). "Inferring Cellular Networks Using Probabilistic Graphical Models," *Science* 303, 799-805.

Friedman, N., Linial, M., Nachman, I. and Pe'er, D. (2000). "Using Bayesian Network to Analyze Expression Data," *Journal of Computational Biology* 7, 601-620.

Gerber, G. K., Joseph, Z.-B., Lee, T. I., Robert, F., Gordon, D. B., Fraenkel, E., Simon, I., Jaakkola, T. S., Young, R. A. and Gifford, D. K. (2003). "Computational Discovery of Gene Modules and Regulatory Networks," in *11th International Conference on Intelligent Systems For Molecular Biology*.

Gollub, J., Ball, C. A., Binkley, G., Sherlock, G. and Al., E. (2003). "The Stanford Microarray Database: Data Access and Quality Assessment Tools," *Nucleic Acids Research* 31, 94-96.

Hartemink, A. and Segal, E. (2004). "Session Introduction," in *Pacific Symposium on Biocomputing*, Pp. 262-263.

Hartemink, A. J., Gifford, D. K., Jaakkola, T. S. and Young, R. A. (2002). "Combining Location and Expression Data for Principled Discovery of Genetic Regulatory Network Models," in *Pacific Symposium on Biocomputing*, Pp. 437-449.

Heckerman, D., Geiger, D. and Chickering, D. H. (1995). "Learning Bayesian Networks: The Combination of Knowledge and Statistical Data," *Machine Learning* 20, 197-243.

Heidtke, K. R. and Schulze-Kremer, S. (1998). "Design and implementation of a Qualitative Simulation Model of Lambda Phage infection," *Bioinformatics* 14, 81-91.

Henrion, M. and Druzdzel, M. J. (1991). "Qualitative Propagation and Scenario based Approaches to Explanation in Probabilistic Reasoning," *Sixth Conference on Uncertainty in Artificial Intelligence*, Pp. 17-32.

Husmeier, D. (2003). "Sensitivity and Specificity of Inferring Genetic Regulatory Interactions from Microarray Experiments with Dynamic Bayesian Networks," *Bioinformatics* 19, 2271-2282.

Imoto, S., Higuchi, T., Goto, T., Tashiro, K., Kuhara, S. and Miyano, S. (2003). "Estimating Gene Networks by Bayesian Networks from Microarrays and Biological Knowledge," in *11th International Conference on Intelligent Systems For Molecular Biology*.

Imoto, S., Higuchi, T., Goto, T., Tashiro, K., Kuhara, S. and Miyano, S. (2004). "Combining Microarrays and Biological Knowledge for Estimating Gene Networks via Bayesian Networks," *Journal of Bioinformatics and Computational Biology* 2, 77-98.

Iossifov, I., Krauthammer, M., Friedman, C., Hatzivassiloglou, V., Bader, J. S., White, K. P. and Rzhetsky, A. (2004). "Probabilistic Inference of Molecular Networks from Noisy Data Sources," *Bioinformatics* 20, 1205-13.

Jansen, R., Greenbaum, D. and Gerstein, M. (2002). "Relating Whole-genome Expression Data with Protein-protein Interactions," *Genome Research* 12, 37-46.

Kanehisa, M., Goto, S., Kawashima, S. and Nakaya, A. (2002). "The KEGG Databases at GenomeNet," *Nucleic Acids Research* 30, 42-46.

Karp, P. D. (1993). "A Qualitative Biochemistry and Its Application to the Tryptophan Operon," in Hunter, L. (Ed), *Artificial Intelligence and Molecular Biology*, AAAI Press, Pp. 289-324.

Karp, P. D. and Mavrovouniotis, M. M. (1994). "Representing, Analyzing, and Synthesizing Biochemical Pathways," *IEEE Expert* 9, 11-22.

Karp, P. D., Riley, M., Saier, M., Paulsen, I. T., Collado-Vides, J., Paley, S. M., Pellegrini-Toole, A., Bonavides, C., & Gama-Castro, S. (2002). "The EcoCyc Database," *Nucleic Acids Research*, 30, 56-58.

Kazic, T. (1993). "Reasoning About Biochemical Compounds and Processes," in *Second International Conference on Bioinformatics, Supercomputing and the Human Genome Project. Singapore,* Pp. 35-49.

Kondor, R. I. and Lafferty., J. (2002). "Diffusion Kernels on Graphs and Other Discrete Input Spaces," in *International Conference on Machine Learning*, Pp. 315-322.

Krull, M., Voss, N., Choi, C., Pistor, S., Potapov, A. and Wingender, E. (2003).

"TRANSPATH: An Integrated Database on Signal Transduction and a Tool for Array Analysis," *Nucleic Acids Res.* 31, 97-100.

Kuipers, B. (1986). "Qualitative Simulation," *Artificial Intelligence* 29, 289-338.

Kuipers, B. and Kassirer, J. (1987). "Knowledge Acquisition by Analysis of Verbatim Protocols," in Kidd, A. (Ed), *Knowledge Acquisition For Expert Systems,* Plenum, Pp. 289-338.

Lanckriet, G. R. G., Cristianini, N., Bartlett, P., Ghaoui, L. E. and Jordan, M. I. (2002). "Learning the Kernel Matrix with Semi-definite Programming," in *19th International Conference on Machine Learning,* Pp. 323-330.

Lanckriet, G. R. G., Deng, M., Cristianini, N., Jordan, M. I. and Noble, W. S. (2004). "Kernel-based Data Fusion and Its Application to Protein Function Prediction in Yeast," in *Pacific Symposium on Biocomputing,* Pp. 300-311.

Lemer, C., Antezana, E., Couche, F., Fays, F. and Al., E. (2004). "The AMAZE LightBench: A Web Interface to a Relational Database of Cellular Processes," *Nucleic Acids Research* 32, D443-D448.

Leroy, G. and Chen, H. (2002). "Filling Preposition-based Templates to Capture Information from Medical Abstracts," in *Pacific Symposium on Biocomputing,* Pp. 350-361.

Li, H., Li, J., Tan, S. H. and Ng, S.-K. (2004). "Discovery of Binding Motif Pairs from Protein Complex Structural Data and Protein Interaction Sequence Data," in *Pacific Symposium on Biocomputing,* Pp. 312-323.

McAdams, H. H. and Shapiro, L. (1995). "Circuit Simulation of Genetic Networks," *Science* 269.

Meyers, S. and Friedland, P. (1984). "Knowledge Based Simulation of Genetic Regulation in Bacteriophage Lambda," *Nucleic Acids Research* 12, 1-9.

Model, F., Adorjan, P., Olek, A. and Piepenbrock, C. (2001). "Feature Selection for DNA Methylation Based Cancer Classification," *Bioinformatics* 17, 157-164.

Nariai, N., Kim, S., Imoto, S. and Miyano, S. (2004). "Using Protein-protein Interactions for Refining Gene Networks Estimated from Microarray Data by Bayesian Networks," *Pacific Symposium on Biocomputing,* Pp. 336-347.

Pe'er, D., Regev, A., Elidan, G. and Friedman, N. (2001). "Inferring Subnetworks from Perturbed Expression Profiles," *Bioinformatics* 17, S215-24.

Platt, J. C. (1998). "Fast Training of Support Vector Machines Using Sequential Minimum Pptimization," in Schölkopf, B., Burges, C., and Smola, A. (Ed), *Advances in Kernel Methods- Support Vector Learning, M*IT Press, Pp. 185-08.

Prakash, A., Blanchette, M., Sinha, S. and Tompa, M. (2004). "Motif Discovery in Heterogeneous Sequence Data," in *Pacific Symposium on Biocomputing, P*p. 348-359.

Salwinski, L., Miller, C. S., Smith, A. J., Pettit, F. K., Bowie, J. U. and Eisenberg, D. (2004). "The Database of Interacting Proteins: 2004 Update," *Nucleic Acids Research* 32, D449-D451.

Segal, E., Barash, Y., Simon, I., Friedman, N. and Koller, D. (2002). "From Promoter Sequence to Expression: A Probabilistic Framework," in *6th International Conference on Research in Computational Molecular Biology.*

Segal, E., Wang, H. and Koller, D. (2003). "Discovering Molecular Pathways from Protein Interaction and Gene Expression Data," *Bioinformatics* 19, i264-i272.

Smith, T. F. and Waterman, M. S. (1981). "Identification of Common Molecular Subsequences," *Journal of Molecular Biology* 147, 195-197.

Somorjai, R. L., Dolenko, B. and Baumgartner, R. (2003). "Class Prediction and Discovery Using Gene Microarray and Proteomics Mass Spectroscopy Data: Curses, Caveats, Cautions," *Bioinformatics* 19, 1484-91.

Spellman, P., Sherlock, G., Zhang, M., Iyer, V., Anders, K., Eisen, M., Brown, P., Botstein,

D. and Futcher, B. (1998). "Comprehensive Identification of Cell Cycle-regulated Genes of the Yeast Saccharomyces Cerevisiae by Microarray Hybridization," *Molecular Biology of the Cell* 9, 3, 273-297.

Takai-Igarashi, T. and Kaminuma, T. (1999). "A Pathway Finding System for the Cell Signaling Networks Database," *Silico Biology* 1, 129-146.

Takusagawa, K. T. and Gifford, D. K. (2004). "Negative Information For Motif Discovery," in *Pacific Symposium on Biocomputing,* Pp. 360-371.

Tamada, Y., Kim, S., Bannai, H., Imoto, S., Tashiro, K., Kuhara, S. and Miyano, S. (2003). "Estimating Gene Networks from Gene Expression Data by Combining Bayesian Network Model with Promoter Element Detection," *Bioinformatics* 19, II227-II236.

Tsuda, K. and Noble, W. S. (2004). "Learning Kernels from Biological Networks by Maximizing Entropy," *Bioinformatics* 20, I326-I333.

Vandenberghe, L. and Boyd, S. (1996). "Semidefinite Programming," *SIAM Review* 38, 49-95.

Vapnik, V. (1995). The *Nature of Statistical Learning Theory.* Springer Verlag.

Weld, D. S. and De Kleer, J. (1990). *Readings in Qualitative Reasoning About Physical Systems.* Morgan Kaufmann.

Wellman, M. P. (1990). "Fundamental Concepts of Qualitative Probabilistic Networks," *Artificial Intelligence* 44, 257-303.

Wu, C. H., Nikolskaya, A., Huang, H., Yeh, L.-S. L. and Natale, D. A. (2004). "PIRSF: Family Classification System At the Protein Information Resource," *Nucleic Acids Research 32,* D112-D114.

Yoo, C., Thorsson, V. and Cooper, G. F. (2002). "Discovery of Causal Relationships in a Gene-regulation Pathway from a Mixture of Experimental and Observational DNA Microarray Data," in *Pacific Symposium on Biocomputing,* Pp. 498-509.

Zanzoni, A., Montecchi-Palazzi, L., Quondam, M., Ausiello, G., Helmer-Citterich, M. and Cesareni, G. (2002). "MINT: A Molecular INTeraction Database," *FEBS Letters* 513, 135-140.

SUGGESTED READINGS

Baldi, P. and S. Brunak. 2001. *Bioinformatics: The Machine Learning Approach,* The MIT Press, Cambridge.
A good introductory book on application of machine learning to biology research.

Buntine, W. 1996. "A guide to the literature on learning probabilistic networks from data," *IEEE Transactions on Knowledge and Data Engineering,* 8(2), 195-210.
An early review paper on regulatory network analysis.

Cheng, J., R. Greiner, J. Kelly, D. A. Bell and W. Liu. 2002. "Learning Bayesian networks from data: an information-theory based approach," *The Artificial Intelligence Journal,* 137, 43-90.
This paper provides a complete presentation of the information-theory based dependency analysis algorithm for learning Bayesian networks.

Chrisman, L., P. Langley, S. Bay and A. Pohorille. 2003. "Incorporating biological knowledge into evaluation of causal regulatory hypotheses," In the Proceedings of Pacific Symposium on Biocomputing, 8, 128-139.
This is one of the early studies that combine biological knowledge with gene expression data to infer regulatory networks.

De Jong, H. 2002. "Modeling and simulation of genetic regulatory systems: a literature review," *Journal of Computational Biology*, 9, 67-103.
This is a comprehensive review of quantitative methods for modeling genetic regulatory networks.

Friedman, N. 2004. "Inferring cellular networks using probabilistic graphical models," *Science*, 303(5659), 799-805.
This is an important recent paper summarizing the major types of regulatory network analysis using probabilistic graphical models.

Hartemink, A. J., D. K. Gifford, T. S. Jaakkola and R. A. Young. 2002. "Combining location and expression data for principled discovery of genetic regulatory network models," In the Proceedings of Pacific Symposium on Biocomputing, 7, 437-449.
This is one of the early studies on joint learning from multiple types of genomic data for regulatory network analysis.

Imoto, S., T. Higuchi, T. Goto, K. Tashiro, S. Kuhara and S. Miyano. 2004. "Combining microarrays and biological knowledge for estimating gene networks via Bayesian networks," *Journal of Bioinformatics and Computational Biology*, 2(1), 77-98.
This is a recent study that combines relatively large-scale biological knowledge from human-curated database with gene expression data for learning regulatory networks.

Lanckriet, G. R. G., M. Deng, N. Cristianini, M. I. Jordan and W. S. Noble. 2004. "Kernel-based data fusion and Its application to protein function prediction in yeast," In the Proceedings of Pacific Symposium on Biocomputing, 9, 300-311.
This paper proposed the kernel-based data fusion model to perform joint learning. One of the case studies in this chapter is based on this paper.

Segal, E., H. Wang and D. Koller. 2003. "Discovering molecular pathways from protein interaction and gene expression data," *Bioinformatics*, 19(Suppl: 1), i264-i272.
This is an early paper exploring joint learning with protein interaction data and gene expression data to enhance regulatory network learning.

Speed, T. 2003. *Statistical Analysis of Gene Expression Microarray Data*, CRC Press.
An introductory book to gene expression data analysis.

Tamada, Y., S. Kim, H. Bannai, S. Imoto, K. Tashiro, S. Kuhara and S. Miyano. 2003. "Estimating gene networks from gene expression data by combining Bayesian network model with promoter element detection," *Bioinformatics*, 19(Suppl 2), II227-II236.
Another joint learning paper exploring the combination of gene expression data and promoter elements.

ONLINE RESOURCES

DOE HGP Genomics Primers:
http://www.ornl.gov/sci/techresources/Human_Genome/publicat/primer/index.shtml

Functional Genomics:
http://www.functionalgenomics.org.uk/sections/resources/

Gene Regulatory Networks:
http://doegenomestolife.org/science/generegulatorynetwork.shtml

On-line Bioinformatics Courses:
http://www.bioinformatik.de/cgi-bin/browse/Catalog/Research_and_Education/Online_Courses_and_Tutorials/

Pacific Symposium on Biocomputing:
http://psb.stanford.edu/

GenBank:
http://www.ncbi.nlm.nih.gov/

SwissProt:
http://au.expasy.org/sprot

TrEMBL:
http://au.expasy.org/sprot/

PIR (Protein Information Resource):
http://pir.georgetown.edu/pirwww

UniProt (Universal Protein Resource):
http://www.expasy.uniprot.org/index.shtml

PDB:
http://www.rcsb.org/pdb/

Stanford Microarray Database:
http://genome-www5.stanford.edu/

ArrayExpress at EBI:
http://www.ebi.ac.uk/arrayexpress/

OPD (Open Proteomic Database):
http://bioinformatics.icmb.utexas.edu/OPD/

DIP:
http://dip.doe-mbi.ucla.edu

MINT:
http://mint.bio.uniroma2.it/mint/

BioCyc:
http://www.biocyc.org/

KEGG:
http://www.genome.ad.jp/kegg/

TransFac & TransPath:
http://www.biobase.de/pages/products/transpath.html

PubMed:
http://pubmed.gov

Nucleic Acids Research (NAR) Database List:
http://www3.oup.co.uk/nar/database/c/

QUESTIONS FOR DISCUSSION

1. What are the advantages of joint learning from multiple types of data? What are the potential problems of learning jointly from multiple types of data?

2. What biological questions, other than those mentioned in this chapter, can be appropriately addressed using joint learning algorithms?

3. What are additional types of data and knowledge can be utilized to improve prediction of the protein function classification? What meaningful kernel representations can be used for these data and knowledge?

4. Which types of genomic or proteomic data can be used for learning regulatory networks? What are the limitations of using microarray data to infer regulatory networks?

5. What are the potential problems of representing known genetic interactions documented in the literature or stored in the genetic interaction databases as binary relations in the form of "gene A activates gene B?" How could this simple representation be enhanced to capture additional information? What computational problems will be introduced by these representational enhancements?

AUTHOR INDEX

SUBJECT INDEX

B

C

D

E

F

G

H

I

J

K

L

M

N

O

P

Q

R

S

T

U

Y

Z